Advances in Conservation Genetics

Advances in Conservation Genetics

Edited by Jason Hendon

SYRAWOOD
PUBLISHING HOUSE
New York

Published by Syrawood Publishing House,
750 Third Avenue, 9th Floor,
New York, NY 10017, USA
www.syrawoodpublishinghouse.com

Advances in Conservation Genetics
Edited by Jason Hendon

International Standard Book Number: 978-1-68286-537-8 (Hardback)

Cataloging-in-Publication Data

Advances in conservation genetics / edited by Jason Hendon.
p. cm.
Includes bibliographical references and index.
ISBN 978-1-68286-537-8
1. Conservation biology. 2. Ecological genetics. 3. Germplasm resources conservation.
I. Hendon, Jason.
QH75 .A38 2018
576.58--dc23

Printed in China.

TABLE OF CONTENTS

PREFACE

This book has been a concerted effort by a group of academicians, researchers and scientists, who have contributed their research works for the realization of the book. This book has materialized in the wake of emerging advancements and innovations in this field. Therefore, the need of the hour was to compile all the required researches and disseminate the knowledge to a broad spectrum of people comprising of students, researchers and specialists of the field.

Conservation genetics falls under the umbrella of population genetics. It helps in understanding causes of extinction and methods to avoid extinction, in order to restore biodiversity. Genetic variability is an important concept of this field as it influences the health and long-term survival of species. Conservation genetics focuses on assessing the genome of species and evaluating the probable causes of extinction. Some of the techniques used in this field are mini satellite, microsatellite, alloenzyme, etc. Most of the topics introduced in this book cover new techniques and the applications of conservation genetics. It includes contribution of experts and scientists which will provide innovative insights into this field.

At the end of the preface, I would like to thank the authors for their brilliant chapters and the publisher for guiding us all-through the making of the book till its final stage. Also, I would like to thank my family for providing the support and encouragement throughout my academic career and research projects.

Editor

Genetic and Ecological Outcomes of *Inga vera* Subsp. *affinis* (Leguminosae) Tree Plantations in a Fragmented Tropical Landscape

Oswaldo Cruz Neto[1], Antonio V. Aguiar[2], Alex D. Twyford[3], Linda E. Neaves[3], R. Toby Pennington[3], Ariadna V. Lopes[1]*

1 Departamento de Botânica, Universidade Federal de Pernambuco, Recife, Pernambuco, Brazil, 2 Department of Wildlife Ecology and Conservation, University of Florida, Gainesville, Florida, United States of America, 3 Royal Botanic Garden Edinburgh, Edinburgh, United Kingdom

Abstract

Planting of native trees for habitat restoration is a widespread practice, but the consequences for the retention and transmission of genetic diversity in planted and natural populations are unclear. Using *Inga vera* subsp. *affinis* as a model species, we genotyped five natural and five planted populations in the Atlantic forest of northeastern Brazil at polymorphic microsatellite loci. We studied the breeding system and population structure to test how much genetic diversity is retained in planted relative to natural populations. We then genotyped seedlings from these populations to test whether genetic diversity in planted populations is restored by outcrossing to natural populations of *I. vera*. The breeding system of natural *I. vera* populations was confirmed to be highly outcrossing ($t = 0.92$; $F_{IS} = -0.061$, $P = 0.04$), with populations showing weak population substructure ($F_{ST} = 0.028$). Genetic diversity in planted populations was 50% less than that of natural populations (planted: $A_R = 14.9$, $H_O = 0.865$ and natural: $A_R = 30.8$, $H_O = 0.655$). However, seedlings from planted populations showed a 30% higher allelic richness relative to their parents (seedlings $A_R = 10.5$, parents $A_R = 7.6$). Understanding the processes and interactions that shape this system are necessary to provide ecologically sensible goals and successfully restore hyper-fragmented habitats. Future restoration plans for *I. vera* must consider the genetic diversity of planted populations and the potential for gene flow between natural populations in the landscape, in order to preserve ecological interactions (i.e. pollination), and promote opportunities for outcrossing.

Editor: Erin Espeland, USDA- ARS, United States of America

Funding: Financial support was provided by the Conselho Nacional de Desenvolvimento Científico e Tecnológico (CNPq) of Brazil (grant numbers 567739/2008-2 and 307209/2010-5 to AVL and 556990/2010-2 to OCN; http://www.cnpq.br), and by the Coordenação de Aperfeiçoamento de Pessoal de Nível Superior (CAPES) of Brazil (grant number 5998/11-5 to OCN; http://www.capes.gov.br). The funders had no role in study design, data collection and analysis, decision to publish, or preparation of the manuscript.

Competing Interests: The authors have declared that no competing interests exist.

* Email: avflopes@ufpe.br

Introduction

The long term survival of a species is critically influenced by the maintenance of genetic variation within populations. Habitat loss and fragmentation, and reduced population sizes, decrease genetic diversity within populations, and may increase the levels of inbreeding and population substructure of plant species [1–3]. Restoration can mitigate these population changes but the benefits will be limited in plantations of native trees that have been sourced from narrow genetic founding stocks, which have low levels of genetic diversity [4,5], and as a consequence may suffer from reduced long-term survival and productivity [6]. One way genetic diversity in such plantations may be restored without human intervention is through outcrossing with native conspecific populations [4].

In tropical forests, habitat loss and fragmentation are a ubiquitous consequence of human population growth and the expansion of agricultural areas [7]. One example is the Atlantic forests of Brazil, which once stretched from Rio Grande do Sul to Rio Grande do Norte states, covering an estimated area of 1.3 million km^2 [8]. The Atlantic rain forest now covers less than 16% of its original extent and occurs in small, isolated fragments [9]. Despite its hyper-fragmented state, the Atlantic rain forest is extremely species-rich with an estimated 13,972 angiosperms species, of which c. 7,000 are endemic [10].

The reported consequences of habitat loss and fragmentation on the reproduction of flowering plants include the reduced range of reproductive attributes of communities, deprived plant-pollinator interactions and decreased gene flow between populations [1,2,11,12,13,14]. To mitigate some of the detrimental effects of habitat loss and fragmentation, an increased number of habitat restoration actions have been conducted in several terrestrial ecosystems [2,4,15]. Forest restoration in Central and South America, including in Brazil [6,16], often involves planting single-species stands of *Inga*, with a high density of individuals. Unfortunately, genetic diversity is neither explicitly measured nor considered in some habitat restoration projects [5,17]. In addition, restored populations are frequently founded with seeds from a limited number of individuals that may only represent a portion of the genetic diversity present in natural populations [4,18,19].

Species of *Inga* present brush type-flowers, crepuscular or nocturnal anthesis, and secrete large amounts of nectar [20,21]. Despite the potentially lower genetic diversity found in restored areas relative to natural areas, planted populations can be considered as a stepping stone for pollinators in animal pollinated plants such as *Inga* spp. [19], and therefore these populations may improve connectivity of natural populations across a fragmented landscape [4]. However, few studies have analyzed the genetic diversity of planted trees in relation to natural populations and the impacts of single species reforestation on gene flow in a fragmented landscape [4,15,19].

The aim of this study is to determine the level of genetic diversity in planted relative to natural populations of *I. vera*, and how plantations influence the transmission of genetic diversity from adults to seeds in a fragmented landscape. We test the hypotheses that (*i*) adult individuals in planted populations have reduced genetic diversity relative to adults from natural populations and (*ii*) outcrossing between planted and natural populations increases genetic diversity in seedlings from planted populations relative to their parents. Microsatellite markers revealed remarkable differences in genetic diversity among adults and seedlings of planted relative to natural populations. Additionally, plantations of *Inga* seem to be connected with the natural remnants by pollen flow.

Materials and Methods

Study site

We conducted this study in Usina Serra Grande (USGA), a private property located on the Borborema Plateau, Alagoas state, northeastern Brazil (8°30′S, 35°50′W). Approval to conduct fieldwork was kindly granted by the landowner (L.A. Bezerra) and one of the property managers (C. Bakker). The total area is 24,000 ha, of which ca. 9,000 ha represents Atlantic forest remnants of different shapes and sizes. These remnants are scattered in an agricultural matrix, dominated by sugar cane monoculture and pastures [11,14] (Fig. 1). In the early 1980's, the landowners reforested areas along river courses as a measure to improve water and soil quality. The reforested patches contain *Inga vera* or *Inga edulis,* trees of which are planted in single species clumps at a density ranging from 20 to 240 (159±89.8) trees per hectare. These monospecific patches are physically isolated from natural areas, and since seed dispersal in *I. vera* is most likely by primates, which are unlikely to cross non-forest areas, there is little chance that any adult *Inga* trees observed in plantations could have arrived via seed dispersal.

The vegetation in the area is classified as moist tropical forest under the Holdridge system, with a dry season from October to February and a rainy season from April to September (climate data for the period 1922–2001, USGA). The richest plant families, in terms of tree species, are Leguminosae (30 spp., among which 14 are subfamily Mimosoideae and eight are *Inga*), Sapotaceae (13 spp.), Lauraceae (11 spp.) and Sapindaceae (8 spp.) [11].

Study species

Inga vera subsp. *affinis* (hereafter *Inga vera*) is distributed from Central America to Uruguay [20]. Its reproductive system is self-incompatible and its fruit set rate is less than 2% in natural conditions [21]. Fruit dispersal is by primates [20]. Though *I. vera* flowers are pollinated by hawkmoths (sphingophilous), they are also visited by bats and hummingbirds [21], (Cruz Neto et al., unpublished). Individuals of *I. vera* flower synchronously and massively at the same period each year [22]. This strategy ensures that the majority of individuals may contribute via pollen and seeds to the seedling cohort. Due to its dependence on biotic pollination, and occurrence in plantations physically isolated from natural remnants, which is a standard restoration practice in northeastern Atlantic forest, *I. vera* is an excellent model species to investigate how planted stands of trees affect gene flow across disturbed landscapes.

Sampling and data gathering

We randomly selected 10 populations of *I. vera*: five planted populations in the restored monospecific stands, and five natural populations in Atlantic forest fragments (Fig. 1C). We defined a population as individuals from either a forest remnant or planted stand, which were physically isolated in the landscape (Figure 1C). To ensure spatial independence, we selected populations at least 300 m away from each other, both within and among the two spatial treatments (natural and planted). For each population, we randomly sampled 10 reproductive trees totalling 100 adult individuals – 50 for each treatment. We collected six to seven fruit per adult individual, totalling 638 fruits, and sterilized all the 6,416 seeds of these fruits in 1% sodium hypochlorite solution prior to planting in a sterile germination substrate. From the 49 (± 6) available seedlings per maternal family, we sampled five seedlings (i.e. one seedling per fruit), after seedling establishment (ca. 4-month old seedlings). In total, 500 seedlings were sampled across populations. For all the four treatment combinations, two spatial (natural *vs* planted) and two generations (adults *vs* seedlings), we collected young leaves from individuals and stored them in silica gel for DNA extraction.

DNA extraction, PCR and genotyping. We extracted DNA from 600 individuals, 100 adults and 500 seedlings, following the Cetyltrimethylammonium Bromide (CTAB) protocol [23]. All individuals were genotyped at four microsatellite loci (Table 1) using primers developed for the related legumes *Pithecelobium elegans* and *I. edulis* [18,24]. Simplex PCR reactions were performed using a Veriti 384-well Thermal Cycler (Applied Biosystems) in a total volume of 20 µl. The reaction mix contained ~ 1.5 ng template DNA, 0.32 mM of each primer (the forward was fluorescently labeled with 6-FAM, NED and PET dyes, Table 1), 4 mM KCL-$MgCl_2$, 3 mM $MgCl_2$, 160 µM of dNTPs, 40 µM of Taq DNA polymerase (Fermentas). We used the following PCR profile: 95°C for 2 min followed by 30 cycles of 95°C for 15 s, 56 or 62°C for 30 s, 72°C for 30 s; 72°C for 15 min (Table 1; Cruz Neto et al., unpublished). PCR products were diluted to 10 ng final concentration with HPLC water. PCR products with each fluorescent dye color were mixed in a single reaction containing 1µl of each diluted PCR product and 6µl of formamide Hi-Di (Applied Biosystems) prior to fragment analysis with a 500-Liz size standard.

PCR fragments were run on an ABI3730 genetic analyzer (Applied Biosystems). Alleles were scored manually using Peak Scanner 1.0 (Applied Biosystems) and binned according to their motif length with Flexibin [25].

Data analysis

We tested for Hardy-Weinberg equilibrium (HWE) and linkage disequilibrium for each population using FSTAT v.2.9.3.2 [26], with Bonferroni adjusted significance. We also tested the presence of null alleles using Micro-checker [27] and where necessary applied the INA correction method in Freena software [28].

Genetic diversity of adults from planted and natural populations. We compared the genetic diversity of *I. vera* trees between planted and natural populations to test our first hypothesis that genetic diversity is lower in planted than in natural populations. The average number of alleles per locus (*A*),

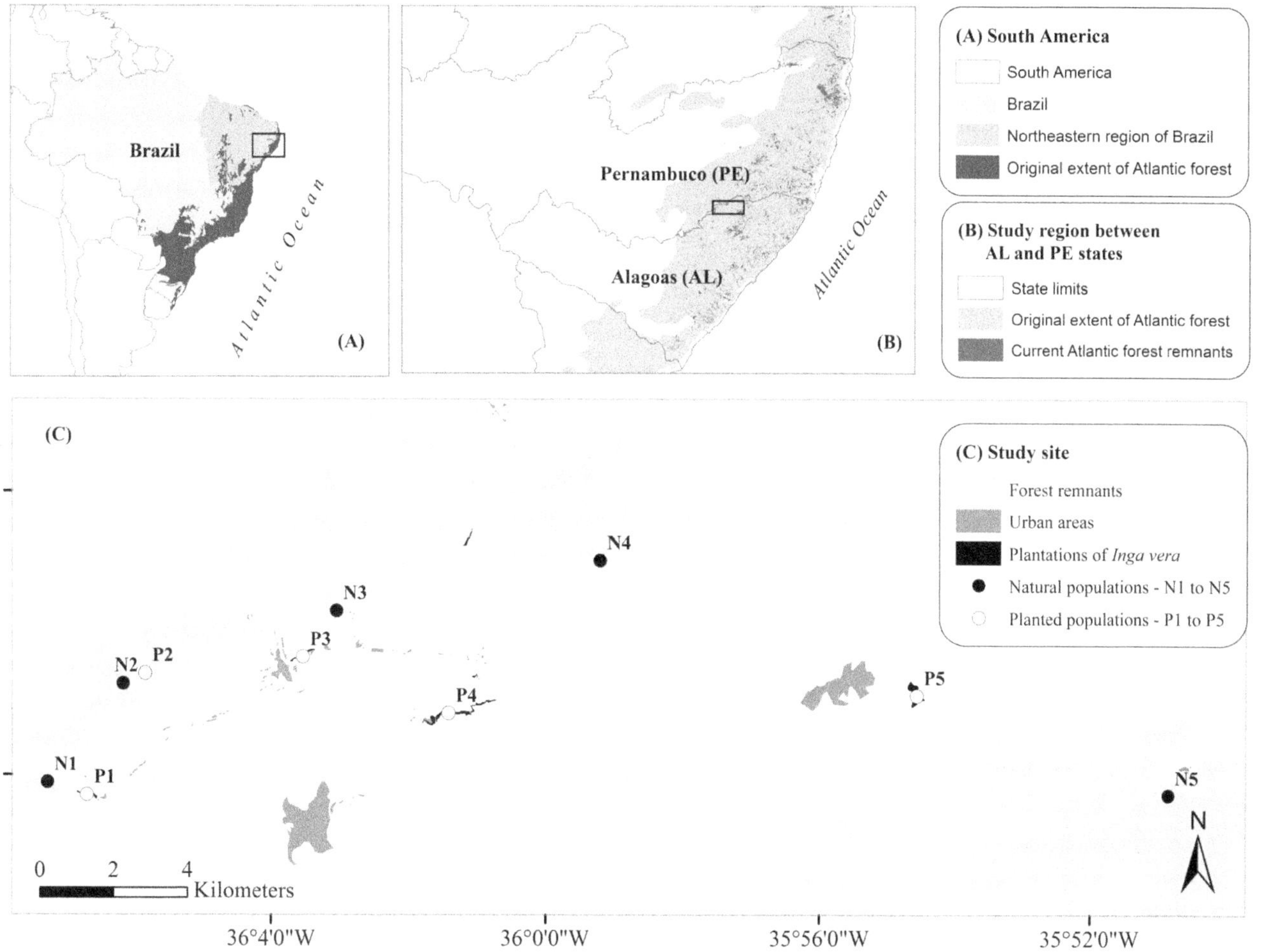

Figure 1. Location of the study area. (A) Map of South America highlighting northeastern region of Brazil and original extent of Atlantic rain forest; (B) Original and remaining distribution of the Atlantic rain forest between Alagoas (AL) and Pernambuco (PE) states; (C) Study site map highlighting remnants of the Atlantic rain forest, *Inga* plantations, and urban areas.

allelic richness rarefied to the minimum sample size of diploid individuals (A_R) [29], and the observed and expected gene diversity (H_O and H_S) [30] values were calculated separately for each planted and natural population using FSTAT. We also scored the number of alleles found exclusively in natural or planted populations using HP-Rare [31].

Genetic diversity of adults and seedlings of *Inga vera* stands. The transmission of genetic diversity in planted and natural populations were assessed by comparing estimates of

Table 1. Genetic diversity per locus in adults and seedlings of *Inga vera* in planted and natural populations.

Locus	Reference	T_a (°C)	Range (bp)	H_O	H_S	F_{ST}
Inga03	Hollingsworth et al. 2005	56	58–90	0.841	0.807	0.015
Inga08	Hollingsworth et al. 2005	56	130–174	0.658	0.78	0.052
Inga33	Hollingsworth et al. 2005	62	216–244	0.583	0.803	0.078
Pel05	Dayanandan et al. 1997	62	179–215	0.807	0.758	0.072

Locus name, reference of the primers, annealing temperature (T_a), range (bp) of the amplified fragments, observed (H_O) and expected (H_S) heterozygosity, and genetic differentiation coefficient (F_{ST}). The primers *Inga03* and *Inga33* were labelled with 6-FAM while the primers *Inga08* and *Pel05* were labelled with NED and PET, respectively.

genetic diversity (A, A_R, H_O and H_S) between adults and seedlings of each population. Differences between adults and seedlings were tested using a hierarchical analysis with 1,000 permutations implemented in FSTAT, followed by Bonferroni correction.

Seedlings from planted populations were scored for the presence of alleles found exclusively in natural adult populations (see above). The presence of these unique alleles would be consistent with pollen-mediated gene flow from natural to planted populations. By considering unique alleles found in all 50 plants in natural populations against the 50 plants in planted populations, we hope to avoid the upward bias of estimates of private alleles that would be expected given the small sample sizes of individuals per population (10 individuals).

Breeding system and population structure. We calculated the inbreeding coefficient (F_{IS}) [32] for adults and seedlings from each natural population, to gain insight into the breeding system of *I. vera*. We used 10 adults and 10 seedlings (i.e. one seedling per adult tree) in each population for inferring F_{IS}. Due to the sensitivity of F_{IS} to small sample sizes, we also calculated the mean across natural and planted stands. Calculations were performed in FSTAT, and significance levels were assessed with 100,000 permutations.

The breeding system of *I. vera* was also inferred through outcrossing rate (t) analysis of natural and planted populations. To estimate the outcrossing rate, we used 10 families per population. Families comprised five seedlings from different fruits per tree (i.e. one seedling per fruit). The maternal parent was also genotyped. We used the Bayesian method implemented in BORICE [33] to calculate t. We choose a chain of 100,000 steps with a burn-in of the first 10,000 steps to reach a stable posterior distribution of t values. The t value per population was obtained after five replicate runs.

Two approaches were used to investigate the substructuring of populations. First we calculated the average F_{ST} [32] across all natural and planted populations. Then, Bayesian clustering analysis in BAPS 5.3 [34] was used to infer the number of genetic clusters for adults from natural populations, and this analysis was repeated for seedlings from natural populations. The software BAPS chooses the most likely number of genetic clusters with a stochastic optimization algorithm. We selected the option 'clustering of groups of individuals' and used 10 iterations per clusters (K), with K values ranging from 1 to 5. The output of this initial analysis was used as the basis for the admixture analysis. We selected the option 'admixture based on mixture clustering' choosing three as the minimum population size, 100,000 iterations, and 5,000 as the reference number of individuals. Five replicates were made for each K value, and the final number of clusters considered that with the highest log likelihood value. We then inspected the BAPS plots displaying only significant admixture between populations ($P<0.05$).

Results

Microsatellite evaluation

All loci were highly polymorphic, containing between 10 and 21 alleles. There was no evidence of null alleles for adult individuals, but there was evidence for null alleles in seedlings of population N4 at locus *Inga 08*. We included this population in subsequent analyses following correction for null alleles, although this had little influence on our results. No evidence for linkage disequilibrium was detected for any planted or natural population, and all populations were in Hardy-Weinberg equilibrium for at least two loci. Measures of genetic diversity (H_O, H_S) were consistent between loci (Table 1).

Genetic diversity in adults of planted populations relative to natural populations

The total number of alleles across loci in adults of natural and planted populations was 63 and 31, respectively. Twelve alleles were only found in natural populations, whereas no alleles were found exclusively in planted populations.

The average allelic richness per natural population of adults was 7.7 (population range from 7.29 to 8.1; Table 2). In contrast, allelic richness per planted population was low ($A_9=3.7$), ranging from 3.42 to 4.26 (Table 2). Comparisons across all planted and natural populations revealed a 50% reduction in allelic richness in adults of planted populations ($A_{38}=7.599$) relative to adults of natural populations ($A_{38}=14.128$, $P=0.008$; Table 3). Similarly, gene diversity was lower ($P=0.013$) in planted ($H_S=0.696$) than in natural populations ($H_S=0.885$; Table 3).

Genetic diversity of seedlings relative to adults

Of the 63 alleles in adults sampled from natural populations, 62 were also present in the seedlings. In contrast, the number of alleles was 38% higher in seedlings (43 alleles) than adults (31) from planted populations. Furthermore, all alleles absent in the adults of planted populations, but present in their seedlings, were found in adults of natural populations. The average allelic richness per population in seedlings from natural stands was 7.06 (population range 6.37 to 7.95), while an average allelic richness of 4.91 (population range 3.96 to 5.66) was found in planted populations (Table 2). The gene diversity in seedlings of planted and natural populations ranged from 0.721 to 0.811 and from 0.831 to 0.897, respectively.

Genetic diversity across planted populations measured as mean allelic richness was higher in seedlings ($A_{38}=10.468$) than in adults ($A_{38}=7.59$, $P=0.013$; Table 3). Furthermore, gene diversity of seedlings ($H_S=0.782$) was higher ($P=0.027$) than that of adults ($H_S=0.696$) in planted populations (Table 3). On the other hand, allelic richness and gene diversity did not differ ($P=0.57$ and $P=0.96$, respectively) between adults and seedlings of natural populations (Table 3).

Breeding system and population structure in *Inga vera*

The population inbreeding coefficient of adults in natural populations was not different from zero ($F_{IS}=-0.061$; $P>0.05$), with population values ranging from -0.106 to 0.019. The inbreeding coefficient of seedlings from natural populations, based on one seedling per maternal tree, differed significantly from zero ($F_{IS}=0.149$, $P<0.05$; Table 3) and ranged from -0.067 to 0.243. Despite higher levels of inbreeding in seedlings relative to adults, the average inbreeding coefficient across populations was low (mean $F_{IS}=0.065$, $P<0.05$). In accordance with the low values of F_{IS}, we detected a high outcrossing rate in natural populations ($t=0.92$, 2.5 percentile = 0.88, 97.5 percentile = 0.95), ranging from 0.83 to 0.98, as would be expected in species with an outcrossing breeding system. Outcrossing rate in planted populations was also high ($t=0.89$, 2.5 percentile = 0.81, 97.5 percentile = 0.93).

Adult planted stands exhibited low inbreeding coefficients, which did not differ from zero, similar to that observed in natural populations ($F_{IS}=-0.084$, $P=0.3$). There was no difference between the inbreeding coefficient of adults from planted and natural populations ($P=0.659$; Table 3). The inbreeding coefficient of seedlings from planted populations was moderate ($F_{IS}=0.303$), and significantly different from zero ($P<0.05$), and higher than F_{IS} of seedlings from natural populations ($F_{IS}=0.149$, $P=0.05$; Table 3).

Table 2. Genetic diversity per population of adults and seedlings of *Inga vera* in planted and natural stands.

Populations	Adults			Seedlings		
	A	A_R	H_S	A	A_R	H_S
Planted						
P1	4.3	3.62	0.612	6	3.96	0.684
P2	4.8	3.98	0.698	5.75	5.35	0.811
P3	4.0	3.42	0.637	5.25	4.01	0.721
P4	4.3	3.45	0.611	6.25	5.58	0.78
P5	4.8	4.26	0.726	6.5	5.66	0.798
Mean (±SD)	4.4 (0.33)	3.7 (0.36)	0.655 (0.05)	6.01 (0.48)	4.91 (0.85)	0.759 (0.05)
Natural						
N1	8.3	8	0.867	7.5	6.64	0.839
N2	7.8	7.29	0.861	8	7.06	0.881
N3	8.0	7.29	0.838	7.25	6.37	0.831
N4	9.0	8.1	0.885	8.25	7.26	0.886
N5	8.5	7.89	0.876	9.5	7.95	0.897
Mean (±SD)	8.3 (0.48)	7.7 (0.4)	0.865 (0.01)	8.1 (0.87)	7.06 (0.6)	0.867 (0.03)

Average number of alleles per locus (*A*), allelic richness rarefied to nine individuals (A_R; see text) and gene diversity (H_S) per population and overall in adults and seeds among planted and natural populations.

The extent of population substructuring was greater in adults of planted populations (F_{ST} = 0.084) compared to adults of natural populations (F_{ST} = 0.028, P = 0.011; Table 3). Low levels of genetic substructuring were also detected in seedlings from planted populations (F_{ST} = 0.041; Table 3). Despite the low values, genetic substructuring in adults and seedlings of planted populations was significantly greater than zero ($P<0.05$; Table 3). The results of the Bayesian assignment analyses performed in BAPS were consistent with this pattern, with a single genetic cluster (K = 1) recovered for all adult individuals from natural populations, and three genetic clusters (K = 3) detected in adults from planted populations, with limited admixture between them. The same difference regarding the genetic clustering was detected in seedlings of natural (K = 1) and planted (K = 3) populations, as the respective adults (Fig. 2).

Discussion

In this study we used *Inga vera* to analyse the retention of genetic diversity in restoration initiatives in the most threatened region of the Atlantic forest of Brazil. We showed that *Inga vera* is an outcrossing species, with high levels of genetic diversity within populations. Comparative analyses revealed that the adults of planted populations have reduced genetic diversity in relation to the adults of natural populations, and also present some genetic structuring, whereas natural populations have no spatial structure. In addition, the seedlings of planted populations have lower genetic diversity than those of natural areas. However, the seedlings of planted populations have higher genetic diversity than the adults, suggesting pollen-mediated gene flow from natural to planted stands. Our results suggest that for self-incompatible species, like *I. vera*, plantations sourced from narrow genetic founding stock may recoup genetic diversity by outcrossing with natural stands. In addition, we hypothesize that plantations of outcrossing species may positively contribute to gene flow through hyper-disturbed landscapes.

Reduced genetic diversity in planted stands of *Inga vera*

Planted stands of *I.vera* had lower allelic richness and expected heterozygosity relative to natural populations. A reduction of

Table 3. Genetic variation for the four categories of *Inga vera* population studies.

Populations	A_R	H_S	F_{IS}	F_{ST}
Planted				
Adults	7.599[a]	0.696[a]	−0.084[a]	**0.084**[a]
Seedlings	10.468[b]	0.782[b]	**0.303**[b]	**0.041**[b]
Natural				
Adults	14.128[c]	0.885[c]	−0.061[a]	0.028[b]
Seedlings	13.195[c]	0.881[c]	**0.149**[c]	0.021[b]

Allelic richness (A_R) per locus rarefied to 38 individuals, gene diversity (H_S), inbreeding coefficient (F_{IS}) and genetic differentiation (F_{ST}). Values in the same column followed by distinct letters were statistically different at $P\leq0.01$ for A_R and $P\leq0.05$ for H_S, F_{IS} and F_{ST}; All the comparisons were based on 1,000 permutations; Values of F_{IS} and F_{ST} in bold were different from zero at $P<0.05$).

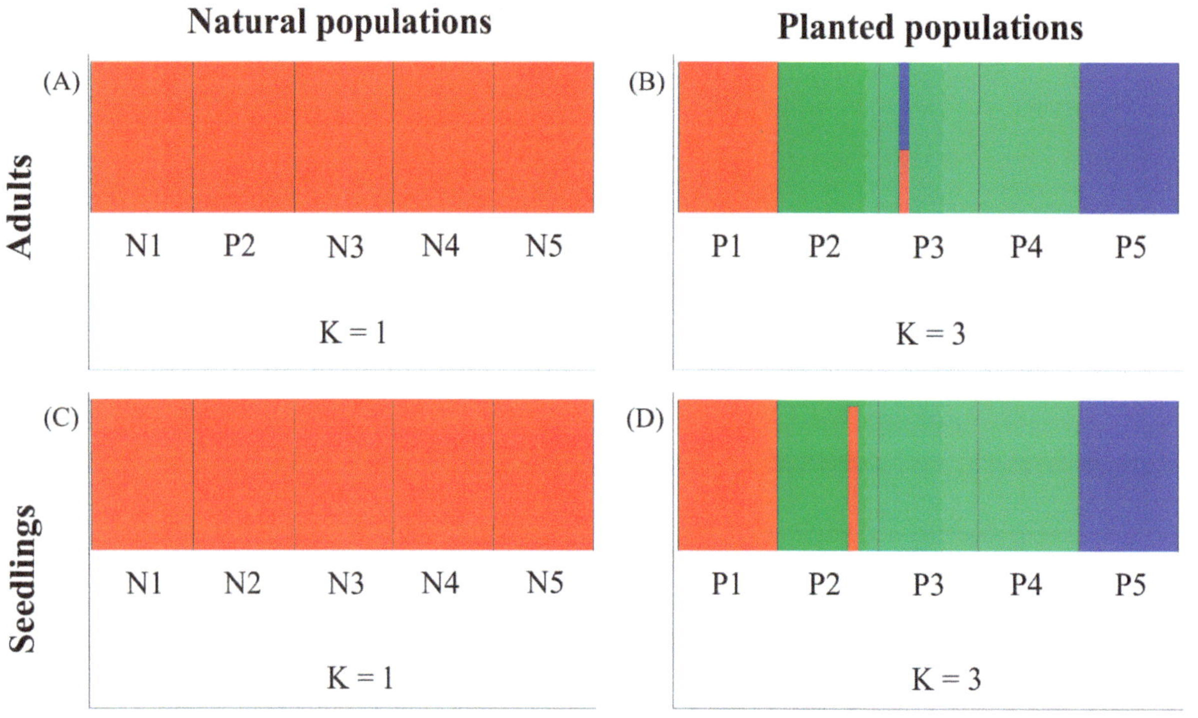

Figure 2. Genetic substructure of planted and natural populations *Inga vera*. Structure plot generated using BAPS 5.3, displaying only significant admixture ($P<0.05$) between planted and natural populations of adults (A and B) and seedlings (C and D). Number of genetic clusters (K) were assessed after five replicates.

genetic diversity in planted relative to natural populations was also found in *I. edulis* for both nuclear and organellar DNA [5,18], and this trend has also been observed in other tree species [15,17]. It is likely that low genetic diversity in *I. vera* planted populations is explained by planting seeds from only a few trees, a standard practice in the northeastern Atlantic forest [16]. However, the increased genetic structure detected in planted populations relative to natural populations, as indicated by higher F_{ST} values and the greater number of clusters in the Bayesian analysis, suggests that not all planted populations are derived from fruits of a single maternal parent.

The clustering of *Inga* pollen in polyads with approximately 20 grains [35] may also influence patterns of genetic diversity. Each pollen grain in a polyad can fertilize one ovule in the same flower. This strategy, where many pollen grains are dispersed as one unit, and where any one of the grains is able to fertilize the ovules, is called "sweepstake" reproduction [36]. Despite the fact that sweepstake reproduction results in a high seed set for a given fruit, the genetic diversity among seeds of the same fruit will be low [36]. Therefore, the planting of seeds from few fruits will restrict the genetic diversity of planted populations of *Inga* species. We emphasize that restoration efforts with species of *Inga* should be done with seeds from different individuals and fruits to avoid close kinship relationships between the seedlings, which may impair the long-term reproductive output (i.e. seed quality, vigour) of planted populations.

Recovery of genetic diversity in planted populations through outcrossing

We found evidence suggesting that planted populations might be outcrossing with natural populations. We detected an increase of 30% in allelic richness in seedlings compared with adults of planted populations, while no change was detected in natural populations. The presence of alleles only found in adults from surrounding natural forest fragments (and not in planted stands) provides some support for pollen flow from natural stands enhancing levels of diversity in planted populations. We also detected an increase in expected heterozygosity from adults to seedlings in planted populations, without a change in natural populations, as found for allelic richness. We do, however, recognize that our experimental design, intended to sample multiple populations and seedlings with low coverage of individuals, is unlikely to detect the full range of allelic diversity in a given population, and therefore cannot comprehensively prove outcrossing to natural populations.

Gene flow from natural to planted stands has been detected for *Pinus canariensis*, which is a wind pollinated tree [4]. In this case, immigration rates into artificially regenerated forest were high, ranging from 0.68 to 0.75, and genetic diversity was significantly increased, indicating genetic recovery in *Pinus canariensis* plantations surrounded by larger natural stands. Our results are also consistent with genetic recovery in planted populations of trees, but in this case in an angiosperm tree species that is insect pollinated in a hyper-fragmented area. This shows that while the planting of trees generally results in populations with lower allelic richness [15,17,18], as in *I. vera*, genetic diversity can, to some extent, be restored by receiving pollen from natural populations.

Breeding system and genetic structure of natural populations

Previous pollination experiments with nine *Inga* species, including *I. vera*, demonstrated that all species were obligate outcrossers (have a xenogamic mating system) and are dependent on cross pollination to set fruits and seeds [35]. The uniformly low values for the inbreeding coefficient across natural populations of *I. vera* reported here are consistent with obligatory xenogamy. Furthermore, we found high genetic diversity (e.g. allelic richness) and outcrossing rate values in adults and seedlings from natural populations, as would be expected in an obligate xenogamic tree species with large effective population sizes. Similar patterns of low inbreeding values and high genetic diversity have also been found in *I. edulis* from Peruvian Amazon forest [18]. Overall, the genetic results presented here, and those from previous studies, support obligate xenogamy as the dominant reproductive system for *Inga* species.

The low genetic structuring detected among populations of *I. vera* is likely to indicate high levels of gene flow between populations. Weak population genetic structure may be a consequence of the pollination system of *I. vera* and also confirms the occurrence of outcrossing in the populations that we studied. The majority of *Inga* species, including *I. vera,* can be considered hawkmoth pollinated, despite occasional vistitation by bat and hummingbirds during the day [21], (Cruz Neto et al., unpublished). Hawkmoths, bats and hummingbirds can fly across large areas, ca. 15 km, during their foraging routes carrying pollen grains to distant individuals in well preserved habitats [2,35,37]. Fruit and seed set in self-incompatible *Inga* species are ensured by the cross pollination associated with vectors that can move long distances [35]. Therefore, pollen flow between distant individuals in different populations, due to pollinator behavior, contributes to the high outcrossing rate and the weak population substructure found in natural populations of *I. vera.*

Curiously, there is an increase in inbreeding of seedlings relative to adults in both natural and planted populations despite the outcrossing breeding system and low population substructuring in adults of *I. vera.* Habitat fragmentation can alter reproductive strategies in plant communities [11] and negatively interferes in plant-pollinator interactions [12] by favouring short distance dispersion of pollen grains [3]. These changes can increase inbreeding by leading to crossing between related individuals [3,13]. Our study site is included in the Pernambuco endemism centre which is the most fragmented region of Atlantic rain forest [40].

Despite their large foraging area, hawkmoth species are negatively affected by habitat disturbance and may become locally extinct, or have reduced foraging routes in hyper-fragmented habitats [38,39]. This could also reduce pollen flow between populations, and increase breeding among potentially related individuals within the same forest fragment. Thus the association between intense forest fragmentation in this area and the dependence of biotic pollinators may contribute to the increase in inbreeding of seedlings of *I. vera* in both natural and planted populations.

Implications for the conservation and restoration of the Atlantic forest

Many habitat restoration plans have been implemented in the Atlantic forest during the last 30 years [16]. The current consensus is that restoration plans must enhance natural succession and preserve ecological interactions, such as pollination services that may be critical for ensuring restoration success [41], and high genetic diversity of planted populations [17]. Plants in tropical biodiversity hotspots are likely to exhibit high levels of biotic dependence for pollination [41,42]. We emphasize that the success of restoration actions in hyper-fragmented ecosystems, such as the Atlantic forest, is likely to involve high number of plant species pollinated by vectors with the ability to fly long distances in their foraging routes [41]. This is the case of species of *Inga* in which the main pollinators are typically hawkmoths, bats and hummingbirds [20,22,35,43]. Pollination in reforested areas may be also facilitated by proximity to natural patches that support pollinator communities [41], such as some of the reforested stands in this study.

The genetic recovery found in this study may also be attributed to the highly synchronous flowering among populations of *I. vera* in the northeastern Atlantic forest [22]. The synchronous flowering events of *I. vera* which are typical of other *Inga* [22,44] species, means that planted populations may be an abundant source of flowers for pollinators supplementing natural populations. Planted stands of *I. vera* with a high density of individuals can increase the availability of floral resource such as nectar and pollen, and therefore favour the attraction of pollinators [45,46], connecting natural fragments with the reforested stands.

Conclusions

Our population genetic data for *Inga vera* reveal that low genetic diversity of planted populations can be recovered by cross-pollination with natural stands, particularly given the outcrossing nature of the study species. Recovery of genetic diversity will, however, be hampered if the genetic diversity of natural stands is reduced by habitat degradation, or if very large stands are planted using a narrow genetic founding stock. Future restoration plans in the Atlantic rain forest should consider the potential of pollen dispersal across large spatial scales, and make an effort to ensure high genetic diversity in planted populations, in order to facilitate successful restoration in this hyper-fragmented ecosystem.

Acknowledgments

We thank to L.G. Dantas, A.P. Harand, J. Silva, M. Woolley, and R. Fernandes (Universidade Federal de Pernambuco) for essential support during the labwork; R. Ennos and M. Ruhsam for useful discussions on patterns of gene flow; E. Espeland (Academic Editor), P. Wolf, and an anonymous reviewer for valuable suggestions on the manuscript. We are also grateful to the Centro de Pesquisas Ambientais do Nordeste and the Projeto Serra Grande for providing logistical support to our research.

Author Contributions

Conceived and designed the experiments: OCN AVA AVL. Performed the experiments: OCN AVA. Analyzed the data: OCN AVA ADT LEN RTP AVL. Wrote the paper: OCN AVA ADT LEN RTP AVL.

References

1. Aguilar R, Quesada M, Ashworth L, Herrerias-Diego Y, Lobo J (2008) Genetic consequences of habitat fragmentation in plant populations: susceptible signals in plant traits and methodological approaches. Mol Ecol 17(24): 5177–5188.
2. Llorens TM, Byrne M, Yahtes CJ, Nistelberger HM, Coates DJ (2012) Evaluating the influence of different aspect of habitat fragmentation on mating patterns and pollen dispersal in the bird-pollinated *Banskia sphaerocarpa* var. *caesia.* Mol Ecol 21: 314–328.
3. Lobo J, Solís S, Fuchs EJ, Quesada M (2013) Individual and temporal variation in outcrossing rates and pollen flow patterns in *Ceiba pentandra* (Malvaceae: Bombacoidea). Biotropica 45(2): 185–194.
4. Navascués M, Emerson BC (2007) Natural recovery of genetic conservation by geneflow in reforested areas of the endemic Canary Island pine, *Pinus canariensis.* Forest Ecol Manag 244: 122–128.
5. Dawson IK, Hollingsworth PM, Doyle JJ, Kresovich S, Weber JC, et al. (2007) Origins and genetic conservation of tropical trees in agroforestry systems: a case study from Peruvian Amazon. Conserv Genet 9: 361–372.

6. Rodrigues RR, Lima RAF, Gandolfi S, Nave AG (2009) On the restoration of high diversity forests: 30 years of experience in the Brazilian Atlantic Forest. Biol Conserv 142: 1241–1251.
7. Wright SJ, Muller-Landau HC (2006) The future of tropical forest species. Biotropica 38: 287–301.
8. Morellato LPC, Haddad CFB (2000) The Brazilian Atlantic Forest. Biotropica 32: 786–792.
9. Ribeiro MC, Metzger JP, Martensen AC, Ponzoni FJ, Hirota MM (2009) The Brazilian Atlantic forest: How much is left, and how is the remaining forest distributed? Implications for conservation. Biol Conserv 142: 1141–1153.
10. Forzza RC, Baumgratz JFA, Bicudo CEM, Carvalho-Junior AA, Costa A, et al. (2010) Catálogo de plantas e fungos do Brasil. Rio de Janeiro: Instituto Jardim Botânico do Rio de Janeiro. 875 p.
11. Girão LC, Lopes AV, Tabarelli M, Bruna EM (2007) Changes in tree reproductive traits reduce functional diversity in a fragmented Atlantic forest landscape. PLoS ONE 2: e908.
12. Aguilar R, Ashworth L, Galetto L, Aizen MA (2006) Plant reproductive susceptibility to habitat fragmentation: review and synthesis through a meta analysis. Ecol Lett 9: 968–980.
13. Kettle CJ, Hollingsworth PM, Affré TJ, Moran B, Ennos RA (2007) Identifying the early genetic consequences of habitat degradation in a highly threatened tropical conifer, *Araucaria nemorosa* Laubenfels. Mol Ecol 16: 3581–3591.
14. Lopes AV, Girão LC, Santos BA, Peres CA, Tabarelli M (2009) Long-term erosion of tree reproductive trait diversity in edge-dominated Atlantic forest fragments. Biol Conserv 142: 1154–1165.
15. Steinitz O, Robledo-Arnuncio JJ, Nathan R (2012) Effects of Forest plantations on the genetic composition of conspecific native Aleppo pine populations. Mol Ecol 21: 300–313.
16. Alves-Costa CP, Lôbo D, Leão T, Brancalion PHS, Nave AG et al. (2008) Implementando reflorestamentos com alta diversidade na Zona da Mata Nordestina: Guia Prático. Recife: J. Luiz Vasconcelos.
17. Lengkeek AG, Jaenicke H, Dawson IK (2004) Genetic bottlenecks in agroforestry systems: results of tree nursery surveys in East Africa. Agroforest Syst 63: 149–155.
18. Hollingsworth PM, Dawson IK, Goodall-Copestake P, Richardson JE, Weber JC, et al. (2005) Do farmers reduce genetic diversity when they domesticate tropical trees? A case study from Amazonia. Mol Ecol 14: 497–501.
19. Lengkeek AG, Mwangi AM, Agufa CAC, Ahenda JO, Dawson IK (2006) Comparing genetic diversity in agroforestry systems with natural forest: a case study of the important timber tree *Vitex fischeri* in central Kenya. Agroforest Syst 67: 293–300.
20. Pennington TD (1997) The genus *Inga*. Royal Botanic Gardens, Kew.
21. Cruz-Neto O, Machado IC, Lopes AV (2007) Ecologia da polinização de *Inga striata* (Benth.) (Leguminosae-Mimosoideae) em um remanescente de Mata Atlântica no Nordeste do Brasil. Rev Bras Biociênc 5: 570–572.
22. Cruz-Neto O, Machado IC, Duarte JA, Lopes AV (2011) Synchronous phenology of hawkmoths (Sphingidae) and *Inga* species (Fabaceae–Mimosoideae): implications for the restoration of the Atlantic forest of northeastern Brazil. Biodivers Conserv 20: 751–765.
23. Ferreira ME, Grattapaglia D (1998) Protocolos: Extração de DNA genômico total de plantas, In: Ferreira ME, Grattapaglia D, editors. Introdução ao uso de marcadores moleculares em análise genética. 2nd edition. Embrapa Cenargen, Brasília, pp. 121–138.
24. Dayanandan S, Bawa KS, Kesseli R (1997) Conservation of microsatellites among tropical trees (Leguminosae). Am J Bot 84: 1658–1663.
25. Amos W, Hoffman JI, Frodsham A, Zhang L, Best S, et al. (2007) Automated binning of microsatellite alleles: problems and solutions. Mol Ecol Notes 7: 10–14.
26. Goudet J (2001) FSTAT, a Program to Estimate and Test Gene Diversities and Fixation Indices, Version 2.9.3.2. Available: http://www2.unil.ch/izea/softwares/fstat.html. Accessed 2012 Jan 16.
27. Van-Oosterhout C, Hutchinson WF, Wills DPM, Shipley P (2004) Micro-checker: software for identifying and correcting genotyping errors in microsatellite data. Mol Ecol Notes 4: 535–538.
28. Chapius MP, Estoup A (2007) Microsatellite null alleles and estimation of population differentiation. Mol Biol Evol 24: 621–631.
29. El Mousadik A, Petit RJ (1996) High level of genetic differentiation for allelic richness among populations of the argan tree [*Argania spinosa* (L) Skeels] endemic to Morocco. Theor Appl Genet 92: 832–839.
30. Nei M (1987) Estimation of average heterozygosity and genetic distance from a small number of individuals. Genetics 89: 583–590.
31. Kalinowski ST (2005) HP-RARE 1.0: a computer program for performing rarefaction on measures of allelic richness. Mol Ecol Notes 5: 187–189.
32. Weir BS, Cockerham CC (1984) Estimating F-statistics for the analysis of population structure. Evolution 38: 1358–1370.
33. Koelling VA, Monhahan PJ, Kelly JK (2012) A Bayesian method for the joint estimation of outcrossing rate and inbreeding depression. Heredity 109: 393–400.
34. Corander J, Marttinen P, Sirén J, Tang J (2008) Enhanced Bayesian modelling in BAPS software for learning genetic structures of populations. BMC Bioinf 9: 539.
35. Koptur S (1984) Outcrossing and pollinator limitation of fruit set: Breeding systems of neotropical *Inga* trees (Fabaceae: Mimosoideae). Evolution 38: 1130–1143.
36. Cruden RW (1977) Pollen-Ovule ratios: A conservative indicator of breeding systems in flowering plants. Evolution 31: 32–46.
37. Elmore LW, Miller DA, Villella FJ (2004) Foraging area size and habitat use by red bats (*Lausirus borealis*) in an intensively managed pine landscape in Missippi. Am Midl Nat 153: 405–417.
38. Kevan P (1999) Pollinators as bioindicators of the state of the environment: species activity and diversity. Agric Ecosyst Environ 74:373–393
39. Kitching IJ, Cadiou JM (2000) Hawkmoths of the world—an annotated and illustrated revisionary checklist(Lepidoptera: Sphingidae). Cornell University Press, New York.
40. Silva JMC, Tabarelli M (2000) Tree species impoverishment and the future flora of the Atlantic Forest of northeast Brazil. Nature 404: 72–73.
41. Dixon KW (2009) Pollination and restoration. Science 325: 571–572.
42. Olesen JM, Jordano P (2002) Geographic patterns in plant-pollinator mutualistic networks. Ecology 83: 2416–2424.
43. Amorim FW, Galetto L, Sazima M (2013) Beyond the pollination syndrome: Nectar ecology and the role of diurnal and nocturnal pollinators in the reproductive success of *Inga sessilis* (Fabaceae). Plant Biol 15: 317–327.
44. Koptur S (1983) Flowering phenology and floral biology of *Inga* (Fabaceae – Mimosoideae). Syst Bot 8: 354–368.
45. Hodges SA (1995) The influence of nectar production on hawkmoth behaviour, self pollination and seed production in *Mirabilis multiflora* (Nyctaginaceae). Am J Bot 82: 197–204.
46. Klinkhamer PGL, De-Jong T, Linnebank LA (2001) Small-scale patterns determine ecological relationships: an experimental example using nectar production rates. Ecol Lett 4: 559–567.

2

Community Turnover of Wood-Inhabiting Fungi across Hierarchical Spatial Scales

Nerea Abrego[1,2]*, Gonzalo García-Baquero[2], Panu Halme[1,3], Otso Ovaskainen[4], Isabel Salcedo[2]

1 Department of Biological and Environmental Science, University of Jyväskylä, Jyväskylä, Finland, **2** Department of Plant Biology and Ecology, University of the Basque Country (UPV/EHU), Bilbao, Spain, **3** Natural History Museum, University of Jyväskylä, Jyväskylä, Finland, **4** Department of Biosciences, University of Helsinki, Helsinki, Finland

Abstract

For efficient use of conservation resources it is important to determine how species diversity changes across spatial scales. In many poorly known species groups little is known about at which spatial scales the conservation efforts should be focused. Here we examined how the community turnover of wood-inhabiting fungi is realised at three hierarchical levels, and how much of community variation is explained by variation in resource composition and spatial proximity. The hierarchical study design consisted of management type (fixed factor), forest site (random factor, nested within management type) and study plots (randomly placed plots within each study site). To examine how species richness varied across the three hierarchical scales, randomized species accumulation curves and additive partitioning of species richness were applied. To analyse variation in wood-inhabiting species and dead wood composition at each scale, linear and Permanova modelling approaches were used. Wood-inhabiting fungal communities were dominated by rare and infrequent species. The similarity of fungal communities was higher within sites and within management categories than among sites or between the two management categories, and it decreased with increasing distance among the sampling plots and with decreasing similarity of dead wood resources. However, only a small part of community variation could be explained by these factors. The species present in managed forests were in a large extent a subset of those species present in natural forests. Our results suggest that in particular the protection of rare species requires a large total area. As managed forests have only little additional value complementing the diversity of natural forests, the conservation of natural forests is the key to ecologically effective conservation. As the dissimilarity of fungal communities increases with distance, the conserved natural forest sites should be broadly distributed in space, yet the individual conserved areas should be large enough to ensure local persistence.

Editor: Rick Edward Paul, Institut Pasteur, France

Funding: This study was partially funded by a PhD student fellowship to NA by the University of the Basque Country (UPV/EHU) (PIF10/2010/PIF10008), by Maj and Tor Nessling Foundation (a grant to PH), and by the Academy of Finland (grant no. 250444 to OO). The funders had no role in study design, data collection and analysis, decision to publish, or preparation of the manuscript.

Competing Interests: The authors have declared that no competing interests exist.

* Email: nerea.abrego@ehu.es

Introduction

Biodiversity is decreasing faster than ever in the history of the earth, largely as a result of human activity [1]. Consequently, conservation and management actions aiming at sustainable use of ecosystems and maintenance of biological diversity have become a priority worldwide. The most commonly applied conservation measures in terrestrial ecosystems have focused on preserving key habitats, species or communities of conservation interest. To date, most conservation strategies have focused on designating protected set aside areas or increasing the areas of existing protected areas. For example, regarding temperate forests, a considerable proportion of the remaining old-growth and semi-natural forests are already conserved, and the protected area is expected to continue increasing [2]. To make conservation efforts cost-effective in an evidence based manner, protection efforts need to be supported by research about the effects of habitat protection [2]. One factor potentially compromising the cost-effectiveness of biodiversity conservation plans is that the patterns and drivers of distribution and turnover of biodiversity across different spatial scales are still poorly understood for many organism groups [3].

A successful design of conservation area networks requires an understanding of the processes determining species composition and persistence at different spatial and temporal scales [4]. For example, if local processes are the most important factor determining the persistence and diversity of the focal species group (as has been observed for several taxa; [5–7]), it may be more effective to aim management actions or site selection procedures towards maintaining the heterogeneity and variability of habitats at the local scale, instead of e.g. increasing the connectivity among protected areas [5]. The influences of local versus regional processes in community assembly often interact with the traits of the species. For example, in a study of forest-dwelling beetles, Gering et al. [8] found that the spatial turnover of rare species was realised across large spatial scales (ecoregions), whereas the spatial turnover of common species was realised across smaller spatial scales. The most commonly used concept in measuring spatial species turnover is beta-diversity. However, this concept has been defined and used in multiple ways, making it difficult to synthesize studies based on this concept, and leading to a debate on its correct use [9,10].

Wood-inhabiting fungi are a diverse and functionally important group of organisms in forest ecosystems. In addition to their central role in nutrient cycling [11], they play a key role on the creation of new habitats and resources for other wood-dwelling species [12]. Furthermore, their dependence on dead wood makes them especially vulnerable to the changes caused by forest management (e.g. [13,14]). At the resource unit level, most species show preferences to a specific substrate type in terms of host species, log type (snag, log, stump…), decay stage and log size [15]. Therefore, the reduction of particular dead wood types (especially of coarse woody debris) in managed forests has directly led to declines and extinctions of certain species [16,17]. At the stand level, fungal communities are highly influenced by microclimatic conditions, which modify e.g. the water content of dead wood and thus can inhibit or facilitate the colonization and fructification of species [18,19]. Further, the size and isolation of forest stands can significantly influence fungal communities: specialist species have been found to be negatively affected by isolation while generalist species can even benefit from fragmented surroundings [15]. At the broadest scale, wood-inhabiting fungi are influenced by regional factors, related e.g. to climatic variation and management history [20].

Only a few studies have examined the importance of environmental factors at different spatial scales in determining the diversity of wood-inhabiting fungi [21,22]. Berglund et al. [23] applied a hierarchical study design and a hierarchical data analysis to show that habitat suitability determines the presence of polypore species at a small spatial scale. Bässler et al. [24] quantified the relative importance of the plot and forest management levels, showing that in spruce forests, species turnover is higher among plots than among management categories. Based on these previous studies, wood-inhabiting fungal species diversity is much affected by small scale processes, though also regional-scale processes seem important especially for specialized and threatened species. However, we are not aware of any previous studies in which the variation in species richness and community composition have been explicitly partitioned to different spatial scales. Such information would be critically needed to design effective conservation areas.

The overall objective of this study is to partition the variation in the assembly of wood-inhabiting fungal communities into different spatial levels. More specifically, our aim is to examine how community turnover of wood-inhabiting fungi is realised at the levels of forest management types, forest sites and plots, and how much of community variability is explained by variation in resource composition and spatial proximity. We utilize the conclusions of the present study to derive recommendations for the design of protection area networks for the conservation of wood-inhabiting fungi in temperate broadleaved forests.

Materials and Methods

Study permits were provided by the Government of Navarre (the local official institution for forest management) and the field studies did not involve endangered or protected species.

1. Study area

The study was carried out in the northern part of Navarre in northern Spain. The study area covers two biogeographical regions with temperate climate: the Atlantic region in the Northwest and the Alpine region in the Northeast. The predominant forest type is beech-dominated forest (with total area of ca. 61,000 ha, covering ca. 14% of the landscape), which also represents the most productive forest type for commercial harvesting in the area.

The northern part of Navarre consists of a valley-dominated landscape with a mountainous topography. Towns and farms are generally located in the lower parts of the valleys and the forests grow in the hillsides. The major part of the forested area has been intensively managed by thinning and selective cutting, but there is also a network of permanently protected areas (see Files S1 and S2). In this study, we examined the difference between two management types, called henceforth managed and natural forests. As natural forests we considered such protected areas in which logging activities have been prohibited, allowing the development of old-growth beech forests. As managed forests we considered such unprotected sites in which commercial harvesting has been continuously conducted (File S1).

2. Study design

2.1. Sampling design. We assembled a list of natural and managed forest sites based on the above definitions. From this list, eight managed and eight natural forest sites (File S1) were randomly selected. Within each site, fungal data (species abundances, defined as number of logs in which each species was observed) were recorded in five randomly placed 100 m^2 sampling plots (File S2). The sampling plots were separated from each other by tens of meters to hundreds of meters, whereas the study sites were separated by hundreds of meters to kilometres (the minimum, maximum and mean distances between plots were 0.07, 114 and 44 km respectively). The sites belonging to the two management types were distributed in space without a strong spatial autocorrelation, i.e. the managed and natural sites did not form their own clusters. The nested design has thus two factors: management (considered as fixed factor; with two levels) and forest site (considered as random factor; with eight levels, nested within management). Variation among study sites represents spatial variation at a regional scale, whereas the residual variation among the sampling plots represents spatial variation at a local scale.

2.2. Data collection. The fieldwork was carried out in the year 2011 during the main period of fungal fruit body production (from late September to early November). All woody debris (called henceforth resource units) with minimum diameter of 1 cm were measured for diameter and decay class, and examined for the presence of fungal species. The resource units were classified according to their diameter (three classes) and decay stage (five classes), thus forming altogether 15 classes. The diameter classes correspond to those used by Abrego & Salcedo [17]: very fine woody debris (VFWD), including resource units with diameter in the range 1–5 cm; fine woody debris (FWD), with 5 cm<diameter ≤10 cm; and coarse woody debris (CWD), with diameter >10 cm. The decay stages were classified into 5 categories based on a modified version of the classification by Renvall [25].

The presence of all saproxylic macromycetes (fungi with visible fruit bodies) was registered at the level of the resource units. Microscopic identification was carried out in the laboratory when necessary. The literature consulted for identifying fungal fruit bodies is summarized in File S3. Each species found on one resource unit was considered as one record, regardless of the number of fruit bodies. Thus, the abundances of the species correspond to the number of woody debris pieces in which each species was found in each sampling plot.

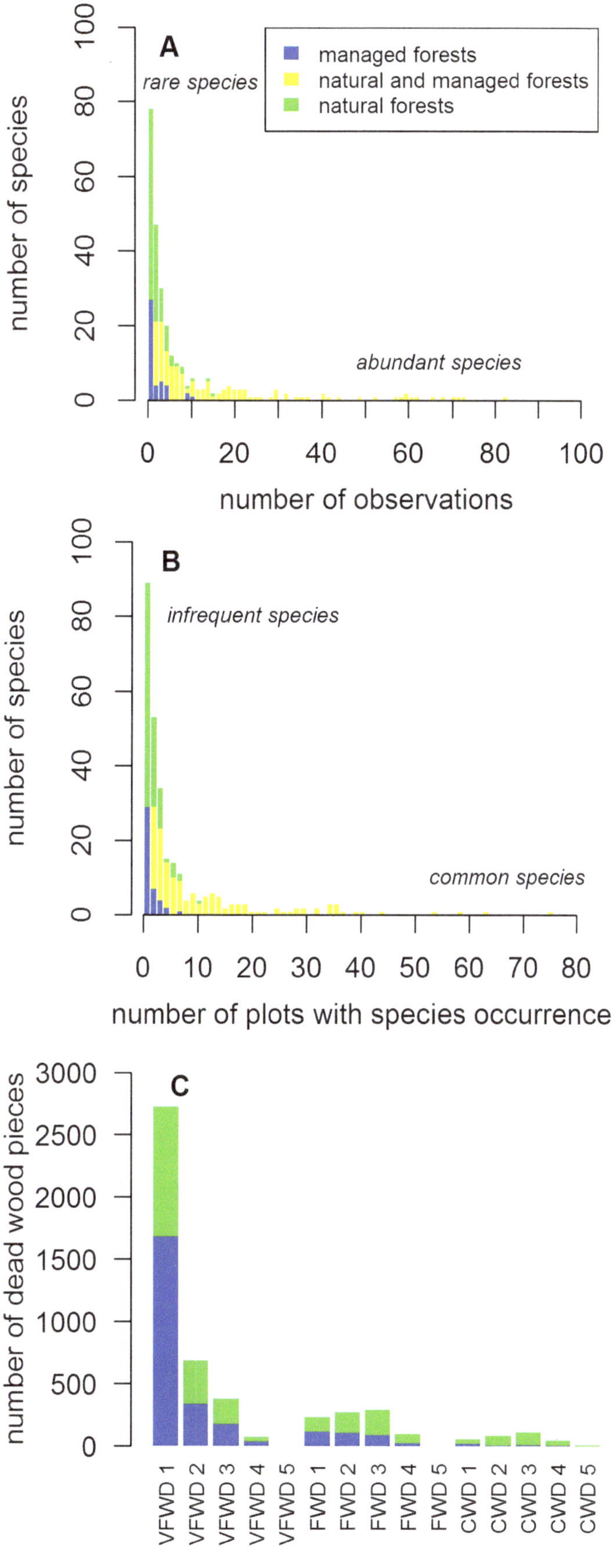
A
rare species
managed forests
natural and managed forests
natural forests
abundant species
number of species
number of observations
B
infrequent species
common species
number of species
number of plots with species occurrence
C
number of dead wood pieces
VFWD 1
VFWD 2
VFWD 3
VFWD 4
VFWD 5
FWD 1
FWD 2
FWD 3
FWD 4
FWD 5
CWD 1
CWD 2
CWD 3
CWD 4
CWD 5

Figure 1. The abundance distributions of species and dead wood types. The total number of occurrences of each species (Panel A), the number of plots in which each species was recorded (Panel B) and the number of occurrences of each dead wood type (Panel C). The species and dead wood pieces that were recorded in managed forests are represented with blue colour, those that were recorded in natural forests with yellow colour, and those species recorded in both forest types with green colour.

3. Statistical analyses

We first conducted a descriptive analysis examining how species richness in fungal communities varies across the three hierarchical scales included in this study. To do so, we constructed accumulation curves in which the total number of species was computed as a function of the number of surveyed plots, assuming that the plots were selected randomly either a) within the same randomly selected managed forest site, b) within the same randomly selected natural forest site, c) among all sites in managed forests, d) among all sites in natural forests, or e) among all sites. We estimated the mean number of species and the interquartile (25%–75%) range for each of these five cases and for each number of sampling plots using 1000 random replicates.

We supplemented the accumulation curves by an additive partitioning in species richness at different hierarchical scales [26,27]. In this methodology, the total diversity γ (here, species

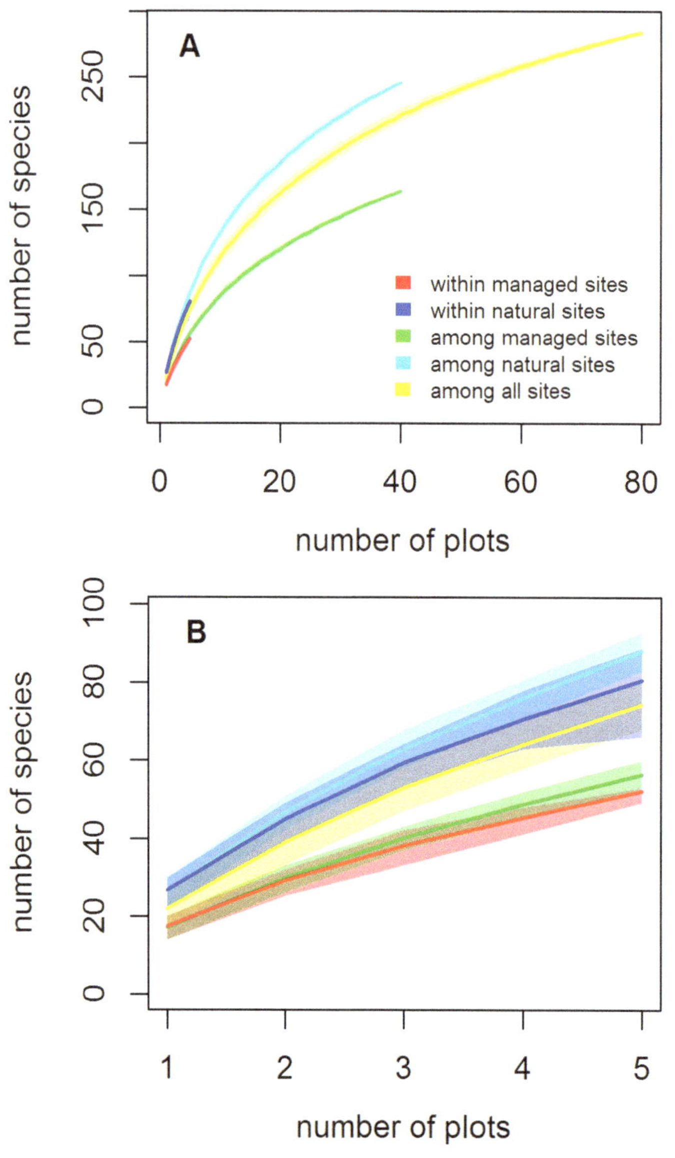

Figure 2. Species accumulation curves showing the total number of species in a given number of plots. The thick lines show means and the shaded regions the interquartile (25%–75%) ranges. The different lines correspond to the five cases which assume that the plots were selected randomly either within the same randomly selected managed forest site (red), the same randomly selected natural forest site (dark blue), among all sites in managed forests (green), among all sites in natural forests (light blue), or among all sites (yellow). Note that the maximum number of plots for which the number of species could be computed is either 5, 40 or 80, depending on the category. Panel A shows the full data, panel B a zoom, in which the number of plots varies from 1 to 5.

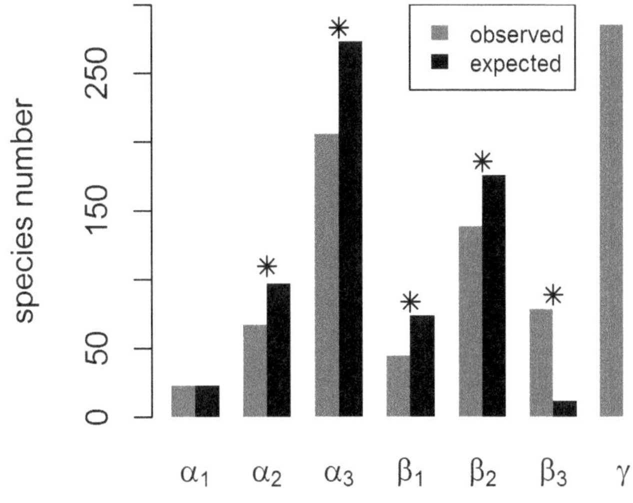

Figure 3. Total species richness (γ) explained by alpha and beta components of diversity at three spatial scales: the mean species number per plot (α_1), the mean species number per site (α_2), the mean species number per management category (α_3), the average addition of species from plots to sites (β_1), the average addition of species from sites to management categories (β_2), the average addition of species from management categories to the total species number (β_3) and the total species number (γ). The grey bars correspond to the observed values and the black bars to the expected values based on a null model. The asterisks indicate the levels of statistical significance (* corresponding to $p<0.001$) for a difference between observed and expected values.

richness) is modelled as a sum of the average diversities within sampling units (α) and among sampling units (β), the latter across the scales included in the hierarchical design. In our case, the partitioning of species richness γ was thus expressed as $\gamma = \alpha_{1(plots)} + \beta_{1(plots)} + \beta_{2(sites)} + \beta_{3(management)}$, where α_1 is the mean species number in the sampling plots, β_1 is the turnover (beta-diversity) between sampling plots, β_2 is the turnover between forest sites, and β_3 is the turnover between managed and natural forests. The additive partitioning and the related significance tests were performed under 999 permutations of the null model r0 of the function *adipart* in the R-package *vegan* [28]. The null model r0 is suited to binary (presence/absence) data, and it randomizes the community matrix in a way that preserves the number of species observed in each site.

We then analysed variation in community composition at the different levels of our hierarchical study design by calculating Bray-Curtis similarity measures [29] for all pairs of sampling plots. We constructed similarity matrices both for the fungal data (square root transformed species abundances, organised as a matrix for 80 sampling plots by 285 species) and for the dead wood data (square root transformed counts of resource unit types, organised as a matrix of 80 sampling plots by 15 resource unit types). We examined how much of the similarity in dead wood composition (SDC) was explained by the two hierarchical factors of management and site, as well as spatial distance (D). We examined how much of the similarity in fungal community composition (SFC) was explained by management and site, as well as by similarity in dead wood composition (SDC) and spatial distance (D). To do so, we constructed matrices measuring similarity in site (SS; with value 1 if the plots belonged to the same site and 0 otherwise), similarity in management (SM; with value 0 if the plots belonged to a different management category, and values 1 or 2 if both plots belonged to the managed or to the natural category, respectively) and log-transformed spatial distance (D) between the sampling plots. We treated SS and SM as fixed factors and SDC and D as continuous covariates. In the models (SDC = SS+SM+D and SFC = SS+SM+SDC+D) we considered also all two-way interactions (except for D). We used backward variable selection to include in the final model only the statistically significant variables. To account for the non-independence of data points that relate to pairs of plots rather than to individual plots, we tested for the statistical significance by comparing the *F-values* related to each factor to the null distribution of *F-values*. To compute the null distribution, we constructed 999 permutations of the species community, in each of which we permuted randomly the identity numbers of the sampling plots. As a complementary approach to quantify the variation on species and resource similarity explained by each scale, we carried out a permutational multivariate analysis of variance (PERMANOVA; [30]). The *pseudo-F* values were used to evaluate which components of the models explained more variability than expected by a null model. This analysis was performed using 999 permutations of residuals under a reduced model with the function *nested.npmanova* of the R-package *BiodiversityR*.

All statistical analyses were performed using R software [31], the code and the original data (the fungal species data matrix, UTM locations of all sampling plots and resource type data matrix) being available in the Supporting Information section (Files S4–S7).

Results

1. Variation in species richness and resource types

We recorded in total 3241 occurrences of 285 species in 2724 dead wood pieces. The species abundance distribution was dominated by rare species (Fig. 1A), with 27% of the recorded species (2.4% of the occurrences) being represented only once in our data. The five most abundant species were simultaneously the five species with highest prevalence at the plot level, comprising over 25% of all occurrences. In general, the species with high abundances were found in both (managed and natural) forest types (Fig. 1, A and B; black bars). Nevertheless, note that in the tail of Fig. 1 (A) there are also some species that were found only in managed forests. These species are *Steccherinum fimbriatum* and *Hyphodermella rosae*, which were recorded in 7 managed plots, 18 and 11 times respectively.

Table 1. Linear models for similarity in fungal and dead wood composition.

Response variable: Similarity in fungal composition (SFC)						
Source of variation	df	Estimate	SSq	MS	F	p
Management (SM)	2	MM: 0 MU: −0.060 UU: −0.047	1.93	0.96	152	<0.001
Site (SS)	1	Different: 0 Same: 0.019	0.71	0.71	112	<0.001
Similarity in dead wood composition (SDC)	1	0.11	1.11	1.11	175	<0.001
Spatial distance (D)	1	−0.02	0.46	0.46	73	<0.001
Residual	3154		20.02	0.01		
Response variable: Similarity in dead wood composition (SDC)						
Source of variation	df	Estimate	SSq	MS	F	p
Management (SM)	2	MM: 0 MU: −0.067 UU: −0.016	3.07	1.54	192	<0.001
Site (SS)	1	Different: 0 Same: 0.019	0.05	0.05	6.58	<0.001
Residual	3156		25.26	0.01		

df = degrees of freedom, SSq = sum of squares, MS = mean squares, MM = pairs of plots from managed forests, MU = pairs of plots belonging to a different management types, UU = pairs of plots from natural (unmanaged) forests. Statistical significance (*p*) based on permutation test.

The dead wood profile was numerically strongly dominated by small diameter classes in early decay stages (Fig. 1C). The coarser dead wood classes in the latest decay stages were the most uncommon, and were mostly found in natural forests.

The species accumulation curves indicate that natural forests were substantially more species rich than managed forests (Fig. 2A). Further, the number of species increased much faster if the sampling plots were located only in natural forests (Fig. 2A; light blue line) than if they were located both in natural and managed forests (Fig. 2A; yellow line), suggesting that the species assemblages in managed forests were largely a subset of those present in natural forests. The total number of species increased only marginally faster if the sampling plots were spread across forest sites than if they were located within a forest site, both in managed (Fig. 2B; green and red lines) and natural (Fig. 2B; light blue and dark blue lines).

The additive partitioning of total species richness (Fig. 3) was in line with the patterns visible in the species accumulation curves. The increase in species number when moving from the plot level to the forest site level (β_1) was lower than would expected by chance, and consequently so was also the number of species within a site, averaged across the sites (α_2). Similarly, the increase in species number when moving from the site level to the management level (β_2) was lower than would expected by chance, and consequently so was also the mean number of species within a management category, averaged across the two management categories (α_3). Both of these results reflect the fact that the most important part of species diversity was realized among the management categories (β_3). This is consistent with Fig. 2, which shows that natural forests were much more species rich than management forests, and that managed forests largely contained a subset of the species found from natural forests. Note that α_3 corresponds to the mean of the light blue and green lines in Fig. 2, which indeed is below the yellow line, which corresponds to a null expectation.

2. Variation in community and resource similarities

After applying the model selection procedure, the model for similarity in fungal community composition (SFC) included as explanatory variables management type similarity (SM), forest site similarity (SS), similarity in dead wood composition (SDC) and spatial distance (D) (Table 1). The model for similarity in dead wood composition (SDC) included management type similarity (SM) and forest site similarity (SS) but not spatial distance (Table 1). In both models none of the two-way interactions were significant (Permutation comparison of *F-values*, $p>0.05$). The variability explained by both models was low, as only the 17% of the total variability was explained in the SFC model and 11% in the SDC model. According to the predicted values of these models, the plots within managed forests showed higher fungal and dead wood similarities than the plots within natural forests, and the lowest fungal and dead wood similarities were found among pairs of plots out which one belonged to a natural and the other one to a managed forest (Fig. 4A). The effect of forest site similarity (SS) on the similarity of both fungal composition (SFC) and dead wood was small but yet significant (Fig. 4, Table 1). Fungal community similarity decreased with increasing spatial distance (Fig. 4B).

The results from the Permanova analysis (Table 2) were in line with the results of the linear model described above: for both fungal and dead wood species composition, both forest management type and forest site level explained a significant amount of variation. For dead wood composition, the variation explained by forest management type was much higher than that explained by forest site level.

Discussion

We found that the assembly of fungal communities was influenced by all scales that we considered in this study. Indeed, the similarity of fungal communities was higher within sites and within management categories than among sites or between the two management categories; it decreased with increasing distance

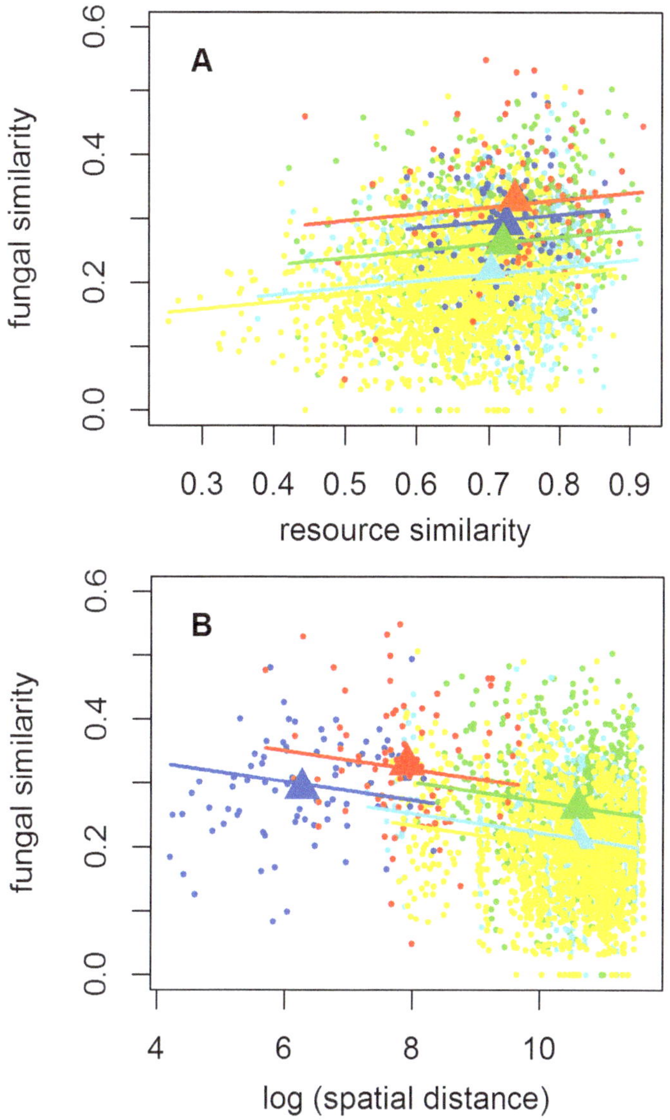

Figure 4. Predicted similarity of the fungal composition (SFC), among sampling plots. When two sampling plots are located either within the same managed forest site are represented by red colour, when located within the same natural forest site are represented by dark blue colour, when located in two different sites in managed forests are represented by green colour, when located in two different sites in natural forests are represented by light blue colour, and when plots are located in sites from different management types are represented by yellow colour. The dots show raw data, the triangles the mean values and the lines the model predictions. When constructing the model prediction as a function of resource similarity (panel A), we normalized spatial distance to its mean value within each category. When constructing the model prediction as a function of spatial distance (panel B), we normalized resource similarity to its mean value within each category.

among the sampling plots and increased with increasing similarity of dead wood resources. However, the majority of the variation remained unexplained, suggesting that biological interactions, neutral mechanisms such as inherent randomness in dispersal, establishment and fruiting [32], as well as environmental constraints not considered here are likely to play an important role in determining local assemblies of wood-inhabiting fungi. This result partly reflects the fact that the wood-inhabiting fungal communities studied here were dominated by locally rare and infrequent species, which directly leads to a high species turnover from one plot to another. This result is in line with a previous study in a coniferous forest where a high variability of wood inhabiting fungal species at small scales was also detected [24].

Bässler et al. [24] partitioned wood-inhabiting fungal richness between plots and management categories, with the result that community turnover was higher between study plots than between management categories. Thus Bässler et al. [24] concluded that the stand level is the most relevant scale for the conservation the wood-inhabiting diversity. In contrast, we found that fungal communities were significantly influenced by all scales that we considered, and in particular by the management category, as natural forests were significantly more diverse than managed forests. The difference between our results and those of Bässler et al. [24] can be at least partially attributed to a difference in the study design, as the study of Bässler et al. [24] involved only a single forest site for each management category.

Table 2. Output from the Permanova analysis for fungal species and dead wood composition models.

	Response variable: Similarity in fungal composition (SFC)					Response variable: Similarity in dead wood composition (SDC)			
Source of variation	**df**	**SS**	**MS**	***Pseudo-F***	***p***	**SSq**	**MS**	***Pseudo-F***	***p***
Management	1	17737	17737	3.53	0.001	8549.2	8549.2	12.45	0.001
Site	14	70343	5024.5	2.08	0.001	9612.1	686.58	1.74	0.002
Residual	64	154670	2416.7			25288	395.13		

df = degrees of freedom, SSq = sum of squares, MS = mean squares.

In general, wood-inhabiting fungal species show preferences for certain resource types in terms of tree species, size and decay class [15], so from a niche differentiation perspective [33], variation in resource type is expected to explain a major part of species variation at a small scale. In our study, although increasing similarity in resource composition (as measured by dead wood decay class and size) significantly increased the similarity of wood-inhabiting fungal communities, the trend was weak. Hence, at a small spatial scale, wood-inhabiting fungal communities seem to be largely influenced by stochastic processes related to species' idiosyncrasy and spatial proximity, rather than habitat factors. Some hierarchically designed studies have reported similar results, and showed that the variation in the occurrence of species among different forest stands [23] or study plots [34] remains largely unexplained even when the most important explanatory variables were considered.

The dispersal capabilities of species can be considered as a species-specific factor limiting potentially their occurrence on suitable habitats [35,36]. One of the covariates that significantly influenced fungal species similarity in our study was the spatial distance, with the result that similarity in fungal composition decreased with increasing distance between the study plots. This result suggests that dispersal limitation is one major factor shaping the local assembly of fungal communities.

Other factors creating unexplained variation between study units are related to the low detectability of wood-inhabiting fungi [37], partly due to their fruiting phenology and its dependence on microclimatic variation [38]. Many fungal species have ephemeral, short living fruit bodies, which makes it difficult to assess fungal diversity by a single fruit body based survey [37]. Further, a large proportion of fungal species can never be detected as fruit bodies [39]. With presence-absence data, low detectability generates a larger amount of random variation on small spatial scales because on larger scale the increasing number of potentially detectable individuals increases the likelihood that at least one is detected [40]. Thus, also in our study, the low general detectability of fungi as well as the climatic dependence of their phenology may have inflated the variation realized at the smallest spatial scale.

Due to the overall high number of species and the large proportion of rare species, the species accumulation curves we constructed were far from saturated. The only way to achieve saturated accumulation curves for hyperdiverse organism groups consists of sampling intensively over a limited area during several years [8,41]. Nevertheless, in spatially extensive studies even unsaturated accumulation curves provide a suitable method for the assessment of the heterogeneity of species diversity [8].

It is well documented that wood-inhabiting fungal communities in natural forests are significantly more species rich and different from those in managed forests (e.g. [16,42,43]), including the communities in beech forests [17]. The main factors behind this difference are related to resource availability and connectivity at forest stand level [15,42]. Our previous study [17] showed that species specialised in utilizing certain dead wood types were absent from managed forests, due to the absence of suitable resource type. Based on the previous [17] and present results, the community differences between different management categories are not related to qualitative differences in species compositions, but to quantitative differences, in the sense that the species from managed forests are a subset from those in natural ones (because their required resource has disappeared). Therefore we cannot say that the natural and managed forests have different species compositions, but rather that natural forests have more species. Indeed, the general trend reported in the literature is that the reduction of particular dead wood pieces caused by forestry practices in managed forests causes extinctions of habitat specialist saproxylic species (e.g. [44–46]). Thus, forest management significantly affects resource composition, and managed forests are more homogeneous in their dead wood and fungal species compositions. Nevertheless we found some species to be more abundant in managed forests than in natural ones, in particular *Steccherinum fimbriatum* and *Hyphodermella rosae*. Consistently with our results, earlier studies have shown that there are species adapted to disturbances, such as those caused by forest management [15,47].

Conservation implications and conclusions

Identifying the spatial scales at which species vary in their diversity and composition is an important research goal in conservation ecology [4,48]. Additive partitioning of diversity is one of the most commonly applied statistical approaches for quantifying the fraction of diversity explained by different scales and for concluding which scales are critical for species conservation (e.g. [5–8]). We found that fungal communities show a high level of turnover at all the studied scales, and that only a small part of the turnover can be explained by factors such as spatial proximity or resource type similarity. The large number of random variation and high number of rare species suggests that a large total area of protected forests is needed to conserve a substantial part of fungal diversity. As managed forests contributed only little to the total number of species, an ecologically effective way to conserve fungal communities necessarily requires conserving natural forests. Nevertheless, leaving dead wood in managed forests can be seen an additional supplementary method, partly because some rare species only appear in managed forests, and partly because increasing the quality of the matrix (here, managed forests) can be essential for ensuring the long-term viability of populations in fragmented landscapes [49]. We found the similarity of fungal communities to decrease with increasing spatial distance, suggesting that protected areas should be distributed thorough the landscape. However, the individual protected areas or local networks of protected areas must simultaneously be kept large enough to ensure the persistence of local populations, which process we did not however consider in the present study.

Supporting Information

File S1 Principal characteristics of the sampled beech forest sites.

File S2 A map of the sampled forest sites and plots in Navarre.

File S3 Literature consulted for the identification of wood-inhabiting species.

File S4 Code for performing the analyses in R.

File S5 Fungal species data.

File S6 Geographical UTM coordinates.

File S7 Dead wood composition data.

Acknowledgments

We are grateful to the Government of Navarre for giving the sampling permissions. We also thank A. Chiarucci and an anonymous reviewer for their valuable comments on the manuscript.

Author Contributions

Conceived and designed the experiments: NA GG-B IS. Performed the experiments: NA. Analyzed the data: NA GG-B OO. Contributed to the writing of the manuscript: NA GG-B PH OO IS.

References

1. Barnosky AD, Matzke N, Tomiya S, Wogan GOU, Swartz B, et al. (2011) Has the earth/'s sixth mass extinction already arrived? Nature 471: 51–57. Available: http://dx.doi.org/10.1038/nature09678.
2. Götmark F. (2013) Habitat management alternatives for conservation forests in the temperate zone: Review, synthesis, and implications. For Ecol Manage 306: 292–307. Available: http://dx.doi.org/10.1016/j.foreco.2013.06.014.
3. Chiarucci A, Bacaro G, Rocchini D. (2008) Quantifying plant species diversity in a natura 2000 network: Old ideas and new proposals. Biol Conserv 141: 2608–2618. Available: http://dx.doi.org/10.1016/j.biocon.2008.07.024.
4. Summerville KS, Boulware MJ, Veech JA, Crist TO. (2003) Spatial variation in species diversity and composition of forest lepidoptera in eastern deciduous forests of north america. Conserv Biol 17: 1045–1057. 10.1046/j.1523-1739.2003.02059.x.
5. Wagner H, Wildi O, Ewald K. (2000) Additive partitioning of plant species diversity in an agricultural mosaic landscape. Landscape Ecol 15: 219–227. Available: http://dx.doi.org/10.1023/A%3A1008114117913.
6. Summerville KS, Crist TO. (2003) Determinants of lepidopteran community composition and species diversity in eastern deciduous forests: Roles of season, eco-region and patch size. Oikos 100: 134–148. 10.1034/j.1600-0706.2003.11992.x.
7. Müller J, Goβner MM. (2010) Three-dimensional partitioning of diversity informs state-wide strategies for the conservation of saproxylic beetles. Biol Conserv 143: 625–633. Available: http://dx.doi.org/10.1016/j.biocon.2009.11.027.
8. Gering JC, Crist TO, Veech JA. (2003) Additive partitioning of species diversity across multiple spatial scales: Implications for regional conservation of biodiversity. Conserv Biol 17: 488–499. 10.1046/j.1523-1739.2003.01465.x.
9. Tuomisto H. (2010) A consistent terminology for quantifying species diversity? yes, it does exist. Oecologia 164: 853–860. Available: http://dx.doi.org/10.1007/s00442-010-1812-0.
10. Anderson MJ, Crist TO, Chase JM, Vellend M, Inouye BD, et al. (2011) Navigating the multiple meanings of β diversity: A roadmap for the practicing ecologist. Ecol Lett 14: 19–28. 10.1111/j.1461-0248.2010.01552.x.
11. Boddy L, Frankland JC, van West P. (2008) Ecology of saprotrophic basidiomycetes. London, United Kingdom: Elsevier Academic Press. Available: http://dx.doi.org/10.1016/S0275-0287(08)80001-X.
12. Stokland JN, Siitonen J, Jonsson BG. (2012) Biodiversity in dead wood. Cambridge: Cambridge University Press.
13. Penttilä R, Siitonen J, Kuusinen M. (2004) Polypore diversity in managed and old-growth boreal picea abies forests in southern finland. Biol Conserv 117: 271–283. 10.1016/j.biocon.2003.12.007.
14. Hottola J, Ovaskainen O, Hanski I. (2009) A unified measure of the number, volume and diversity of dead trees and the response of fungal communities. J Ecol 97: 1320–1328. 10.1111/j.1365-2745.2009.01583.x.
15. Nordén J, Penttilä R, Siitonen J, Tomppo E, Ovaskainen O. (2013) Specialist species of wood-inhabiting fungi struggle while generalists thrive in fragmented boreal forests. J Ecol 101: 701–712. 10.1111/1365-2745.12085.
16. Küffer N, Senn-Irlet B. (2005) Influence of forest management on the species richness and composition of wood-inhabiting basidiomycetes in swiss forests. Biodivers Conserv 14: 2419–2435. 10.1007/s10531-004-0151-z.
17. Abrego N, Salcedo I. (2013) Variety of woody debris as the factor influencing wood-inhabiting fungal richness and assemblages: Is it a question of quantity or quality? For Ecol Manage 291: 377–385. 10.1016/j.foreco.2012.11.025.
18. Heilmann-Clausen J. (2001) A gradient analysis of communities of macrofungi and slime moulds on decaying beech logs. Mycol Res 105: 575–596. Available: http://dx.doi.org/10.1017/S0953756201003665.
19. Siitonen P, Lehtinen A, Siitonen M. (2005) Effects of forest edges on the distribution, abundance, and regional persistence of wood-rotting fungi. Conserv Biol 19: 250–260. 10.1111/j.1523-1739.2005.00232.x.
20. Ódor P., Heilmann-Clausen J., Christensen M., Aude E., Dort K., et al. (2006) Diversity of dead wood inhabiting fungi an bryophytes in semi-natural beech forest in Europe. Biol Conserv. 131: 58–71. Available: http://dx.doi.org/10.1016/j.biocon.2006.02.004.
21. Junninen K, Komonen A. (2011) Conservation ecology of boreal polypores: A review. Biol Conserv 144: 11–20. 10.1016/j.biocon.2010.07.010.

22. Sverdrup-Thygeson A, Gustafsson L, Kouki J. (2014) Spatial and temporal scales relevant for conservation of dead-wood associated species: Current status and perspectives. Available: http://dx.doi.org/10.1007/s10531-014-0628-3.
23. Berglund H, Hottola J, Penttilä R, Siitonen J. (2011) Linking substrate and habitat requirements of wood-inhabiting fungi to their regional extinction vulnerability. Ecography 34: 864–875. 10.1111/j.1600-0587.2010.06141.x.
24. Bässler C, Müller J, Svoboda M, Lepšová A, Hahn C, et al. (2012) Diversity of wood-decaying fungi under different disturbance regimes–a case study from spruce mountain forests. Biodivers Conserv 21: 33–49. 10.1007/s10531-011-0159-0.
25. Renvall P. (1995) Community structure and dynamics of wood-rotting Basidiomycetes on decomposing conifer trunks in northern Finland. Karstenia 35: 1–51.
26. Lande R. (1996) Statistics and partitioning of species diversity, and similarity among multiple communities. Oikos 76: 5–13. Available: http://www.jstor.org/stable/3545743.
27. Crist TO, Veech JA, Gering JC, Summerville KS. (2003) Partitioning species diversity across landscapes and regions: A hierarchical analysis of α, β, and γ diversity. Am Nat 162: 734–743. Available: http://www.jstor.org/stable/10.1086/378901.
28. Oksanen J, Blanchet FG, Roeland R, Legendre P, Minchin PR, et al. (2013) Community ecology package "vegan", R package version 2.0–7.
29. Bray JR, Curtis JT. (1957) An ordination of the upland forest communities of southern wisconsin. Ecol Monogr 27: 325–349.
30. Anderson MJ. (2001) A new method for non-parametric multivariate analysis of variance. Austral Ecol 26: 32–46. 10.1111/j.1442-9993.2001.01070.pp.x.
31. R Development Core Team. (2013) R: A language and environment for statistical computing. Vienna, Austria: R Foundation for Statistical Computing.
32. Hubbell SP. (2001) The unified neutral theory of biodiversity and biogeography. Princeton, N. J.: Princeton Univ. Press.
33. Chase JM, Leibold MA. (2003) Ecological niches: Linking classical and contemporary approaches. Chicago, Illinois, USA: University of Chicago Press.
34. Abrego N, Salcedo I. (2014) Response of wood-inhabiting fungal community to fragmentation in a beech forest landscape. Fungal Ecol 8: 18–27. Available: http://dx.doi.org/10.1016/j.funeco.2013.12.007.
35. Edman M, Gustafsson M, Stenlid J, Ericson L. (2004) Abundance and viability of fungal spores along a forestry gradient - responses to habitat loss and isolation? Oikos 104: 35–42. 10.1111/j.0030-1299.2004.12454.x.
36. Norros V, Penttilä R, Suominen M, Ovaskainen O. (2012) Dispersal may limit the occurrence of specialist wood decay fungi already at small spatial scales. Oikos 121: 961–974. 10.1111/j.1600–0706.2012.20052.x.
37. Halme P, Kotiaho JS. (2012) The importance of timing and number of surveys in fungal biodiversity research. Biodivers Conserv 21: 205–219. 10.1007/s10531-011-0176-z.
38. Moore D, Gange AC, Gange EG, Boddy L. Chapter 5 fruit bodies: Their production and development in relation to environment. In: Anonymous British Mycological Society Symposia Series: Academic Press. pp.79–103. Available: http://dx.doi.org/10.1016/S0275-0287(08)80007-0.
39. Halme P, Heilmann-Clausen J, Rämä T, Kosonen T, Kunttu P. (2012) Monitoring fungal biodiversity – towards an integrated approach. Fungal Ecol 5: 750–758. Available: http://dx.doi.org/10.1016/j.funeco.2012.05.005.
40. MacKenzie DI, Nichols JD, Royle JA, Pollock KH, Bailey LL, et al. (2006) Occupancy estimation and modeling: Inferring patterns and dynamics of species occurrence. San Diego, USA: Elsevier.
41. Novotný V, Basset Y. (2000) Rare species in communities of tropical insect herbivores: Pondering the mystery of singletons. Oikos 89: 564–572. 10.1034/j.1600-0706.2000.890316.x.
42. Penttilä R, Lindgren M, Miettinen O, Rita H, Hanski I. (2006) Consequences of forest fragmentation for polyporous fungi at two spatial scales. Oikos 114: 225–240. 10.1111/j.2006.0030-1299.14349.x.
43. Juutilainen K, Mönkkönen M, Kotiranta H, Halme P. (2014) The effects of forest management on wood-inhabiting fungi occupying dead wood of different diameter fractions. For Ecol Manage 313: 283–291. Available: http://dx.doi.org/10.1016/j.foreco.2013.11.019.
44. Ódor P, Standovár T. (2001) Richness of bryophyte vegetation in near-natural and managed beech stands: The effects of management-induced differences in dead wood. Ecol Bull: 219–229. Available: http://www.jstor.org/stable/20113278.
45. Siitonen J. (2001) Forest management, coarse woody debris and saproxylic organisms: Fennoscandian boreal forests as an example. Ecol Bull: 11–41. Available: http://www.jstor.org/stable/20113262.
46. Stokland J, Kauserud H. (2004) Phellinus nigrolimitatus–a wood-decomposing fungus highly influenced by forestry. For Ecol Manage 187: 333–343. Available: http://dx.doi.org/10.1016/j.foreco.2003.07.004.
47. Stokland JN, Larsson K. (2011) Legacies from natural forest dynamics: Different effects of forest management on wood-inhabiting fungi in pine and spruce forests. For Ecol Manage 261: 1707–1721. 10.1016/j.foreco.2011.01.003.
48. Dray S, Pélissier R, Couteron P, Fortin M-, Legendre P, et al. (2012) Community ecology in the age of multivariate multiscale spatial analysis. Ecol Monogr 82: 257–275. 10.1890/11–1183.1. Available: http://dx.doi.org/10.1890/11-1183.1.
49. Jules ES, Shahani P. (2003) A broader ecological context to habitat fragmentation: Why matrix habitat is more important than we thought. J Veg Sci 14: 459–464. 10.1111/j.1654-1103.2003.tb02172.x.

3

Identity and Specificity of *Rhizoctonia*-Like Fungi from Different Populations of *Liparis japonica* (Orchidaceae) in Northeast China

Rui Ding, Xu-Hui Chen*, Li-Jun Zhang, Xiao-Dan Yu, Bo Qu, Ru Duan, Yu-Feng Xu

College of Bioscience and Biotechnology, Shenyang Agricultural University, Shenyang, Liaoning, People's Republic of China

Abstract

Mycorrhizal association is known to be important to orchid species, and a complete understanding of the fungi that form mycorrhizas is required for orchid ecology and conservation. *Liparis japonica* (Orchidaceae) is a widespread terrestrial photosynthetic orchid in Northeast China. Previously, we found the genetic diversity of this species has been reduced recent years due to habitat destruction and fragmentation, but little was known about the relationship between this orchid species and the mycorrhizal fungi. The *Rhizoctonia*-like fungi are the commonly accepted mycorrhizal fungi associated with orchids. In this study, the distribution, diversity and specificity of culturable *Rhizoctonia*-like fungi associated with *L. japonica* species were investigated from seven populations in Northeast China. Among the 201 endophytic fungal isolates obtained, 86 *Rhizoctonia*-like fungi were identified based on morphological characters and molecular methods, and the ITS sequences and phylogenetic analysis revealed that all these *Rhizoctonia*-like fungi fell in the same main clade and were closely related to those of *Tulasnella calospora* species group. These findings indicated the high mycorrhizal specificity existed in *L. japonica* species regardless of habitats at least in Northeast China. Our results also supported the wide distribution of this fungal partner, and implied that the decline of *L. japonica* in Northeast China did not result from high mycorrhizal specificity. Using culture-dependent technology, these mycorrhizal fungal isolates might be important sources for the further utilizing in orchids conservation.

Editor: Zhong-Jian Liu, The National Orchid Conservation Center of China; The Orchid Conservation & Research Center of Shenzhen, China

Funding: This research was funded by the National Natural Science Foundation of China (No. 31100299 and No. 31200011), (URL: http://www.nsfc.gov.cn/), and China Postdoctoral Science Foundation (No. 2012M510835) (URL: http://res.chinapostdoctor.org.cn/Program/Main.html). The funders had no role in study design, data collection and analysis, decision to publish, or preparation of the manuscript.

Competing Interests: The authors have declared that no competing interests exist.

* Email: chenxh017@163.com

Introduction

The mycorrhizal association is ubiquitous but very important symbiosis in nature, which plays an essential role in the maintenance of most terrestrial ecosystems [1]. Over 90% of all plant species can form mycorrhizas with different kinds of fungi, and the existence of mycorrhizal fungi can confer to their hosts many adaptive advantages via improved water and nutrient/minerals uptake from the soil [2,3,4], enhanced plant growth [5,6], reduced toxic element accumulation [1,7], and increased resistence to pathogen damage [8]. Nowadays, based on their essential role in plant species and the whole ecosystems, more attention has been given to the potential of exploiting fungi for effective use [9,10]. Although a large number of mycorrhizal fungi existed in plants, only a fraction has been described and explored to date, and their function in ecological systems remains indistinct.

The Orchidaceae, which is one of the largest and most diverse plant families, is distributed worldwide [11]. However, many orchid species have suffered dramatic declines in distribution and some species have become rare and endangered in recent decades [12]. Mycorrhizal association is known to be important to orchids because they depend on the presence of suitable fungal partners for seed germination and seedling development [13,14]. Therefore, a complete understanding of the mycorrhizal fungi of the many threatened orchid species is required for conservation action plans [15].

Although molecular taxonomic identification of the endophytic fungi of orchid species has now revealed that the diversity of orchid associates is much complex [16], the study of the earliest-diverging orchid lineages and distribution of mycorrhizal fungal associates across orchid phylogeny supported that the ancestral state is an association to the *Rhizoctonia*-like fungi lineages [17], and orchid mycorrhizas are predominantly represented by associations between photosynthetic plants and *Rhizoctonia*-like fungi [15]. The *Rhizoctonia*-like fungi includes members of the Ceratobasidiaceae, Sebacinales and Tulasnellaceae [15,17]. However, the taxonomy, diversity and distribution patterns of mycorrhizal fungi may vary considerably among different orchid species or different populations within the same species, ranging from very narrow specificity to little specificity [18–26]. Therefore, knowing the identity of the fungi that form mycorrhizas with orchids and the specificity of the relationships are of crucial importance for orchid ecology and conservation [27].

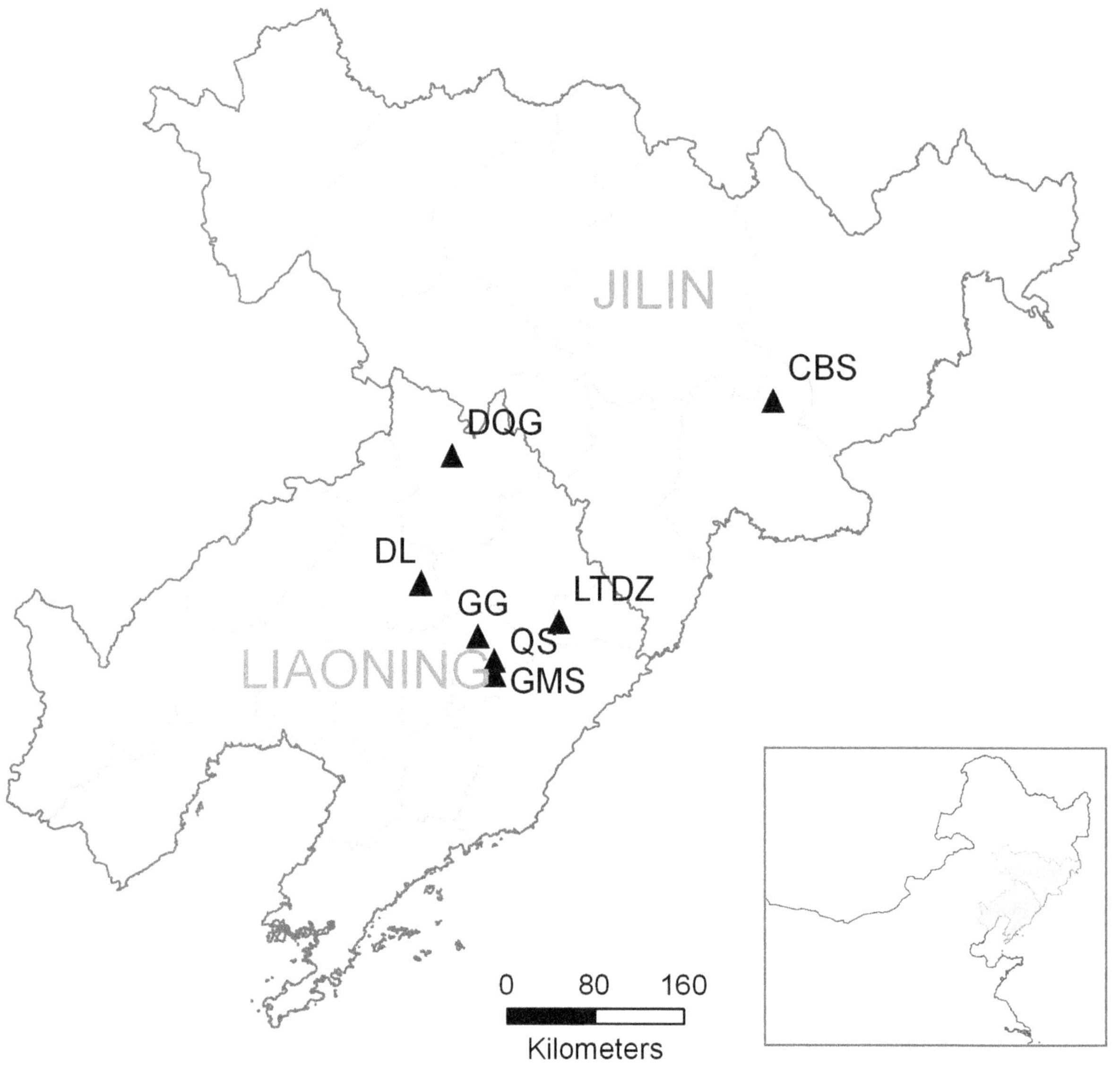

Figure 1. Locations of sampling sites of *Liparis japonica* in Northeast China.

Liparis japonica, which is a terrestrial photosynthetic orchid species, has a wide geographic distribution in the tropical and subtropical areas such as China, Korea, Japan and Russia [28]. Historical record indicated that this species had a relatively wide distribution in Northeast China. Unfortunately, the number of *L. japonica* populations has declined sharply during the last several decades and many populations recorded previously has disappeared [29]. In a previous study, we assessed the genetic diversity and population genetic structure of *L. japonica* in Northeast China, and found potential restricted gene flow existed among populations due to habitat destruction and fragmentation [29]. To conserve surviving wild populations and reintroduce plants into declining populations requires a more complete understanding on the mycorrhizal fungal partners of *L. japonica* species.

In this study, we investigated in detail the taxonomy and diversity of culturable mycorrhizal fungi associated with *L. japonica* in Northeast China using morphological and ITS phylogenetic analysis. Seven different populations of *L. japonica* species were examined in order to evaluate how fungi diversity varied within and between habitats. Using a culture-dependent method, the fungal isolates obtained will be important sources for deep study and further utilizing in orchid conservation.

Materials and Methods

Study species and sampling

185 individuals of *L. japonica* belonging to seven natural populations were collected in Northeast China at the flowering stage in June 2010 and July 2011, and the distances between these populations varied from 25 to 460 km. Details of locations and number of sampled individuals of each population were listed in Table 1 and Fig. 1. Plants were excavated with a clod of their surrounding soil and transferred to the laboratory. The field studies did not involve endangered or protected species, and no specific permissions were required for these locations.

Table 1. Location of the seven populations of *Liparis japonica* and the corresponding fungal isolates.

Population Code	Location	Longitude (E)	Latitude (N)	Altitude (m)	Sample size	No. of fungal isolates
DL	Dongling park, Shenyang, Liaoning	123°34′39″	41°49′53″	81	51	25
GG	Gaoguan, Benxi, Liaoning	124°02′32″	41°20′19″	274	30	30
GMS	Guanmenshan forest park, Benxi, Liaoning	124°09′30″	41°07′41″	438	18	27
CBS	Changbaishan nature reserve, Jilin	127°47′55″	42°30′35″	741	10	18
LTDZ	Laotudingzi nature reserve, Benxi, Liaoning	124°53′56″	41°18′09″	817	24	21
DQG	Daqinggou nature reserve, Fuxin, Liaoning	122°12′03″	42°43′55″	194	26	37
QS	Qianshan nature reserve, Anshan, Liaoning	123°07′48″	41°01′21″	163	26	43
Total					185	201

Fungal isolation and morphological identification

Potential mycorrhizal fungi were isolated from the orchid plants and identified from pure culture. *L. japonica*, unlike many other temperate, terrestrial orchids which have thick roots and often produce abundant pelotons, has few active pelotons suitable for isolation. Thus, endophytic fungi were isolated from single hyphal

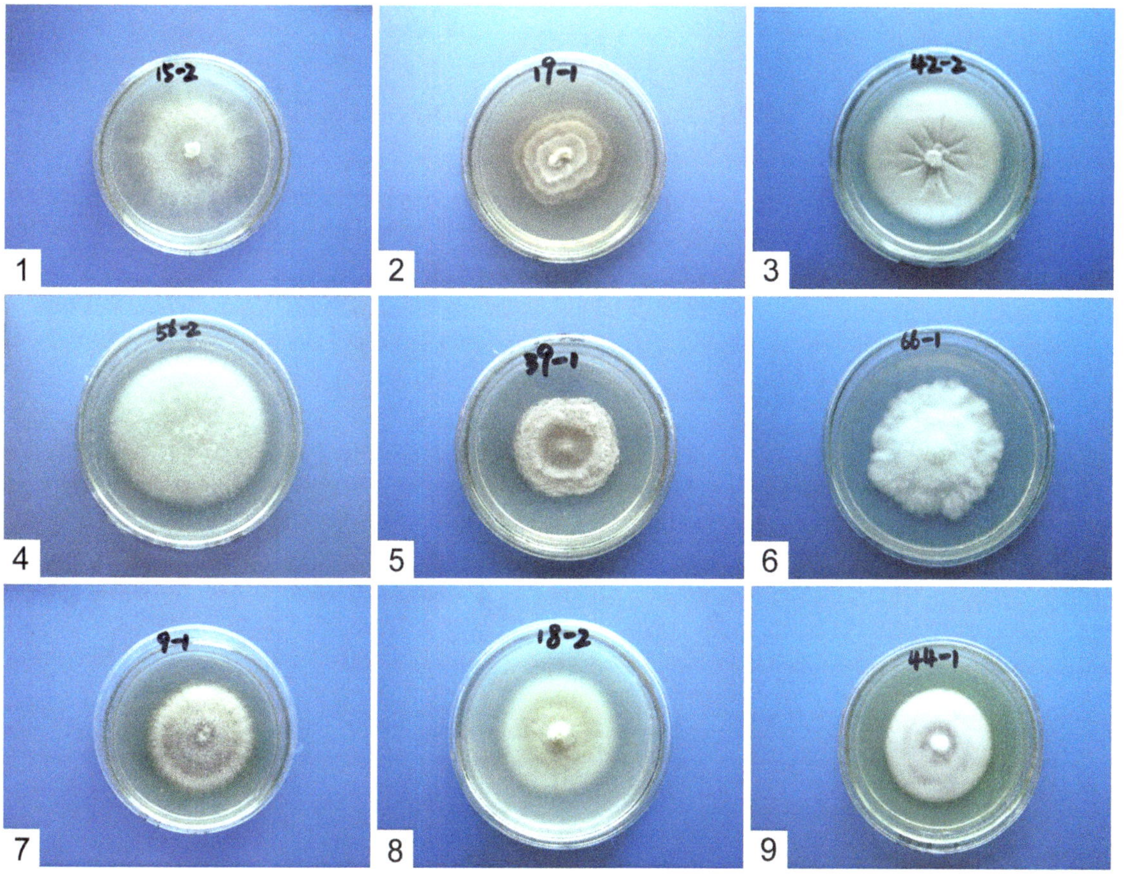

Figure 2. The nine morphotypes of endophytes isolated from *Liparis japonica* at seven populations. (morphotype I: *Rhizoctonia* sp., morphotype II: *Phomopsis* sp., morphotype III: *Verticillium* sp., morphotype IV: *Fusarium* sp., morphotype V: *Chaetomium* sp., morphotype VI: *Gliocladium* sp., morphotype VII: *Cylindrocarpon* sp., morphotype VIII: *Phialophora* sp., morphotype IX: *Paecilomyces* sp.).

Table 2. Distribution of 201 fungal isolates in different morphotypes.

Morphotype	No. of isolates in different morphotype							Total	Percentage (%)
	DL	GG	GMS	CBS	LTDZ	DQG	QS		
I	7	8	9	12	7	21	22	86	42.8
II	3	5	14	3	2	5	6	38	18.9
III	5	5	1	0	3	4	6	24	11.9
IV	6	3	1	0	5	3	3	21	10.4
V	2	5	2	2	3	0	4	18	9.0
VI	1	2	0	0	1	2	1	7	3.5
VII	0	2	0	0	0	0	1	3	1.5
VIII	0	0	0	1	0	2	0	3	1.5
IX	1	0	0	0	0	0	0	1	0.5
Total	25	30	27	18	21	37	43	201	

tips emerging from sterilized root portions as in the isolation method described by [30]. Three roots per plant were carefully cleaned from the soil under running water, surface-sterilized in 0.1% (*v/v*) Mercuric chloride and 75% (*v/v*) ethanol for 3 min and 5 s respectively, and subsequently washed three times in sterile distilled water. Root sections of 3–5 mm thickness were obtained by cutting and placed in a petri dish with potato dextrose agar (PDA). Petri dishes were incubated at 25°C in the dark and observed for fungi growing every 2 days for at least 3 weeks. The growing colonies were separated onto fresh media for purity and this process was repeated three times. Classification of the endophytic fungi was based on their growth rate and morphological characteristics, including colonial morphology, production of conidiogenous cells, conidial size and dimension on PDA medium [31], and similar isolates were grouped into one morphotype. The *Rhizoctonia*-like fungal endophytes were recognized by the following characteristics: hyphas hyaline with constricted branch points, 2.5–7 μm diam; submerged growth in PDA; ellipsoid, globose or irregular monilioid cells 10–20×10–20 μm; colony creamy white to pale tan or orange, rubbery or leathery in appearance and texture [32–34].

DNA extraction, amplification and sequencing

Genomic DNA was extracted from each *Rhizoctonia*-like fungi isolate as described by [35]. ITS sequences of rDNA were amplified using the following primer pairs: ITS1-OF/ITS4-OF [36]. PCRs were performed in 50-μL volumes containing approximately 10 μg of fungal genomic DNA, 1×PCR buffer (Sangon Biotech Co. Ltd), 1.5 mM $MgCl_2$, 200 μM each of dNTP (dATP, dCTP, dGTP, and dTTP), 100 nM of each primer and 2.5 unit of *Taq* DNA polymerase (Sangon Biotech Co. Ltd). PCRs were carried out with a BIO-RAD T100^{tm} Thermalcycler with an initial incubation at 95°C for 5 min, followed by 35 cycles with denaturation at 95°C for 30 s, annealing at 60°C for 1 min and extension at 72°C for 1 min, and a final extension reaction at 72°C for 10 min. The PCR products were purified using the TaKaRa Agarose Gel DNA Purification Kit (TaKaRa, Dalian, China) and sequenced using the BigDye terminator v3.1 on an ABI 3730xl DNA Analyzer (Applied Biosystems, Forster City, California, USA). New ITS sequences from this study have been deposited in GenBank (the accession numbers is from KF537635 to KF537658).

Data analysis

ITS sequences were analyzed with BLAST [37] against the NCBI sequence database to find the closest sequence matches in the GenBank database (http://www.ncbi.nlm.nih.gov). Genus and species of the database match were accepted whenever identity between our sequence and that of the database was greater than 97% [38]. Sequences used to construct phylogenetic trees were from organisms considered to be closely related to the mycorrhizal fungus of *L. japonica* based on BLAST searches, and additional orchid mycorrhizal fungi sequences (mainly based on [19]) were also chosen for phylogenetic analysis. Selected sequences were aligned using the program Clustal_X 2.1 [39]. Alignments were checked for ambiguities and adjusted manually when necessary and the alignment gaps were treated as missing information.

Phylogenetic analyses were estimated using maximum parsimony (MP), Neighbor-joining (NJ) and Bayesian inference (BI) methods in PAUP* version 4.0b10 [40], MEGA version 4 [41] and MrBayes version 3.0b4 [42], respectively. For MP analysis, the bootstrap support values were estimated by 1000 replicates with 10 random sequence additions and tree-bisection-reconnection (TBR) branch swapping. For NJ analysis, genetic distances

Table 3. Morphological characters of the 9 morphotypes of fungal isolates from *Liparis japonica.*

Morphotype	Colony color	Colony texture	Growth rate (mm day^{-1})	Aerial mycelium	Conidial shape	Conidia Size (μm)
I	off-white	leathery, compact	11.9±1.5	absent	–	–
II	pink to tan	leathery, radial draped	4.9±0.5	absent	elliptic cylindrical	6–8×3–4
III	white or buff	cottony, slightly loose	5.7±0.4	scarce	fusiform	5×1–1.5
IV	white	lanose, loose	11.8±0.5	abundant	ovate, cylindrical	cylindrical: 15–20×5–10 ovale: 4×3
V	white at first, then olive-green	pulvinate, compact	4.8±0.5	absent	limoniform	8–9×4–5
VI	white	cottony, slightly loose	5.0±0.2	abundant	elliptic	6–8×3–5
VII	pale tan	cottony, compact	5.2±0.2	abundant	clavate, elliptic	40×5–10
VIII	Pale taupe	felted, compact	3.4±0.4	scarce	ovate	3–5×2.5
IX	lilac	felted, compact	3.3±1.7	abundant	elliptic	4–5×2

were calculated using the Kimura two-parameter model, and the bootstrap support values were estimated by 1000 replicates. Prior to the Bayesian analysis, the Akaike information criterion (AIC) in Modeltest version 3.7 [43] was used to select the best-fit model of molecular evolution for each dataset. For BI analysis, four chains of the Markov Chain Monte Carlo (MCMC) chains were run, sampling one tree every 1000 generations for 3,000,000 generations, starting with a random tree. Bayesian posterior probabilities (PP) were calculated from the majority rule consensus of the tree. The sequence of *Septobasidium carestianum* (GenBank accession number DQ241448) was selected as outgroup [21].

Results

Fungal distribution and morphological diversity

After isolation and purification, a total of 201 isolates of endophytic fungi from 185 plants were obtained from seven *L. japonica* populations. According to their morphological characters and growth rate on PDA medium, fungal isolates were classified into nine morphotypes with four to seven morphotypes in each population (Table 2). The isolates in morphotypes I and II were observed in all populations studied and represented the majority of the fungal community (61.7%), while isolates from other morphotypes were not common and occurred only at one or a few populations (Table 2). Morphological characters and detailed descriptions of the nine morphotypes were given in Fig. 2 and Table 3, and according to the colony and micro-morphological characters, we identified morphotype I as *Rhizoctonia* sp., morphotype II as *Phomopsis* sp., and morphotype III to IX as *Verticillium* sp., *Fusarium* sp., *Chaetomium* sp., *Gliocladium* sp., *Cylindrocarpon* sp., *Phialophora* sp., and *Paecilomyces* sp., respectively. Only the *Rhizoctonia*-like isolates in morphotype I were given codes and subjected to further collection of mycelium for DNA extraction and phylogenetic analysis.

Of the 185 individuals of *L. japonica* samples, each of 180 samples gave rise to one *Rhizoctonia*-like isolate, whereas the other five samples gave rise to two *Rhizoctonia*-like isolates, and no sample gave rise to three or more *Rhizoctonia*-like isolates.

Phylogenetic analysis

About 640-bp portions of the ITS were sequenced for the 24 representative isolates of morphotype I, with 3 to 4 isolates within each population. All sequences identified by BLAST in NCBI were from Tulasnellaceae strains. Thus, the representative *Tulasnella* spp. that have been identified were selected for phylogenetic analysis with the sequences obtained in this study, and a total of 54 different sequences were used.

The overall topology of the trees produced with Maximum parsimony (MP), Bayesian inference (BI) and Neighbour-joining (NJ) methods were similar based on our ITS data (Fig. 3). The MP strict consensus tree (Fig. 3) showed that fungal associates of *L. japonica* all belonged to the genus *Tulasnella*, and all isolates belonged to a single major clade. Some mycorrhizal isolates from other orchid species which have been identified as *T. calospora*, and a mycorrhizal isolate (*Epulorhiza* sp., AM040890; [44]) from *L. loeselii* species in Hungary also fell in this major clade.

Discussion

Compared with the direct DNA amplification technique, identification of orchid mycorrhizal fungi with the pure culture technique has been criticized because of the bias toward fast-growing and easily cultured fungi [15]. However, the culture-dependent technique is still used in a vast majority of orchid mycorrhizal fungi studies, mainly because fungi partners of these plants were usually easy to isolate and no significant differences were found between the pure culture technique and direct DNA amplification technique [20,23,45–47]. Unlike many other temperate, terrestrial orchids which have thick roots and often produce abundant pelotons, *L. japonica* has few active pelotons suitable for isolation and direct PCR identification, so we used standard culture method for fungi identification in this study, and

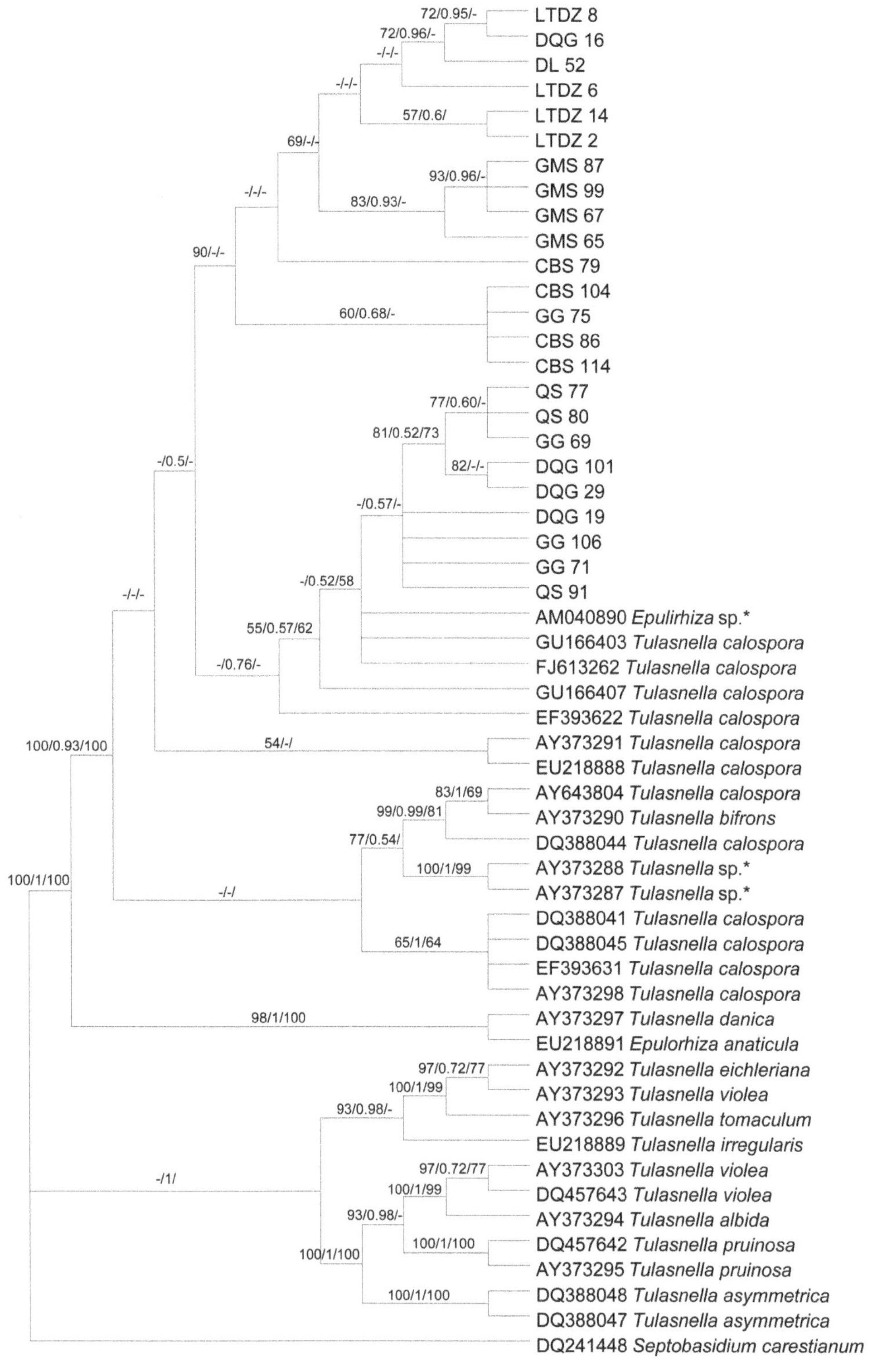

Figure 3. Strict consensus tree of Maximum parsimony (MP) based on ITS sequences of representative isolates of *Rhizoctonia*-like fungi isolated from *Liparis japonica* and identified *Tulasnella* spp. and *Epulorhiza* spp. Numbers at the nodes are MP bootstrap percentages, Bayesian posterior probabilities and Neighbour-joining support percentages (≥50%), respectively. "-" indicates node not supported in Bayesian inference and Neighbour-joining analyses. An asterisk indicates fungi isolated from *Liparis lilifolia* and *L. loeselii* (other *Liparis* species). The *Septobasidium carestianum* sequence was designated as outgroup for rooting the trees. Sequences obtained from GenBank were shown with accession numbers.

the results were credible by the striking observation of the same fungal clade isolated across such a wide geographical distribution of *L. japonica*. Moreover, although the molecular method is powerful for fungi identification, isolation and cultivation of the mycorrhizal fungi isolates are still needed for deep study and further utilizing in orchids conservation.

In the present study, various endophytic fungi were isolated from roots of *L. japonica* and identified by morphological and the rDNA ITS sequences, and most of the identified endophytes of *L. japonica* are Basidiomycetes and Ascomycetes. In general, endophytic fungi in orchid plants are very abundant, but it seems likely that only *Rhizoctonia*-like fungi species may form mycorrhizal associations with host [48]. In this study, 86 endophyte fungi isolates belonging to the *Tulasnella* genus were isolated (42.8%) based on morphological characters and phylogenetic relationships reconstructed from ITS sequence, and they existed in every *L. japonica* population. Although some endophytes have been shown to be able to increase the tolerance of some plant species [49], little is known about the role of endophytes for orchid performance, and the non-*Rhizoctonia*-like endophytic fungi also deserve additional study.

The specificity of orchid mycorrhizal associations has important implications for orchid biology, conservation and restoration of orchid populations [50]. Specific fungal partners may lead to enhanced seed germination rates, and the efficiency of nutrient exchange between partners may be heightened with specific plant-fungus combinations [45]. Our results showed that all mycorrhizal fungi isolates obtained belonged to a single clade of *Tulasnella*, and the molecular analysis showed that they are all representatives of the same species, *T. calospora*, which indicates that the relationship between *L. japonica* and its mycorrhizal fungus is very specific regardless of habitats, at least in Northeast China. There are more and more reports showed that the degree of specificity can vary among species independent on orchid types [19,22]. There is also evidence that a preference exists among some orchids for specific fungal partners, and some photosynthetic orchids, even when sampled over a wide range, have a single dominant mycorrhizal fungus, for example, *Spiranthes sinensis* [33], *Goodyera pubescens* [51], and *Pterostylis nutans* [52]. Our results also supported that high specificity in widespread and photosynthetic orchids may be more common than previously thought. Moreover, the phenomenon that multiple fungal taxa existed in a single root system was detected in some photosynthetic orchids which indicated that multiple fungi in a single root system may occur more often in photosynthetic than in mycoheterotrophic orchids [26]. While our results showed that this characteristic is not universal in orchid mycorrhizal systems.

From another perspective, the mycorrhizal specificity we observed might not be accurate for the reason that our results were based on a single collection of root tissue in adult *L. japonica* plants. However, the diversity and identify of orchid mycorrhizal fungi may vary according to different life stages of plant. Specificity during germination and early development has been detected in orchids, whereas the adult stage of plant growth in the same species can have less specific fungal partners [13,51,53]. Accordingly, fungi required for germination and recruitment of *L. japonica* might be different from those associated with adult plants, and this will be the next logical steps to investigate the pattern of mycorrhizal association.

Although *L. japonica* had a strong preference for *T. calospora* species across the sampled range, the fungus did not have an equivalent specificity. *T. calospora* has been shown to be a very common mycorrhizal fungus of many terrestrial orchid species [15], and the fact that *T. calospora* stains were present in different populations of *L. japonica* in Northeast China in our study and many other orchids in other regions of the world indicates its wide distribution in different ecological environments. Because of this, we considered that the decline of *L. japonica* in Northeast China was not resulted from fungal specificity, although there exists a point that a dependence on narrowly specific interactions with fungi and pollinator may predispose many orchids to become rare [15,45,54].

Acknowledgments

We thank Zhi-hui Zhao (College of Environment and Resources, Dalian Nationalities University, China) for morphological identifying the orchid mycorrhizal fungal. We also thank Xiao-guo Xiang (State Key Laboratory of Systematic and Evolutionary Botany, Institute of Botany, Chinese Academy of Sciences, China) for her great help on the phylogenetic analyses.

Author Contributions

Conceived and designed the experiments: XHC LJZ. Performed the experiments: R. Ding. Analyzed the data: R. Ding XHC XDY. Contributed reagents/materials/analysis tools: BQ R. Duan YFX. Contributed to the writing of the manuscript: R. Ding XHC.

References

1. Smith SE, Read DJ (2008) Mycorrhizal Symbiosis. New York: Academic Press. 19–41.
2. Gardes M (2002) An orchid-fungus marriage: physical promiscuity, conflict and cheating. New Phytol 154: 4–7.
3. Allen MF (2007) Mycorrhizal fungi: highways for water and nutrients in arid soils. Vadose Zone J 6: 291–297.
4. Leigh J, Hodge A, Fitter AH (2009) Arbuscular mycorrhizal fungi can transfer substantial amounts of nitrogen to their host plant from organic material. New Phytol 181: 199–207.
5. Artursson V, Finlay RD, Jansson JK (2006) Interactions between arbuscular mycorrhizal fungi and bacteria and their potential for stimulating plant growth. Environ Microbiol 8: 1–10.
6. Porras-Soriano A, Soriano-Martn ML, Porras-Piedra A, Azcón R (2009) Abruscular mycorrhizal fungi increased growth, nutrient uptake and tolerance to salinity in olive trees under nursery conditions. J Plant Physiol 166: 1350–1359.
7. Fomina MA, Alexander IJ, Colpaert JV, Gadd GM (2005) Solubilization of toxic metal minerals and metal tolerance of mycorrhizal fungi. Soil Biol Biochem 37: 851–866.
8. Yao MK, Tweddell RJ, Désilets H (2002) Effect of two vesicular-arbuscular mycorrhizal fungi on the growth of micropropagated potato plantlets and on the extent of disease caused by *Rhizoctonia solani*. Mycorrhiza 12: 235–242.
9. Harms H, Schlosser D, Wick LY (2011) Untapped potential: exploiting fungi in bioremediation of hazardous chemicals. Nat Rev Microbiol 9: 177–192.
10. Zeng Y, Guo LP, Chen BD, Hao ZP, Wang JY, et al. (2013) Arbuscular mycorrhizal symbiosis and active ingredients of medicinal plants: current research status and prospectives. Mycorrhiza 23: 253–265.
11. Dressler RL (1993) Phylogeny and classification of the orchid family. Portland: Dioscorides Press.
12. Swarts ND, Dixon KW (2009) Perspectives on orchid conservation in botanic gardens. Trends Plant Sci 14: 590–598.
13. Bidartondo MI, Read DJ (2008) Fungal specificity bottlenecks during orchid germination and development. Mol Ecol 17: 3707–3716.
14. Rasmussen HN (2002) Recent developments in the study of orchid mycorrhiza. Plant Soil 244: 149–163.
15. Dearnaley JDW, Martos F, Selosse M-A (2012) Orchid mycorrhizas: molecular ecology, physiology, evolution and conservation aspects. In: Hock B, editors. Fungal Association, 2nd edn. Berlin: Springer-Verlag. 207–230.
16. Tondello A, Vendramin E, Villani M, Baldan B, Squartini A (2012) Fungi associated with the southern Eurasian orchid *Spiranthes spiralis* (L.) Chevall. Fungal Biol 116: 543–549.
17. Yukawa T, Ogura-Tsujita Y, Shefferson RP, Yokoyama J (2009) Mycorrhizal diversity in *Apostasia* (Orchidaceae) indicates the origin and evolution of orchid mycorrhiza. Am J Bot 96: 1997–2009.
18. Chen J, Zhang LC, Xing YM, Wang YQ, Xing XK, et al. (2013) Diversity and taxonomy of endophytic Xylariaceous fungi from medicinal plants of *Dendrobium* (Orchidaceae). PLOS ONE 8: e58268.

19. McCormick MK, Whigham DF, O'Neill J (2004) Mycorrhizal diversity in photosynthetic terrestrial orchids. New Phytol 163: 425–438.
20. Otero JT, Flanagan NS, Herre EA, Ackerman JD, Bayman P (2007) Widespread mycorrhizal specificity correlates to mycorrhizal function in the neotropical, epiphytic orchid *Ionopsis utricularioides* (Orchidaceae). Am J Bot 94: 1944–1950.
21. Nontachaiyapoom S, Sasirat S, Manoch L (2010) Isolation and identification of *Rhizoctonia*-like fungi from roots of three orchid genera, *Paphiopedilum*, *Dendrobium*, and *Cymbidium*, collected in Chiang Rai and Chiang Mai provinces of Thailand. Mycorrhiza 20: 459–471.
22. Jacquemyn H, Deja A, De hert K, Bailarote BC, Lievens B (2012) Variation in mycorrhizal associations with Tulasnelloid fungi among populations of five *Dactylorhiza* species. PLOS ONE 7: e42212.
23. De Long JR, Swarts ND, Dixon KW, Egerton-Warburton LM (2013) Mycorrhizal preference promotes habitat invasion by a native Australian orchid: *Microtis media*. Ann Bot 111: 409–418.
24. Xing X, Ma X, Deng Z, Chen J, Wu F, et al. (2013) Specificity and preference of mycorrhizal associations in two species of the genus *Dendrobium* (Orchidaceae). Mycorrhiza 23: 317–324.
25. Oliveira SF, Bocayuva MF, Veloso TGR, Bazzolli DMS, da Silva CC, et al. (2014) Endophytic and mycorrhizal fungi associated with roots of endangered native orchids from the Atlantic Forest, Brazil. Mycorrhiza 24: 55–64.
26. Pandey M, Sharma J, Taylor DL, Yadon VL (2013) A narrowly endemic photosynthetic orchid is non-specific in its mycorrhizal associations. Mol Ecol 22: 2341–2354.
27. McCormick MK, Jacquemyn H (2014) What constrains the distribution of orchid populations? New Phytol 202: 392–400.
28. Editorial Committee of the Flora of China (1992) Flora of China. Beijing: Science Press. 18: 62–64.
29. Chen XH, Guan JJ, Ding R, Zhang Q, Ling XZ, et al. (2013) Conservation genetics of the endangered terrestrial orchid *Liparis japonica* in Northeast China based on AFLP markers. Plant Syst Evol 299: 691–698.
30. Curtis JT (1939) The relation of specificity of orchid mycorrhizal fungi to the problem of symbiosis. Am J Bot 26: 390–399.
31. Wei YK, Gao YB, Zhang X, Su D, Wang YH, et al. (2007) Distribution and diversity of *Epichloë/Neotyphodium* fungal endophytes from different populations of *Achnatherum sibiricum* (Poaceae) in the Inner Mongolia Steppe, China. Fungal Divers 24: 329–345.
32. Currah RS, Sherburne R (1992) Septal ultrastructure of some fungal endophytes from boreal orchid mycorrhizas. Mycol Res 96: 583–587.
33. Masuhara G, Katsuya K (1994) *In situ* and *in vitro* specificity between *Rhizoctonia* spp. and *Spiranthes-sinensis* (Persoon) Ames var. *amoena* (M. Bieberstein) Hara (Orchidaceae). New Phytol 127: 711–718.
34. Ma M, Tan TK, Wong SM (2003) Identification and molecular phylogeny of *Epulorhiza* isolates from tropical orchids. Mycol Res 107: 1041–1049.
35. Cenis JL (1992) Repid extraction of fungal DNA for PCR amplification. Nucleic Acids Res 20: 2380.
36. Taylor DL, McCormick MK (2008) Internal transcribed spacer primers and sequences for improved characterization of basidiomycetous orchid mycorrhizas. New Phytol 177: 1020–1033.
37. Altschul SF, Madden TL, Schaffer AA, Zhang J, Zhang Z, et al. (1997) Gapped BLAST and PSI-BLAST: A new generation of protein database search programs. Nucleic Acids Res 25: 3389–3402.
38. Sánchez MS, Bills GF, Zabalgogeazcoa I (2008) Diversity and structure of the fungal endophytic assemblages from two sympatric coastal grasses. Fungal Divers 33: 87–100.
39. Thompson JD, Gibson TJ, Plewniak F, Jeanmougin F, Higgins DG (1997) The Clustal_X windows interface: flexible strategies for multiple sequence alignment aided by quality analysis tools. Nucleic Acids Res 25: 4876–4882.
40. Swofford DL (2002) PAUP*: phylogenetic analysis using parsimony (* and other methods), version 4.0b10. Sinauer, Sunderland.
41. Tamura K, Dudley J, Nei M, Kumar S (2007) MEGA4: molecular evolutionary genetics analysis (MEGA) software version 4.0. Mol Biol Evol 24: 1596–1599.
42. Ronquist F, Huelsenbeck JP (2003) MrBayes 3: Bayesian phylotenetic inference under mixed models. Bioinformatics 19: 1572–1574.
43. Posada D, Crandall KA (1998) Modeltest: testing the model of DNA substitution. Bioinformatics 14: 817–818.
44. Illyés Z, Halász K, Rudnóy S, Ouanphanivanh N, Garay T, et al. (2009) Changes in the diversity of the mycorrhizal fungi of orchids as a function of the water supply of the habitat. J Appl Bot Food Qual 83: 28–36.
45. Bonnardeaux Y, Brundrett M, Batty A, Dixon K, Koch J, et al. (2007) Diversity of mycorrhizal fungi of terrestrial orchids: compatibility webs, brief encounters, lasting relationships and alien invasions. Mycol Res 111: 51–61.
46. Roche SA, Carter RJ, Peakall R, Smith LM, Whitehead MR, et al. (2010) A narrow group of monophyletic *Tulasnella* (Tulasnellaceae) symbiont lineages are associated with multiple species of *Chiloglottis* (Orchideacea): implications for orchid diversity. Am J Bot 97: 1313–1327.
47. Kohout P, Těšitelová T, Roy M, Vohník M, Jersáková J (2013) A diversity fungal community associated with *Pseudorchis albida* (Orchidaceae) roots. Fungal Ecol 6: 50–64.
48. Suárez JP, Weiβ M, Abele A, Garnica S, Oberwinkler F, et al. (2006) Diverse tulasnelloid fungi form mycorrhizas with epiphytic orchids in an Andean cloud forest. Mycol Res 110: 1257–1270.
49. Rodriguez RJ, Henson J, Volkenburgh EV, Hoy M, Wright L, et al. (2008) Stress tolerance in plants via habitat-adapted symbiosis. ISME J 2: 404–416.
50. Dearnaley JDW (2007) Further advances in orchid mycorrhizal research. Mycorrhiza 17: 475–486.
51. McCormick MK, Whigham DF, Sloan D, O'Malley K, Hodkinson B (2006) Orchid-fungus fidelity: a marriage meant to last? Ecology 87: 903–911.
52. Irwin MJ, Bougoure JJ, Dearnaley JDW (2007) *Pterostylis nutans* (Orchidaceae) has a specific association with two *Ceratobasidium* root-associated fungi across its range in eastern Australia. Mycoscience 48: 231–239.
53. Zellter LW, Piskin KA (2011) Mycorrhizal fungi from protocorms, seedlings and mature plants of the eastern prairie fringed orchid, *Platanthera leucophaea* (Nutt.) Lindley: A comprehensive list to augment conservation. Am Midl Nat 166: 29–39.
54. Swarts NS, Sinclair EA, Francis A, Dixon KW (2010) Ecological specialization in mycorrhizal symbiosis leads to rarity in an endangered orchid. Mol Ecol 19: 3226–3242.

4

Genetic Structure of Chinese Indigenous Goats and the Special Geographical Structure in the Southwest China as a Geographic Barrier Driving the Fragmentation of a Large Population

Caihong Wei[1]⁹, Jian Lu[2]⁹, Lingyang Xu[1], Gang Liu[2], Zhigang Wang[2], Fuping Zhao[1], Li Zhang[1], Xu Han[2], Lixin Du[1]*, Chousheng Liu[2]*

1 National Center for Molecular Genetics and Breeding of Animal, Institute of Animal Sciences, Chinese Academy of Agricultural Sciences, Beijing, People's Republic of China, **2** National Center of Preservation & Utilization of Genetic Resources of Animal, National Animal Husbandry Service, Beijing, People's Republic of China

Abstract

Background: China has numerous native domestic goat breeds, however, extensive studies are focused on the genetic diversity within the fewer breeds and limited regions, the population demograogic history and origin of Chinese goats are still unclear. The roles of geographical structure have not been analyzed in Chinese goat domestic process. In this study, the genetic relationships of Chinese indigenous goat populations were evaluated using 30 microsatellite markers.

Methodology/Principal Findings: Forty Chinese indigenous populations containing 2078 goats were sampled from different geographic regions of China. Moderate genetic diversity at the population level (H_S of 0.644) and high population diversity at the species level (H_T value of 0.737) were estimated. Significant moderate population differentiation was detected (F_{ST} value of 0.129). Significant excess homozygosity (F_{IS} of 0.105) and recent population bottlenecks were detected in thirty-six populations. Neighbour-joining tree, principal components analysis and Bayesian clusters all revealed that Chinese goat populations could be subdivided into at least four genetic clusters: Southwest China, South China, Northwest China and East China. It was observed that the genetic diversity of Northern China goats was highest among these clusters. The results here suggested that the goat populations in Southwest China might be the earliest domestic goats in China.

Conclusions/Significance: Our results suggested that the current genetic structure of Chinese goats were resulted from the special geographical structure, especially in the Western China, and the Western goat populations had been separated by the geographic structure (Hengduan Mountains and Qinling Mountains-Huaihe River Line) into two clusters: the Southwest and Northwest. It also indicated that the current genetic structure was caused by the geographical origin mainly, in close accordance with the human's migration history throughout China. This study provides a fundamental genetic profile for the conservation of these populations and better to understand the domestication process and origin of Chinese goats.

Editor: Qin Zhang, China Agricultrual University, China

Funding: This work was supported by grant 2011BAD28B05-2 from the National Key Technology R&D Program of China and Ministry of Agriculture 2013 Protection Project of Animal Germplasm Resources (No. [2013]36). The funders had no role in study design, data collection and analysis, decision to publish, or preparation of the manuscript.

Competing Interests: The authors have declared that no competing interests exist.

* E-mail: lxdu@263.net (LD); liuchousheng@sina.com (CL)

⁹ These authors contributed equally to this work.

Introduction

The archaeological and genetic evidence suggests that goat (Capra hircus) was probably first domesticated in the Fertile Crescent region of the Near East ≈10,000 years ago [1,2]. The domestic goat has always played an important role in providing a full range of useful products to human society, such as meat, milk, pelts and fiber, especially in China and other development countries [3]. Goats are the most adaptable and geographically wide spread livestock species, ranging from the mountains of Siberia to the deserts and reopics of Africa [4,5]. To date, there are about 1,000 goat breeds, and more than 875 million goats are kept around the world, according to the statistics of the UN Food and Agriculture Organization (http://www.fao.org/corp/statistics/en/). In China, there are about 58 native domestic goat breeds, distributed in different environmental areas including the north pastoral region, the Qinghai-Tibet plateau region, the mixed pastoral-agricultural region, and the north and south agricultural region (Animal Genetic Resources in China: Sheep and Goats, 2011). During this decade, these domestic breeds represent important genetic resources and have been aroused us more great attention, therefore, conservation of the domestic animal diversity is essential to meet the future needs. Considering the special economic and ecological characteristics, China recently has become concerned about these indigenous animal populations and initiated the conservation programs, such as setting the conservation area, conservation farm and the genebank of genetic resource for special breeds.

The study of genetic diversity within and across breeds provides insight into population structure and relationships, and is essential for the conservation of these indigenous breeds [6,7]. Studies on the genetic diversity of domestic goats using mitochondrial DNA, microsatellites and single nucleotide polynorphism (SNP) markers have been conducted over few years in the world [4,8,9,10,11]. These studies have demonstrated that the domestic goat showed a weak phylogeographic structure. Recently, Kijas et al also used the goat genome-wide SNP array to assess the genetic diversity and polledness in divergent goat populations [12]. It is suggested that the insight on breed development and introgression provided by the different types of genetic markers is complementary. The nuclear microsatellite genetic markers have been extensively used to assess genetic diversity and inbreeding levels of populations, introgression from other genetic groups, genetic differentiation and population structure, even reflect the consequences of genetic drift, founder effects and population admixture [9,13,14,15].

In the Western region of China, the Hengduan Mountains, as a natural geographical barrier, block the Qinghai-Tibet Plateau and the Yunnan-Guizhou Plateau, and have the most complex river systems and a profoundly complex and dynamic geological history [16]. In addition, the Qinling Mountains represent an important geographic barrier in Eastern Asia that divides the current mainland of China into southern and northern and temperate and semi-tropical regions [17]. These mountains had divided the West China into Northwest China and Southwest China, and results to the different climatic characteristic and geographical structure.

Since China is a vast subcontinent with the complex geographical situation, Chinese goat breeds exhibit enormous variations in fecundity; production of meat, milk, and fiber; draughtability; disease resistance; and heat tolerance. In addition, many previous studies of Chinese goats were conducted in the limited number of samples within fewer breeds and regions [8,9,15,18,19]. The objective of the present study was to use 30 nuclear microsatellite markers to analyze the genetic relationships and genetic differentiation of 40 Chinese indigenous populations containing 2078 goats sampled from different geographic regions of China. Furthermore, the bottleneck, the origins and evolutionary trajectories of Chinese domestic goats also had been investigated. Our results demonstrated that Chinese goat populations could be subdivided into at least the following four genetic clusters: Southwest China, South China, Northwest China and East China, and the genetic diversity of Northern China goats was highest among these clusters. It was also suggested that the current genetic structure of Chinese goats was caused by the geographical origin mainly, in close accordance with the human's migration history throughout China. This information may be valuable for the conservation and maintaining diversity and genetic uniqueness of the genetic resource of these indigenous goat breeds, is better to understand the domestication process and origin of Chinese goats.

Materials and Methods

Biological Samples Collection

In total, 2078 animals representing 40 goat breeds were studied (Additional file: Table S1). Goats were sampled from 23 provinces or autonomous regions in China, including almost all of the geographical regions, from high latitude to plains (Figure 1). The sampling protocol for each goat breed was below: coming from the main producing area, meeting the characteristics and features of the breed, without relationship in the three generations, healthy, 30 males and 30 females. The blood samples were collected by the owner of the animal or a collaborator and shipped to this laboratory for processing. All experimental procedures were approved by the Law of Animal Husbandry in People's Republic of China (Dec 29, 2005). The whole study protocols for collection of the blood samples of experimental individuals were reviewed and approved by the Biological Studies Animal Care and Use Committee of National Animal Husbandry Service, Beijing, People's Republic of China. The genomic DNA was isolated from cryopreserved blood samples using DNeasy Blood & Tissue Kit (Qiagen, Valencia, CA). After removing the poor quality DNA samples, the number of samples in this study was 2078 finally.

Microsatellite Markers

A common set of 30 microsatellites were selected from a panel of 31 markers recommended for genetic diversity studies by the International society for Animal Genetics (ISAG)/Food and Agriculture Organization of the United Nations (FAO) working group: CSRD247, ETH10, MAF209, OARAE54, SRCRSP15, SRCRSP3, SRCRSP5, TGLA53, DRBP1, ILSTS087, INRA-BERN172, MAF065, MCM527, OARFCB20, SPS113, SRCRSP8, ILSTS029, INRA023, INRA063, INRABERN185, P19(DYA), SRCRSP23, SRCRSP9, TCRVB6, BM6444, ILSTS011, ILSTS005, SRCRSP7, OARFCB48, MAF70.

The 30 microsatellite markers were amplified in multiplex polymerase chain reactions (PCRs) using fluorescence-labelled primers. PCR products had been run on PAGE-SDS gel to check the real size, and separated using an ABI-PRISM 3100-Avant Genetic Analyzer (Applied Biosystems, Fster City, CA) according to manufacturer recommendations. Allele sizes relative to an internal size standard (GS400HD) were determined using GENEMAPPER 3.5 (Applied Biosystems, Fster City, CA).

Statistical Analysis

For each microsatellite locus, the mean number of alleles, observed heterozygosities (H_O), expected (H_E) heterozygosities average genetic diversity (H_S) both within populations and among populations and total gene diversity (H_T) were obtained from the FSTAT 2.9.3 software [20]. Wright's F-statistics were estimated from FSTAT 2.9.3 using Weir and Cockerham method [21]. Deviations from Hardy-Weinberg equilibrium (HWE) were assessed with GENEPOP v 4.2 software [22]. Unbiased estimates of exact P-values were obtained using the Markov Chain Monte Carlo algorithm of 10000 dememorization steps, 100 batches, and 5000 iterations. The linkage disequilibrium between each pair of loci was assessed from GENEPOP software [23].

To estimate whether of the reduction in the effective population size, the Wilcoxon's signed-rank tests (one-tailed) with the null hypothesis, $H_E > H_{EQ}$, and the alternative hypothesis, $H_E < H_{EQ}$, were used under the assumption of mutation-drift equilibrium in the infinite-alleles model (IAM), two-phase model of mutation (TPM) and the stepwise mutation model (SMM), and the allele frequency distribution mode shift analysis were performed using BOTTLENECK version 1.2.02 [24].

According to the geographic origin of these populations, a hierarchical analysis of variance was performed to partition the total genetic variance into components due to within-individuals, among-populations and among individuals within populations' differences. Variance components were used to compute the fixation indexes and their significance was tested using a non-parametric

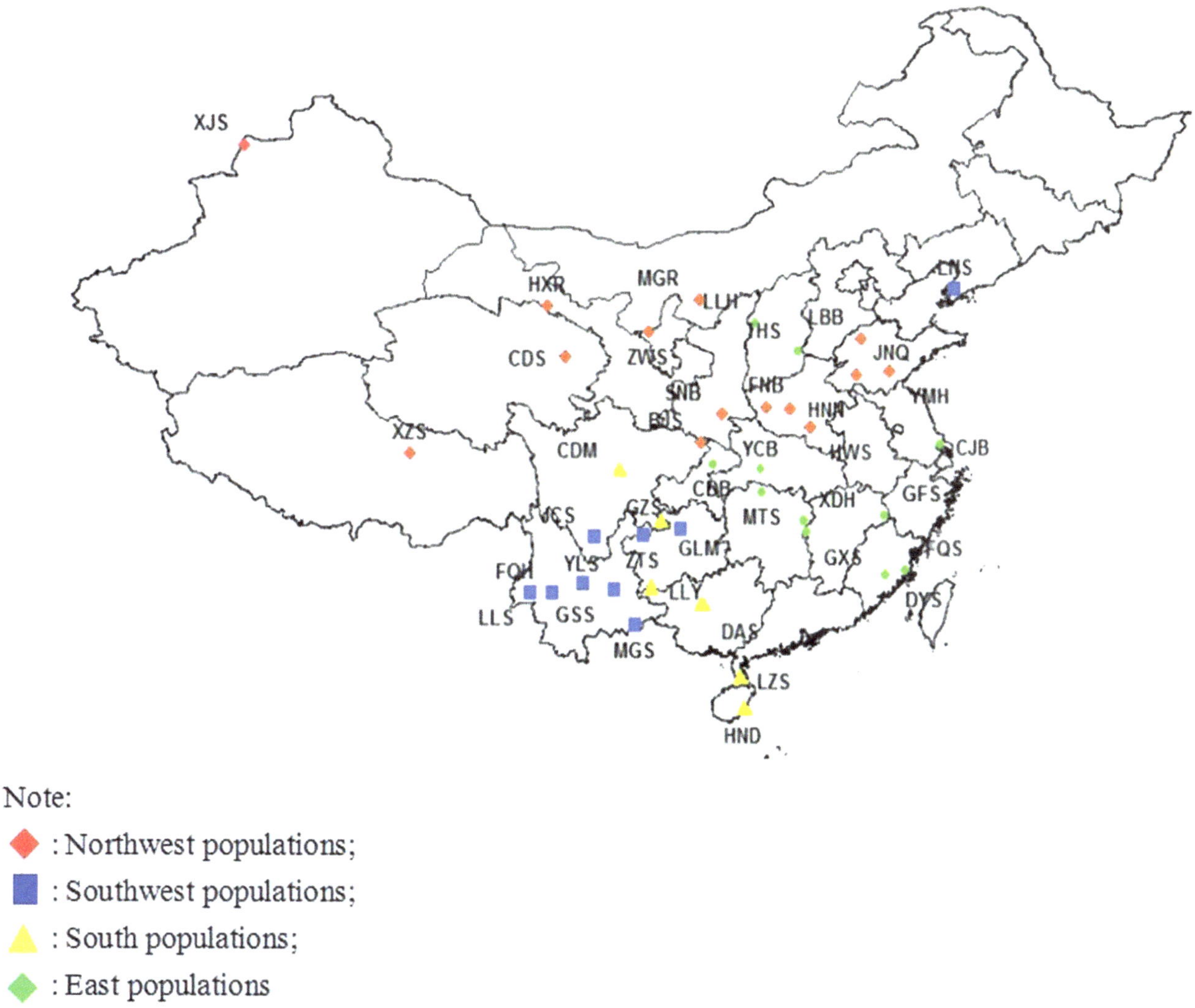

Figure 1. Geographic distribution of the 40 goat populations in China.

permutation approach [25]. The analysis of molecular variance (AMOVA) module of ARLEQUIN 3.5.1.2 software was used to perform the computation [26].

To illustrate the genetic divergence among the breeds, the Reynolds distances were estimated by the PHYLIP package to reconstruct the neighbor-joining consensus tree, with the boot-strapping of 1,000 replicates, and the dendrogram was depicted using the software package TreeView version 1.4.0 [27,28]. Additionally, the pattern of genetic variation among populations was also assessed by principal components analysis (PCA) which was based on allele frequencies using the Multivariate Statistical Package (MVSP) version 3.13b (Kovach Computing Services, Anglesey, Wales, UK).

To study population genetic structure and to detect the most likely number of clusters (K) in the data set, the STRUCTURE 2.3 software was performed [29]. The Structure algorithm assumes K populations, each of which is in Hardy-Weinberg and linkage equilibrium and characterized by a set of allele frequencies at each locus. To choose the appropriate number of inferred clusters to model the data, 2 to 40 inferred clusters were performed with 10 independent runs each. All analyses were performed with a burn-in length of 100,000 followed by 100,000 Markov Chain Monte Carlo iterations for each of K. To find optimal alignments of independent runs, the computer program CLUMPP 1.1 was used; and the output obtained was used directly as input by the cluster visualization program Distruct 1.1 software [30,31].

Results

Genetic Diversity

In the present study, a set of 30 microsatellite markers listed in Table 1 was used to analysis samples of the 40 goat populations (Table S1). The result of microsatellite markers amplification analyze showed that the total number of alleles was 471, and the largest number of alleles was found at locus BM6444 (35) and the smallest at locus MAF 209 (4), with an average of 15.7 alleles per locus (Table 1). The average genetic diversity (H_S) among the locus was 0.644 and ranging from 0.237 to 0.820. Additionally, the average overall genetic diversity (H_T) estimate among the populations was 0.737, ranging from 0.274 to 0.907. As shown in Table 2, the genetic variability of each population was studied. The highest mean number of alleles (MNA) was 8.27 in the FNB goat and the lowest was 4.33 in the GFS goat. Among these populations, the allelic richness (AR) based on 23 diploid individuals per population ranged from 3.89 in DYS goat to 7.35 in FNB. The estimated H_E per population ranged from 0.5070 in DYS to 0.7378 in FNB. H_O ranged from 0.4336 in DYS to 0.6730 in HNN. It is observed that the FNB population was the highest one in MNA, AR and H_E among all the 40 populations, showing the higher genetic variation. In contrast, the DYS population showed the lower genetic variation than other populations. Additionally, it is obtained that H_O was lower than H_E in all the populations analyzes in this study. As shown in Table 2, the F_{IS} over all the loci was found to be significant in thirty-eight populations except CDM and LLH, whereas the F_{IS} within populations was significant for twenty loci (CSRD247, OARAE54, SRCRSP15, SRCRSP3, SRCRSP5, TGLA53, DRBP1, ILSTS087, INRABERN172, MAF065, OARFCB20, SPS113, SRCRSP8, INRA023, INRA063, P19(DYA), TCRVB6, BM6444, ILSTS011, SRCRSP7) (Table 1). These results suggested a significant deviation from the Hardy-Weinberg equilibrium.

According to Wilcoxon's signed-rank test performed using BOTTLENECK, thirty-six populations (LLS, MGS, YLS, ZTS, JCS, GSS, FQH, GZS, LNS, CDM, GLM, LLY, LZS, HND, DAS, MGR, XZS, HXR, ZWS, SNB, FNB, HNN, HWS, YMH, LBB, LLH, THS, CJB, CDB, GFS, GXS, MTS, YCB, XDH, FQS, DYS) showed a significant deviation (excess heterozygosity) from the mutation-drift equilibrium under the SMM (after sequential Bonferroni correction), whereas under the IAM and TPM, all the 40 goat populations showed a significant deviation from the mutation-drift equilibrium, suggesting that decreased in the population size had occurred due to a bottleneck effect (Table S2). The mode shift test also detected that 26 populations showed the distortion of allele frequency, whereas another 14 populations showed a typical property of a population in equilibrium were ZTS, GSS, GZS, GLM, CDS, XJS, BJS, HNN, HWS, JNQ, YMH, LBB, LLH and CDB (Table S2).

Genetic Differentiation

As shown in Table 1, the population differentiation was significant within each locus, with the average F_{ST} value of 0.129 detected, ranging from 0.066 (INRABERN172) to 0.273 (INRA023). The measure of the relative genetic differentiation for the pairwise F_{ST} values between populations was found to be significant ($P<0.05$, Table S3), except for four pairwise populations (YLS-ZTS, CDS-XJS, ZWS-HXR and YCB-MTS).

F-statistics and their 95% confidence intervals obtained with 10000 bootstraps over 30 loci were as follow: $F_{IS} = 0.105$ (−0.007–0.552), $F_{IT} = 0.223$ (−0.028–0.601) and $F_{ST} = 0.132$ (0.072–0.293). Breeds estimated of F_{IS} were computed using FSTAT (Goudet, 2000) and ranged from 0.018 to 0.21 (Table 1). Breeds with the highest F_{IS} values were YMH, HWS, GSS and BJS. The breeds with the lowest F_{IS} value were CDM, LLH, JCS and MGS.

The AMOVA results indicated that 77.72% of the total variance was partitioned within individuals, whereas 13.15% and 9.12% of the total variance were explained by differences among the populations and among the individuals within populations, respectively ($P<0.001$, Table 3).

Population Structure

The neighbor-joining tree constructed on the Reynolds distances was used to portray the degree of the genetic relationships among the goat populations. The tree showed that the 40 populations were clustered into four major clades in China (Figure 2). As shown in the Figure 2, the Southwest China clade comprised ZTS, LLS, YLS, MGS, JCS, GSS, GZS, LNS and FQH, whereas the South China clade comprised LLY, CDM, GLM, DAS, HND and LZS. Similarly, the Northwest China clade comprised HXR, MGR, XJS, ZWS, XZS, CDS, SNB, BJS, HNN, FNB, JNQ and HWS, whereas the East China comprised YMH, LBB, CDB, THS, LLH, CJB, GFS, GXS, YCB, MTS, XDH, FQS and DYS. As shown in the tree, it was suggested that each population had been clustered to its geographical distribution and origin (Figure 1), not to the administrative division of China. For example, the population JCS was located in the Hengduan mountains, south of Sichuan province, therefore, which had divided into the Southwest cluster. Similar situation was also happened in BJS which is located in the northeast of Sichuan province, nearby the Qingling Mountains, but had been divided into the Northwest cluster, not to the Southwest cluster or South cluster. It was also observed that the populations (YMH, LBB, JNQ and HWS) in the Shandong province and Henan province located in the North China Plain had been grouped to the East China cluster, but admixed with the North China cluster. The cause of this situation maybe is that there is no mountain to restrict the gene flow between the two clusters: East China and Northwest China.

The population-based PCA was performed using allele frequencies of the 30 microsatellite markers in 40 goat populations. The first principal component accounts for 21.09% of the total variation and especially separated the whole populations into the West and the East. The second principal component condenses 13.42% of the variation, which separated the 40 populations into the South and North. The PCA clustered these populations into four groups, as shown in the Figure 3, which corresponded to the four gene pools (Figure 2). However, there was some ambiguity regarding the placement of the individuals of populations LBB, JNQ, YMH, HHS and LNS, as the individuals of these populations represented a mixture of the other groups. The results are mostly coincided with the topology of the neighbor-joining tree. The populations are grouped on the basis of geographical distribution and origin generally.

The Bayesian clustering model-based method allowed for assessment of the genetic structure and admixture among the populations. The results of Bayesian-based clustering revealed that the inference of the number of gene pools (K) was not straightforward because log-likelihood values for the data conditional on K, and $\ln P(x|K)$, increased progressively as K increased, as shown in Figure 4. When the number of populations varied from $K=2$ to 40, the largest change in the log of the likelihood function (ΔK) was when $K=4$. The ΔK value obtained was 19.90 at $K=3$, 120.50 at $K=4$, 0.78 at $K=5$, 2.59 at $K=6$, and 0.52 at $K=7$. Changing the assumptions of an "equal alpha for each population" and "correlated allele frequencies" could not change the final result. The proportion of individuals in each population assigned into four clusters (or gene pools) is shown in Figure 4, and the corresponding 40 populations were plotted onto the map, as shown in Figure 1. The results of Bayesian clustering for $K=2$ indicated that there was a clear transition from the West China populations (green) to the East China populations (red). At $K=3$, the West China populations had been divided into two clusters, the Southwest China populations (green) and the Northwest China populations (blue). At $K=4$, the ΔK value reached a maximum, and the Southwest China populations subsequently had been divided into the Southwest China populations (green) and South China populations (yellow). It was suggested that gene pool 1(Southwest China) included nine populations: LLS, MGS, YLS, ZTS, JCS, GSS, FQH, GZS and LNS, and that this pool was most abundant in the Yunnan-Guizhou Plateau, except LNS in the Liaoning Province. Similar results were also found in the NJ-tree

Table 1. Genetic diversity measures estimated at each 30 loci across the 40 Chinese indigenous goat populations.

Locus	Location	TA	Allele size	H_O	H_S	H_T	F_{IT}	F_{ST}	F_{IS}
CSRD247	14	13	220–247	0.667	0.723	0.804	0.174***	0.105***	0.078***
ETH10	5	11	200–210	0.376	0.309	0.379	0.012	0.189***	−0.218
MAF209	17	4	100–104	0.246	0.237	0.274	0.104***	0.138***	−0.039
OARAE54	25	15	115–138	0.582	0.607	0.722	0.200***	0.161***	0.047***
SRCRSP15	unknown	11	172–198	0.507	0.542	0.626	0.192***	0.134***	0.067***
SRCRSP3	10	10	109–123	0.527	0.594	0.714	0.260***	0.167***	0.111***
SRCRSP5	21	14	156–178	0.670	0.693	0.786	0.146***	0.121***	0.029**
TGLA53	16	18	126–160	0.358	0.622	0.703	0.503***	0.116***	0.438***
DRBP1	23	24	195–229	0.358	0.758	0.838	0.574***	0.097***	0.528***
ILSTS087	6	13	135–155	0.702	0.764	0.853	0.182***	0.105***	0.085***
INRABERN172	26	12	234–256	0.570	0.607	0.650	0.123***	0.066***	0.061***
MAF065	15	18	116–158	0.709	0.733	0.792	0.103***	0.074***	0.031**
MCM527	5	12	165–187	0.603	0.610	0.669	0.097***	0.088***	0.010
OARFCB20	2	15	93–112	0.589	0.652	0.746	0.208***	0.132***	0.088***
SPS113	10	12	134–158	0.671	0.697	0.768	0.129***	0.096***	0.037**
SRCRSP8	unknown	16	215–255	0.560	0.743	0.830	0.314***	0.104***	0.235***
ILSTS029	3	20	148–170	0.661	0.653	0.776	0.156***	0.163***	−0.009
INRA023	3	19	196–215	0.489	0.635	0.863	0.437***	0.273***	0.226***
INRA063	18	23	164–186	0.462	0.663	0.828	0.438***	0.205***	0.293***
INRABERN185	18	18	261–289	0.504	0.514	0.674	0.256***	0.244***	0.016
P19(DYA)	23	18	160–196	0.636	0.690	0.771	0.178***	0.105***	0.081***
SRCRSP23	unknown	18	81–119	0.714	0.694	0.811	0.117***	0.146***	−0.034
SRCRSP9	12	17	99–135	0.747	0.743	0.839	0.115***	0.118***	−0.003
TCRVB6	10	24	217–255	0.590	0.722	0.797	0.263***	0.098***	0.183***
BM6444	2	35	118–200	0.541	0.820	0.907	0.411***	0.099***	0.347***
ILSTS011	14	13	256–294	0.648	0.675	0.747	0.126***	0.098***	0.031*
ILSTS005	10	13	172–218	0.657	0.614	0.704	0.071***	0.130***	−0.068
SRCRSP7	6	7	117–131	0.493	0.569	0.642	0.237***	0.116***	0.137***
OARFCB48	17	12	149–173	0.731	0.715	0.786	0.067***	0.092***	−0.027
MAF70	4	14	134–168	0.708	0.722	0.807	0.124***	0.109***	0.017
	All	471		0.576	0.644	0.737	0.220***	0.129***	0.105***

Note: *Location*, the locus located on the chromosomes; T_A, total number of alleles; H_O, observed heterozygosity; H_S, gene diversity; H_T, overall gene diversity; F_{IS}, F_{ST} and F_{IT}, measures of the F-statistics.
*$P<0.05$;
**$P<0.01$;
***$P<0.001$.

Table 2. Genetic diversity measures estimated using 30 microsatellite loci in each of 40 goat populations.

No.	Breed	n	TNA	MNA	AR	H_E	H_O	F_{IS}
1	LLS	52	180	6.00	5.21	0.5882	0.4908	0.167
2	MGS	50	160	5.33	4.68	0.5981	0.5729	0.043
3	YLS	47	180	6.00	5.36	0.6283	0.5517	0.123
4	ZTS	50	178	5.93	5.16	0.5925	0.5299	0.107
5	JCS	60	172	5.73	4.95	0.6000	0.5754	0.041
6	GSS	59	230	7.67	6.56	0.6797	0.5490	0.194
7	FQH	43	137	4.57	4.23	0.5584	0.5178	0.073
8	GZS	50	169	5.63	4.91	0.5823	0.5465	0.062
9	LNS	58	193	6.43	5.69	0.6635	0.6198	0.066
10	CDM	58	178	5.93	5.09	0.6060	0.5952	0.018
11	GLM	60	195	6.50	5.66	0.6492	0.5965	0.082
12	LLY	52	169	5.63	5.13	0.6395	0.6003	0.062
13	LZS	56	164	5.47	4.71	0.5826	0.5393	0.075
14	HND	58	178	5.93	5.18	0.5944	0.5078	0.147
15	DAS	36	178	5.93	5.60	0.6601	0.5574	0.157
16	MGR	48	208	6.93	6.19	0.6614	0.5556	0.161
17	CDS	59	237	7.90	6.87	0.7090	0.6530	0.080
18	XJS	59	247	8.23	7.03	0.6891	0.6155	0.108
19	XZS	53	221	7.37	6.60	0.6846	0.6116	0.108
20	HXR	43	243	8.10	7.27	0.7171	0.6339	0.117
21	ZWS	25	206	6.87	6.80	0.6941	0.6218	0.106
22	SNB	56	194	6.47	5.83	0.6830	0.6010	0.121
23	BJS	58	202	5.73	5.70	0.6815	0.5532	0.190
24	FNB	45	248	8.27	7.35	0.7378	0.6186	0.163
25	HNN	60	220	7.33	6.31	0.7076	0.6730	0.049
26	HWS	50	221	7.37	6.75	0.7294	0.5882	0.195
27	JNQ	59	219	7.30	6.56	0.7369	0.6461	0.124
28	YMH	53	235	7.83	6.73	0.7043	0.5573	0.210
29	LBB	48	230	7.67	6.72	0.7069	0.6279	0.113
30	LLH	58	193	6.43	5.52	0.6618	0.6476	0.022
31	THS	60	194	6.47	5.59	0.6569	0.6151	0.064
32	CJB	42	173	5.77	5.32	0.6440	0.6234	0.032
33	CDB	60	194	6.47	5.39	0.6273	0.5536	0.118
34	GFS	44	130	4.33	4.02	0.5269	0.5051	0.042
35	GXS	43	144	4.80	4.35	0.5836	0.5375	0.080

Table 2. Cont.

No.	Breed	n	TNA	MNA	AR	H_E	H_O	F_{IS}
36	MTS	60	161	5.37	4.78	0.6251	0.5689	0.091
37	YCB	52	178	5.93	5.19	0.6549	0.6227	0.050
38	XDH	52	161	5.37	4.74	0.6080	0.5493	0.097
39	FQS	55	147	4.90	4.30	0.5692	0.4745	0.168
40	DYS	47	132	4.40	3.89	0.5070	0.4336	0.146
	Average		190	6.31	5.60	0.6433	0.5760	

Note: n, sample size; TNA, total number alleles; MNA, mean number of alleles; AR, allelic richness, H_E, heterozygosity estimates, H_O, heterozygosity observed; F_{IS}, fixation index.

and PCA analysis. The gene pool 2 (South China) surrounded the Yunnan-Guizhou Plateau and reached the southernmost of China along the Pearl River, and that this pool included six populations: CDM, GLM, LLY, LZS, HND and DAS. Gene pool 3(Northwest China) was predominant on the northern of Qingling Mountains – Huaihe River Line which represented the geographical boundaries of North and South in China, especially Eastern China, and encompassed fourteen populations: MGR, CDS, XJS, XZS, HXR, ZWS, SNB, BJS, FNB, HNN, HWS, JNQ, YMH and LBB. Gene pool 4 (East China) was predominant on Yangtze River hills and encompassed twelve populations: LLH, THS, CJB, CDB, GFS, GXS, MTS, YCB, XDH, FQS and DYS. However, the populations: YMH and LBB, which located at the Shandong Province, showed a mixture of the four ancestral genetic resources (Figure 1 and Figure 4). With the exception of LNS, it was obvious that the four gene pools were almost in accordance with the natural distribution of goat in China. The three clustering results from the three different analysis methods were clearly in total agreement with each other.

Discussion

Moderate Genetic Diversity and High Genetic Differentiation of the Goat Populations

In the current study, we attempted to estimate the genetic diversity and differentiation of the 40 China indigenous goat populations using 30 microsatellite markers. It was detected that the moderate genetic diversity at the population level ($H_S = 0.644$, $H_E = 0.6433$) and high population diversity ($H_T = 0.737$) at the species level. Our results showed that the genetic variation estimated by MNA and AR in Chinese indigenous populations was lower than that of European breeds which was 5.2–9.1 MNAs, with an average of 7.1 and 6.1–7.9 AR values, but was higher than that of EastAsian indigenous populations which was 2.5–7.6, with an average of 5.8 and 2.3–6.89 AR values [10,32]. In the study of Nomura et al, the genetic diversity of the populations in Mongolia and Bangladesh was higher than that of other countries, such as Japan, Korea and Indonesia [32]. Similar result also occurred in our study. In this investigation, genetic diversity of the goats in the Northwest populations (such as XJS, MGR and FNB goat) was comparatively higher than that in others clusters. Although, the populations were different, the similar result also occurred in the study of Li et al, that the diversity of the North China populations, such as Taihang goat and Neimonggol goat, was also higher than that of other populations (such as Small-xiang goat, Chuangdong White goat and Nanjiang Brown goat) [9]. The reason of the high genetic diversity maybe is that the Northwest China populations are nearby the center of Fertile Crescent region of Near East, and the gene flow was converged in the region.

Significant population differentiation was indicated by the high F_{ST} values (averaging 0.129), similar to the EastAsian indigenous populations with the mean value of 0.13, but higher than that of the European breeds, which had a mean value of 0.07 [10,32]. The Eastern China populations showed the higher pairwise F_{ST} values than the mean, suggesting that these goat populations showed higher genetic differentiation than other populations. The regions of these Eastern populations were located farthest from the goat domestication center of West Asia.

In the present study, Bottlenecks occurred in thirty-six populations, except CDS, XJS, BJS and YMH, which were located in North of Qinling-Huaihe Line. Our results indicated that population bottleneck had happened over the entire distribution of Chinese goats. It is the first report to analysis the bottleneck in Chinese indigenous goats using 30 microsatellite

Table 3. Results of analysis of molecular variance (AMOVA) for 40 populations of Chinese goats.

Source of variation	d.f	Sum of squares	Variation components	Variations (%)
Among populations	40	6405.542	1.46086	13.15127***
Among individuals within populations	2038	21923.663	1.01360	9.12486***
Within individuals	2073	18106	8.63368	77.72387***

d.f., degrees of freedom.
***P<0.001, P-values were obtained by 20000 permutations.

markers. The results of Bottleneck analysis of Chinese goats would help us to understand the developing progress of these populations, and provide new sights to establish the conservation programs to maintain the high genetic diversity of these indigenous goat populations.

In the present study, the results of the F_{IS} value over all the loci (Table 2) and the F_{IS} within individual populations per locus (Table 1) indicated that there was a significant deviation from the Hardy-Weinberg (HW) equilibrium in the 40 goat populations. The main reasons of the significant positive values of the F_{IS} are the existence of inbreeding and excess homozygosity.

Distinct Genetic Structure of the Chinese Goat Populations

According to the results of Bayesian model-based clustering and the NJ tree based on Nei's genetic distance, it was detected that there was the same clear population structure of Chinese domestic goat populations, comprising four major genetic distinct groups, the Southwest China, the South China, the East China and the Northwest Chinas. The analysis of PCA was also consistent to the result of clustering, suggesting that the clustering of these Chinese goat populations was efficient and reliable in this experiment. In the studies of Zhang et al, the Nanjiang Yellow goat and the Leizhou goat in South China had also been clustered into one group, whereas the other three breeds in North China had been clustered into one group, consisting to our results [33]. Furthermore, it was observed that these goat populations had not been clustered by their purpose by the artificial selection, such as meat, cashmere or fur, on the contrary, the genetic relationships among the populations of goat reflected the natural geographical locations of the populations. The cause of the results here might be due to the short time of artificial selection in China. We also found that the populations with the close geographical distance were characterized by low genetic differentiation, but as the distance among the populations had been increased, the genetic differentiation among the populations also increased. Similarly, the results of AMOVA also supported the observed patterns of population differentiation, which indicated significant molecular variance both among populations (13.15%) and among individuals within populations (9.12%) (Table 3).

In the three clustering methods, we found that the current genetic structure of Chinese goats has resulted from the special geographical structure, especially in the Western China. It was observed that the Hengduan Mountains located at the middle of Qinghai-Tibet Plateau and Yunnan-Guizhou Plateau, which restricted the gene flow between XZS in Tibet and other goat populations (such as LLS, YLS, ZTS and JCS) in the Yunnan province.

In addition, as a geographical boundary in China, the Qinling Mountains-Huaihe River Line which divides the regions of China into the South and the North, and generates the difference between the South and the North: the climate, agricultural forms, geographical features as well as the human's habits. In the present study, our results clearly demonstrated that the Qinling Mountains-Huaihe River Line also had separated the West goat populations into two clusters: the Southwest cluster and the Northwest cluster. Although with some admixture to the Northwest China population, our results indicated that the LNS located in the north of China had been clustered into the Southwest China populations, inconsistent to other results [33]. The reasons caused the difference are following: firstly, there is 30 microsatellite loci selected for the analysis in our study, not the limit numbers; on the other hand, the number of goat populations in the genetic structure analysis is 40, sampling from almost all of the geographical regions, from high latitude to plains.

The Origin of Goat in China

Previous genetic studies and archaeological data had revealed goats were probably first domesticated in the Fertile Crescent region of Near East around 10,000 years ago [1,2]. Some studies hint that a second independent domestication in Pakistan had given rise to the Cashmere-like goat breeds [1,34]. However, the question that where is the Chinese goats from, has not been elaborated completely, although there are so many studies focus on it [8,35,36,37]. Phylogenetic analysis of the first hypervariable region of mitochondrial DNA control region had suggested that the goats has multiple maternal origins with a possible center of origin in Aisa, as well as in the Fertile Crescent [8]. Li et al studied the mtDNA RFLP of 18 native goat populations in China, indicating that the native goat populations in China might originate from two different maternal ancestors [35]. In addition, it was proven that at least one subclade of lineage B was originated from eastern Asia, especially in southwest China [36]. Some studies also support the hypothesis that the southwest of China might be one of the domestication centers [37,38].

In our study, we estimated that the goat populations in Southwest China might be the earliest domestic goats in China, even in Asia. The reasons contributed to the result are as following. Firstly, the results of genetic structure analysis by STRUCTURE demonstrated that, the Northwest China populations combining with the Southwest populations had been clustered into the West populations, when K was 2. However, the East China populations had been clustered into a single clade. Similar result was also observed in the NJ-tree. In other words, if the goat populations in Southwest China were not the earliest domestic goats, the Northwest China populations would be an independent cluster, while the Southwest China and East China populations would be another cluster, however, it is inconsistent with the present results. The observation that the genetic diversity of Northwest China populations was higher than the Southwest and the East China populations provided additional evidence to support this hypothesis.

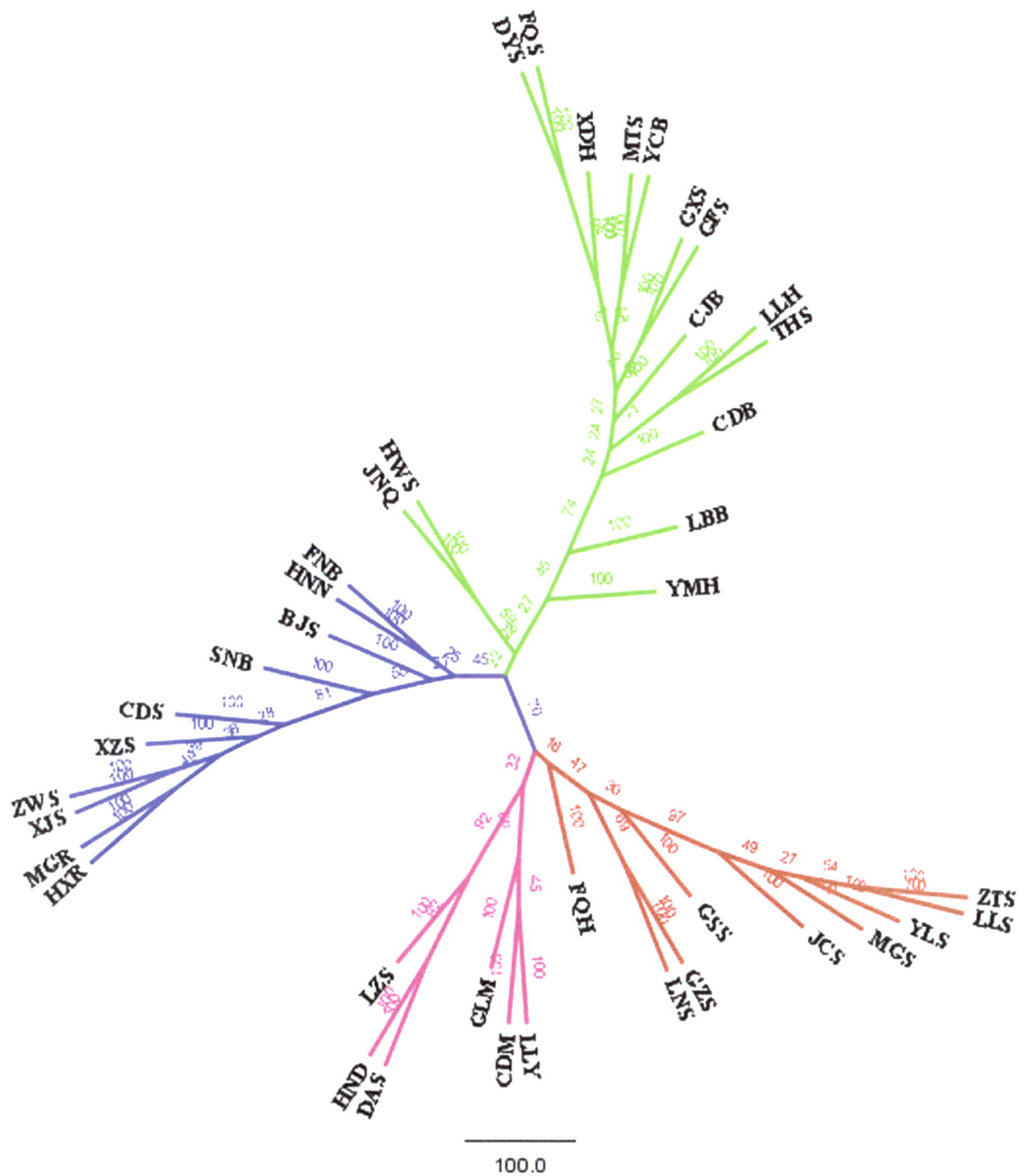

Figure 2. The neighbor-joining tree of 40 goat populations in China was constructed based on Nei's genetic distance (D_A). The population abbreviations are shown in Table 1. The numbers on the nodes indicate the bootstrap values (%) obtained from 1000 replications.

Secondly, the domestication process of livestock is closely associated to the human activities [5]. It is generally recognized that the modern humans in East Asia have a common origin in Africa, and the first entry of modern humans into the southern part of East Asia occurred about 18,000–60,000 years ago and was followed by a north-ward migration that coincided with the receding glaciers in that area [39,40]. The evidence of archaeology also indicated that the modern Chinese humans were originated in the South China [41]. Therefore, the migratory routes of modern humans in China help us to better understand the domestication process of the goats.

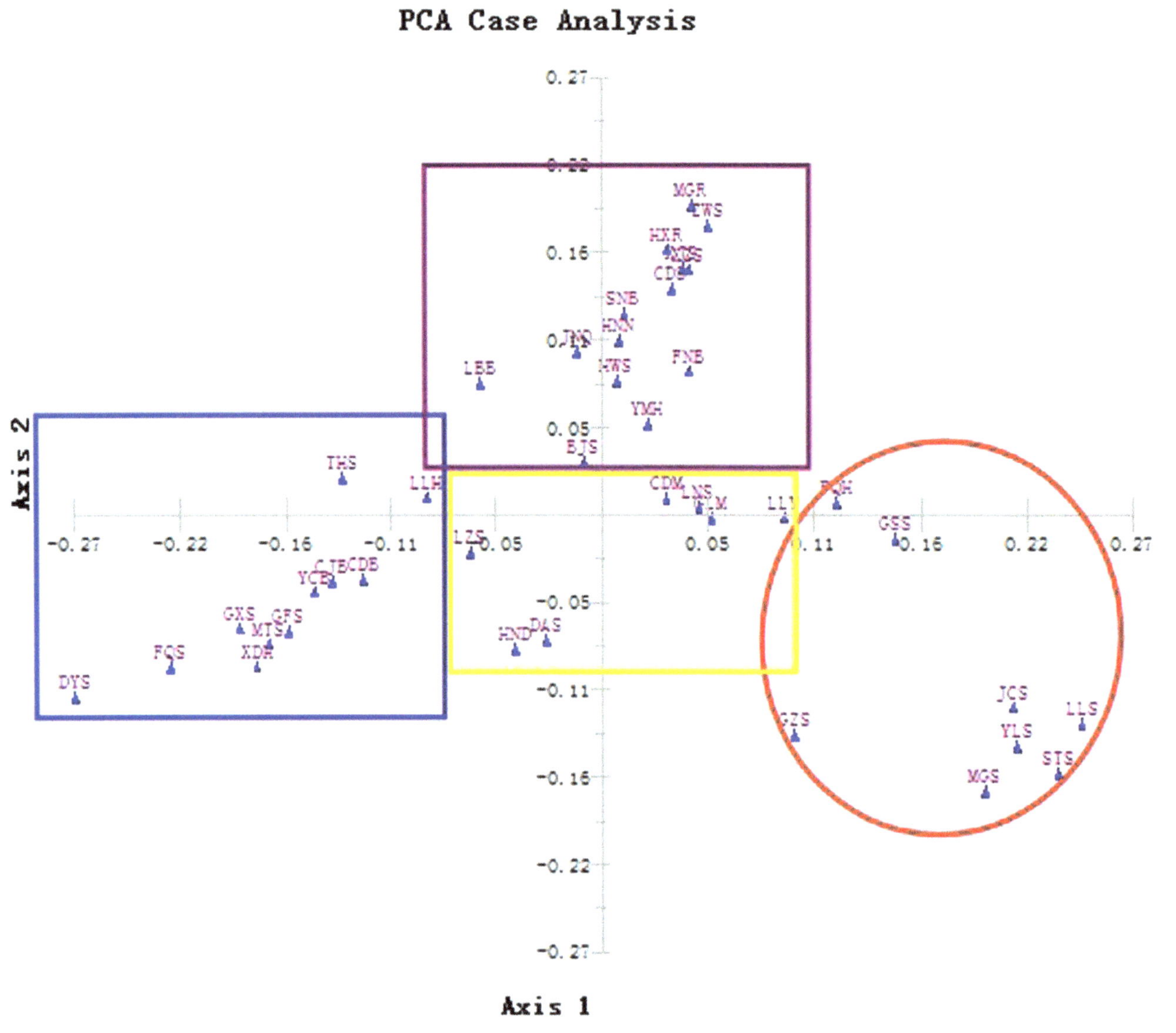

Figure 3. PCA analysis.

Thirdly, the Southwest China has the most complex river systems and a profoundly complex and dynamic geological history, and it also is rich of biodiversity [16,17]. The special geographic structures of the region, such as the Hengduan Mountains and Qingling-Huaihe River Line, had restricted the gene flow of the livestock between the North and Southwest China goat populations. In this region, it is appear that multiple origin events documented as common phenomena in domestic animals such as cattle, chickens and pigs, suggesting that the southwest of China might be one of the domestication centers [42,43,44,45]. Moreover, the mitochondrial DNA analysis of the goats in five Asian countries around of China (India, Pakistan, Mongolia, Malaysia and Laos) had revealed that the mtDNA lineage B were higher in Chinese goats than those of mixed goat populations in the five countries, suggesting that mtDNA lineage B of goats probably originated from China [37]. Combing with our results, it is inferred that the goat populations in Southwest China might be the earliest domestic goats in China, even in Asia.

Certainly, the results of microsatellite markers don't have the convincing evidence to support the hypothesis completely, due to the low density of microsatellite markers in the DNA nuclear. Therefore, it is necessary to explore the origin of Chinese goats through a detailed study of goats in Asia using the archeological records, ancient DNA, wild goat samples from more locations, and more accurate molecular methods, such as the genome-wide SNP array and mitochondrial DNA.

In this study, we had analyzed the diversity of forty Chinese domestic goat breeds which were come from all the regions of China, from Qinghai-Tibet plateau region to the north and south agricultural region, and these breeds also had special purpose, such as meat, cashmere, fur and so on. The results here provide valuable information to help understanding the genetic structure of these important indigenous goats for future genetic improvement and setting the conservation program in China. The information of this study also would help to formulate the species-specific breeding programs to maintain the genetic characteristics and to reduce unnecessary inbreeding or gene flow among populations. The present work presents the first substantial analysis of the diversity of Chinese domestic goat microsatellite markers and provides information about the genetic structure of

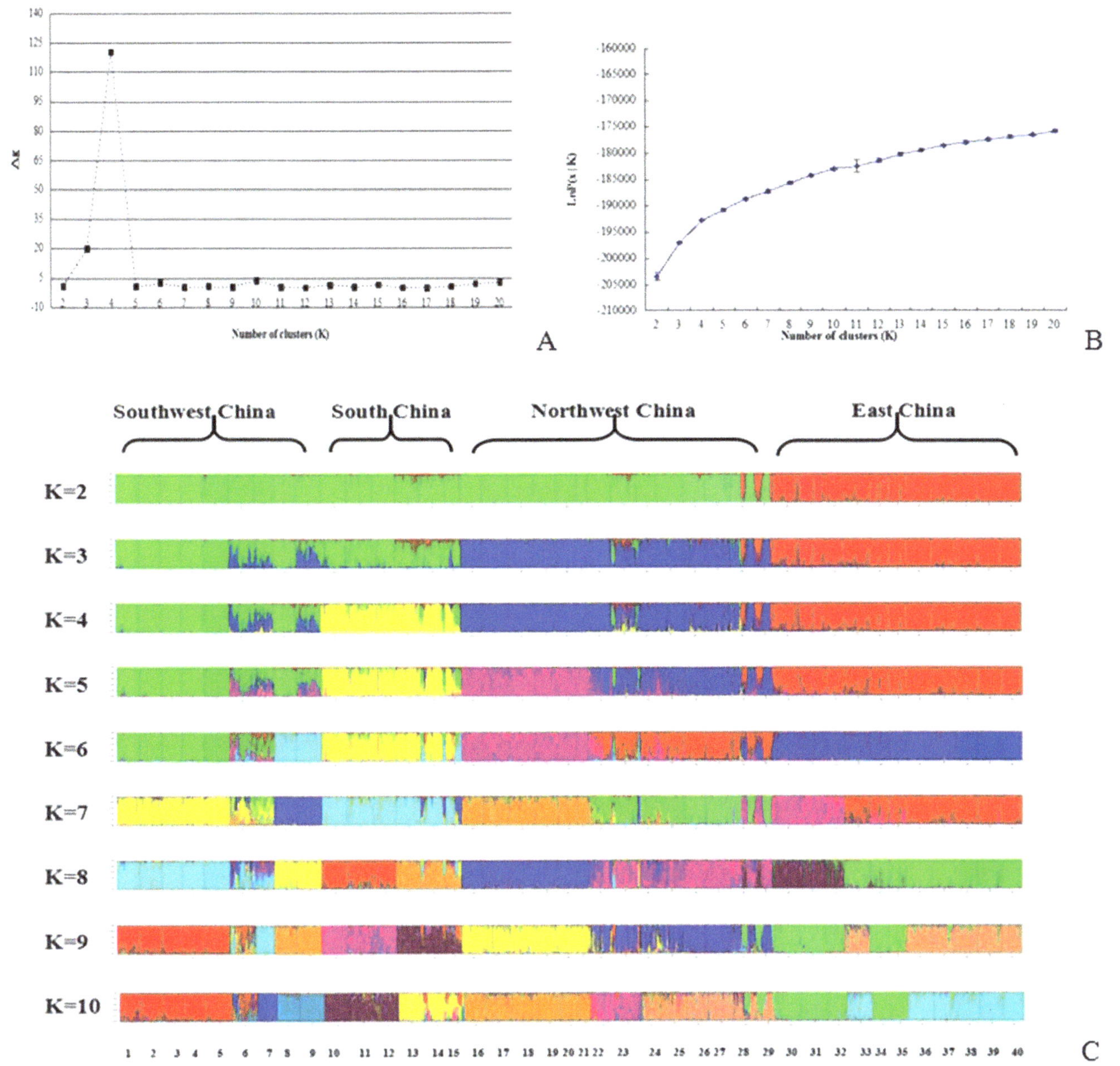

Figure 4. Population structure of 40 goat breeds based on 30 microsatellite loci using STRUCTURE. *K* was estimated using (A) the posterior probability of the data given each *K* (10 replicates) and (B) the distribution of Δ*K*, and (C) the colored clusters represented *K* from 2 to 10 were detected from STRUCTURE analysis. The number given below was corresponding to the name of each breed (seen in the Table 2).

goat breeds within this important region, and thus insights into their genetic history. With the introduction of foreign species, these indigenous goat populations have been threatened, some of which have nearly disappeared or admixed with the exotic species. Assessment of the genetic status of Chinese indigenous goat population is essential for the establishment of the conservation programs.

Supporting Information

Table S1 The information of the 40 Chinese indigenous goat populations. The information included the name, number and phenotypic characteristic of these goat populations, even the geographical location in China.

Table S2 Bottleneck analysis of the 40 Chinese indigenous goat populations.

Table S3 Pairwise F_{ST} values (upper diagonal) and D_A genetic distance (lower diagonal) between 40 goat populations.

Acknowledgments

We are grateful to the anonymous reviewers and the editor for their critical review; and their comments substantially contributed to the revision and improvement of this work. We also thank Dr. Xiaotian Qiu for assistance with sample collections.

Author Contributions

Conceived and designed the experiments: CHW LXD CSL. Performed the experiments: XH GL. Analyzed the data: JL LYX FPZ. Contributed reagents/materials/analysis tools: ZGW LZ. Wrote the paper: CHW JL.

References

1. Porter V (1996) Goats of the world: Farming Press. Available: http://agris.fao.org.
2. Pringle H (1998) Reading the signs of ancient animal domestication. Science 282: 1448–1448. doi: 10.1126/science.282.5393.1448.
3. Joshi MB, Rout PK, Mandal AK, Tyler-Smith C, Singh L, et al. (2004) Phylogeography and origin of Indian domestic goats. Mol Biol Evol 21: 454–462. doi: 10.1093/molbev/msh038.
4. Luikart G, Gielly L, Excoffier L, Vigne J-D, Bouvet J, et al. (2001) Multiple maternal origins and weak phylogeographic structure in domestic goats. Proc Natl Acad Sci U S A 98: 5927–5932. doi: 10.1073/pnas.091591198.
5. MacHugh DE, Bradley DG (2001) Livestock genetic origins: goats buck the trend. Proc Natl Acad Sci U S A 98: 5382–5384. doi: 10.1073/pnas.111163198.
6. Glowatzki-Mullis M L, Muntwyler J, Bäumle E, Gaillard C (2009) Genetic diversity of Swiss sheep breeds in the focus of conservation research. J Anim Breed Genet 126: 164–175. doi: 10.1111/j.1439-0388.2008.00768.x.
7. Dalvit C, De Marchi M, Zanetti E, Cassandro M (2009) Genetic variation and population structure of Italian native sheep breeds undergoing in situ conservation. J Anim Sci 87: 3837–3844. doi:10.2527/jas.2008-1682.
8. Chen SY, Su YH, Wu SF, Sha T, Zhang YP (2005) Mitochondrial diversity and phylogeographic structure of Chinese domestic goats. Mol Phylogenet Evol 37: 804–814. http://dx.doi.org/10.1016/j.ympev.2005.06.014.
9. Li MH, Zhao SH, Bian C, Wang HS, Wei H, et al. (2002) Genetic relationships among twelve Chinese indigenous goat populations based on microsatellite analysis. Genet Sel Evol 34: 729–744. doi:10.1186/1297-9686-34-6-729.
10. Canon J, Garcıa D, Garcıa-Atance MA, Obexer-Ruff, GLenstra JA, et al. (2006) Geographical partitioning of goat diversity in Europe and the Middle East. Anim Genet 37: 327–334.: doi:10.1111/j.1365-2052.2006.01461.x.
11. Naderi S, Rezaei H-R, Taberlet P, Zundel S, Rafat S-A, et al. (2007) Large-scale mitochondrial DNA analysis of the domestic goat reveals six haplogroups with high diversity. PLoS One 2: e1012. DOI: 10.1371/journal.pone.0001012.
12. Kijas JW, Ortiz JS, McCulloch R, James A, Brice B, et al. (2012) Genetic diversity and investigation of polledness in divergent goat populations using 52 088 SNPs. Anim Genet 44(3): 325–335. doi: 10.1111/age.12011.
13. Kim KS, Yeo JS, Lee JW, Kim JW, Choi CB (2002) Genetic diversity of goats from Korea and China using microsatellite analysis. Asian Austral J Anim 15: 461–465.
14. Poissant J, Shafer ABA, Davis CS, Mainguy J, Hogg JT, et al. (2009) Genome-wide cross-amplification of domestic sheep microsatellites in bighorn sheep and mountain goats. Mol Ecol Resour 9: 1121–1126. doi: 10.1111/j.1755-0998.2009.02575.x.
15. Di R, Farhad Vahidi SM, Ma YH, He XH, Zhao QJ, et al. (2011) Microsatellite analysis revealed genetic diversity and population structure among Chinese cashmere goats. Anim Genet 42(4): 428–431. doi: 10.1111/j.1365-2052.2010.02072.x.
16. Liu Q, Chen P, He K, Kilpatrick CW, Liu SY, et al. (2012) Phylogeographic study of Apodemus ilex (Rodentia: Muridae) in southwest China. PLoS One 7: e31453. doi: 10.1371/journal.pone.0031453.
17. Yuan JH, Cheng FY, Zhou SL (2012) Genetic structure of the tree Peony (Paeonia rockii) and the Qinling Mountains as a geographic barrier driving the fragmentation of a large population. PLoS One 7: e34955. doi: 10.1371/journal.pone.0034955.
18. Li XL, Valentini A (2004) Genetic diversity of Chinese indigenous goat breeds based on microsatellite markers. J Anim Breed Genet 121: 350–355. doi: 10.1111/j.1439-0388.2004.00465.x.
19. Li JY, Chen H, Lan XY, Kong XJ, Min LJ (2008) Genetic diversity of five Chinese goat breeds assessed by microsatellite markers. Czech J Anim Sci 53: 315–319. http://journals.uzpi.cz:8050/uniqueFiles/01923.pdf.
20. Goudet J (2001) FSTAT, a program to estimate and test gene diversities and fixation indices (version 2.9. 3). http://www.unil.ch/izea/softwares/fstat.html.
21. Weir BS, Cockerham CC (1984) Estimating F-statistics for the analysis of population structure. Evolution: 1358–1370. http://www.jstor.org/stable/2408641.
22. Raymond M, Rousset F (1995) GENEPOP (version 1.2): population genetics software for exact tests and ecumenicism. J Hered 86: 248–249. http://hal.archives-ouvertes.fr/halsde-00186383/.
23. Cornuet J-M, Piry S, Luikart G, Estoup A, Solignac M (1999) New methods employing multilocus genotypes to select or exclude populations as origins of individuals. Genetics 153: 1989–2000. http://www.genetics.org/content/153/4/1989.short.
24. Nei M, Maruyama T, Chakraborty R (1975) The bottleneck effect and genetic variability in populations. Evolution: 1–10. http://www.jstor.org/stable/2407137.
25. Excoffier L, Smouse PE, Quattro JM (1992) Analysis of molecular variance inferred from metric distances among DNA haplotypes: application to human mitochondrial DNA restriction data. Genetics 131: 479–491. http://www.genetics.org/content/131/2/479.short.
26. Excoffier L, Laval G, Schneider S (2005) Arlequin (version 3.0): an integrated software package for population genetics data analysis. Evol Bioinform Online 1: 47. http://cmpg.unibe.ch/software/arlequin3.
27. Felsenstein J (2008) Phylogeny Inference Package (PHYLIP). Genomes Sciences, Department of Genetics, University of Washington, Seattle, WA. http://evolution.gs.washington.edu/phylip.html.
28. Page RDM (1996) TREEVIEW: An application to display phylogenetic trees on personal computers. Comput Appl Biosci 12: 357–358. http://www.ncbi.nlm.nih.gov/pubmed/8902363?dopt=Abstract&holding=f1000,f1000m,isrctn.
29. Pritchard JK, Stephens M, Donnelly P (2000) Inference of population structure using multilocus genotype data. Genetics 155: 945–959. http://www.genetics.org/content/155/2/945.short.
30. Evanno G, Regnaut S, Goudet J (2005) Detecting the number of clusters of individuals using the software STRUCTURE: a simulation study. Mol Ecol 14: 2611–2620. doi: 10.1111/j.1365-294X.2005.02553.x.
31. Rosenberg NA (2004) DISTRUCT: a program for the graphical display of population structure. Mol Ecol Notes 4: 137–138. doi: 10.1046/j.1471-8286.2003.00566.x.
32. Nomura K, Ishii K, Dadi H, Takahashi Y, Minezawa M, et al. (2012) Microsatellite DNA markers indicate three genetic lineages in East Asian indigenous goat populations. Anim Genet 43: 760–767. doi: 10.1111/j.1365-2052.2012.02334.x.
33. Zhang LY, Yang QF, Li XL, Zhou RY, Li LH (2012) Genetic diversity of five goat breeds in China based on microsatellite markers. Afr J Biotechnol 11: 11671–11677.
34. Meadow RH (1993) Animal domestication in the Middle East: a revised view from the eastern margin. Harappan Civilization: 295–320. edited by G. Possehl. Oxfird abd IBH, New Dehli.
35. Li XL, Zhang YP, Chen S, Zeng F, Qiu X (1999) Study on the origin and differentiation of the goat based on the mtDNA diversity. Acta Veterinaria et Zootechnica Sinica 30: 313–319.
36. Wu YP, Guan WJ, Zhao QJ, He XH, Pu YB, et al. (2009) A fine map for maternal lineage analysis by mitochondrial hypervariable region in 12 Chinese goat breeds. Anim Sci J 80: 372–380. doi: 10.1111/j.1740-0929.2009.00659.x.
37. Zhong T, Zhao QJ, Niu LL, Wang J, Jin PF, et al. (2013) Genetic phylogeography and maternal lineages of 18 Chinese black goat breeds. Trop Anim Health Pro: 1–5. doi: 10.1007/s11250-013-0432-1.
38. Lin BZ, Odahara S, Ishida M, Kato T, Sasazaki S, et al. (2013) Molecular phylogeography and genetic diversity of East Asian goats. Anim Genet 44(1): 79–85. doi: 10.1111/j.1365-2052.2012.02358.x.
39. Jin L, Su B (2000) Natives or immigrants: modern human origin in East Asia. Nat Rev Genet 1: 126–133. doi:10.1038/35038565.
40. Chu JY, Huang W, Kuang SQ, Wang JM, Xu JJ, et al. (1998) Genetic relationship of populations in China. Proc Natl Acad Sci U S A 95(20): 11763–11768. doi: 10.1073/pnas.95.20.11763.
41. Kong QP, Sun C, Wang HW, Zhao M, Wang WZ, et al. (2011) Large-scale mtDNA screening reveals a surprising matrilineal complexity in East Asia and its implications to the peopling of the region. Mol Biol Evol 28: 513–522. doi: 10.1093/molbev/msq219.
42. Lai S, Chen S, Liu Y, Yao Y (2007) Mitochondrial DNA sequence diversity and origin of Chinese domestic yak. Anim Genet 38: 77–80. doi: 10.1111/j.1365-2052.2007.01555.x.
43. Wu GS, Yao YG, Qu KX, Ding ZL, Li H, et al. (2007) Population phylogenomic analysis of mitochondrial DNA in wild boars and domestic pigs revealed multiple domestication events in East Asia. Genome Biol 8: R245. doi: 10.1186/gb-2007-8-11-r245.
44. Jin L, Zhang M, Ma J, Zhang J, Zhou C, et al. (2012) Mitochondrial DNA evidence indicates the local origin of domestic pigs in the upstream region of the Yangtze river. PLoS One 7: e51649. doi: 10.1371/journal.pone.0051649.
45. Liu YP, Wu GS, Yao YG, Miao YW, Luikart G, et al. (2006) Multiple maternal origins of chickens: out of the Asian jungles. Mol Phylogenet Evol 38: 12–19. http://dx.doi.org/10.1016/j.ympev.2005.09.014.

5

Biogeography of Speciation of Two Sister Species of Neotropical *Amazona* (Aves, Psittaciformes) Based on Mitochondrial Sequence Data

Amanda V. Rocha[1], Luis O. Rivera[2], Jaime Martinez[3], Nêmora P. Prestes[3,4], Renato Caparroz[1]*

1 Laboratório de Genética e Biodiversidade, Instituto de Ciências Biológicas, Universidade de Brasília, Brasília, Distrito Federal, Brazil, **2** Centro de Investigación y Transferencia de Jujuy, Universidad Nacional Jujuy - CONICET, Jujuy, Argentina, **3** Projeto Charão - Associação Amigos do Meio Ambiente, Carazinho, Rio Grande do Sul, Brazil, **4** Instituto de Ciências, Universidade de Passo Fundo, Passo Fundo, Rio Grande do Sul, Brazil

Abstract

Coalescent theory provides powerful models for population genetic inference and is now increasingly important in estimates of divergence times and speciation research. We use molecular data and methods based on coalescent theory to investigate whether genetic evidence supports the hypothesis of *A. pretrei* and *A. tucumana* as separate species and whether genetic data allow us to assess which allopatric model seems to better explain the diversification process in these taxa. We sampled 13 *A. tucumana* from two provinces in northern Argentina and 28 *A. pretrei* from nine localities of Rio Grande do Sul, Brazil. A 491 bp segment of the mitochondrial gene cytochrome c oxidase I was evaluated using the haplotype network and phylogenetic methods. The divergence time and other demographic quantities were estimated using the isolation and migration model based on coalescent theory. The network and phylogenetic reconstructions showed similar results, supporting reciprocal monophyly for these two *taxa*. The divergence time of lineage separation was estimated to be approximately 1.3 million years ago, which corresponds to the lower Pleistocene. Our results enforce the current taxonomic status for these two Amazon species. They also support that *A. pretrei* and *A. tucumana* diverged with little or no gene flow approximately 1.3 million years ago, most likely after the establishment of a small population in the Southern Yungas forest by dispersion of a few founders from the *A. pretrei* ancestral population. This process may have been favored by habitat corridors formed in hot and humid periods of the Quaternary. Considering that these two species are considered threatened, the results were evaluated for their implications for the conservation of these two species.

Editor: Michael Knapp, Bangor University, United Kingdom

Funding: This work is supported by graduate scholarships for A. V. Rocha by Coordenação de Aperfeiçoamento de Pessoal de Nível Superior - Capes. The funders had no role in study design, data collection and analysis, decision to publish, or preparation of the manuscript.

Competing Interests: The authors have declared that no competing interests exist.

* Email: renatocz@yahoo.com.br

Introduction

Speciation remains an important question in science, even after nearly two centuries of research and intense debate owing to the evolutionary theories of Darwin and Wallace [1]. Three models have been proposed to explain the origin of new species of animals with sexual reproduction: allopatric, parapatric and sympatric (for review, see [2]). Allopatric speciation can occur basically in two models: vicariant speciation [3] and dispersion/founder effect [4]. Under the first model, a widely distributed species becomes subdivided into two or more relatively large populations due to extrinsic barriers, and genetic differences begin to accumulate between isolates as each population responds to its own array of selective forces and tracks its ever-changing environment. Under the second model, species evolve through the establishment of small demes often peripherally isolated by dispersing a few founders from an ancestral population.

There are numerous studies of phylogenetic analyses based on molecular data that support vicariance (e.g., [5]), dispersal (e.g., [6,7]), or both (e.g., [8]) as causes of distribution patterns in Neotropical birds. Similarly, empirical studies have increasingly found dispersion/colonization as a force behind the most common allopatric speciation events [9–13]. Estimates of divergence times based on the molecular clock can be used to correlate the time of cladogenesis of species with paleogeographical vicariant events. Depending on the relative ages of species divergence and vicariant events, assessments can be made of whether a dispersal hypothesis (barrier older than species divergence) or vicariant hypothesis (barrier age and species divergence similar) better explains the observed distributions [6]. In many cases, it has been shown that the study group is too young to have been affected by the alleged vicariant barrier, suggesting a recent history of dispersal instead of vicariance of the *taxa* [6,14–16]. In recent years, methods based on coalescent theory have been applied to estimate divergence times among populations and/or species, enabling pure isolation with subsequent independent evolution to be distinguished from isolation with subsequent gene flow [17].

The Spectacled Amazon (*Amazona pretrei*) and the Tucuman Amazon (*A. tucumana*) are Neotropical parrots which were shown to be sister species in phylogenetic reconstructions [18]. However, the Tucuman Amazon was previously regarded as a race of *A.*

pretrei [19], but was separated as a full species based on a (most likely erroneous [20]) record of sympatry in the Misiones province of Argentina. These two forms seem to differ in their adult and juvenile plumage, with *A. pretrei* being sexually dimorphic, while the sexes are alike in *A. tucumana* [21,22]. Different feeding requirements also tend to support the view that these two amazons should be treated as two species, although other evidence is needed to confirm their true relationship [23].

Currently, the *A. pretrei* distribution area comprises the state of Rio Grande do Sul and southern Santa Catarina in Brazil, with some records in the Misiones province in Argentina (Fig. 1). This species is observed mainly at altitudes between 500 and 1,000 meters in environments marked by the presence of Araucaria (*Araucaria angustifolia*) forests and savannas with Araucaria [24]. *A. tucumana* is restricted to the Southern Yungas forests, a montane forest of the Andes of northwestern Argentina and eastern Bolivia, particularly in areas 1,500–2,200 meters in altitude [25]. These two forest formations are currently separated by approximately 700 km of xerophytic Chaco woodland. The latter is an alluvial plain formed by sediments from the Andes [26] and is part of diagonal open biomes in South America, covering an area of approximately 840,000 km^2 in northern Argentina, western Paraguay, southeastern Bolivia and the extreme west of Mato Grosso do Sul, Brazil [27]. The flora of this region is considered a relict of the Pliocene or lower Pleistocene, stabilized on salinic soils left after two events of marine transgressions [28], which occurred between 15 and 5 million years ago [29]. Currently, it has a semi-arid climate with intensely hot summers and severe frosts in winter [26]. However, during the interglacial periods of the Quaternary, the Chaco area may have been more humid than today [30–32].

According to Nores [33], forests advanced from the southern Yungas and the Paranense region along the Bermejo and Pilcomayo rivers deep enough to form a continuous forest bridge between both regions. Furthermore, Nores [33] noted that some bird species presumably expanded to form a continuous distribution along this continuous forest bridge and became isolated again in the subsequent drier period. This connection-interruption process most likely occurred several times during the Pleistocene and Holocene periods [30,34–36].

Having considered the biogeographical scenario described above and several pairs of species and subspecies with disjunct populations in the Southern Yungas forests and Paraná state (Brazil), Nores [33] described four different levels of bird speciation currently observable in both forests. These different levels of speciation were most likely related to the time elapsed since the species crossed the forest bridge between these two regions. The author classified *A. pretrei* and *A. tucumana* as belonging to an allospecies group, in which differentiation most likely began very early in the wet-dry cycles, although it is possible that isolation and speciation began before the Quaternary period. However, the author noted that future studies applying new methodologies might assist in the review of this classification.

In the present study, we use molecular data and methods based on coalescent theory to investigate whether genetic evidence supports the hypothesis of *A. pretrei* and *A. tucumana* as separate species, and whether genetic data allow us to assess which allopatric model seems to better explain the diversification process of *A. pretrei* and *A. tucumana*. Based on allopatry by vicariance, we should expect that separation of these species occurred before the Quaternary climatic fluctuations, possibly due to the formation of vegetation in the Chaco province estimated to have started during the Pliocene. In allopatry via the dispersion/founder effect or peripheral isolation [4], we should expect that separation occurred subsequently to the formation of the Chaco by a dispersal event of a few individuals which could have been facilitated by the forest corridor, estimated to have occured during the Pleistocene. Finally, considering that the two species in this study are considered to be threatened at a global level [37] or endangered at a national levels [38,39], the results have been evaluated for their implications for the conservation of these species. For endangered species, knowledge of these historical processes is critical for correct interpretation of current patterns of genetic variation. Furthermore, this information should guide the assignment of conservation priorities and the implementation and elaboration of subsequent conservation strategies.

Materials and Methods

Taxon sampling

We sampled 13 *A. tucumana* from two provinces located in northern Argentina and 28 *A. pretrei* from nine localities of Rio Grande do Sul, Brazil (Table 1, Fig. 1). All samples were obtained from nestlings found in natural nests with the permission of the respective environmental agencies (DPMAyRN license n° 082/2006 and APN license n° 04/2008 in Argentina, and SISBIO license n° 25694-3 in Brazil). Only one nestling from each nest was considered for genetic analysis. Three to four drops of blood were taken from each nestling via puncture of the brachial wing vein using needles and glass capillaries, subsequently stored in microtubes containing absolute ethanol. All the samples were stored frozen at the Laboratory of Molecular Genetics and Bird Evolution at the Institute of Biosciences of the University of São Paulo.

Sample size

For evaluation of whether sequence samples were representative of the genetic constitution of the species, we estimated $P = [(k-1)/(k+1)]$, which represents the probability that a sample of size k and the whole population share the most recent ancestor [40]. P can be interpreted as the probability of the sample being representative of the genetic diversity of the population.

DNA extraction and sequencing

Total genomic DNA was extracted using a standard proteinase K/SDS and phenol–chloroform protocol [41]. Using a Lambda DNA marker (Biosciences) as a reference, the DNA samples were quantified in a 1.5% agarose gel stained with ethidium bromide.

A segment of the mitochondrial gene cytochrome c oxidase I (COI) was amplified using COIA and COIF primers [42] by polymerase chain reaction (PCR) as follows: final volume of 10 µL containing 0.5 mM of each primer, 0.5 U of Taq DNA polymerase (Phoneutria, CA), 1 mM dNTP, 1× buffer (10 mM Tris-HCl pH 8.3, 50 mM KCl, 1.5 mM $MgCl_2$) and 20 to 50 ng DNA. The reactions were subjected to a cycle with a temperature of 95°C for the initial denaturation, followed by 35 cycles at 95°C for denaturation, 53°C for primer annealing and extension at 72°C, ending with one cycle at 72°C for 10 minutes for the final extension. The samples were purified by enzymatic reaction with the enzymes exonuclease I (EXO, USB) and Shrimp Alkaline Phosphatase (SAP, USB). The purified PCR products were sequenced using the DYEnamic ET Terminator kit (Amersham Bioscience) in an automated sequencer ABI 3100 (Applied Biosystems) according to manufacturer's recommendations. The sequences obtained were evaluated for quality and checked for ambiguities using SeqScape v.2.1 (Applied Biosystems). The alignment of the sequences was performed with ClustalW algorithm [43] using the MEGA 5.0 software [44]. The COI

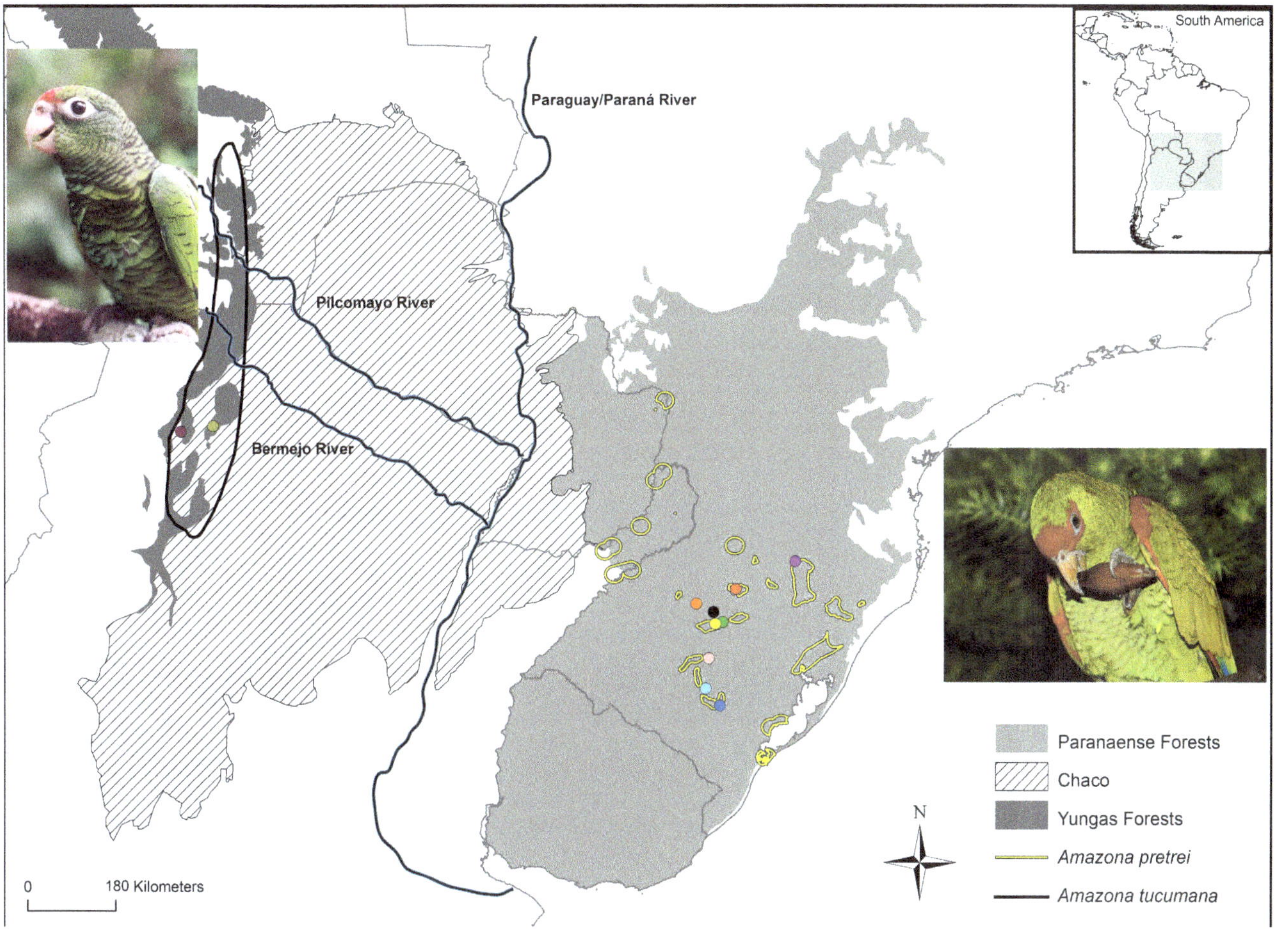

Figure 1. Geographic distribution of *Amazona pretrei* (black circle) and *A. tucumana* (yellow circles). The geographical distribution maps of species were kindly donated by IUCN [71]). Collection localities of samples of *Amazona pretrei* and *A. tucumana* analyzed in this study were indicated by colored dots. For more details of the collection sites, see Table 1.

haplotype sequences generated and used in this work were deposited in the GenBank (accession numbers KF697704–KF697713).

Sequence statistics, genetic diversity, and phylogenetic analysis

The nucleotide and haplotype diversity for each species were estimated using the DNAsp 5.10.01 program [45]. The relationship between distinct genetic sequences or haplotypes was evaluated by the haplotype network via median joining method [46] using Network 4.1 (Fluxus Technology Ltd). To verify whether the sequences conformed to neutral expectations of nucleotide substitution, we computed Tajima's D- [47], Fu and Li D* and F* values [48] for each gene using DNAsp 5.10.01.

Phylogenetic relationships among sequences were reconstructed based on the maximum parsimony using MEGA 5.0 [44] and on the Bayesian analyses by Markov Chain Monte Carlo (MCMC) using MrBayes 3.2 [49]. Maximum parsimony analysis was performed considering all the character states as unordered and equally weighted. A heuristic search was implemented with ten random addition sequence replicates, tree bisection-reconnection branch swapping, and a maximum of 100 trees saved per round. To evaluate relative robustness of the clades found in the most parsimonious trees, the bootstrap analysis [50] employed 1,000 replicates using the same heuristic search settings and a maximum of 100 trees were saved per round. Bayesian inference analysis was performed using HKY models. The best-fit model of nucleotide evolution was selected with the Akaike information Criterion (AIC) in JModeltest 2.1.4 [51]. Four Markov chains were run for 2,000,000 generations (sampling every 100 generations) to allow adequate time for convergence. Posterior probabilities of the nodes were computed for all Bayesian analyses across the sampled trees after burn-in. The number of generations required to reach stationary of the posterior distribution was determined by examining marginal probabilities plotted as a time series in Tracer v1.5 [52]. The burn-in period was set to 10%. All MCMCs were run twice to confirm consistent approximation of the posterior parameter distributions. In both analyses, a homologous sequence of *Amazona vinacea* (Genbank AY301462) was used as outgroup. This species was chosen as outgroup based on phylogeny of Russello and Amato [18].

Divergence time and other demographic quantities of the two species were estimated using the isolation and migration model described by Nielsen and Wakeley [17] employing IMa2 [53]. This method uses a Markov Chain Monte Carlo to estimate jointly the posterior distribution of the model parameters and the demographic quantities: θ ($N_e\mu$, where N_e is the effective population size of females and μ is the mutation rate per sequence

Table 1. *Taxa* used in this study: species, sample number, sampled locality, haplotype name and GenBank accession number for COI sequences.

Species	Sample	Country	State/Province	Locality	Coordinates	Haplotype	Genbank
A. pretrei	3388	Brazil	Rio Grande do Sul	Fortaleza dos Valos	28°48′30.32″S/53°15′34.28″W	pre 7	KF697712
	3390						
	1249			Santana da Boa vista	30°52′58.74″S/53°07′11.97″W	pre 1	KF697709
	1248					pre 6	KF697708
	1234			Jacuizinho	29°01′58.74″S/53°03′04.75″W	pre 3	KF697710
	2257					pre 5	KF697707
	2261						
	1255			Restinga seca	29°50′37.26″S/53°22′15.26″W	pre 5	KF697707
	2257			Carazinho	28°18′13.63″S/52°45′30.35″W	pre 1	KF697709
	2259						
	2260						
	3380						
	3382						
	4699						
	4700						
	4846						
	4701					pre 2	KF697713
	3383					pre 4	KF697711
	3384					pre 5	KF697707
	2267						
	5683						
	5676			Cruz alta	28°37′39.83″S/53°39′00.38″W	pre 5	KF697707
	5693			Caçapava do sul	30°30′40.26″S/53°26′25.26″W	pre 5	KF697707
	2252			Salto do Jacuí	29°04′18.18″S/53°13′21.64″W	pre 1	KF697709
	2253						
	2255						
	4702						
	5675			Barracão	27°41′57.32″S/51°25′32.59″W	pre 1	KF697709
A. tucumana	UFG442	Argentina	Salta	Parque Nacional El Rey	24°42′27.56″S/64°37′09.37″W	tuc 1	KF697704
	UFG446						
	UFG440					tuc 2	KF697706
	UFG450					tuc 3	KF697705
	UFG413		Jujuy	Santa Barbara	24°50′06.55″S/65°21′10.19″W	tuc 1	KF697704
	UFG416						
	UFG419						

Table 1. Cont.

Species	Sample	Country	State/Province	Locality	Coordinates	Haplotype	Genbank
	UFG420						
	UFG424						
	UFG435						
	UFG430						
	AMCC110751						AY301459
	UFG428					tuc 2	KF697706

per generation), M (Nem, where m is the proportion of effective number of migrants per generation) and t (time since population splitting) and TMRCA (time to the most recent common ancestor). For mitochondrial COI, we used the substitution rate of 0.772%/lineage/myr, based on the mean for the Neoaves clade derived in a mitogenomic timescale for birds [54].This rate was converted to 3.8×10^{-6} substitutions/locus/year by multiplying by the number of base pairs of the locus (491), and transforming from millions of years to years. Generation time was set as five years [55]. Several preliminary runs were performed in IMa2 with different priors and heating schemes to find the conditions that allowed proper mixing among chains to avoid local optimum parameter values. When good conditions were achieved, three runs were performed to verify whether the estimated parameters were converging to similar results. Final IMa2 analyses were run for 10,000,000 generations after a burn-in of 10,000 steps using geometric heating, with high heating parameters ($g1 = 0.9$ and $g2 = 0.95$), and 20 chains.

Finally, we also evaluated the historical demography by calculating the R2 statistic [56] in DNAsp version 4.0. Significant and low values of R2 suggest demographic expansion if neutrality was not rejected.

Results

Genetic variation, haplotype network and phylogenetic analysis

We analyzed a segment of 491 bp of the COI for all individuals. Translation of the protein-coding gene COI did not reveal stop codons or frame-shift mutations, and third codon positions were more variable than first and second codon positions, as expected in functional genes. In the haplotype network, two haplotype groups were clearly identified, which correspond to the haplotypes of the two taxonomically recognized species (Fig. 2).Ten unique haplotypes defined by seven variable sites (five are parsimony informative) were identified in *A. pretrei*, while only three haplotypes defined by two variable sites (only one parsimony informative) were identified in *A. tucumana*. These two species did not share haplotypes (Fig. 2). Singleton COI haplotypes differed from one another by one to four mutational steps in *A. pretrei*, and by one or two mutational steps in *A. tucumana*. The number of mutational steps separating these species was nine. Considering the genetic variability, *A. pretrei* presented haplotype diversity ($D_H = 0.68$) greater than in *A. tucumana* ($D_H = 0.41$), as well as an average genetic distance (D_M) three times greater compared to *A. tucumana* ($D_M = 0.003$ in *A. pretrei* and $D_M = 0.001$ in *A. tucumana*). The average genetic distance interspecific was $D_M = 0.022$. When compared to the outgroup (*A. vinacea*), the average genetic distance was $D_N = 0.057$. The COI sequences have an 85% and 93% probability of having correctly sampled the size of the gene genealogy of the *A. tucumana* and *A. pretrei*, respectively. Therefore, our gene samples are considered suitable for the subsequent studies.

The phylogenetic reconstructions based on maximum parsimony (data not shown) and Bayesian analysis (Fig. 3) showed similar results to those obtained by the haplotype network, showing reciprocal monophyly for these two species, with 100% of bootstrap support value and posterior probability of 1.

Coalescent analyses

Sequence data sets for both species had nonsignificant values ($p>0.10$) of Tajima's D values and Fu and Li's D* and F* statistics, suggesting that the COI gene is selectively neutral. Therefore, it is appropriate to use it to estimate gene flow and other parameters with neutral coalescent methods.

The effective population size of females for *A. pretrei* and *A. tucumana* was approximately 357,000 and 94,000 individuals, respectively (Table 2). The effective size of the ancestral population could not be estimated properly with the analyzed COI fragment. The divergence between lineages occurred with an absence or low level of gene flow because the effective number of female migrants among populations was estimated to be less than one individual per generation (Table 2). The divergence time of lineage separation was estimated to be approximately 1.3 million years ago (Table 2), which corresponds to the lower Pleistocene. In relation to the time of the most recent common ancestor, it was estimated to be 1.6 million years ago (ranging between 562,016 and 1,996,124 years ago). Assuming a rate of 0.772%/lineage/myr, and based on 5.7% genetic divergence between *A. vinacea* and *A. pretrei/A. tucumana* lineages, we estimate that the separation between these lineages occurred at least 3.96 million years ago.

According to the analysis of demographic expansion (R2), the values were not statistically significant ($R2 = 0.861$, $p > 0.05$), suggesting population stability for both species.

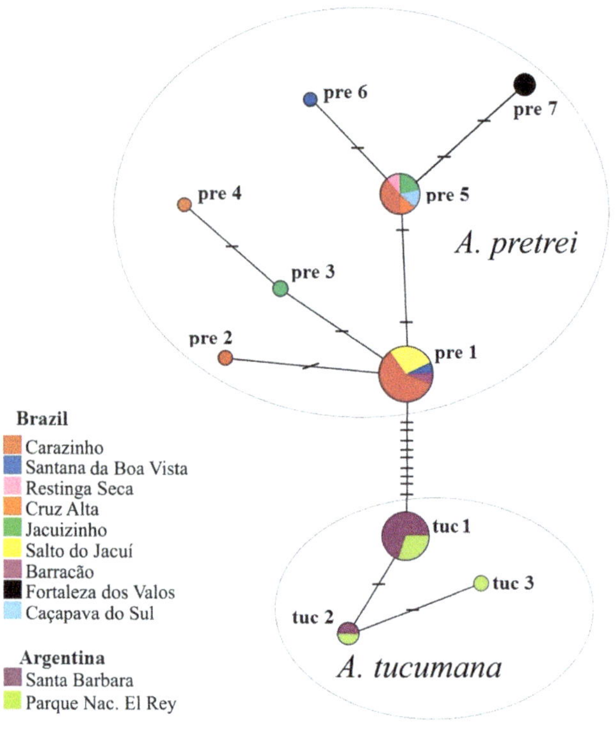

Figure 2. Relationships among the seven *Amazona pretrei* and three *A. tucumana haplotypes* based on the median-joining network of 491 bp of mitochondrial DNA COI sequences. Sizes of circles are proportional to the haplotype frequency. The number of variable sites between haplotypes is indicated by dashes between them.

Discussion

Species delimitation

Our results indicate that the two amazon species studied here are clearly genetically differentiated. The average genetic distance between these two amazon species was seven fold of those observed within them, including nine fixed nucleotide substitutions. Sequences from the COI fragment indicated a clear subdivision with haplotypes confined to each clade in both network and phylogenetic analysis, indicating reciprocal monophyly for these two species. Coalescent analysis suggested that divergence with little or no gene flow occurred between the two groups in the Lower Pleistocene.

The unified species concept [57] suggests treating independently evolving population lineages as different species if independent lines of evidence are available to support their reciprocal isolation [20]. Thus, the current taxonomic status for these two amazon species is reinforced, since genetic evidence and differences in morphology and ecology (see Introduction) indicate *A. tucumana* lineage is evolving independently from *A. pretrei* lineage.

Speciation and biogeography

It is not always easy to identify whether two sister species have resulted from vicariance or dispersal events. The vicariance model is supported if the sister species were separated by an intermittent barrier or a barrier that is younger than the estimated age of the ancestral population. In contrast, the dispersion/founder effect model is supported if the sister species were differentiated by means of a permanent barrier, which is considered to be older than the pair of species [58].

According to our analysis, *A. pretrei* and *A. tucumana* diverged with little or no gene flow when the current barrier (xerophytic Chaco vegetation) that separates these two species had already formed (see Introduction). Most likely, this divergence occurred after the establishment of a small population in the Southern Yungas forest through dispersion of a few founders from the *A. pretrei* ancestral population. Thus, this scenario suggests that the allopatry by the dispersion/founder effect model (peripheral isolation) seems to be more likely than by the vicariance model to explain the diversification of these two Amazon species, as discussed below.

According to the phylogeny proposed by Russello and Amato [18], *Amazona vinacea* is the sister species of *A. pretrei/A. tucumana* lineage. Considering that *A. vinacea* such as *A. pretrei* occupy the Araucaria forests of southern Brazil [59], it is acceptable to assume that the *A. pretrei/A. tucumana* lineage arose in this region. The divergence between *A. vinacea* and *A. pretrei/A. tucumana* lineages should have occurred at least 3.96 million years ago when the vegetation of the Chaco was most likely still in training on salty soil, due to the last marine transgression in South America [29,60]. This non-forest vegetation may have limited the spread of the *A. pretrei/A. tucumana* lineage in this period.

Based on our results, the divergence between *A. tucumana* and *A. pretrei* occurred approximately 1.3 million years ago, possibly when the xerophytic vegetation of the Chaco was already similar to current vegetation, limiting the occupation of this area by a typical forest species. It is possible that the formation of more suitable habitat corridors, such as those proposed along the banks of the Bermejo and Pilcomayo rivers across the Chaco during interglacial periods of the Quaternary [33], allowed the passage of amazons through the Chaco arid zone, reaching the wetlands of the tropical forests of the Yungas forest. The isolation between those lineages most likely occurred after the disappearance of these corridors, establishing a condition of isolation between the *A. pretrei* ancestral population and the *A. tucumana* peripheral population. Willis [61] previously also raised the subject of the use of corridors, commenting on the possible use of a south-Andean route (Andean-south route) by *A. pretrei* and *A. tucumana*.

The haplotype diversity observed in *A. tucumana* is lower than in *A. pretrei*, and the relationship pattern among the *A. tucumana*

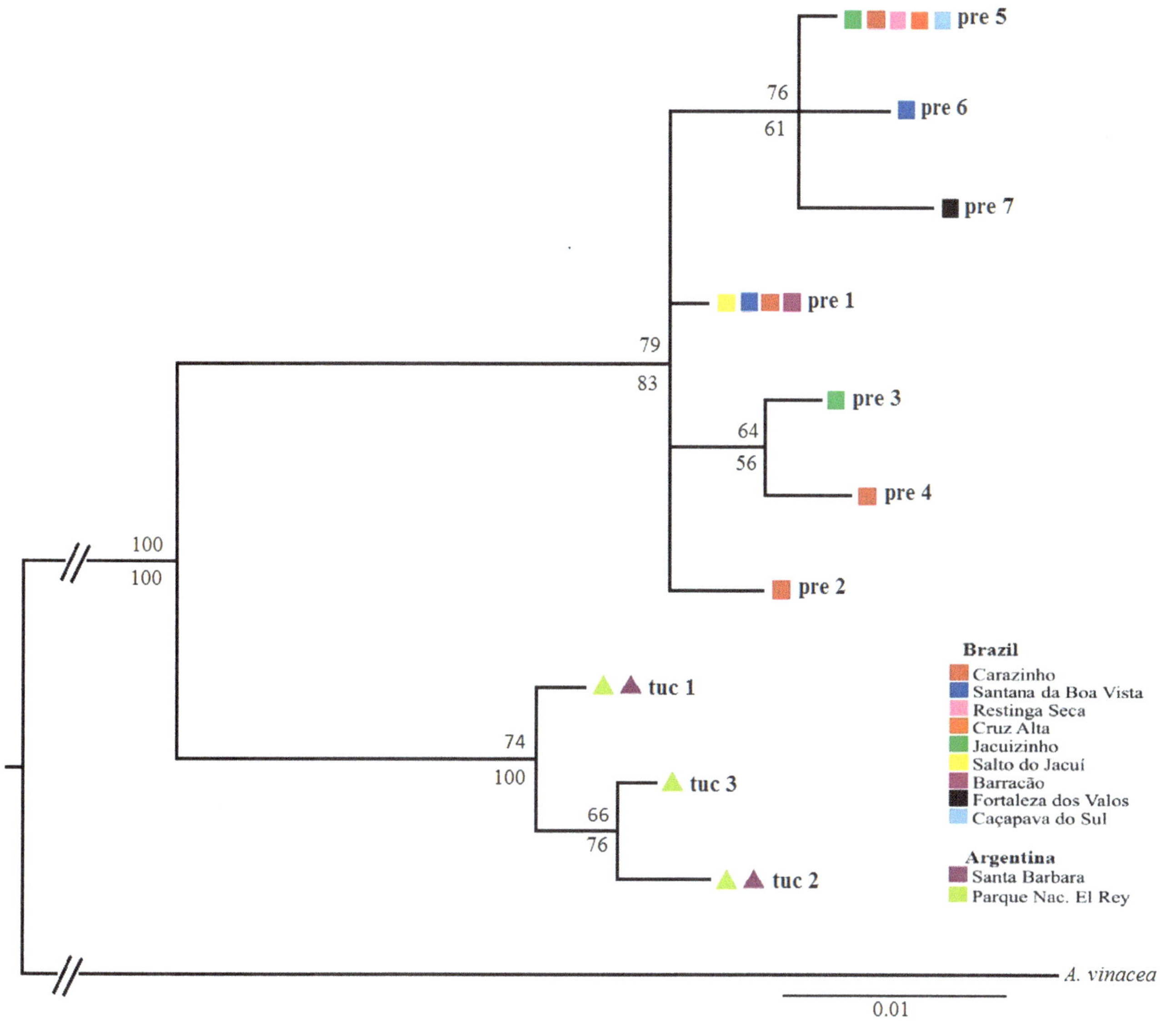

Figure 3. Bayesian tree of 491 bp alignment of mitochondrial DNA COI haplotypes of *Amazona pretrei* (colored squares) and *Amazona tucumana* (colored triangle). Tree is a 50% majority-rule consensus of 18,000 trees. The numbers indicate Bayesian posterior probabilities (above the branches) and maximum parsimony bootstrap (below the branches) values.

haplotypes in the network, composed by a frequent and internal haplotype and other derivative thereof, can be interpreted as indicative of a founder effect, supporting the peripheral isolation model. However, there is a difference between the sample sizes: the smallest sample has at least 85% genetic variability, moreover, when comparing equivalent numbers of individuals of the two

Table 2. Demographic and population parameters for *A. pretrei* and *A. tucumana* based on analysis of 491 bp fragment of the mitochondrial gene cytochrome oxidase I (COI) using IMa2 program.

Species	Ne_F (CI)	m1>m2 (CI)	m2>m1 (CI)	*t* (CI)
A. pretrei	94,913	0.00		1,339,474
(m1)	(36,192 to 220,785)	(0.00–4.38)		(265,504 to 1,735,876)
A. tucumana	27,471		0.00	
(m2)	(2,762 to 135,901)		(0.00–1.86)	

Ne_F - effective number of females; CI - confidence interval of 95%; m1>m2 - effective number of females migrants/generation from m1 to m2; m2>m1 - effective number of females migrants/generation from m2 to m1; *t* - divergence time between *A. pretrei* and *A. tucumana*.

species, variability is still higher in *A. pretrei*. As an example, considering only the 13 spectacled Amazons (same number of *A. tucumana*) from Carazinho, RS, a wider range of haplotypes ($D_H = 0.64$, four haplotypes) was still found in this locality.

Ecological observation also supports the origin of *A. tucumana* in the Pleistocene. *A. pretrei* seems to be dependent on *Podocarpus lambertii* seeds during the breeding season [62] mainly in southern Rio Grande do Sul (savannah with Araucaria), while in the Southern Yungas forests of northwestern Argentina, *A. tucumana* feed their nestlings mainly with *Podocarpus parlatorei* seeds, with 95% of the crop content consisting of seeds (L. O. Rivera, personal observation). Moreover, an increase in nest density was found in years with a high production of *P. parlatorei* fruits. All the populations of *A. tucumana* recorded along its distribution range are found in forests of *P. parlatorei*; this tree is a very good predictor of the presence of *A. tucumana* (L. O. Rivera, personal communication). Ledru et al. [63] found high genetic homogeneity in *P. lambertii*, which occurs mainly in the southernmost latitudes of Brazil, and also noted that such high genetic homogeneity most likely reflects the recent expansion of this genus in the area. This observation, and the affinity between southeastern *Podocarpus* populations and *P. parlatorei* from Argentina and Bolivia, indicates an ancient connection with other South American regions. They suggested that this finding supported the hypothesis of a wider pre-Quaternary distribution of *P. lambertii* on the South American continent, most likely before the formation of the Andes. Thus, it could be possible that a western portion of the continuous extension of *P. lambertii* forest became isolated before the Quaternary and that this forest was then colonized in the Quaternary by the dispersion of some *A. pretrei* individuals, giving origin to *A. tucumana*.

In conclusion, our genetic data indicate that the *taxa* studied here should be considered both distinct and worthy of conservation efforts. However, the scenario of speciation described in this paper was constructed based on a single marker (mitochondrial DNA) with maternal inheritance. Several deviations related to the estimated divergence time and levels of gene flow between populations during this process may be different if other independent markers are observed (see review in Nichols [64]). Thus, it is important to assess this scenario based on other markers to corroborate our biogeographical hypothesis of speciation for these two *taxa*.

Conservation implications

The conservation of genetic diversity within populations is always desirable to ensure that they are genetically sustainable, particularly in response to future environmental changes. Furthermore, genetic diversity is broadly correlated with population size, hence conservation efforts should seek to maintain or create large populations. The main factors affecting the two amazon species studied here are natural habitat destruction and fragmentation, as well as the illegal trafficking of nestlings and adults [24,25]. Although these two species are suffering similar human pressures, *A. pretrei* seems to have a better conservation *status* than its sister species.

Systematic monitoring, which began in 1991, indicates that the *A. pretrei* population is stable at approximately 19,000 amazons [65]. Additionally, the genetic variability observed in this species ($D_H = 0.68$) is similar to the one observed ($D_H = 0.75$) in the non-endangered amazon *Amazona aestiva* [55]. This can also be interpreted by comparative analysis made from the analysis of microsatellite loci [65], where *A. pretrei* also shows similar genetic variability to other non-endangered parrots. Thus, the high genetic variability observed in these species can be seen as indicative that anthropogenic interferences have not yet drastically affected its genetic variability. Based on this, it is possible to suggest that the conservation measures to be adopted for this species should focus on the conservation of the remaining forest and restoration of degraded areas, thus ensuring conditions to maintain the current population size and hence genetic variability. In this direction, several actions were proposed in the National Plan of Action for the Conservation of Atlantic Parrots [66] to ensure the conservation of this species, among which we highlight the need for the conservation of remaining forests for nesting and feeding by amazons.

In contrast, *A. tucumana* has a smaller population size in relation to its sister species, with population estimates have identified approximately 8,000 individuals, 80% of which are in Argentina in the northern and central distribution [25,67]. The low genetic variability observed in this species ($D_H = 0.41$) when compared with its sister species may be mainly the result of its evolutionary history, due to the founding event in the early formation of this species. Furthermore, perhaps in recent centuries, it could have also been influenced by the different negative impacts of human actions, making the species more genetically vulnerable than its sister species does. Thus, conservation efforts for the Tucuman Amazon should prioritize the northern and central regions of Southern Yungas in Argentina, where 80% of the remaining Tucuman Amazon population occurs, with high priority given to the creation of public and private reserves [25]. Furthermore, this species may require actions aimed at increasing the population size, such as eliminating nest poaching and retaining nest and food trees in forestry operations [68,69]. Finally, the low genetic variability and genetic distinctiveness in *A. tucumana* revealed in our study can be seen as very valuable information given the recent change of the categorization of this species on the IUCN Red List from Near Threatened to Vulnerable [70].

Acknowledgments

We thank the collaboration of Juliana Gondim de Albuquerque Lima and Vivian Ribeiro in performing some laboratory analyses, Cassia Alves Lima for her help in designing the figures, and Natan Medeiros, Lilian Giugliano and two anonymous referees for providing important suggestions for the manuscript. We are grateful to the National Park Administration of Argentina and the Direction of Natural Resources of the Jujuy Province that granted the research permits.

Author Contributions

Conceived and designed the experiments: AVR LOR JM NPP RC. Performed the experiments: AVR RC. Analyzed the data: AVR LOR JM NPP RC. Contributed reagents/materials/analysis tools: AVR LOR JM NPP RC. Contributed to the writing of the manuscript: AVR LOR JM NPP RC.

References

1. Darwin CR, Wallace AR (1858) Proceedings of the meeting of the Linnean Society held on July 1st, 1858. J Proc Linn Soc Zool 3: 54–56. (John van Wyhe editor) 2002. The complete work of Charles Darwin online. Available: http://darwin-online.org.uk/.
2. Bush GL (1975) Modes of animal speciation. Annu Rev Ecol Evol Syst 6: 339 364.
3. Dobzhansky T (1970) Genetic of the evolutionary process. New York: Columbia university press. 505 p.
4. Mayr E (1942) Systematics and the origin of species. New York: Columbia University Press. 334 p.

5. Quintero E, Ribas CC, Cracraft J (2013) The Andean *Hapalopsittaca* parrots (Psittacidae, Aves): an example of montane-tropical lowland vicariance. Zool Scr 42(1): 28–43.
6. Voelker G (1999) Dispersal, vicariance and clocks: historical biogeography and speciation in a cosmopolitan passerine genus (Anthus: Motacillidae). Evolution 53: 1536–1552.
7. Voelker G (2002) Systematics and historical biogeography of wagtails (Aves: Motacilla): dispersal versus vicariance revisited. Condor 104: 725–739.
8. Weir JT, Price M (2011) Andean uplift promotes lowland speciation through vicariance and dispersal in *Dendrocincla* woodcreepers. Mol Ecol 20(21): 4550–4563.
9. Cunnningham CW, Collins TM (1998) Beyond area relationships: extinction and recolonization in molecular marine biogeography. In: Schierwater B, Streit B, Wagner GP, Desalle R, editors. Molecular Ecology and Evolution: Approaches and Applications: Birkhauser, Basel, Switzerland. pp. 297–322.
10. Waters JM, Dijkstra LH, Wallis GP (2000a) Biogeography of a southern hemisphere freshwater fish: how important is a marine dispersal? Mol Ecol 9: 1815–1821.
11. De Queiroz K (2005) The resurrection of oceanic dispersal in historical biogeography. Trends Ecol Evol 20: 68–73.
12. Noonam BP, Chippindale PT (2006) Vicariant origin of Malagasy reptiles supports late cretaceous Antarctic landbrige. Am Nat 168: 730–774.
13. Yoder AD, Nowak MD (2006) Has vicariance or dispersal been the predominant biogeographic force in Madagascar? Only time will tell. Annu Rev Ecol Evol Syst 37: 405–431.
14. Waters JM, Lopez A, Wallis GP (2000b) Molecular phylogenetics and biogeography of galaxiid fishes (Osteichthyes: Galaxiidae): dispersal, vicariance and the position of Lepidogalaxias salamandroides. Syst Biol 49: 777–795.
15. Sanmartín I, Enghoff H, and Ronquist F (2001) Patterns of animal dispersal, vicariance and diversification in the Holarctic. Biol J Linn Soc 73: 345–390.
16. Johnson NK, Cicero C (2002) The role of ecological diversification in sibling speciation of *Empidonax flycatchers* (Tyrannidae): multigene evidence from mtDNA. Mol Ecol 11: 2065–2081.
17. Nielsen R, Wakeley J (2001) Distinguishing migration from isolation: a Markov Chain Monte Carlo approach. Genetics 158: 885–896.
18. Russello MA, Amato G (2004) A molecular phylogeny of *Amazona*: implications for neotropical parrot biogeography, taxonomy, and conservation. Mol Phylogenet Evol 30: 421–437.
19. Peters JL (1937) Checklist of birds of the world. Cambridge: Harvard University Press. 318 p.
20. Ridgely RS (1981) A guide to the birds of Panama. Princeton: Princeton University Press. 412 p.
21. Forshaw JM (1977) Parrots of the World. 3rd ed, Willoughby: Lansdowne Press. 584 p.
22. Diefenbach KH, Goldhammer SP (1986) Biologie und okologie der rotzchwanzamazone *Amazona brasilienis*. Trochilus 7: 72–78.
23. Collar NJ, Gonzaga LP, Krabbe N, Madroño Nieto A, Naranjo LG, et al. (1992) Threatened birds of the Americas: the ICBP/IUCN Red Data Book. International Council for Bird Preservation, Cambridge, U.K.
24. Martinez J, Prestes NP (2008) Biologia da conservação: estudo de caso com papagaio-charão e outros papagaios brasileiros.Passo Fundo:Editora UPF. 287 p.
25. Rivera L, Politi N, Bucher EH (2007) Decline of the Tucuma' n parrot *Amazona tucumana* in Argentina: present status and conservation needs. Oryx 41: 101–105.
26. Prado D (1993) What is the Gran Chaco vegetation in South America?. II. A redefinition. Contribution to the study of flora and vegetation of the Chaco. Candollea 48: 615–629.
27. Pennington RT, Prado DE, Pendry CA (2000) Neotropical seasonally dry forestsand Quaternary vegetation changes. J Biogeogr 27: 261–273.
28. Iriondo M (1993) Geomorphology and late Quaternary of the Chaco (South America). Geomorphology 7: 289–303.
29. Hernández RM, Jordan TE, Farjat AD, Echavarría L, Idleman BD, et al. (2005) Age, distribution, tectonics, and eustatic controls of the Paranense and Caribbean marine transgressions in southern Bolivia and Argentina. J S Am Earth Sci 19: 495–512.
30. Vuilleumier F (1965) Relationships and evolution within the Cracidae. Bull of the Mus Comp Zool 134: 1–27.
31. Short LL (1975) A zoogeographical analysis of the South American Chaco avifauna. Bull Amer Mus Nat Hist 154: 163–352.
32. Simpson BB (1979) Quaternary biogeography of the righ montane regions of South America. In: Duelman WE editor. The South American Herpetofauna: its origin, evolution, and dispersal. Monograph of the Museum of Natural History. Laurence Kansas: University of Kansas. pp. 157–188.
33. Nores M (1992) Bird speciation in subtropical South America in relation to forest expansion and retraction. Auk 109: 346–357.
34. Van der Hammen T (1974) The Pleistocene changes of vegetation and climate in tropical South America. J Biogeogr 1: 3–26.
35. Van der hammen T (1983) The palaeoecology and palaeogeography of savannas. In: Bourliere F, editor. Tropical savannas. Amsterdam: Elsevier. pp. 19–35.
36. Fitzpatrick JW (1980) Some aspects of speciation bin South American flycatchers. In: Short LL, editor. Symposium on speciation in South American birds. Actis 17 Congressus Internationalis Ornithologici. Berlin: Deutsche Ornithologen-gesellschaft. pp. 1273–1279.
37. IUCN (2013) The IUCN Red List of Threatened Species. v. 2013.1. Available: http://www.iucnredlist.org. Accessed 2 July 2013.
38. Machado AB, Drummond GM, Paglia AP (org.) (2008) Livro vermelho da fauna brasileira ameaçada de extinção. Ministério do Meio Ambiente, Brasília, DF, v.2. 1420 p.
39. López-Lanús B, Grilli P, Coconier E, Di Giacom A, Banchs R (2008) Categorización de las aves de la Argentina según su estado de conservació-n.Informe de Aves Argentinas/AOP y Secretaría de Ambiente y Desarrollo Sustentable. Buenos Aires, Argentina.
40. Hein J, Schierup MH, Wiuf C (2005) Gene genealogies, variation and evolution: a primer in coalescent theory. Oxford: Oxford University Press.
41. Brudford MW, Hannote O, Brookfield JFY, Burke T (1992) Single locus and multilocus DNA fingerprinting. In: Hoezel CAR, editor. Molecular genetic analyses of populations: a pratical approach. New York: Oxford University Press. pp. 225–269.
42. Palumbi SR (1996) Nucleic acids II: The polymerase chain reaction. In: Hillis DM, Moritz DM, Mable BK, editors. Molecular systematics. Sunderland: Sinauer Associates pp. 205–247.
43. Thompson JD, Gibson TJ, Plewniak F, Jeanmougin F, Higgins DG (1997) The clustalx windows interface: flexible strategies for multiple sequence alignment aided by quality analysis tools. Nucleic Acids Res 24: 4876–4882.
44. Tamura N, Peterson D, Peterson N, Skecher S, Nei M, et al. (2011) MEGA 5: Molecular Evolutionary Genetics Analysis using Maximum Likelihood, Evolutionary Distance, and Maximun Parsimony methods. Mol Biol Evol 28: 2731–2739.
45. Rozas J, Sánchez-De I, Barrio JC, Messeguer X, Rozas R (2003) DNAsp, DNA polymorphism analyses by the coalescent and other methods. Bioinformatics 19: 2496–2497.
46. Bandelt HJ, Forster P, Röhl A (1999) Median-joining networks for inferring intraspecific phylogenies. Mol Biol Evol 16: 37–48.
47. Tajima F (1989) Statistical method for testing the neutral mutation hypothesis. Genetics 123: 585–595.
48. Fu Y, Li W (1993) Statistical tests of neutrality of mutations. Genetics 133: 693–709.
49. Huelsenbeck JP, Ronquist F (2001) MRBAYES: Bayesian inference of Phylogenetic trees. Bioinformatics 17: 754–755.
50. Felsenstein J (1985) Confidence limits on phylogenies: an approach using the bootstrap. Evolution 39: 783–791.
51. Darriba D, Taboada GL, Doallo R, Posada D (2012) jModelTest 2: more models, new heuristics and parallel computing. Nat Methods 9(8): 772.
52. Rambaut A, Drummond AJ (2007). TRACER v1.5. Available: http://beast.bio.ed.ac.uk/Tracer.
53. Hey J, Nielsen R (2004) Multilocus methods for estimating population sizes, migration rates and divergence time, with applications to the divergence of *Drosophila pseudoobscura* and *D. persimilis*. Genetics 167: 747–760.
54. Pereira SL, Baker AJ (2006) A mitogenomic timescale for birds detects variable phylogenetic rates of molecular evolution and refutes the standard molecular clock. Mol Biol Evol 23: 1731–1740.
55. Caparroz R, Seixas GHF, Berkunsky I, Collevatti RG (2009) The role of demography and climatic events in shaping the phylogeography of *Amazona aestiva* (Psittaciformes, Aves) and definition of management units for conservation. Diversity Distrib 15: 459–468.
56. Ramos-Onsins SF, Rozas J (2002) Statistical properties of new neutrality test against population growth. Mol Biol Evol 19: 2092–2100.
57. De Queiroz K (2007) Species concepts and species delimitation. Syst Biol 56: 879–886.
58. Newton I (2003) The speciation and biogeography of birds. London: Academic Press. 668 p.
59. Mata JRR, Erize F, Rumboll M (2006) Birds of South America non-passerines: Rheas to Woodpeckers. Oxford: Princeton University Press. 376 p.
60. Hulka C, Gräfe KU, Sames CE, Uba C, Heubeck C (2006) Depositional setting of the middle to Late Miocene Yecua formation of the Chaco Foreland Basin, southern Bolivia. J S Am Earth Sci 21: 135–150.
61. Willis EO (1992) Zoogeographical origins of eastern Brazilian birds. Ornitol Neotrop 3: 1–15.
62. Martinez J, Prestes NP (2002) Ecologia e conservação do papagaio-charão *Amazona pretrei*. In: Galetti M, Pizo MA, editors. Ecologia e conservação de psitacídeos no Brasil. Belo Horizonte: Melopsittacus Publicações Científicas. pp. 173–192.
63. Ledru M-P, Salatino MLF, Ceccantini G, Salatino A, Pinheiro F, et al. (2007) Regional assessment of the impacto f climatic change on the distribution of a tropical conifer in the lowlands of South America. Diversity Distrib 13: 761–771.
64. Nichols R (2011) Gene trees and species trees are not the same. Trends Ecol Evol 16 (7): 358–364.
65. Prestes NP, Martinez J (2011) Papagaio- charão. In: Schunck F, Samenzari M, Lugarini R, Soares ES, editors. Plano de ação nacional para a conservação dos papagaios da Mata Atlântica. ICMBio/MMA Série Especial (20). pp. 34–40.
66. Schunck F, Samenzari M, Lugarini R, Soares ES (org) (2011) Plano de ação nacional para a conservação dos papagaios da Mata Atlântica. ICMBio/MMA Série Especial (20). 128 p.

67. Rivera L, Llanos RR, Politi N, Hennessey B, Bucher EH (2010) The Near Threatened Tucumán Parrot *Amazona tucumana* in Bolivia: insights for a global assessment. Oryx 44: 110–113.
68. Rivera L, Politi N, Bucher EH (2012) Nesting habitat of the Tucuman Parrot *Amazona tucumana* in an old-growth cloud-forest of Argentina. Bird Conserv Internat 22: 398–410.
69. Rivera L, Politi N, Bucher EH, Pidgeon A (2013) Nesting success and productivity of Tucuman Parrots (*Amazona tucumana*) in high-altitude forests of Argentina: do they differ from lowland *Amazona* parrots? Emu 114(1): 41–49.
70. IUCN (2012) IUCN Red List of Threatened Species. Version 2012.1. Available: http://www.iucnredlist.org. Downloaded on 15 November 2013.
71. Birdlife International (2013) Species factsheet: Tucuman Amazon *Amazona tucumana*. Available: http://www.birdlife.org/datazone/speciesfactsheet. Php?Id=1667. Accessed 25 October 2013.

6

Taxonomic Status, Phylogenetic Affinities and Genetic Diversity of a Presumed Extinct Genus, *Paraisometrum* W.T. Wang (Gesneriaceae) from the Karst Regions of Southwest China

Wen-Hong Chen[1,5], Yu-Min Shui[1]*, Jun-Bo Yang[2], Hong Wang[1]*, Kanae Nishii[3], Fang Wen[4], Zhi-Rong Zhang[2], Michael Möller[3]*

1 Key Laboratory for Plant Diversity and Biogeography of East Asia, Kunming Institute of Botany, Chinese Academy of Sciences, Kunming, Yunnan, China, **2** Plant Germplasm and Genomics Center, Germplasm Bank of Wild Species, Kunming Institute of Botany, Chinese Academy of Sciences, Kunming, Yunnan, China, **3** Science Division, Royal Botanic Garden Edinburgh, Edinburgh, Scotland, United Kingdom, **4** Guangxi Institute of Botany, Guangxi Zhuang Autonomous Region and Chinese Academy of Sciences, Guilin, Guangxi, China, **5** University of the Chinese Academy of Sciences, Beijing, China

Abstract

Background: The karst regions in South China have an abundance of endemic plants that face high extinction risks. The Chinese Gesneriaceae endemic *Paraisometrum mileense* (= *Oreocharis mileensis*), was presumed extinct for 100 years. After its re-discovery, the species has become one of five key plants selected by the Chinese forestry government to establish a new conservation category for plants with extremely small populations. For conservation purposes, we studied the phylogenetic and population genetic status of *P. mileense* at the three only known localities in Guangxi, Guizhou and Yunnan.

Methodology/Principal Findings: We collected 64 samples (52 species) of *Oreocharis* and 8 samples from three provinces of *P. mileense* and generated molecular phylogenies, and inferred that *P. mileense* represents a relatively isolated and derived taxonomic unit within *Oreocharis*. Phylogeographic results of 104 samples of 12 populations of *P. mileense* indicated that the populations in Yunnan have derived from those in Guangxi and Guizhou. Based on AFLP data, the populations were found to harbor low levels of genetic diversity ($He = 0.118$), with no apparent gradient across the species' range, a restricted gene flow and significant isolation-by-distance with limited genetic differentiation among the populations across the three provinces ($F_{ST} = 0.207$, $P < 0.001$). The 10 populations in Yunnan were found to represent two distinct lineages residing at different altitudes and distances from villages.

Conclusion/Significance: The low levels of genetic diversity found in *P. mileense* are perhaps a consequence of severe bottlenecks in the recent past. The distribution of the genetic diversity suggests that all populations are significant for conservation. Current *in situ* and *ex situ* measures are discussed. Further conservation actions are apparently needed to fully safeguard this conservation flagship species. Our work provides a model of an integrated study for the numerous endemic species in the karst regions with extremely small populations.

Editor: Igor Mokrousov, St. Petersburg Pasteur Institute, Russian Federation

Funding: This work was supported by the National Natural Science Foundation of China (grant no. 31000258, 31470306), the National Key Basic Research Program of China (no. 2014CB954100), the Open Project of the Key Laboratory for Plant Diversity and Biogeography of East Asia, Kunming Institute of Botany (grant no. KLBB201304) and the project of Integrated Scientific Surveys to Daweishan National Park in China. The funders had no role in study design, data collection and analysis, decision to publish, or preparation of the manuscript.

Competing Interests: The authors have declared that no competing interests exist.

* Email: ymshui@mail.kib.ac.cn (YMS); wanghong@mail.kib.ac.cn (HW); m.moeller@rbge.ac.uk (MM)

Introduction

Paraisometrum mileense W.T.Wang was until relatively recently regarded as extinct in the wild because it had not been recollected for more than 100 years since its type specimen collection in 1906 [1,2]. In 2006, the plant was rediscovered in Shilin county, next to Mile county, Yunnan, where the type specimen was collected [1,3]. It was described as critically endangered (CR), possessing only 101–1000 individuals in a single population [4]. Interestingly, already in 2009, a second locality had been discovered in Longlin county in Guangxi, the province neighbouring Yunnan to the East [5], and soon afterwards, a third location, in Xingyi, in Guizhou province to the North of Guangxi, had been found [6] (Fig. 1), tripling the number of occurrence points. The total number of plants was estimated at >30,000 individuals [7], but up to 2010, detailed fieldwork by some of the authors (YMS and WHC) estimated the number to be significantly lower, with 630 mature plants in Yunnan, 150 in Guangxi, and 60 in Guizhou (Table 1). Incidentally, approximately 70–80% of the seedlings and young individuals died, and 50–60% of mature individuals sustained damage due to an extreme drought that occurred from 2010 to 2011 in the karst region in Yunnan where *P. mileense* grows (Fig. 2). This might partially explain the discrepancy in reported plant numbers, and illustrates the great vulnerability of the species to even short term climate fluctuations. Until recently, only the occurrence point in Guangxi was located in a provincial nature reserve (established in 2005), while the other two were outside protected areas. In 2011, however, the Chinese government set up a small reserve in Shilin, Yunnan, to protect the species at this primary rediscovery point (Fig. 2). The species is also one of five selected key plants to be used to establish a new category of protected areas in China for plant species with extremely small populations (PSESP) [4].

Until 2011, *Paraisometrum* W.T.Wang was regarded as a monotypic genus, but was then included in an enlarged genus *Oreocharis* Benth. as *O. mileensis* (W.T.Wang) Mich.Möller & A.Weber [8]. Irrespective of this recent inevitable taxonomic change, the species represents a highly threatened taxon. Furthermore, under its old name, it has received considerable attention as a strong flagship species for plant conservation in China, and we therefore use its original name here in this context, for consistency with current conservation initiatives, such as the establishment of the above new conservation category in China [4].

To be able to devise meaningful conservation strategies, knowledge of the taxonomic status, closest congeners and the level and distribution of genetic diversity within a taxon is essential [9–11]. Even though *P. mileense* has been included in phylogenetic analyses previously, its exact phylogenetic affinities in *Oreocharis* are still unclear. While it was clearly shown in the phylogenetic analyses that the species has evolved from within the enlarged *Oreocharis*, it fell on a polytomy with species of the *hitherto* genera *Ancylostemon* Craib, *Briggsia* Craib, *Isometrum* Craib, *Opithandra* B.L.Burtt and *Tremacron* Craib [8,12,13]. The generic characters to establish *Paraisometrum* are the presence of four upper corolla lobes, and one lower lobe of the pentamerous flowers [2]. Since this characteristic also occurs in other species of the newly defined *Oreocharis* (e.g. *O. saxatilis* (Hemsl.) Mich. Möller & A.Weber = *Ancylostemon saxatilis* Hemsl.) [14–17], the present work will address the species delineation of *P. mileense*. Furthermore, the plants in Guangxi and Guizhou appear to possess some floral features different from those in Yunnan (pers. obs. YMS and WHC). However, only two samples of *P. mileense*, one from Yunnan and one from Guangxi, were included in the most comprehensive study of *Oreocharis* to date [12]. This is not enough to test the taxon coherence in the light of modern approaches such as DNA barcoding [18–20]. Clearly, more molecular work was needed to include samples from all three occurrence points and from diverse species within the enlarged *Oreocharis*.

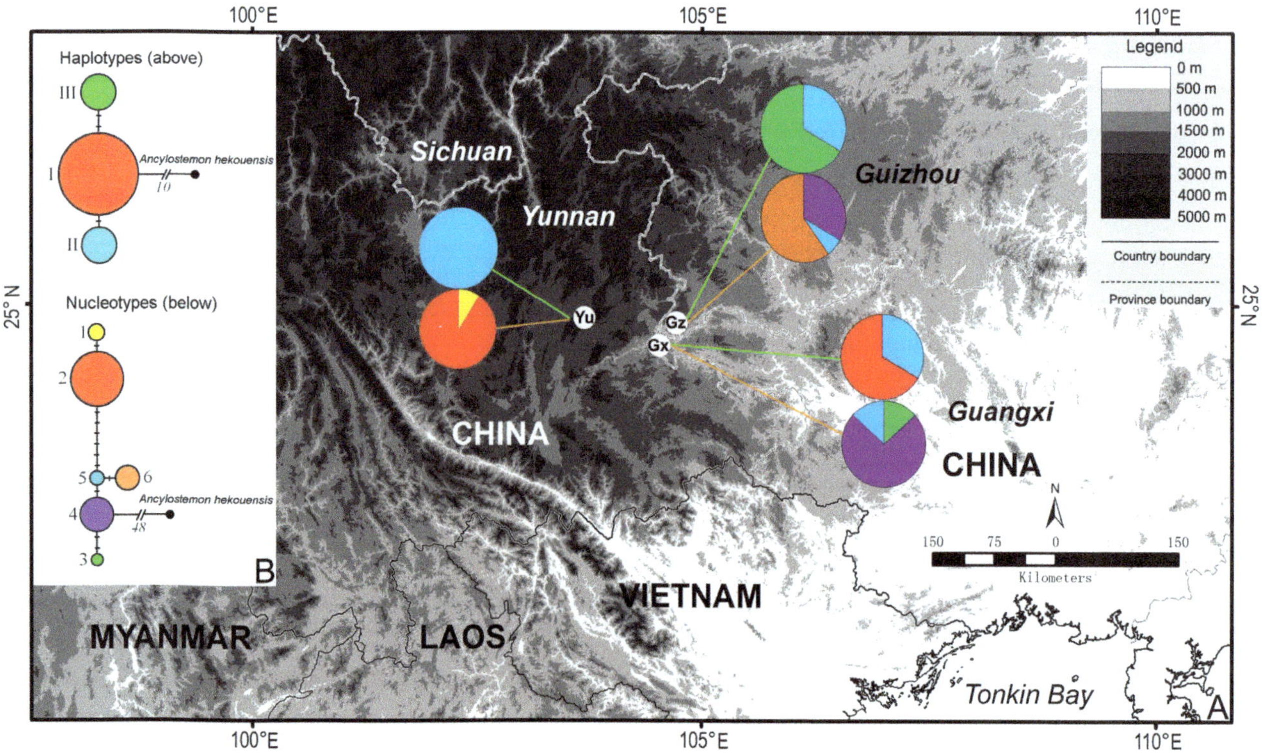

Figure 1. Map of localities of *Paraisometrum mileense* in Yunnan, Guangxi and Guizhou. A. Pie charts for haplotypes (above) and nucleotypes (below) are given for the three main geographical distribution areas. B. Median-joining networks of chloroplast haplotypes based on *trn*LF and *mat*K sequences (above) and of ITS nucleotypes (below) for *P. mileense* samples rooted on its closest relative *Ancylostemon hekouensis*.

Table 1. Detailed locality information for the 12 populations of *Paraisometrum mileense* used for AFLP analysis.

Population code	Locality name	Forest type	Position on slope	Altitude (m)	Distance (m) from village	Population size*	Area size (ha)	Sample number	Individual sample no
Yu1	Yunnan, Shilin, Guishan	secondary	lower	2001	25	46	0.40	5	Yu1a-e
Yu2	Yunnan, Shilin, Guishan	secondary	lower	2083	40	50	0.20	5	Yu2a-e
Yu3	Yunnan, Shilin, Guishan	secondary	lower	2112	20	52	0.25	5	Yu3a-e
Yu4	Yunnan, Shilin, Guishan	secondary	lower	1970	60	100	0.65	15	Yu4a-n
Yu5	Yunnan, Shilin, Guishan	primary	lower-middle	2149	800	76	0.25	8	Yu5a-e
Yu6	Yunnan, Shilin, Guishan	primary	lower-middle	2018	40	84	0.26	5	Yu6a-f
Yu7	Yunnan, Shilin, Guishan	primary	lower-middle	2077	600	120	1.20	5	Yu7a-e
Yu8	Yunnan, Shilin, Guishan	primary	middle-upper	2349	1000	32	0.20	5	Yu8a-e
Yu9	Yunnan, Shilin, Guishan	primary	upper	2412	1200	36	0.15	5	Yu9a-e
Yu10	Yunnan, Shilin, Guishan	primary	upper	2496	1500	34	0.16	10	Yu10a-i
Gx11	Guangxi, Longlin, Yacha	primary	upper	1183	1500	150	0.80	19	Gx11a-s
Gz12	Guizhou, Xingyi, Jingnan	primary	middle-upper	1405	1000	60	0.60	17	Gz12a-q

* - approximate number of mature plants per population based on data from August 2010.

There are many Gesneriaceae species endemic to the karst region in China, typically with small populations [7], such as *Primulina tabacum* Hance. This species occurs in only four populations distributed in Guangdong and Hunan with less than 1,000 plants in each [21]. Despite the relatively low plant number, a surprisingly high genetic diversity was found within the populations. Additionally, because of the long inter-population distances, a disruption of gene flow resulted in high population differentiation [10]. From a conservation perspective, the situation in *P. tabacum* was seen as 'a window of opportunity' to preserve a high level of extant genetic variation in the species. To determine whether a similar situation is present in *P. mileense*, we generated population genetic data using amplified fragment length polymorphisms (AFLPs) for individuals from the three localities in Guangxi, Guizhou and Yunnan. AFLP is a powerful tool for generating data from multiple loci for the detection of genetic variation without the need for pre-existing knowledge of genomic sequences [22,23], and have been successfully applied to small and relict populations (e.g. [10,24–27]), and a range of suitable analytical packages are available for these dominant markers (e.g. [27]).

Thus, our main aims were fourfold, to phylogenetically analyse sequence data, from the chloroplast intron–spacer sequences of *trn*LF, and the nuclear ribosomal internal transcribed spacer regions (nrITS), to a) test the taxonomic status of the species, b) to determine the phylogenetic position and relationships of *P. mileense* within the enlarged *Oreocharis*, c) to reconstruct the phylogeographic history of the *P. mileense* populations using *trn*LF, *mat*K, and ITS sequence data for individuals from 12 populations (or sub-populations) from the only three localities in Guangxi, Guizhou and Yunnan, and to acquire and analyse AFLP data for population samples to d) determine the levels of genetic diversity and differentiation of this species to assess its conservation requirements.

Materials and Methods

Ethics Statement

Of all sampled species, the target species *Paraisometrum mileense* is listed as one of five selected key plants to be used to establish a new category of protected areas in China for plant species with extremely small populations (PSESP). Approvals and permission for field studies were obtained from the Yunnan Forestry Bureau, China (permit no. [2011]–115). The GPS data are sensitive and are not provided for the protection of the plants according to the Yunnan forestry administrative request in the permit. However, we gave their approximate location in Fig. 1.

Study Materials

For the phylogenetic study, we expanded the ingroup sampling from 38 *Oreocharis* species plus one variety (41 samples) [12] to 52 species plus 2 varieties (73 samples), among which 11 species covered at least two population samples each. Of *P. mileense*, we included three samples each from Guangxi and Guizhou and two samples from Yunnan (Table S1). Thus, the *Oreocharis* ingroup sampling covered about 2/3 of the genus (53 out of about 80 species) [12]. The outgroup included 14 samples across six genera of straight-fruited advanced Asiatic and Malesian Gesneriaceae of subtribe Didymocarpinae (*sensu* [28]), with two *Didymocarpus* Wall. samples to root the trees [8,12].

The *Paraisometrum* materials for the phylogeographic analyses included 43 individuals from Yunnan (from 10 populations or sub-populations), 15 from the Guangxi population and 15 from the Guizhou population, and for the AFLP analysis 68 individuals

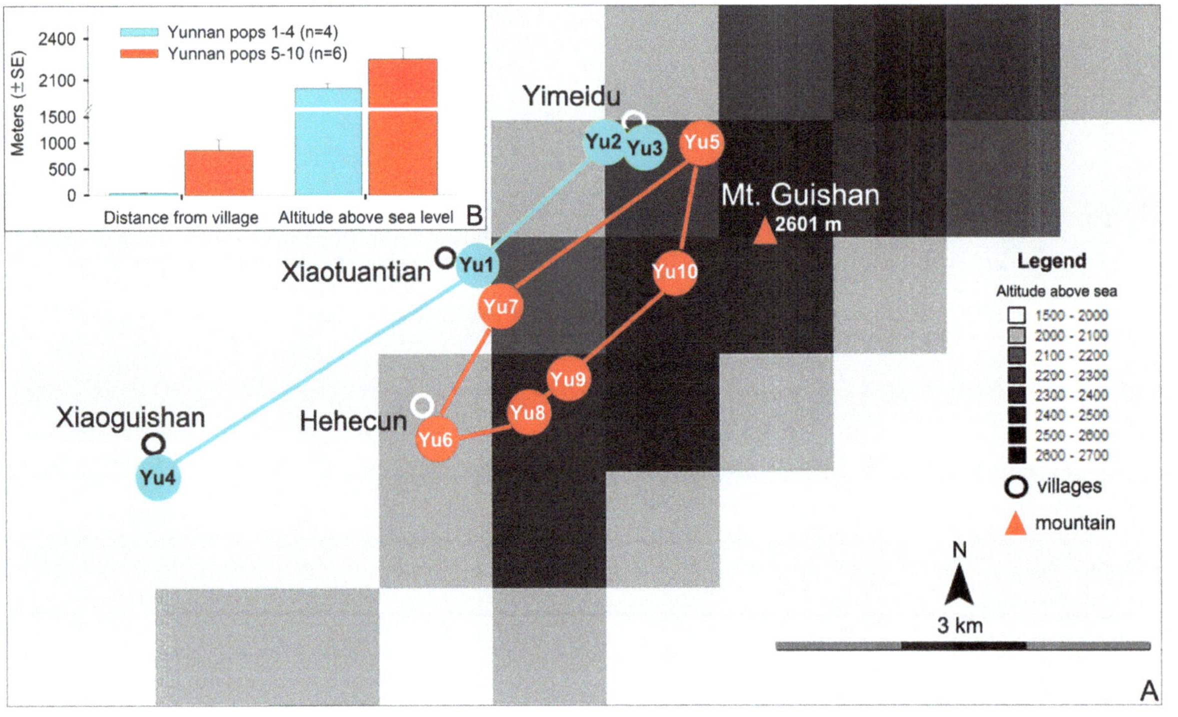

Figure 2. Populations of *Paraisometrum mileense* in Yunnan. A. Distribution of the ten populations analysed. B. Graphs of distance from village (left) and the average altitude (right) of populations 1–4 (light blue) and 5–10 (red).

from Yunnan (from 10 populations, 5–15 samples per population), 19 from the Guangxi population and 17 from the Guizhou population (Fig. 1; Table 1).

Molecular Methods

Genomic DNA was extracted from silica-gel-dried leaves collected in the field following a modified CTAB method [29,30]. The PCR primers for ITS were 'ITS5' and 'ITS4' [31], for *trn*LF 'C' and 'F' [32], and for *mat*K '3F' and '1R' [30]. PCR was performed in reactions containing 30–50 ng genomic DNA, 0.3 µl of each primer (5 µM/µl), 10 µl 2×*Taq* PCR MasterMix (Tiangen Biotech Co., Ltd, Beijing China: 0.1 U *Taq* polymerase/µl, 0.5 nM of each dNTP, 20 mM Tris-HCl (pH 8.3), 100 mM KCl, 3 mM $MgCl_2$) and ddH_2O to make up 20 µl. PCR amplifications were conducted under the following profile: 95°C for 3 min followed by 35 cycles at 94°C for 30 s, at the annealing temperature specific for each primer pair for 30 s (ITS: 55°C; *trn*LF and *mat*K: 52°C), 72°C for 1 min, and a final extension step at 72°C for 5 min. After PCR amplicon purification, Sanger sequencing was carried out in 6 µl reactions containing 0.5 µl PCR product, 0.3 µl primer (5 µM/µl), 1.05 µl SeqBuffer, 0.3 µl BigDye Terminator Mix (Applied Biosystems, Foster City, USA) and 3.85 µl distilled H_2O. Sequencing reactions were cycled under the following conditions: 32 cycles at 96°C for 10 s, 50°C for 5 s, and 60°C for 3 min, and the products analysed on an ABI 3730×l sequencer (Applied Biosystems, Foster City, USA). The newly acquired sequences have been submitted to GenBank.

AFLP was performed based on Vos et al. [23]. Restriction endonuclease enzyme digestion and link reactions were performed in 20 µl reactions containing 4 µl template DNA (50 ng/µl), 1 µl Adaptor, 2 µl *Pst*I/*Mse*I, 2.5 µl 10× reaction buffer, 2.5 µl ATP (10 mM), 1 µl T4 Ligase, 7 µl distilled H_2O. The solution was centrifuged for a few minutes after stirring, and incubated at 37°C for 5 h, then at 8°C for 4 h and then at 4°C overnight. Pre-amplification reactions were performed containing 2 µl template DNA, 1 µl Pre-ampmix, 0.5 µl dNTPs (Table 2), 2.5 µl 10× PCR buffer, 0.5 µl *Taq* polymerase, and 18.5 µl ddH_2O. PCR amplifications were conducted with the profile: 94°C for 2 min, followed by 30 cycles at 94°C for 30 s, 56°C for 30 s, 72°C for 80 s, with a final extension at 72°C for 5 min, and terminated at 4°C. Eight selective amplification primers were used for each *Pst*I and *Mse*I (Table 2). The selective amplification reactions were performed in 25 µl reactions containing 2 µl 1:20 diluted preamplification product, 2.5 µl 10× PCR buffer, 0.5 µl dNTP, 1 µl *Pst*I primer, 1 µl *Mse*I primer, 0.5 µl *Taq* polymerase, and 17.5 µl ddH_2O. PCR amplifications were conducted as follows: 94°C for 30 s, 65°C for 30 s, 72°C for 80 s for the first cycle, followed by 12 cycles with progressively decreasing annealing temperature by 0.7°C each cycle starting from 65°C to 56.6°C, which was followed by 23 cycles of 94°C for 30 s, 55°C for 30 s, 72°C for 80 s, with a final extension step at 72°C for 5 min, and then cooled to 4°C. The AFLP primers, *Taq* polymerase and dNTPs for PCR were purchased from Dingguochangsheng Biotechnology Co. Ltd. (Beijing, China). The amplified fragments were separated and detected with an ABI Prism 377 sequencer (Applied Biosystems, Foster City, USA). Due to ambiguous banding patterns or obvious PCR failures, 19 additional samples (GX11a-s) collected exclusively from Guangxi province were included (Table 1). The scoring error rate was about 1.3%, determined by replicating AFLP runs for 16 individuals (ca. 15% of all samples analysed) (*cf.* [33]).

Phylogenetic Analyses

The phylogenetic analyses were conducted on a matrix containing 86 ITS and *trn*LF sequences. ITS and *trn*LF sequences for 37 samples were newly acquired (Table S1), and for 49 samples

Table 2. Name and DNA sequences of primers and adaptors used in the AFLP experiments on *Paraisometrum mileense* samples.

	Primer name	Primer sequence
Adaptors	*PstI* 1	5′ -CTC GTA GAC TGC GTA CAT GCA
	PstI 2	5′ -TGT ACG CAG TCT AC
	MseI 1	5′ -GAC GAT GAG TCC TGA G
	MseI 2	5′ -TAC TCA GGA CTC AT
Pre-amplification primers	*PstI*	5′ -GAC TGC GTA CAT GCA G
	MseI	5′ -GAT GAG TCC TGA GTA A C
Selective amplification primers		
PstI primers (5 ng/μl)	*PstI*-1	5′ -GAC TGC GTA CAT GCA GAA
	PstI-2	5′ -GAC TGC GTA CAT GCA GAC
	PstI-3	5′ -GAC TGC GTA CAT GCA GAG
	PstI-4	5′ -GAC TGC GTA CAT GCA GAT
	PstI-5	5′ -GAC TGC GTA CAT GCA GTA
	PstI-6	5′ -GAC TGC GTA CAT GCA GTC
	PstI-7	5′ -GAC TGC GTA CAT GCA GTG
	PstI-8	5′ -GAC TGC GTA CAT GCA GTT
MseI primers (30 ng/μl)	*MseI*-1	5′ -GAT GAG TCC TGA GTA ACA A
(marked FAM)	*MseI*-2	5′ -GAT GAG TCC TGA GTA ACA C
	MseI-3	5′ -GAT GAG TCC TGA GTA ACA G
	MseI-4	5′ -GAT GAG TCC TGA GTA ACA T
	MseI-5	5′ -GAT GAG TCC TGA GTA ACT A
	MseI-6	5′ -GAT GAG TCC TGA GTA ACT C
	MseI-7	5′ -GAT GAG TCC TGA GTA ACT G
	MseI-8	5′ -GAT GAG TCC TGA GTA ACT T

retrieved from GenBank. The newly acquired DNA sequences were assembled and trimmed in Sequencher 4.1.4 (Gene Codes, Ann Arbor, MI, USA), and added to, and manually aligned with, the existing matrices. Maximum parsimony (MP) and Bayesian inference (BI) analyses followed Möller et al. [8,12] and Weber et al. [34], using PAUP* 4.0b10 [35] and MrBayes 3.2.2 [36,37], respectively.

The partition-homogeneity test in PAUP* (incongruence length difference test [38,39], on 1,000 replicates of repartitioning with tree bisection-reconnection (TBR) indicated no significant incongruence between the two datasets ($P = 0.28$) and the ITS and *trn*LF data sets were analysed combined. Parsimony was implemented on unordered and unweighted characters and through heuristic tree searches on 10,000 random starting trees with both TBR swapping and MulTrees on. Clade support was obtained as bootstrap indices with 10,000 heuristic replicates of random additions with TBR on and MulTrees off.

The most suitable substitution model for the BI analysis was obtained under the AIC criterion in MrModeltest [40], and was GTR+G for both, *trn*LF and the ITS spacers, and SYM+G for the 5.8S gene. Two independent runs of four MCMC chains of two million generations were run, sampled each 1,000th generation. The first 200 trees (10%) were discarded as burn-in (generations prior to stationarity of likelihood values), and the posterior probabilities (PP) obtained from 50% majority rule consensus trees obtained using the 'sumt' command in MrBayes. A low value of the average standard deviation of split frequencies (0.009025), the high correlation of the PP support values and the parallel distribution of variance values found between the two parallel runs of the Bayesian analysis, and the level runs of the split posteriors indicated a good convergence of the MCMC runs (Table S2; Figs. S1–S3).

Phylogeographic Analyses

Chloroplast haplotypes were determined using combined chloroplast *trn*LF and *mat*K sequence data, and nucleotypes using ITS (GenBank accession numbers *trn*LF: KM062935–KM062942, KM06301–KM063145; *mat*K: KM063008–KM063080; ITS: KM062943–KM063007; KM063175–KM063182). Phylogeographic networks were reconstructed with NETWORK 4.6.1.1 (available at http://www.fluxus-engineering.com). *Ancylostemon hekouensis* Y.M.Shui & W.H.Chen (= *Oreocharis hekouensis* (Y.M.Shui & W.H.Chen) Mich.Möller & A.Weber) was used to root the networks, based on finding of the phylogenetic analyses here.

Population Genetic Analyses

The ABI trace files were analysed in GeneScan 3.1 (Applied Biosystems), and only intensive bands between 70 and 500 base pairs in size converted to a binary matrix (1 = band presence, 0 = band absence) (Table S3). Genetic diversity parameters were obtained in GenAlEx 6.5 [41], and included percentage of polymorphic loci (*P*), number of different alleles (*Na*), number of effective alleles (*Ne*), Shannon's Information index (*SI*), expected heterozygosity (*He*) and unbiased expected heterozygosity (*uHe*). To investigate for patterns among the *Paraisometrum* populations, we used STRUCTURE 2.3.4. [42] to assign individuals to genotypically distinct groups. We used the admixture model and the option for correlated allele frequencies as recommended by Falush et al. [43]. The program was run 14 times for each cluster (K) from $K1$ to $K10$. Each run of 100,000 iterations was preceded by 10,000 iterations as burn-in when convergence was achieved. We plotted the mean likelihood for each cluster $L(K)$ against the cluster number (K). To establish the optimal number of clusters, the relationship between K and ΔK, the second order rate of change of the likelihoods, was plotted [44]. The distribution of genetic variation within and between populations and between regions was analysed in a hierarchical AMOVA in GenAlEx.

To test an isolation-by-distance scenario, a Mantel test was performed on the geographic and *ln* geographic distance *versus* Nei's genetic distance (D) in GenAlEx. For a 3D illustration of spatial relationships of the populations, a principal coordinate analysis (PCoA) was performed using the Jaccard distance in R-pack Le Progiciel R.4.0d10 [45]. To illustrate the genetic relationships between the populations, an unrooted Neighbor Joining (NJ) tree was reconstructed using Nei and Li's restriction site distances in PAUP*. Branch support was obtained from 1000 NJ bootstrap replicates in PAUP*.

Results

Phylogeny

The MP analysis recovered 12 most parsimonious trees of 1538 steps length with a CI of 0.5566 and RI of 0.6982. The genus *Oreocharis* in its new definition formed a strongly supported clade in the strict consensus tree (BS = 100%). *Paraisometrum* was nested deeply within this clade in a derived position with the *P. mileense* samples forming a strongly supported clade (BS = 100%) (Fig. S4). Sister to this clade were two samples of *A. hekouensis*, though with no branch support.

The BI trees showed the same phylogenetic positions with view to the *Paraisometrum* and *A. hekouensis* samples (Fig. 3), the sister-relationship of the two clades received only low support (PP = 0.63). The BI tree was more resolved both within the enlarged *Oreocharis* and within the *Paraisometrum* clade. *Paraisometrum* was nested deeply in a derived position within the clade dominated by species with yellow tetrandrous flowers (Fig. 3). The populations from Guizhou and Guangxi appeared intermixed in two groups not reflecting their origins (PP = 0.62), while the samples from Yunnan were most similar to each other (PP = 1.0), and somewhat distant to those from the other provinces.

Phylogeography

Among the 1676 bases in the combined *trn*LF and *mat*K matrix, three positions were variable among the *Paraisometrum* samples, resulting in three chloroplast haplotypes (Fig. 1; Table 3). Haplotype I was the only one in Yunnan and present in the other two provinces, while haplotype II was unique to Guangxi, and haplotype III was a private haplotype in the population from Guizhou (Fig. 1A). The median-joining network, rooted on *A. hekouensis*, placed haplotype I in a central position with haplotypes II and III as peripherals (Fig. 1B).

Among the 643 bases of the ITS region, eleven positions were variable, giving six different ITS nucleotypes among the samples with types 1 and 2 exclusive to Yunnan, type 4 and 5 shared among the Guangxi and Guizhou populations, while type 3 was private in Guangxi, and type 6 private in Guizhou (Fig. 1A; Table 3). The median-joined network placed type 4 present in Guangxi and Guizhou in the centre, from which first types only present in these two provinces have evolved; the ITS types 1 and 2 present in Yunnan were in a derived position (Fig. 1B).

Genetic Diversity

At the population level, the AFLP data suggested a low level of genetic diversity of the populations, particularly in Yunnan (Table 4). This is likely a result of the low number of individuals included per populations. At the region level, the data suggested that the Yunnan populations harboured an overall higher level of diversity. For example, the percentage of polymorphic loci was 90.88% in Yunnan, compared to 61.18% in Guangxi and 66.99% in Guizhou. Other genetic diversity indices were also slightly higher, such as the Shannon's Information index (Yunnan: 0.247, Guangxi: 0.219, Guizhou: 0.235), though the number of effective alleles and heterozygosity indices were very similar in the three provinces.

Genetic Structure

The AMOVA indicated that most genetic diversity resided within the populations of *P. mileense* (79%) and significant genetic differences existed among populations (21%, $F_{ST} = 0.207$, $P < 0.001$) (Table 5). When structured for regions, a similar genetic differentiation was observed between the three provinces (12%, $F_{CT} = 0.120$, $P < 0.001$) and among populations (12%, $F_{SC} = 0.144$, $P < 0.001$) (Table 6).

The optimal number of clusters for the STRUCTURE analysis was determined as $K = 4$ (Fig. 4 A and B). The analysis revealed a strong separation between the Yunnan populations 1–4 and 5–10 (Fig. 4C). The Guangxi population (Pop 11, Gx) was very homogeneous and distinct. The Guizhou population (Pop 12, Gz) was a mix of two types, one shared with the Guangxi population and the other individuals similar to a great degree to samples of the Yunnan populations 8 and 9. With $K = 3$, the patterns did not change greatly, except that the latter genotypes were clustering with the populations 1–4 from Yunnan. With $K = 5$, no further change in clustering was observed, the fifth genetic cluster scattering in small proportions among all populations. The Mantel test was significant using the geographic distance ($r = 0.398$; $P = 0.01$) and *ln* geographic distance ($r = 0.383$; $P = 0.03$).

The PCoA (1st axis 12% variance; 2nd axis = 7.7%; 3rd axis = 6.6%) separated the samples of the three regions strongly in the first axis, including the Yunnan populations 1–4 and 5–10 (Fig. 5A). Samples from Guangxi clustered very tightly together, while those from Guizhou and Yunnan were widely scattered. The samples from Guizhou were also scattered and almost overlapped with individuals from Guangxi at one end of their distribution, and with those from Yunnan, populations 5–10, at the other (Fig. 5A).

In the unrooted NJ tree, the samples clustered into four groups according to their province of origin including the split between Yunnan populations 1–4 and 5–10 (Fig. 5B). The Guangxi and Yunnan (populations 1–4) clusters received a high bootstrap support of 93%. These clusters were also characterized by short and uniform branch lengths. The same was the case for the cluster of Yunnan populations 5–10, except for three individuals, Yu8e, Yu9c and Yu9e. These had extremely long branches and were the same samples that seemed to be categorized as belonging to the Guizhou population in the STRUCTURE analysis (see Fig. 4C). The Guizhou population cluster was characterized by a loose clustering and relatively uniformly long branches (Fig. 5B).

Discussion

Taxonomic Status and Phylogenetic Affinities

The present study aimed to address several important issues surrounding this enigmatic, and until recently, monotypic genus that was thought to be extinct for 100 years. It was recently

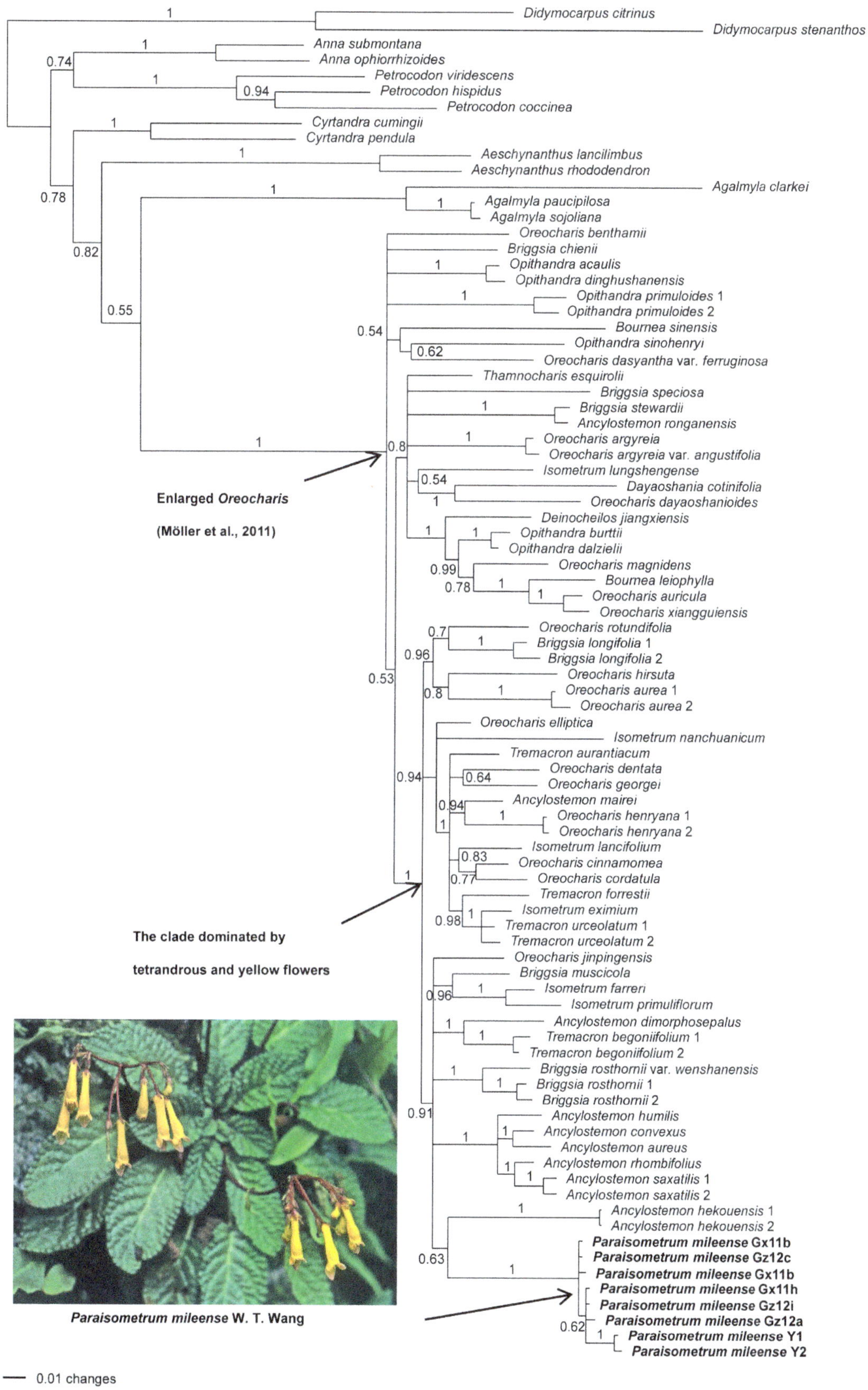

Figure 3. Bayesian inference analysis placing *Paraisometrum mileense* in *Oreocharis*. Tree based on combined ITS and *trn*LF sequence data with average branch lengths and posterior probabilities. Photo of *P. mileense* by Yu-Min Shui.

Table 3. Haplotypes and nucleotypes found among the samples of *Paraisometrum mileense*.

Regions	N	Haplotypes			Nucleotypes					
		I	II	III	1	2	3	4	5	6
Yunnan (Pops1–10)	43	43	-	-	4	39	-	-	-	-
[Yunnan (Pops1–4)]	22	22	-	-	2	20	-	-	-	-
[Yunnan (Pops5–10)]	21	21	-	-	2	19	-	-	-	-
Guangxi (Pop11)	15	5	10	-	-	-	2	11	2	-
Guizhou (Pop12)	15	5	-	10	-	-	-	5	1	9
Total	73	-	-	-	-	-	-	-	-	-

reported to occur only in one population in Shilin, Yunnan [4], but we were able to sample from two further populations, from two different provinces. Since all *Paraisometrum* samples in the phylogenetic study fell in a single strongly supported clade separated from other *Oreocharis* taxa by a long branch, they can be regarded as a single taxonomic unit (Figs. 3, 4). Unlike previous phylogenetic analysis where relationships of this genus to other congeners were unresolved [8,12,13], we found indications that the closest species to *Paraisometrum mileense* is *Ancylostemon hekouensis*, a recently described species endemic to the karst region in Southwest China [46], with a similar leaf and flower morphology to *Paraisometrum*, but without the distinct 4-lobed upper corolla lip. The species with the most similar corolla shape to *P. mileense*, *Ancylostemon saxatilis* Hemsl., was found not closely related to *P. mileense* (Figs. 3, 4). This morphological homoplasy, is in line with previous findings of high levels of parallelism in the evolution of floral morphology in *Oreocharis* [12].

Phylogeographic History

Both, haplotype and nucleotype distribution, suggest a scenario of a close relationship between the Guangxi and Guizhou populations (shared nucleotypes), and a derived Yunnan population (lower haplo- and nucleotype diversity, derived nucleotypes) (Fig. 1, Table 3), and suggests a migration of *Paraisometrum* westward into Yunnan [9].

The significance in the Mantel test on the AFLP data indicates an isolation-by-distance scenario [47], and suggests a limited dispersal across the landscape. This may be a consequence of *Paraisometrum* being pollinated by bees, which have a relatively limited flying distance (e.g. [48,49]), and the seeds of *Paraisometrum*, that, though small (ca. 0.6 mm long), are not dust-like, as in orchids [50,51], to be dispersed far by wind, and do not have special dispersal aids, e.g. hooks or appendages (e.g. [52]) for long distance dispersal [53]. The significant genetic structure detected among the populations and regions (Tables 5, 6), further indicates a breakdown of their genetic connectivity. The ITS data may be used to give an indication of the divergence time between the *Paraisometrum* localities. Taking the average rate of evolution of the ITS spacers of 11 herbaceous plants (4.13×10^{-9} substitutions per site per year) [54,55], the Yunnan populations might have separated from the other populations around 1.6 million years ago (±0.26 SE), at the beginning of the Pleistocene. This is a period that would cover repeated glacial-interglacial cycles during which limited secondary contact may have occurred (see below). This could explain the higher levels of within-population genetic variation observed here.

Our study provides a first insight into the history of a plant considered extinct in the wild. To fully address the evolution of the species, and to untangle historic from contemporary events, an increased sampling is required, additional markers employed such as microsatellites, combined with population demographic analyses (e.g. [56,57]), and, with view to conservation, ecological niche modelling (e.g. [58]).

Genetic Diversity and Differentiation

The STRUCTURE analysis indicated several noteworthy aspects of the *Paraisometrum* populations; firstly, that the Yunnan populations represented two distinct lineages with at least three gene pools and can be divided into two main populations, populations 1–4, and populations 5–10 with some admixture (Fig. 4). This bipartition of the Yunnan populations was also seen in the PCoA clustering analysis (Figs. 5A, S5) and the NJ tree (Fig. 5B). Both groups have the same haplotypes which indicates

Table 4. Genetic diversity indices based on AFLP data among the 12 populations of *Paraisometrum mileense* analysed.

Population	*N*	*P*	*Na*	*Ne*	*SI*	*He*	*uHe*
Yu1	5	33.37	0.744 (0.023)	1.160 (0.007)	0.155 (0.006)	0.100 (0.004)	0.111 (0.004)
Yu2	5	36.37	0.762 (0.024)	1.172 (0.007)	0.170 (0.006)	0.109 (0.004)	0.121 (0.004)
Yu3	5	33.01	0.712 (0.023)	1.161 (0.007)	0.156 (0.006)	0.100 (0.004)	0.112 (0.004)
Yu4	15	55.85	1.130 (0.024)	1.195 (0.007)	0.200 (0.006)	0.124 (0.004)	0.128 (0.004)
Yu5	8	44.34	0.938 (0.024)	1.202 (0.008)	0.195 (0.006)	0.125 (0.004)	0.133 (0.005)
Yu6	5	28.60	0.660 (0.022)	1.154 (0.007)	0.142 (0.006)	0.093 (0.004)	0.103 (0.004)
Yu7	5	32.95	0.727 (0.023)	1.174 (0.007)	0.162 (0.006)	0.106 (0.004)	0.118 (0.005)
Yu8	5	36.01	0.760 (0.024)	1.194 (0.008)	0.179 (0.006)	0.117 (0.004)	0.130 (0.005)
Yu9	5	43.66	0.891 (0.024)	1.197 (0.007)	0.199 (0.006)	0.126 (0.004)	0.140 (0.004)
Yu10	10	49.17	1.017 (0.024)	1.224 (0.008)	0.214 (0.006)	0.137 (0.004)	0.144 (0.005)
Gx	19	57.99	1.190 (0.024)	1.213 (0.008)	0.212 (0.006)	0.133 (0.004)	0.137 (0.004)
Gz	17	66.99	1.344 (0.023)	1.220 (0.007)	0.235 (0.006)	0.143 (0.004)	0.148 (0.004)
Total	8.7(0.036)	43.19 (3.48)	0.906 (0.007)	1.189 (0.002)	0.185 (0.002)	0.118 (0.001)	0.127 (0.001)
Mean over Loci for each region							
Yunnan	68	90.88	1.818 (0.014)	1.231 (0.007)	0.247 (0.006)	0.149 (0.004)	0.150 (0.004)
Guangxi	19	60.75	1.240 (0.024)	1.219 (0.008)	0.219 (0.006)	0.137 (0.004)	0.141 (0.004)
Guizhou	17	67.42	1.352 (0.023)	1.227 (0.007)	0.238 (0.006)	0.146 (0.004)	0.151 (0.004)
Grand Mean over Loci and pops							
Total	34.667(0.337)	73.01(9.14)	1.470(0.013)	1.226(0.004)	0.235(0.003)	0.144(0.002)	0.147(0.002)

Values are means (and SE).
N = no. of samples; *P* = percentage of polymorphic loci; *Na* = no. of different alleles; *Ne* = no. of effective alleles; *SI* = Shannon's Information index; *He* = expected heterozygosity; *uHe* = unbiased expected heterozygosity.

Table 5. Results of an unstructured hierarchical AMOVA on AFLP data of 12 populations of *Paraisometrum mileense*.

Source	df	SS	MS	Est. Var.	Var. (%)	*F* statistics	*P*
Among Pops	11	4914.428	446.766	36.513	21	$F_{ST}=0.207$	<0.001
Within Pops	92	12907.486	140.299	140.299	79	–	–
Total	103	17821.913	–	176.812	100	–	–

Table 6. Results of a structured hierarchical AMOVA on AFLP data of 12 populations of *Paraisometrum mileense*, with three regions, Yunnan (10 pops), Guangxi (1 pop), Guizhou (1 pop).

Source	df	SS	MS	Est. Var.	Var. (%)	*F* statistics	*P*
Among Regions	2	2243.796	1121.898	22.405	12	$F_{CT}=0.120$	<0.001
Among Pops	9	2670.632	296.737	23.581	13	$F_{SC}=0.144$	<0.001
Within Pops	92	12907.486	140.299	140.299	75	$F_{ST}=0.247$	<0.001
Total	103	17821.913	–	186.286	100	–	–

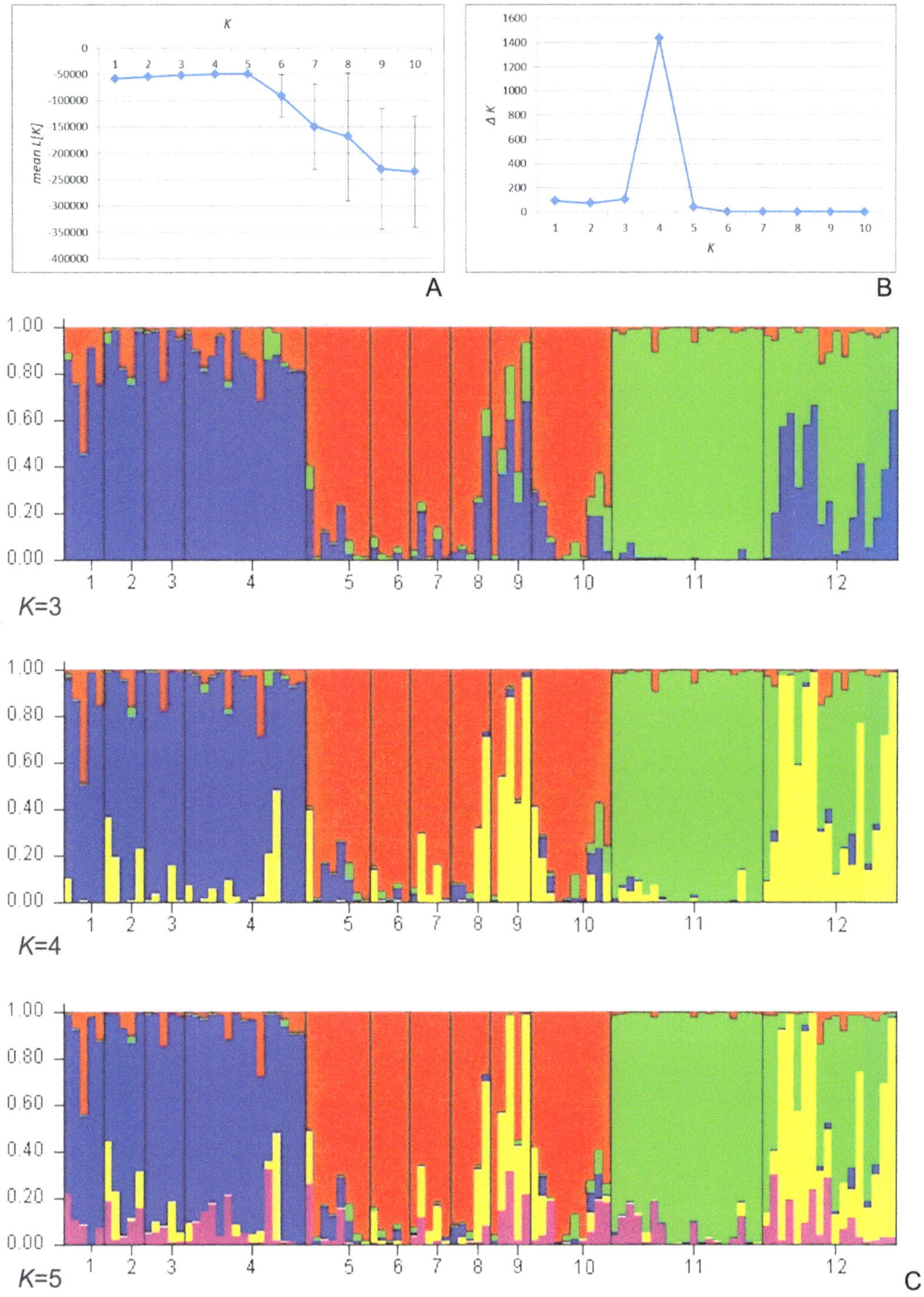

Figure 4. Results of the Bayesian inference STRUCTURE analysis on AFLP data of 12 *Paraisometrum mileense* populations. A. Plot of *K*-clusters *versus* mean (±SD) likelihoods (L[K]). B. *K* plotted against the second order rate of change of the likelihoods (Δ*K*). C. STRUCTURE clustering results for *K* = 3 to 5 as suggested in B. Numbers refer to populations in Table 1. 1–10 = Yunnan, 11 = Guangxi, 12 = Guizhou.

their common ancestry, and the same composition of ITS nucleotypes (Table 3), which suggests fragmentation of a previously more continuously distributed population. Populations 1–4 occur near villages, in the foothills of Mt. Guishan in secondary open forests with frequent disturbance due to human activities, while populations 5–10 grow mostly undisturbed further away from villages, above the foothills almost to the summit of Mt. Guishan, and in more dense and primary forests (Fig. 2B, Table 1). Whether the bipartition is linked to the condition of the habitats surrounding the population groups, being disconnected from each other by disturbed forest, or has an older origin, would require further research. Secondly, the population in Guangxi was distinct and very homogeneous, also seen in the tight clustering in the PCoA (Fig. 5A) and the uniformly short branches

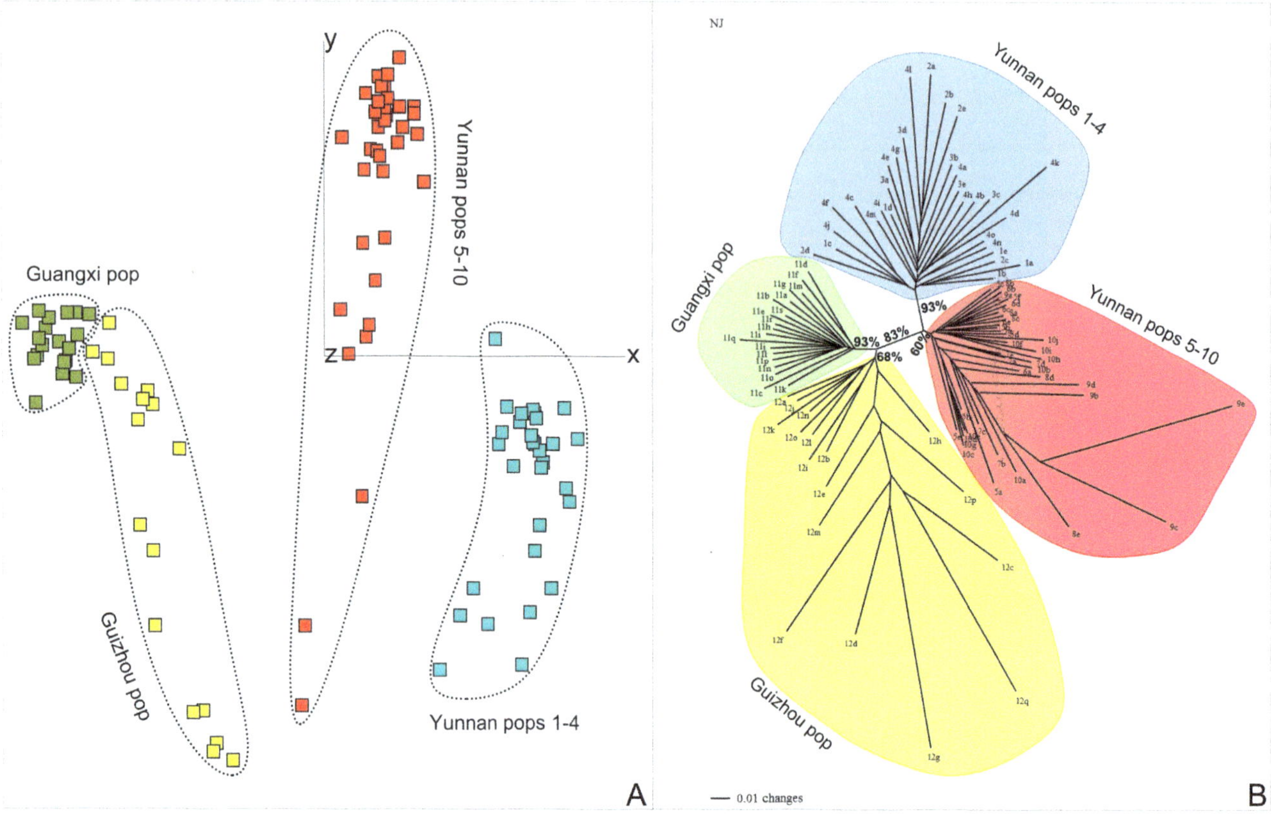

Figure 5. Four genetic lineages exist in *Paraisometrum mileense*. A. PCoA scatter plot based on AFLP data using the Jaccard distance on 12 populations of *P. mileense*. Axis 1(x) and 2(y). B. Unrooted NJ-tree based on Nei and Li's restriction site distances. Bootstrap values only given for internal branches (1000 Neighbor-Joining replicates).

in NJ tree (Fig. 5B). This might indicate that while the plants contain similar genetic diversity levels as other regions, the plants were closer related to each other possibly due to consanguineous matings, perhaps as a side effect of their small population sizes and limited area of distribution of less than a hectare (Table 1), and long distance to the other distribution localities with no apparent intervening populations known (Fig. 1). Thirdly, on the contrary, the internal and terminal branches of the population from Guizhou were long, and the STRUCTURE analysis suggested that some plants had a genetic makeup greatly similar to plants of the Yunnan populations 8 (plant Yu8e) and 9 in particular (plants Yu9c, Yu9e). This was not reflected in a mixed clustering of the respective Yunnan samples among the Guizhou samples, though their branches were unusually long (Fig. 5B). This might reflect genetic links and allele exchanges between these populations, perhaps due to habitat expansion during glacial-interglacial cycles that brought the populations in closer proximity (e.g. [58]). They are currently only some 120 km apart (Fig. 1).

The genetic diversity found at the population level was very low, although this might have been a consequence of the low number of individuals (often 5) included for analysis of the Yunnan populations. However, even when calculated within the three regions, the AFLP diversity levels were relatively low, with no marked difference between the regions (Table 4). Even at the species level, the genetic diversity of *Paraisometrum* was comparatively low ($He = 0.144$), and much lower compared to *Primulina tabacum* ($He = 0.339$), another herbaceous Gesneriaceae species from China [11]. *Primulina tabacum* and *P. mileense* have similar distribution sizes though in different provinces (*P. tabacum* in Guangdong and Hunan, *Paraisometrum* in Guangxi, Guizhou and Yunnan), but both occur in limestone karst areas, though at different altitudes (*P. tabacum* below 300 m, *Paraisometrum* between 1,180–2,500 m) and in different aspects (*P. tabacum* grows around entrances of limestone caves, *Paraisometrum* in limestone forests). The high genetic diversity levels found in *P. tabacum* populations were explained by their refugial history and/or breeding system [11]. The more restricted habitats in cave entrances of *P. tabacum*, as opposed to the more open forest habitats of *Paraisometrum* could be a factor that may have allowed a more continuous distribution and genetic connectivity of *Paraisometrum* populations. However, it is more likely that the greater population sizes of *P. tabacum* (1,000 individuals per population, as opposed to 30–150 individuals in *Paraisometrum* populations) has allowed *P. tabacum* to retain higher diversity levels. The similarly low levels of AFLP diversity compared to *Paraisometrum* found in other herbaceous perennials, such as *Trollius europaeus* L. ($He = 0.158$–0.229) [24], *Silene otitis* St.-Lag. ($He = 0.167$–0.240) [25], and *Draba aizoides* L. ($He = 0.07$–0.15) [26], were linked to effects of population fragmentation. Thus, the low genetic diversity in *Paraisometrum* is likely a result of a combination of fragmentation and small population size. Intriguingly though, in *Draba* L., the low genetic diversity levels were not correlated with a limited reproductive fitness, as suggested by their high germination rates. This may have important conservation implications and could be a field of study in *Paraisometrum* in the future.

Implications for Conservation

Our work provides an example of an integrated study for endemic species in the karst regions in South China with extremely small populations. It is well known that the karst region is characterized by a limestone topography with an abundance of endemic plants with extremely small populations [7,59,60]. In these regions, localised endemics occur often restricted to one or few limestone hills, which are usually isolated by non-limestone topography and the plants effectively occur on islands with high risks of extinction [7,14]. The three distribution points of *Paraisometrum* are isolated in different limestone forests, and the occurrence nearby villages is not unusual for endemic plants of the karst region [59,61].

The villages are commonly closely associated with characteristic forests on the limestone hills, the Feng Shui Forests. These are important for the water supply of the villages, but are recently influenced by activities of local people, such as free range poultry and goat keeping [62], which negatively affect the forest habitats.

With a view to conservation of *Paraisometrum*, the only three distribution points known to date appear to contain a significant amount of diversity within their populations with only a limited level of differentiation. Though, analysing a larger number of individuals per population might show some more resolution. Furthermore, the STRUCTURE analysis shows that they seem to have experienced some gene flow between Yunnan and Guizhou populations and are thus not genetically greatly isolated (Table 1). However, the calculated levels of genetic diversity were quite low, even when estimated across the regions including Yunnan (and not based on the small 'sub-populations' there). This is likely linked to the relatively small population sizes. The low genetic diversity might hinder adaptation to rapid climatic changes, as shown in the exceptionally dry season of 2011 during which they contracted significantly, due to the death of many immature individuals, and hardly any flower was produced in the Yunnan populations, their inflorescences withered (SYM, pers. observation). The relatively uniform distribution of genetic diversity in the three provinces of *Paraisometrum* does not immediately allow a prioritization of regions or populations for conservation efforts. Even though little differences in the genetic make-up were detected between the two Yunnan lineages, the one including populations 1–4, may deserve stronger protection due to their close proximity to human habitation and the negative effects these can bring (Fig. 2). For all of the reasons given above, conservation efforts should thus include all populations and include *in situ* as well as *ex situ* measures to safeguard this enigmatic species.

In situ measures relevant to *Paraisometrum* have already been implemented in Yunnan on a small scale at Shilin populations and involved the removal of *Ageratina adenophora* (Spreng.) R. M. King & H. Robinson, an invasive plant alien to China [63], the exclusion of goats through fencing in the plants, and by involving the local villagers to safeguard these plants that have a high potential for eco-tourism. Future *in situ* measures may also involve forest restoration [64], particularly for populations near villages. However, these may not be sufficient and *ex situ* conservation measures are required alongside *in situ* measures in the light of the observed strong detrimental effects of recent rainfall variation on the populations. This may, for the moment preclude transfer of plants from threatened habitats to new localities in the wild. Though, the plants are easily cultivated and propagated vegetatively and through seeds. Currently, about 56 plants are cultivated at the Kunming Botanical Garden, Yunnan, China, and seeds of three populations have been deposited at the Germplasm Bank of Wild Species, Plant Germplasm and Genomics Center, Kunming Institute of Botany, Yunnan, China. Thus, the immediate and medium term survival of the species seems ensured. However, seeds in germplasm banks are prone to genetic erosion during seed rejuvenation cycles even of inbreeding species [65]. Currently research is underway studying the pollinators to elucidate the reproductive strategy and breeding system to better devise a tailored *ex situ* strategy for *Paraisometrum mileense*.

Conclusions

The erstwhile monotypic genus *Paraisometrum mileense* (now *Oreocharis mileensis*) represents an isolated and independent taxonomic unit, and appears closest related to the erstwhile *Ancylostemon hekouensis* (now *Oreocharis hekouensis*). The populations in Yunnan seem to have derived from those in Guangxi and Guizhou with some evidence of limited gene transfer between Yunnan and Guizhou. The occurrence points in Yunnan were divided into two quite distinct lineages, while the population in Guangxi is relatively homogeneous. Overall, the populations of *P. mileense* contain relatively low levels of genetic diversity with no apparent gradient across the species' range. Several conservation measures are currently implemented, but additional actions are needed. From a conservation genetic perspective, all populations seem equally important for *in situ* and *ex situ* protection.

Supporting Information

Figure S1 Comparison of posterior probability values between run 1 versus run 2 (10% burn-in) of the Bayesian inference analysis.

Figure S2 Posterior probabilities of splits at selected increments over MCMC run1 (A) and 2 (B) of the Bayesian inference analysis.

Figure S3 Comparisons of topological differences within and among MCMC runs of the Bayesian inference analysis.

Figure S4 Maximum parsimony strict consensus tree of 12 parsimonious trees of 1538 steps, based on combined ITS and *trn*LF sequence data (CI = 0.5566; RI = 0.6982). Numbers above branches are bootstrap values (x denotes branches receiving <50% support).

Figure S5 PCoA scatter plot based on AFLP data using the Jaccard distance on 12 populations of Paraisometrum mileense. Axes 1 (x) and 3 (z).

Table S1 Collection details and GenBank numbers of *Oreocharis* (including *Paraisometrum mileense*) and outgroup samples used for the phylogenetic analyses.

Table S2 Characteristics of the Bayesian inference analysis of 86 samples of combined ITS and *trn*LF data.

Table S3 Binary AFLP data for 12 populations of Paraisometrum mileense from Yunnan (Yu), Guizhou (Gz) and Guangxi (GX).

Acknowledgments

We thankfully acknowledge Prof. Yi-Gang Wei, Dr. Qi Gao, Mr. Wen-Bin Xu, Guangxi Institute of Botany, Mr. Xiao-Qing Luo, Institute of Subtropical Crops of Guizhou Province and Mr. Bo Xiao, Malipo Forestry Bureau, and Dr. Zong-Xin Ren, Kunming Institute of Botany, Chinese Academy Sciences, for providing support in the field. We also like to thank two anonymous reviewers for their constructive comments on the manuscript. The Royal Botanic Garden Edinburgh is supported by the Rural and Environment Science and Analytical Services division (RESAS) in the Scottish Government.

Author Contributions

Conceived and designed the experiments: WHC HW MM. Performed the experiments: WHC JBY KN ZRZ. Analyzed the data: MM HW YMS. Contributed reagents/materials/analysis tools: YMS MM FW. Contributed to the writing of the manuscript: WHC MM YMS.

References

1. Smith L (2007) 'Extinct' plant found flowering in China. The Times website. Available: http://www.thetimes.co.uk/tto/news/world/article1973186.ece. Accessed 2014 Sep 1.
2. Weitzman AL, Skog LE, Wang WT, Pan KY, Li ZY (1997) New taxa, new combinations, and notes on Chinese Gesneriaceae. Novon 7: 423–435.
3. Shui YM, Cai J (2007) "100-years-lost" plant reemerges in Yunnan, China. Samara 12: 3.
4. Ma YP, Chen G, Grumbine RE, Dao ZL, Sun WB, et al. (2013) Conserving plant species with extremely small populations (PSESP) in China. Biodivers Conserv 22: 803–809.
5. Xu WB, Pan B, Huang YS, Ye XX, Liu Y (2009) *Paraisometrum* W.T.Wang, a newly recorded genus of Gesneriaceae from Guangxi, China. Guihaia 29: 581–583.
6. Gao Q, Xu WB (2011) *Paraisometrum* W. T. Wang, a newly recorded genus of Gesneriaceae from Guizhou, China. Acta Bot Boreal-Occid Sin 31 858–860.
7. Wei YG, Wen F, Möller M, Monro A, Zhang Q, et al. (2010) Gesneriaceae of South China. Nanning: Guangxi Science and Technology Publishing House. 777 p.
8. Möller M, Forrest A, Wei YG, Weber A (2011) A molecular phylogenetic assessment of the advanced Asiatic and Malesian didymocarpoid Gesneriaceae with focus on non-monophyletic and monotypic genera. Plant Syst Evol 292(3–4): 223–248.
9. Lowe A, Harris S, Ashton P (2007) Chapter 3. Genetic diversity and differentiation. In: Ecological Genetics: Design, Analysis, and Application. Hongkong: Blackwell Publishing. pp. 52–100.
10. Ni XW, Huang YL, Wu L, Zhou RC, Deng SL, et al. (2006) Genetic diversity of the endangered Chinese endemic herb *Primulina tabacum* (Gesneriaceae) revealed by amplified fragment length polymorphism (AFLP). Genetica 127: 177–183.
11. Wang CN, Möller M, Cronk QCB (2004) Population genetic structure of *Titanotrichum oldhamii* (Gesneriaceae), a subtropical bulbiliferous plant with mixed sexual and asexual reproduction. Ann Bot 93: 201–209.
12. Möller M, Middleton DJ, Nishii K, Wei YG, Sontag S, et al. (2011) A new delineation for *Oreocharis* incorporating an additional ten genera of Chinese Gesneriaceae. Phytotaxa 23: 1–36.
13. Tan Y, Wang Z, Sui XY, Hu GW, Motley T, et al. (2011) The systematic placement of the monotypic genus *Paraisometrum* (Gesneriaceae) based on molecular and cytological data. Pl Diversity Resources 33: 465–476.
14. Li ZY, Wang YZ (2004) Plants of Gesneriaceae in China. Zhengzhou: Henan Science and Technology Publishing House. 14–79 p.
15. Pan KY (1987) Taxonomy of the genus *Oreocharis* (Gesneriaceae). Acta Phytotax Sin 25: 164–293.
16. Wang WT, Pan KY, Li ZY (1990) Gesneriaceae. In: Wang WT, editor. Fl. Reipubl. Popularis Sin. 69: 190–203. Science Press, Beijing.
17. Wang WT, Pan KY, Li ZY, Weitzman AL, Skog LE (1998) Gesneriaceae. In: Wu ZY, Raven PH (eds) Flora of China. 18: 268–272. Beijing: Science Press & St Louis: Missouri Botanical Garden Press.
18. Jiao LJ, Shui YM (2013) Evaluating candidate DNA barcodes among Chinese *Begonia* (Begoniaceae) species. Pl Diversity Resources 35: 715–724.
19. Liu J, Möller M, Gao LM, Zhang DQ, Li DZ (2011) DNA barcoding for the discrimination of Eurasian yews (*Taxus* L., Taxaceae) and the discovery of cryptic species. Mol Ecol Res 11: 89–100.
20. Liu ML, Yu WB, Wang H (2013) Rapid identification of plant species and iflora: application of DNA barcoding in a large temperate genus *Pedicularis* (Orobanchaceae). Pl Diversity Resources 35: 707–714.
21. Ren H, Peng SL, Zhang JX, Jian SG, Wei Q, et al. (2003) The ecological and biological characteristics of an endangered plant, *Primulina tabacum* Hance. Acta Ecol Sin 23: 1012–1017.
22. Rafalski JA, Vogel JM, Morgante M, Powell W, Andre C, et al. (1997) Generating and using DNA makers in plants. In: Birren B, Lai E, editors. Non-mammalian Genome Analysis: A practical Guide. New York: Academic Press, pp. 75–134.
23. Vos P, Hogers R, Bleeker M, Reijans M, van de Lee T, et al. (1995) AFLP: a new technique for DNA fingerprinting. Nucleic Acids Res 23: 4407–4414.
24. Despres L, Loriot S, Gaudeul M (2002) Geographic pattern of genetic variation in the European globeflower *Trollius europaeus* L. (Ranunculaceae) inferred from amplified fragment length polymorphism markers. Mol Ecol 11: 2337–2347.
25. Lauterbach D, Ristow M, Gemeinholzer B (2012) Population genetics and fitness in fragmented populations of the dioecious and endangered *Silene otites* (Caryophyllaceae) Plant Syst Evol 298: 155–164.
26. Vogler F, Reisch C (2013) Vital survivors: low genetic variation but high germination in glacial relict populations of the typical rock plant *Draba aizoides*. Biodivers Conserv 22: 1301–1316.
27. Bonin A, Ehrich D, Manel S (2007) Statistical analysis of amplified fragment length polymorphism data: a toolbox for molecular ecologists and evolutionists. Mol Ecol 16: 3737–3758.
28. Weber A, Clark JL, Möller M (2013) A new formal classification of Gesneriaceae. Selbyana 31(2): 68–94.
29. Doyle JJ, Doyle JL (1987) A rapid DNA isolation procedure for small quantities of fresh leaf tissue. Phytochem Bull 19: 11–15.
30. Doyle JJ, Doyle JL (1990) Isolation of plant DNA from fresh tissue. Focus 12: 13–15.
31. White TL, Bruns T, Lee S, Taylor J (1990) Amplification and direct sequencing of fungal ribosomal RNA genes for phylogenetics. In: Innis MA, Gelfand D, Sninsky JJ, White TJ, editors. PCR protocols: a guide to methods and applications. San Diego: Academic Press, pp. 315–322.
32. Taberlet P, Gielly L, Pautou G, Bouvet J (1991) Universal primers for amplification of three non-coding regions of chloroplast DNA. Plant Mol Biol 17: 1105–1109.
33. Bonin A, Bellemain E, Bronken Eidesen P, Pompanoni F, Brochmann C, et al. (2004) How to track and assess genotyping errors in population genetics studies. Mol Ecol 13: 3261–3273.
34. Weber A, Wei YG, Puglisi C, Wen F, Mayer V, et al. (2011) A new definition of the genus *Petrocodon* (Gesneriaceae). Phytotaxa 23: 49–67.
35. Swofford DL (2002) PAUP*: phylogenetic analysis using parsimony (*and other methods), version 4. Sinauer, Sunderland, Massachusetts, USA.
36. Huelsenbeck JP, Ronquist F (2001) MRBAYES: Bayesian inference of phylogenetic trees. Bioinformatics 17: 754–755.
37. Ronquist F, Teslenko M, van der Mark P, Ayres D, Darling A, et al. (2011) MrBayes 3.2: Efficient Bayesian phylogenetic inference and model choice across a large model space. Syst Biol 61: 539–542.
38. Farris J, Källersjö SM, Kluge AG, Bult C (1994) Constructing a significance test for incongruence. Syst Biol 44: 570–572.
39. Farris J, Källersjö SM, Kluge AG, Bult C (1995) Testing significance of incongruence. Cladistics 10: 315–319.
40. Nylander JA, Wilgenbusch JC, Warren DL, Swofford DL (2008) AWTY (are we there yet?): a system for graphical exploration of MCMC convergence in Bayesian phylogenetics. Bioinformatics 24(4): 581–583.
41. Peakall R, Smouse PE (2006) GenAlEx 6.5: genetic analysis in Excel. Population genetic software for teaching and research-an update. Mol Ecol Notes 6: 288–295.
42. Pritchard JK, Stephens M, Donnelly P (2000) Inference of population structure using multilocus genotype data. Genetics 155: 945–959.
43. Falush D, Stephens M, Pritchard JK (2003) Inference of population structure using multilocus genotype data: linked loci and correlated allele frequencies. Genetics 164: 1567–1587.
44. Evanno G, Regnaut S, Goudet J (2005) Detecting the number of clusters of individuals using the software STRUCTURE: a simulation study. Mol Ecol 14: 2611–2620.
45. Casgrain P, Legendre P, Vaudor A (2005) The royal package for multivariate and spatial analysis, Version 4.0 (development release 10). Available: http://adn.biol.umontreal.ca/~numericalecology/old/R/index.html. Accessed 2014 Sep 1.
46. Chen WH, Shui YM (2006) *Ancylostemon hekouensis* (Gesneriaceae), a new species from Yunnan, China. Ann Bot Fenn 43: 448–450.
47. Wright S (1943) Isolation by distance. Genetics 28: 114–138.
48. Pasquet RS, Peltier A, Hufford MB, Oudin E, Saulnier J, et al. (2008) Long-distance pollen flow assessment through evaluation of pollinator foraging range suggests transgene escape distances. P Natl Acad Sci USA 105(36): 13456–13461.
49. Eckert JE (1933) The flight range of the honeybee. J Agric Res 47: 257–285.
50. Dressler RL (1993) Phylogeny and classification of the orchid family. Cambridge: Cambridge University Press, 314 p.
51. Murren CJ, Ellison AM (1998) Seed dispersal characteristics of *Brassavola nodosa* (Orchidaceae). Am J Bot 85(5): 675–680.
52. Kokubugata G, HirayamaY, Peng CI, Yokota M, Möller M (2011) Phytogeographic aspects of *Lysionotus pauciflorus* sensu lato (Gesneriaceae) in

the China, Japan and Taiwan regions: phylogenetic and morphological relationships and taxonomic consequences. Plant Syst Evol 292(3–4): 177–188.
53. Chen WH (2013) Reproductive Biology of *Paraisometrum* W. T. Wang, an endemic to SW China– with special emphasis on its phylogenetic position and genetic diversity. Ph.D. Thesis, Kunming Institute of Botany, Chinese Academy of Sciences, Kunming, China.
54. Kay KM, Whittall JB, Hodges SA (2006) A survey of nuclear ribosomal internal transcribed spacer substitution rates across angiosperms: an appropriate molecular clock with life history effects. BMC Evol Biol 6: 36.
55. Puglisi C, Wei YG, Nishii K, Möller M (2011) *Oreocharis* × *heterandra* (Gesneriaceae): a natural hybrid from the Shengtangshan Mountains, Guangxi, China. Phytotaxa 38: 1–18.
56. Ribeiro RA, Lemos-Filho JP, Ramos ACS, Lovato MB (2011) Phylogeography of the endangered rosewood *Dalbergia nigra* (Fabaceae): insights into the evolutionary history and conservation of the Brazilian Atlantic Forest. Heredity 106: 46–57.
57. Liu J, Möller M, Provan J, Gao LM, Poudel RC, et al. (2013) Geological and ecological factors drive cryptic speciation of yews in a biodiversity hotspot. New Phytol 199: 1093–1108.
58. Poudel RC, Möller M, Gao LM, Ahrends A, Baral SR, et al. (2012) Using morphological, molecular and climatic data to delimitate yews along the Hindu Kush-Himalaya and adjacent regions. PLoS ONE 7(10): e46873.
59. Fang RZ, Bai PY, Huang GB, Wei YG (1995) A floristic study on the seed plants from tropics and subtropics of Dian-Qian-Gui. Acta Bot Yunnan Suppl. 5: 111–150.
60. Sun H (2013) Phytogeographical regions of China. In: Hong DY, Blackmore S, editors. Plants of China: A companion to the Flora of China. Beijing: Sciences Press. pp. 176–204.
61. Chen WH, Möller M, Shui YM, Wang H, Wen F, et al. (2014) Three new species of *Petrocodon* (Gesneriaceae), endemic to the limestone areas of Southwest China, and preliminary insights into the diversification patterns of the genus. Syst Bot 39(1): 316–330.
62. Xu HG, Zhou F (2011) Biodiversity in the karst area of Southwest Guangxi. Beijing: Encyclopedia of China Publishing House. 143 p.
63. Zhu L, Sun OJ, Sang WG, Li ZY, Ma KP (2007) Predicting the spatial distribution of an invasive plant species (*Eupatorium adenophorum*) in China. Landscape Ecol 22(8): 1143–1154.
64. Lamb D, Erskine PD, Parrotta JA (2005) Restoration of degraded tropical forest landscapes. Science 310 (5754): 1628–1632.
65. Parzies HK, Spoor W, Ennos RA (2000) Genetic diversity of barley landrace accessions (*Hordeum vulgare* ssp. *vulgare*) conserved for different lengths of time in *ex situ* gene banks. Heredity 84: 476–486.

Genetic Variability and Population Structure of *Disanthus cercidifolius* subsp. *longipes* (Hamamelidaceae) Based on AFLP Analysis

Yi Yu[1], Qiang Fan[1], Rujiang Shen[1], Wei Guo[3], Jianhua Jin[1], Dafang Cui[2]*, Wenbo Liao[1]*

1 Guangdong Key Laboratory of Plant Resources and Key Laboratory of Biodiversity Dynamics and Conservation of Guangdong Higher Education Institutes, School of Life Sciences, Sun Yat-Sen University, Guangzhou, China, **2** College of Forestry, South China Agriculture University, Guangzhou, China, **3** Department of Horticulture and Landscape Architecture, Zhongkai University of Agriculture and Engineering, Guangzhou, China

Abstract

Disanthus cercidifolius subsp. *longipes* is an endangered species in China. Genetic diversity and structure analysis of this species was investigated using amplified fragments length polymorphism (AFLP) fingerprinting. Nei's gene diversity ranged from 0.1290 to 0.1394. The AMOVA indicated that 75.06% of variation was distributed within populations, while the between-group component 5.04% was smaller than the between populations-within-group component 19.90%. Significant genetic differentiation was detected between populations. Genetic and geographical distances were not correlated. PCA and genetic structure analysis showed that populations from East China were together with those of the Nanling Range. These patterns of genetic diversity and levels of genetic variation may be the result of *D. c.* subsp. *longipes* restricted to several isolated habitats and "excess flowers production, but little fruit set". It is necessary to protect all existing populations of *D. c.* subsp. *longipes* in order to preserve as much genetic variation as possible.

Editor: Ting Wang, Wuhan Botanical Garden, Chinese Academy of Sciences, Wuhan, China

Funding: Support for this study was provided by a grant by the National Natural Science Foundation of China (Project 31170202) to WL, a grant by the National Natural Science Foundation of China (Project 30670141) to WL, a grant by the National Natural Science Foundation of China (Project 31100159) to QF, a grant by the National Natural Science Foundation of China (Project 40672017) to JJ. The funders had no role in study design, data collection and analysis, decision to publish, or preparation of the manuscript.

Competing Interests: The authors have declared that no competing interests exist.

* Email: lsslwb@mail.sysu.edu.cn (WL); cuidf@scau.edu.cn (DC)

Introduction

Disanthus Maxim. (Hamamelidaceae) is a monotypic genus endemic to China and Japan [1]. *Disanthus cercidifolius* subsp. *longipes* is the only subspecies in *Disanthus* which is distributed in China (southern Zhejiang, central and northwestern Jiangxi, and southern Hunan), while its sister, *D. c.* subsp. *cercidifolius*, is endemic to Japan [2]. *D. c.* subsp. *longipes* was first reported by Cheng [3] in 1938, and then revised by Chang [4] in 1948. In a systematic study of *Disanthus* [2], the author believed that the Chinese species of *Disanthus* was a sister population of *D. c.* subsp. *cercidifolius*, which is distributed in warm and humid forests of the Cathayan Land, and also was a Tertiary relic. *D. c.* subsp. *longipes* was characterized by its morphology and preference for humid, acid soils and shady habitats. The inflorescences are paired, axillary; Capitula are 2-flowers, purple; Leaves are heart shaped, green then turning to purple, orange, and red. It is usually a small tree, 2–3 m high, occasionally reaching heights of 6–8 m in forests when growing along streams. Because of severe habitat fragmentation that caused population decline, *Disanthus* was listed in the 1992 Red List of Endangered Plant Species of China [5], the 1994 IUCN Red List of Threatened Species (www.iucnredlist.org), the Key Wild Plants under State Protection [6] and the 2004 China Species Red List as bring a species at high risk of extinction in the wild [7].

Although study of *Disanthus* has attracted many investigators [8–22], little attention has been paid to its genetic analysis and population structure, except Xiao [23]. Xiao investigated the genetic diversity of *D. c.* subsp. *longipes* based on nine allozyme loci and found a higher genetic variation within populations as well as significantly lower variation among populations. However, sampling of Xiao's study was limited to part of the distribution area of *D. c.* subsp. *longipes* only. So a comprehensive study of the populations' genetic structure at different geographic scales is still needed.

Preserving the genetic diversity of endangered species is one of the primary goals in conservation planning. Because survival and evolution of species depended on the maintenance of sufficient genetic variability within and among populations to accommodate new selection pressures caused by environmental changes [24,25]. For endemic endangered species, intraspecific variation is a prerequisite for any adaptive changes or evolution in the future, and have profound implications for species conservation [26,27]. The knowledge of the levels and patterns of genetic diversity is important for designing conservation strategies for threatened and endangered species [28,29]. So identifying variations with molecular markers has provided the abundant information concerning genetic diversity in plant species [30–36]. Amplified fragment length polymorphism (AFLP) [37] is a PCR-based technique which has been successfully applied to the identification and estimation of molecular genetic diversity and population structure [32–34,38–41]. This technique can generate information on multiple loci in a single assay without prior sequence knowledge, and [42]. Using AFLP markers, we would know (1) the degree of genetic diversity within and among populations; (2) which factors might explain genetic variation; and (3) how to apply this information to develop recommendations for management of this endangered species.

Materials and Methods

Ethics Statement

Field studies were approved by Hunan Provincial Bureau of Forestry for collection in Xinning County (1XN), the Dupanglin National Nature Reserve (2DP), the Mangshan National Nature Reserve (3MS), and approved by Guangdong Provincial Bureau of Forestry for collection in Nanling National Nature Reserve (4NL), and approved by Zhejiang Provincial Bureau of Forestry for collection in QianJiang Source National Forestry Park (5QJ) and Zhulong town, Longquan City (6ZL), and approved by Jiangxi Provincial Bureau of Forestry for collection in Guanshan National Nature Reserve (7GS), and Mount Sanqingshan National Park (8SQ).

Specimen collection

The specimens were collected from eight populations from April to June of 2008. All populations but one grew in evergreen and deciduous broad-leaved mixed forests (often lived with species such as in genus *Cyclobalanopsis* and *Sorbus*, *etc.*) at altitudes of 450–1200 m. The exception (3MS) was the one growing in bamboo forests in the Mangshan Mountains of Yizhang. For molecular analysis, 10–11 individuals per population were sampled. The locations and information of populations are provide in Table 1 and Fig. 1. Before DNA extraction, all the dried leaves were preserved in silica gel [43]. All the voucher specimens are deposited at the Herbarium of Sun Yat-sen University (SYS).

DNA extraction and AFLP reactions

Genomic DNA was extracted from silica gel-dried leaves using the cetyl trimethylammonium bromide (CTAB) method [44]. The extracted DNA was dissolved in 100 μL of Tris-hydrochloride (TE) buffer [10 mmol/l Tris-HCl (pH 8.0), 1 mmol/EDTA (pH 8.0)] and used as a template for the polymerase chain reaction (PCR).

AFLP reactions were performed following the method reported by [37] with the following modifications. The restriction digest and ligation steps were done as separate reactions. For the digestion, approximately 500 ng of genomic DNA was incubated at 37°C (EcoRI) or 65°C (MseI) for 2 h in a 20 μL volume reaction containing 10× H Buffer (TOYOBO, Shanghai) and 10 U restriction enzymes EcoRI or MseI. For the ligation, 20 μL of a ligation mix consisting of 10× T4 DNA Ligase (TOYOBO, Shanghai), 1 μL EcoRI-adapter, 1 μL MseI-adapter, and 2 U T4 DNA Ligase was added to the sample and kept at 22°C for 3 h. After ligation, the samples were diluted 10-fold with sterile deionized water (dH_2O). A pre-selective polymerase chain reaction (PCR), using PTC-200 thermocycler (MJ research, Waltham, MA) was done using primer pairs with a single selective nucleotide extension. The reaction mix (total volume 20 μL) consisted of 4 μl template DNA from the restriction/ligation step, 1 μL primer (EcoRI/MseI), and 15 μL AFLP Core Mix (13.8 μL dH_2O, 1.6 μL $MgCl_2$, 1.6 μL dNTPs (2.5 mM), 1 U Taq DNA polymerase, and 10× H buffer). After an initial incubation at 94°C for 2 min, 20 cycles at 94°C for 20 s, 56°C for 30 s, and 72°C for 2 min, with a final extension at 60°C for 30 min, were performed. The PCR products of the amplification reaction were diluted 10-fold with dH_2O and used as a template for the selective amplification using two AFLP primers, each containing three selected nucleotides.

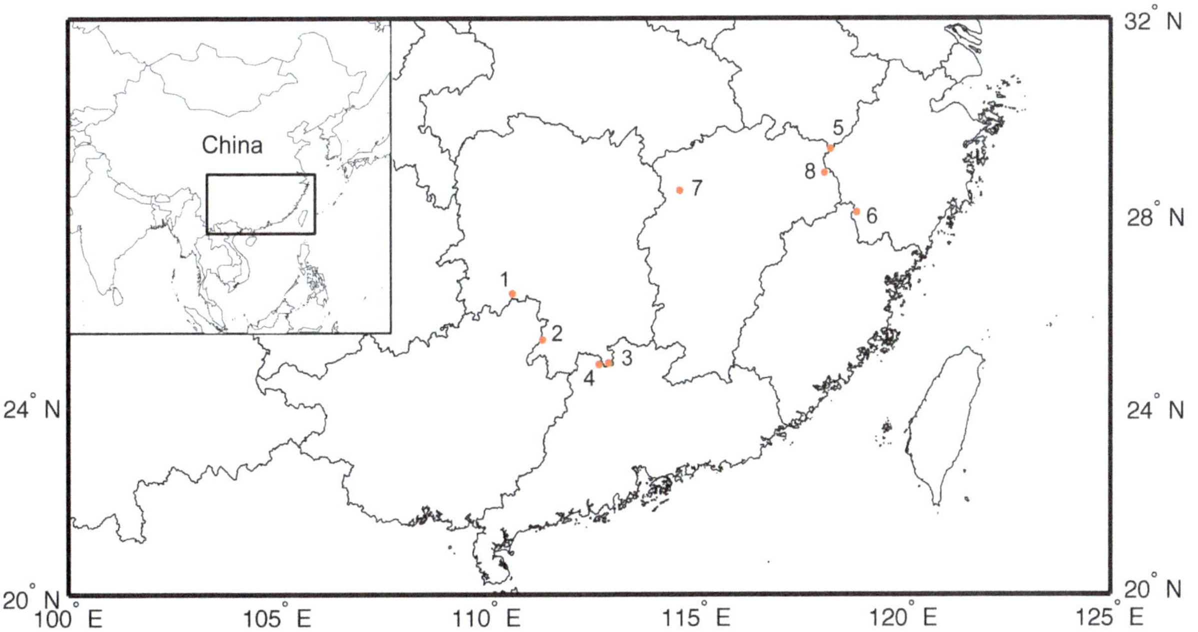

Figure 1. Location of eight populations in two groups sampled in a study of genetic diversity of *Disanthus cercidifolius* subsp. *longipes*. Populations are represented by black dots and located as Table 1. Note: 1:1XN, 2:2DP, 3:3MS, 4:4NL, 5:5QJ, 6:6ZL, 7:7GS, 8:8SQ.

Table 1. Information of location of populations.

ID	GPS co-ordinates	Location of populations	Location type	Altitude (m)	Collection voucher
1XN	26°25′08″N, 110°36′31″E	Xinning county, Hunan Province	mountain slope on the edge of forests	690–740	SHEN Rujiang and GUO Wei
2DP	25°27′43″N, 111°20′20″E	Dupanglin National Nature Reserve, Hunan Province	in bushes by stream	700–850	SHEN Rujiang and GUO Wei
3MS	24°58′12″N, 112°53′16″E	Mangshan National Nature Reserve, Hunan Province	in bamboo forests by stream	680–923	SHEN Rujiang and GUO Wei
4NL	24°56′18″N, 112°39′40″E	Nanling Naitional Nature Reserve, Guangdong Province	mountain slope	680–700	SHEN Rujiang
5QJ	29°23′59″N, 118°12′58″E	Qianjiang Source National Forestry Park, Zhejiang Province	mountain slope	650–700	SHEN Rujiang and GUO Wei
6ZL	28°06′28″N, 118°51′27″E	Zhulong town, Longquan City, Zhejiang Province	mountain slope by streams	1040	SHEN Rujiang and GUO Wei
7GS	28°33′22″N, 114°35′42″E	Guanshan National Nature Reserve, Jiangxi Province	mountain slope	560–600	CHEN Lin
8SQ	28°54′57″N, 118°03′52″E	Mount Sanqingshan National Park, Jiangxi Province	mountain slope in forests	620–1260	Observation team of Sun Yat-sen University

Nine primer combinations labeled with fluorescent 6-carboxy fluorescein (6-FAM) were probed for selective amplification, and only primer combinations with the greatest number of bands (EcoRI/MseI: AAC/CTG; ACA/CTG and ACT/CAT) were selected. Selective amplification was done using the following touchdown PCR conditions: 94°C for 2 min first, then 10 cycles at 94°C for 20 s, and 66°C for 30 s, with a 1°C decrease per cycle then extension at 72°C for 2 min; followed by 20 cycles at 94°C for 20 s, 56°C for 30 s, and 72°C for 2 min. After amplification, 3 μL of the samples were diluted 3-fold with sterile deionized water (dH_2O), and mixed with 10 μL formamide and 0.2 μL Size standard-600 (Beckman Coulter, Fullerton, CA). The mix was used for sequence analysis.

Raw data were collected on a CEQ8000 Sequencer (Beckman Coulter). AFLP products were resolved using a Beckman Coulter CEQ8000 genetic analyzer. Semi-automated fragment analysis was performed using the fragment analysis software of the CEQ8000. The chromatograms of fragment peaks were scored as present (1) or absent (0), and a binary qualitative data matrix was constructed. A total of 82 individuals were run twice with all primer combinations.

Data analysis

As a measure of population diversity, the binary data matrix was input to POPGENE version 1.32 [45], assuming Hardy–Weinberg equilibrium. The following indices were used to quantify the amount of genetic diversity within each population examined: the number of AFLP fragments ($Frag_{tot}$), the percentage of polymorphic fragments ($Frag_{poly}$), Nei's (1973) gene diversity (H), and Shannon's information index (I). In addition, the number of unique fragments ($Frag_{uni}$) and DWs, frequency-down-weighted marker values [46], were calculated as measures of divergence. For each population, the number of occurrences of each AFLP marker in that population was divided by the number of occurrences of that particular marker in the total dataset, then these values were summed up [46,47]. The value of DW was expected to be high in long-term isolated populations where rare markers should accumulate due to mutations, whereas newly established populations were expected to exhibit low values [46]. To even out the unequal sample sizes, DWs were calculated using five randomly chosen individuals. The value of DW was calculated by AFLPdat [48], which is based on R version 2.9.0 (www.r-project.org).

Nei's genetic distance [49] of *D. c.* subsp. *longipes* populations were calculated using the software POPGENE version 1.32 [45], then cluster analysis was performed based on the unweighted pair group method with arithmetic averaging (UPGMA) [50] using NTSYS pc version 2.1e [51]. In addition, based on the genetic distance index devised by [52], the UPGMA dendrogram of individuals was drawn using TREECON version 1.3b [53]. The robustness of the branches was estimated by 1000 bootstrap replicates.

ARLEQUIN version 3.0 [54] was used to perform an analysis of molecular variance (AMOVA) [55] to assess the hierarchical genetic structure among populations and within populations. The AFLPdat program [48], based on R version 2.9.0 (www.r-project.org), was used to convert the AFLP data matrix to the ARLEQUIN input format. The AMOVA was performed by partitioning genetic variation among and within populations regardless of their geographic distribution. In this study, the traditional F-statistics [56] cannot be used in dominant marker AFLP, therefore Φ-statistics [55] replace F-statistics. The significance of Φ values was tested by 1000 permutations. Φ-statistics are computed from a matrix of Euclidean squared distances between every pair of individuals [40]. Two models of the eight populations were tested to investigate regional relationships. Firstly, we treat all populations as a single group to obtain a value for Φ_{st} as an overall measure of population divergence (a two-level analysis), and then we divided the populations into two groups, the East China region and the Nanling Range (three-level hierarchical analyses).

A distance matrix of Φ_{st} between every pair of populations was calculated in ARLEQUIN as a measure of interpopulation genetic differentiation, from which 1000 bootstrapped replicate matrices were then computed, so gene flow (Nm) based on Φ_{st} [56] could be calculated. Isolation-by-distance was investigated by computing the correlation between geographic distance and genetic distance (Φ_{st}) between every pair of populations and applying the Mantel test, using NTSYSpc version 2.1e [51]. The Mantel z-statistic value was tested non-parametrically by creating a null distribution of z using 1000 random permutations and comparing the observed z value [40].

Table 2. Number of loci evaluated for each of three AFLP primer combinations utilized in assays of 82 individuals of *Disanthus cercidifolius* subsp. *longipes* from eight populations.

Primer combination	Number of loci
EcoRI-AAC/MseI-CTG	144
EcoRI-ACA/MseI-CTG	152
EcoRI-ACT/MseI-CAT	157
Total	453

The AFLP data were also subjected to a principal components analysis (PCA), which may help reveal unexpected relationships among a large number of variables, reducing them to two or three new uncorrelated variables so they retain most of the original information [57]. We chose Jaccard's similarity coefficient [50] to calculate the eigenvalue and eigenvector. The standardized data were projected onto the eigenvectors of the correlation matrix and represented in a two-dimensional scatter plot [57]. Plots of samples in relation to the first three principal components were constructed with populations designated as either populations of the East China region or populations of the Nanling Range. The data from the PCA was analyzed using the computer program NTSYSpc version 2.1e [51]. A two-dimensional representation of genetic relationships among *D. c.* subsp. *longipes* genotypes was carried out using SPSS version 16.0 [58].

STRUCTURE version 2.3.1 [59] was used to investigate structure at the individual level. In this study, we inferred that each individual of *D. c.* subsp. *longipes* comes purely from one of the populations. It was applied using a "no admixture" model, 100 000 burn-in period, and 50 000 MCMC replicates after burn-in. The approach requires that the number of clusters K be predefined, and the analysis then assigns the individuals to clusters probabilistically [40]. We performed ten runs for each value of K (1 to 10). To determine the K value, we used both the LnP(D) value and Evanno's ΔK [60]. LnP(D) is the log likelihood of the observed genotype distribution in K clusters and can be output by STRUCTURE simulation [59]. Evanno's ΔK took consideration of the variance of LnP(D) among repeated runs and usually can indicate the ideal K. The suggested $\Delta k = M(|L(k+1)-2L(k)+L(k-1)|)/S[L(k)]$, where L(k) represents the kth LnP(D), M is the mean of 10 runs, and S their standard deviation[60,61]. The output uses color coding to show the assignments of individuals in each population to the clusters.

Population structure was also investigated by HICKORY version 1.1 [62], in order to assess the importance of inbreeding in the data and the assumption of Hardy-Weinberg equilibrium. HICKORY makes it possible to evaluate departures from Hardy-Weinberg equilibrium in dominant as well as co-dominant markers [40]. The program AFLPdat [48] was used to convert the AFLP data matrix to the HICKORY input format. The deviance information criterion (*DIC*) criteria, *Dbar*, *Dhat*, and *pD*, for assessing the importance of inbreeding were computed using the default values: Burn-in 5000; sample 100 000; thin 20 [62].

Results

Population genetic diversity

In this study, three of the nine AFLP primer combinations were used (Table 2). The number of fragments for each primer combination (with percentage of polymorphisms within parenthesis), were: EcoRI-AAC/MseI-CTG: 144 (98.61%), EcoRI-ACA/MseI-CTG: 152 (99.34%), and EcoRI-ACT/MseI-CAT: 157 (96.18%). The length of the fragments varied from 64 bp to 560 bp. The three AFLP primer combinations produced a total of 453 fragments in 82 individuals, of which 444 (98.01%) were polymorphic. The total number of fragments per population ($Frag_{tot}$) varied between 198 (7GS) and 227 (5QJ). The percentage of polymorphism ($Frag_{poly}$) across the eight populations ranged from 43.71% (7GS) to 50.11% (5QJ). The number of fragments that only occur in one population ($Frag_{uni}$) varied between 6 (7GS) and 16 (8SQ). The DW ranged from 56.85 (5QJ) to 111.19 (1XN), mean 76.88, and SD = 21.11. The diversity within a population (H) ranged from 0.1290 (2DP) to 0.1394 (5QJ) (Table 3).

AMOVA

The two-level AMOVA in ARLEQUIN gave a Φ_{st} value of 0.2328 ($P<0.001$), with of 23.28% variation among populations and 76.72% within populations (Table 4). The three-level hierarchical AMOVA analyses of the two-group models shows that three-quarters of the variation (75.06%) was concentrated within populations, while the between group component (5.04%) was less than the between populations-within-group component (19.90%). All three Φ values were significant based on a 1000 permutation test (Table 4).

Pairwise genetic distance and gene flow

The pairwise genetic distance (Φst) and gene flow (*Nm*) matrix were used to establish the level of genetic divergence among the populations (Table 5). Estimates of pairwise genetic distance using AFLP date ranged from 0.1364 ($P<0.001$) for the most closely related populations (4NL and 6ZL), to 0.3428 ($P<0.001$) in the most divergent populations (3MS and 8SQ). The gene flow (*Nm*) among populations ranged from 0.4792 to 1.5830. Except for one population in Nanling National Nature Reserve (4NL), most of populations' Φst were above 0.2, and the corresponding *Nm* were below 1. These results indicate that the genetic distance of *D. c.* subsp. *longipes* is not close, yet nor dependent on geographic distance.

Relationship between geographic distance and genetic distance (The Mantel tests)

A Mantel test showed no correlation between the geographic distance and the Φst genetic distance ($r=0.38917$, $P=0.9920$, Fig. 2). This result implies that *D. c.* subsp. *longipes* does not demonstrate a historical pattern of isolation-by-distance.

Principal components analysis

A PCA of the AFLP-based distance data was performed to examine relationships among the populations of *D. c.* subsp. *longipes*. The PCA showed that populations from East China and the Nanling Range clearly do not cluster separately (Fig. 3). The

Table 3. Region, population (ID), sample size (N), total number of fragments per population ($Frag_{tot}$), percentage of polymorphic fragments ($Frag_{poly}$), number of fragments that only occur in one population ($Frag_{uni}$), frequency-down-weighted marker values (DW), Nei's (1973) gene diversity (H), Sannon's index (I) sampled in a study of *Disanthus cercidifolius* subsp. *longipes* (H. T. Chang) K. Y. Pan populations.

Region	ID*	N	$Frag_{tot}$	$Frag_{poly}$	$Frag_{uni}$	DW	H	I
Nanling range	1XN	10	216	47.68	12	111.19	0.1357±0.1779	0.2106±0.2570
	2DP	11	203	44.81	11	56.96	0.1290±0.1790	0.1991±0.2580
	3MS	10	204	45.03	10	93.96	0.1373±0.1839	0.2100±0.2652
	4NL	10	215	47.46	9	58.09	0.1383±0.1710	0.2136±0.2608
East China region	5QJ	11	227	50.11	10	56.85	0.1394±0.1778	0.2172±0.2564
	6ZL	10	209	46.14	12	77.45	0.1393±0.1822	0.2141±0.2631
	7GS	10	198	43.71	6	64.85	0.1353±0.1852	0.2062±0.2662
	8SQ	10	223	49.23	16	95.68	0.1373±0.1775	0.2138±0.2560
All	-	82	453	98.01	-	-	0.1781±0.1633	0.2918±0.2223

Table 4. Analysis of molecular variance (AMOVA) in *Disanthus cercidifolius* subsp. *longipes* for 82 individuals from eight populations.

Model	Source of variation	df	Sum of squares	Variance component	Percentage variance %	Φ-statistics	*P*
Two levels	among populations	7	1055.95	11.14	23.28	$\Phi_{st}=0.23283$	$P<0.001$
	within populations	74	2716.05	36.7	76.72		
	Total	81	3772	47.84			
three levels	between groups	1	237.66	2.46	5.04	$\Phi_{ct}=0.05040$	$P<0.05$
	between populations-within-groups	6	818.29	9.73	19.90	$\Phi_{sc}=0.20955$	$P<0.001$
	within populations	74	2716.05	36.70	75.06	$\Phi_{st}=0.24939$	$P<0.001$
	Total	81	3772	48.89			

Computed with ARLEQUIN version 3.0. The *P* values represent the probability of obtaining an equal or more extreme value by chance, estimated from 1000 permutations.

Table 5. Pairwise genetic distance (Φst, lower diagonal, $P<0.001$) and gene flow (*Nm*, upper diagonal) between eight populations of *Disanthus cercidifolius* subsp. *longipes* based on AFLP data.

	1XN	2DP	3MS	4NL	5QJ	6ZL	7GS	8SQ
1XN	-	1.4673	0.7122	1.1283	0.7637	0.7610	0.8506	0.6140
2DP	0.1456	-	0.8055	1.1586	0.6833	0.7813	0.7398	0.5624
3MS	0.2598	0.2369	-	1.4639	0.6874	0.7389	0.8037	0.4792
4NL	0.1814	0.1775	0.1459	-	1.0727	1.5830	1.2885	0.6644
5QJ	0.2466	0.2679	0.2667	0.1890	-	1.1441	0.7897	1.3148
6ZL	0.2473	0.2424	0.2528	0.1364	0.1793	-	1.1783	0.6917
7GS	0.2272	0.2526	0.2373	0.1625	0.2405	0.1750	-	0.5466
8SQ	0.2893	0.3077	0.3428	0.2734	0.1598	0.2655	0.3138	-

Significance levels are based on 1000 iterations and indicate that the probability that random Φst values are higher than the observed values. $Nm = (1 - Fst)/4*Fst$ [56].

natural divide is separated into three clusters: cluster 1 (1XN, 2DP), cluster 2 (5QJ, 8SQ), and cluster 3, a mixture that includes some individuals from cluster 1 or cluster 2. The first and second principal component axes, PC1 and PC2, accounted for 25.46% and 25.17% of the total variation, respectively.

Hierarchical and cluster analysis

Nei's genetic distances [49] between populations showed two main clusters of populations (Fig. 4). The cluster that comprises 5QJ and 8SQ is clearly differentiated from the others. Populations 1XN and 2DP are another cluster, and population 4NL and 6ZL are grouped with 7GS, although the UPGMA dendrogram of the 82 individuals based on Nei and Li's genetic distance is not clearly grouped (Fig. 5). Although most individuals in the same population cluster together, some individuals in different populations (except 1XN and 3MS) also are clustered together.

In the genetic structure analysis (STRUCTURE), the highest estimate of the likelihood of the data, given the number of clusters chosen, was k = 10 (ten clusters). We can't get a clear knee in Lnp(D) (Fig. 6). Evanno's ΔK took consideration of the variance of LnP(D) among repeated runs and usually can indicate the best k. See the Fig. 6, we can find that when the k = 6 and we get the highestΔK value. So the best k = 6. The diagram (Fig. 7) showing assignment of individuals to the clusters revealed a clinal structure to the data. In the six-cluster model (k = 6), gene pool 1 (including 5QJ, 8SQ) and gene pool 2 (including 1XN, 2DP) are restricted with a few from another pool, only being distributed in the southwest (gene pool 1) and northeast (gene pool 2). Gene pool 3 (including 3MS, 4NL), while Gene pool 4, pool 5 and pool 6, consist of 6ZL and 7GS. These 4 pools have the most widespread pattern.

The HICKORY results showed that there is inbreeding in populations of *D. c.* subsp. *longipes* (Table 6), because the *DIC* and *Dbar* parameter were lower in "*Full model*" than in other models. This pattern of results allows the "*Full model*" to be considered best, according to the HICKORY manual [62].

Discussion

Genetic diversity in endemic and endangered species

Accurate estimates of genetic diversity are useful for optimizing sampling strategies aiming at the conservation and management of genetic resources [26,63,64]. According to Hamrick and Godt (1989), there are strong associations between geographic range and genetic diversity [65]. Allozyme analyses concluded that endemic and geographically limited plant species generally possess less genetic variation, due to genetic drift and restricted gene flow [66–68]. However, in our study, the percentage of polymorphic fragments ($Frag_{poly}$) and Shannon's information index (I) are higher than those of endemic species based on allozyme [65]. This may account for the new technique can generate more genetic diversity information. Historical events have also been shown to be responsible for variations in genetic diversity [68]. Numerous allozyme studies and an increasing number of cpDNA and mtDNA studies now provide substantial evidence that putative refugee plant populations harbor higher levels of genetic diversity relative to their likely descendant populations [69]. *Disanthus cercidifolius* subsp. *longipes* is a Tertiary relic, which is distributed in warm and humid forests of the Cathayan Land in the Tertiary [2,70]. Lots of present endemics, several of which inhabit Pleistocene refugia during the Quaternary glacial period, were able to maintain higher levels of diversity because of population stability during the glacial cycle. *Disanthus cercidifolius* subsp. *longipes* may be one of them and preserved mutation during the

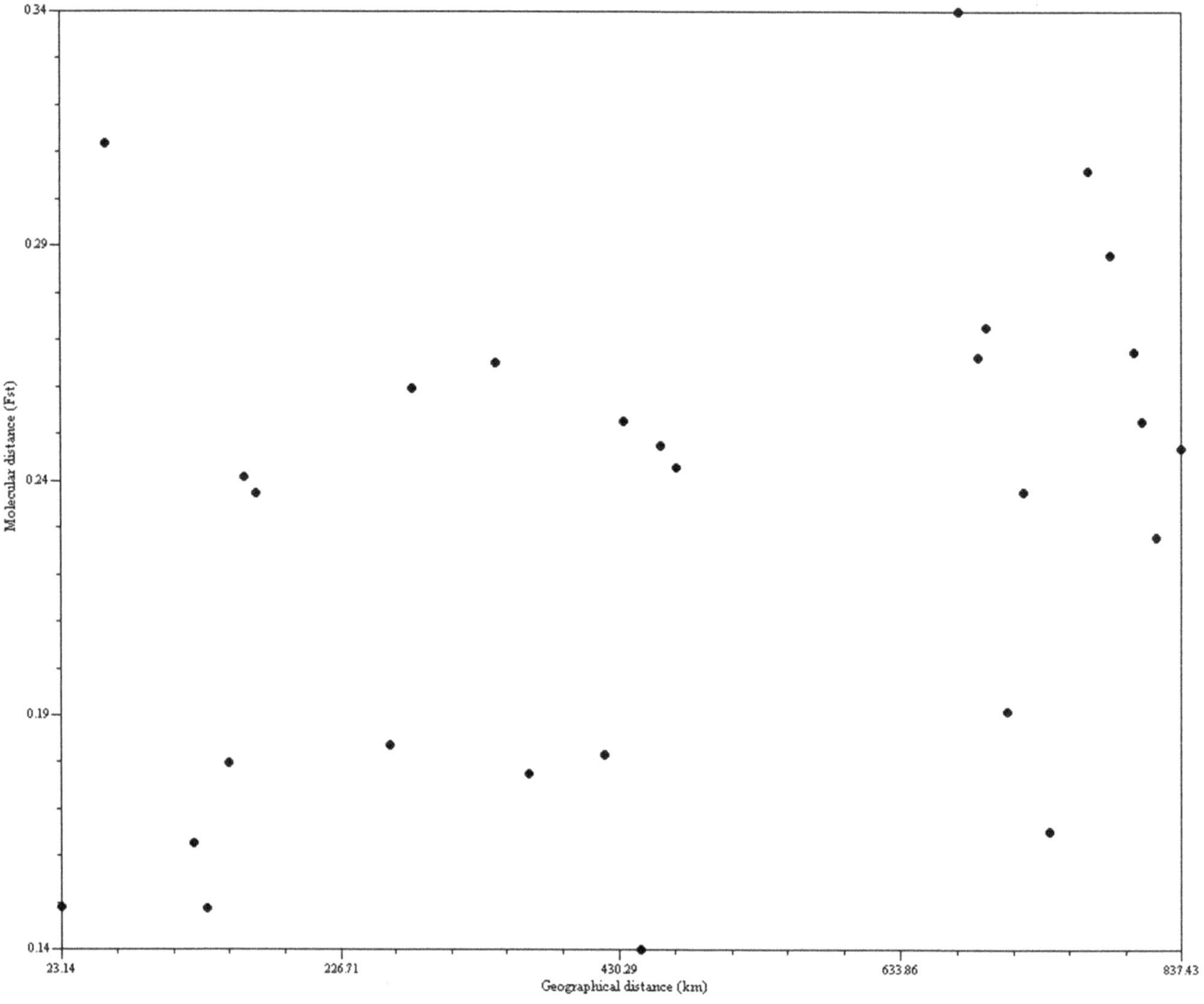

Figure 2. *Disanthus cercidifolius* subsp. *longipes*, bivariate plot showing no correlation between matrices of pair-wise geographic (km) and genetic distance (*Φst*) in eight populations, comprising 82 individuals. Computed with NTSYSpc ver. 2.1e.

relatively long glacial period. So it shows high genetic diversity [69,71]. Compared by AFLP, RAPD, and ISSR markers, the genetic diversity indices in this study revealed an intermediate level of genetic diversity of *D. c.* subsp. *longipes* compared to other endangered species [34,41,66,72,73].

Results of studies of genetic diversity within populations show that all populations were on the similar level, except for the Dupanglin National Nature Reserve (2DP) which is shown to be lower than the others. The lack of genetic diversity resulting from the total existing individual in population 2DP is limited (fewer than 100 individuals). In practice, a larger population often has higher genetic variation [72,74].

Genetic variation and gene flow

In this study AMOVA showed that the largest portion (76.72%) of genetic variance is contributed by genetic variation within populations (Φ_{st} value 0.23283) and only a small portion (23.28%) is due to differences among groups. Some other studies of endemic and endangered species show a similar pattern [34,41,66,72,73]. Moderate or high diversity and low population partitioning in rare plants have previously been attributed to a number of factors [66,75–77], including insufficient length of time for genetic diversity to be reduced following a natural reduction in population size and isolation; adaptation of the genetic system to small population conditions; recent fragmentation (human disturbance) of a once-continuous range, i.e., genetic system; and extensive gene flow due to the combination of bird pollination and high outcrossing rates [66,78]. The direct estimate of gene flow (*Nm*) based on Φ_{st} 0.23283 was 0.82374, which means that the number of migrants per generation is lower than one. The present distribution of *D. c.* subsp. *longipes* is restricted to several isolated habitats [5,7]. However, fossil and palynological evidences suggested that, *Disanthus* was wide spread in warm and humid forests in the Tertiary [2,70]. Accordingly, a reasonable hypothesis is that the modern range of *Disanthus* species was the result of population fragmentation and contraction after the Quaternary glacial cycles. Considering that *D. c.* subsp. *longipes* has poor success in pollination, "excess flowers production, but little fruit set" [79], and fragmented habitat, inbreeding, which generally leads to decreased fitness, or inbreeding depression, of a population, some of the above causes result in the endangered status of *D. c.* subsp. *longipes*.

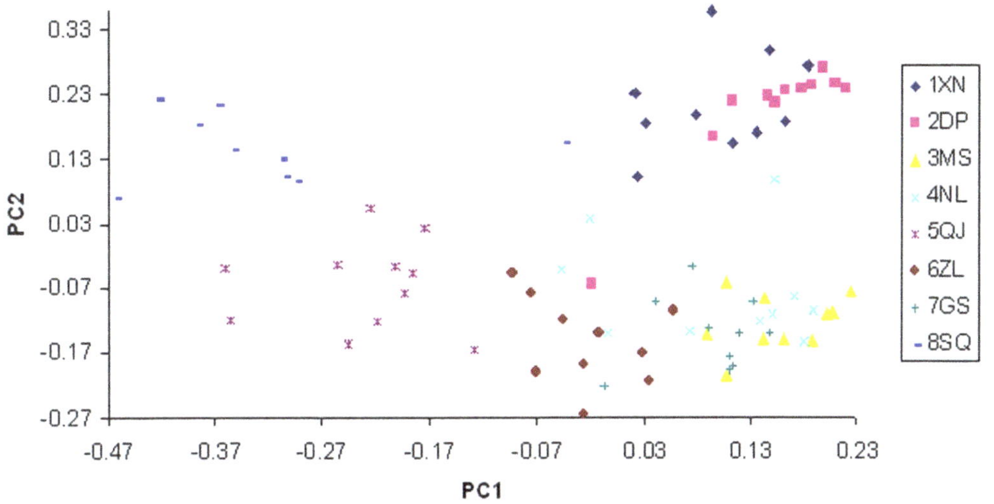

Figure 3. Principal coordinates analysis of 82 individuals from eight populations, based on dissimilarity matrix (Jaccard's coefficient). Accessions are plotted according to the values of first (x axis) and the second (y axis) components and with different symbols according to population. Principal coordinate axes shown (pc1 and pc2) represent 25.46% and 25.17% of respective variance in the dissimilarity matrix.

Genetic structure at different hierarchical levels

Although AFLP markers are dominant, they provide no information on heterozygote frequencies, and our investigation provides no direct information on the reproductive strategy of *D. c.* subsp. *longipes*. In general, outcrossing and long-lived seed plants maintain the most genetic variation within populations, while predominantly selfing, short-lived species harbor comparatively higher variation among populations [65,80]. According to data based primarily on allozyme analysis, Φ_{st} = 0.2 for outcrossing species and 0.5 for inbreeders [65]. The overall Φ_{st} value 0.23283 and the gene flow value for *D. c.* subsp. *longipes* are similar to other species that outcross [81–84]. Permutation tests of the fixation index (Φ_{st} = 0.23) indicated significant genetic structuring. The populations of *D. c.* subsp. *longipes* spread from east to west (Fig. 1). And in the field investigation of our study, we

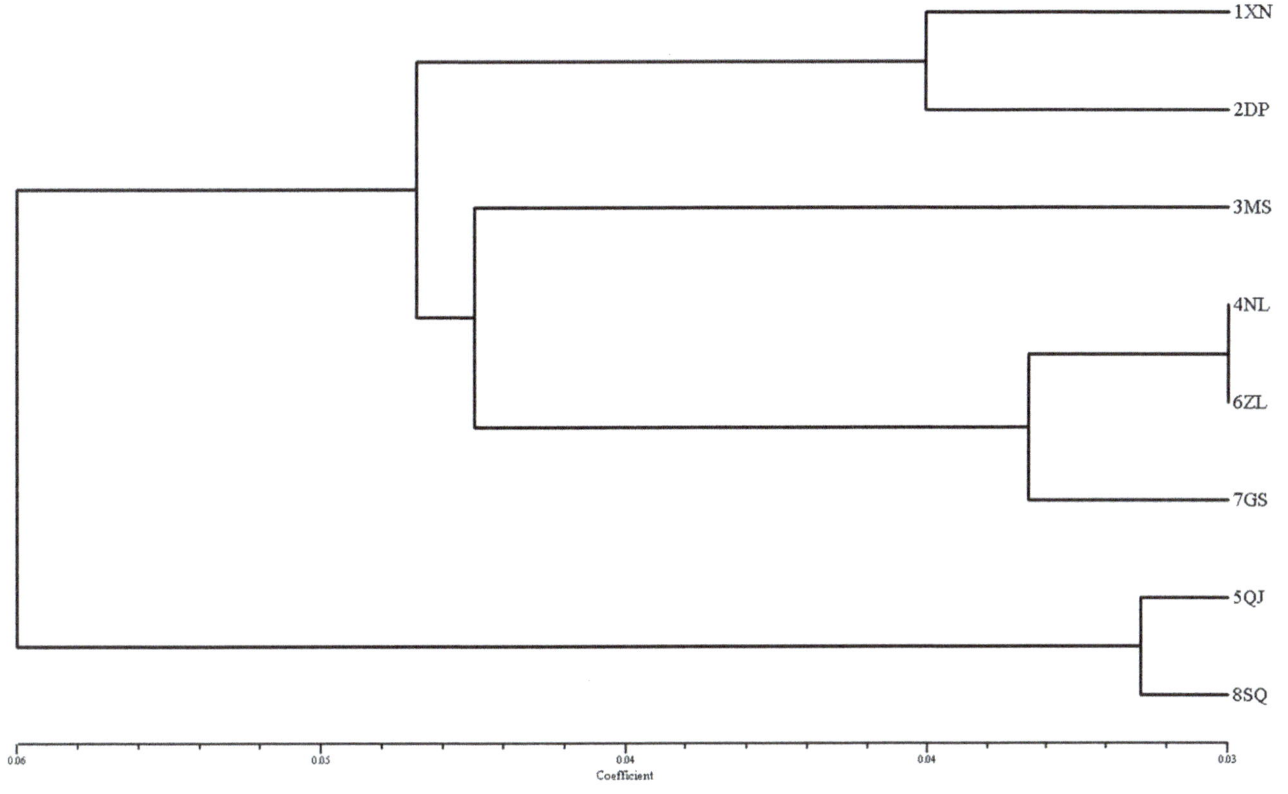

Figure 4. UPGMA dendrogram based on genetic distance. Note: A = 1XN, B = 2DP, E = 3MS, G = 4NL, H = 5QJ, I = 6ZL, J = 7GS, K = 8SQ.

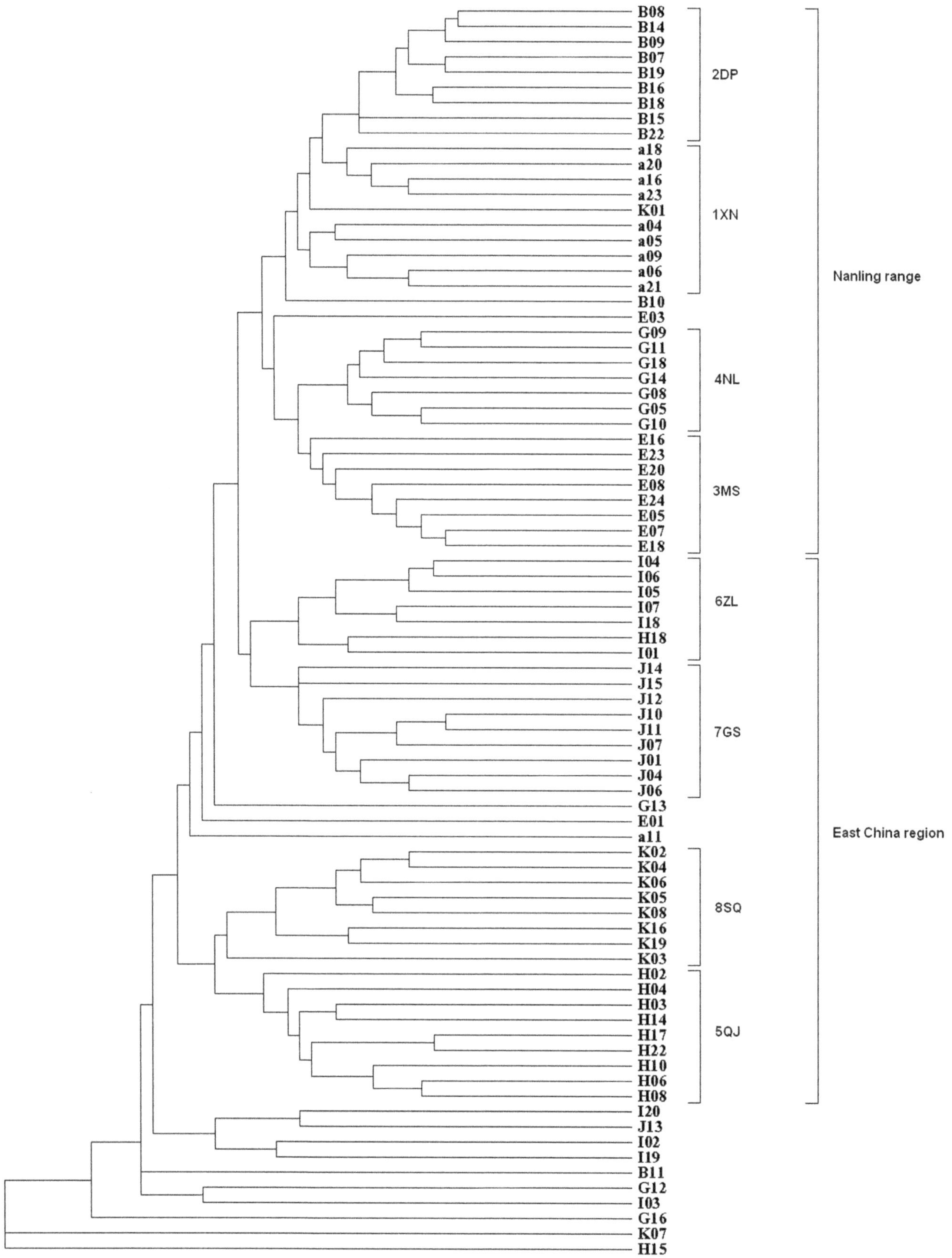

Figure 5. The UPGMA dendrogram based on Nei & Li's genetic distance.

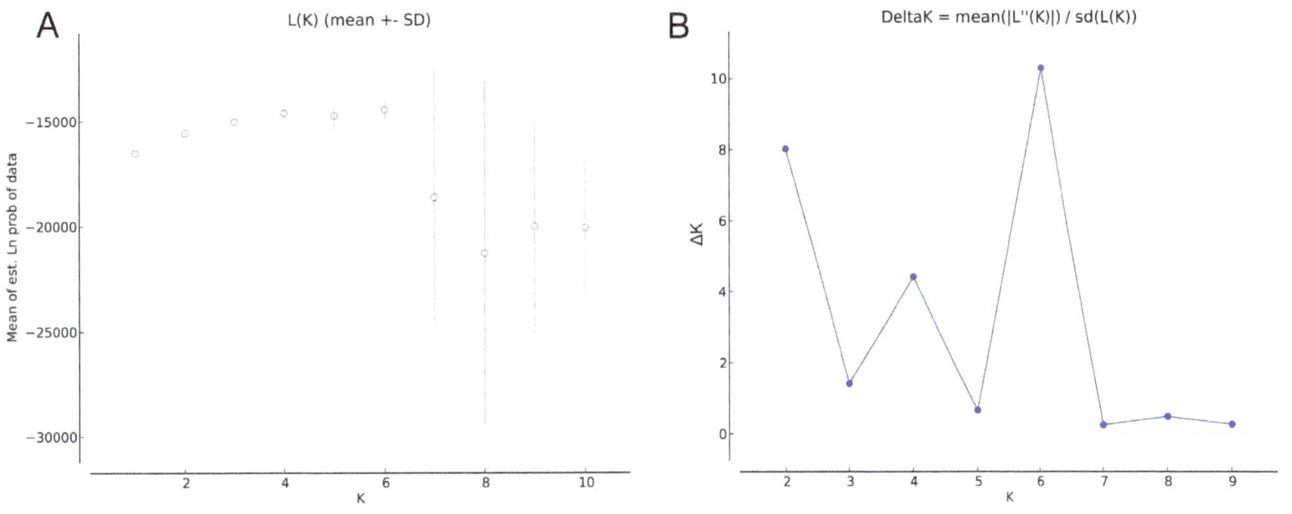

Figure 6. The mean LnP(D) and ΔK over 10 repeats of STRUCTRUE simulations. A. Mean lnP(D) value with k = 1–10. B. ΔK value with k = 2–9.

found that *D. c.* subsp. *longipes* is one of the dominant species in its habitat.

HICKORY software analysis suggests that *D. c.* subsp. *longipes* populations are not genetically differentiated. There is inbreeding in populations of *D. c.* subsp. *longipes* (Table 6), because the *DIC* and *Dbar* parameter were lower in "*Full model*" than that in other models [62]. While inbreeding generally leads to decreased fitness (inbreeding depression) of a population, it also can be advantageous, allowing plants to adapt to disadvantageous conditions [85–88]. *D. c.* subsp. *longipes* originally had an outcrossing system, but under some conditions (e.g., poor efficiency in wind or insect pollination), it would adopt inbreeding or mixed systems to reduce the risk of reproductive failure. This result is consistent with Xiao's conclusions [89] from his study of the reproductive ecology of *D. c.* subsp. *longipes*. If we use co-dominant markers, it most likely would provide an affirmative result and better understanding of the processes.

Hierarchical cluster analysis (Fig. 5) revealed that most individuals of a specific population were grouped together, despite the mixing-in of some individuals from different populations. Most geographically close populations tended to cluster together. The DW values show that the two close populations are clearly divergent, that one is long-term isolated (DW value higher) and the other is newly established (DW value lower). The DW value of 7GS is close to average, but its site is isolated. It might be that there once had been other populations between 6ZL and 7GS. In fact, there is a population of *D. c.* subsp. *longipes* in the Junfeng Mountains, Jiangxi Province, according to Pan's data [2], but we have not obtained any specimens for analysis since 2008. Moreover, although *D. c.* subsp. *longipes* in the Jinggang Mountains, Jiangxi Province, might have been transplanted from Guanshan National Nature Reserve based on the data [90], we found a large population area of *D. c.* subsp. *longipes* in the Jinggang Mountains in October 2009, evidence that the present distribution of *D. c.* subsp. *longipes* appears to be relic from a once extensive range and population.

Genetic structure analysis of the individual samples using STRUCTURE shows that six gene pools are represented in the data, and each gene pool is relatively independent except a few mixtures from other clusters. 1XN and 2DP are the first cluster separated from the other populations; 5QJ and 8SQ form the distinct cluster 2 from the others; 3MS and 4NL also cluster with each other and become distinct from rest populations. These three clusters show high correlation with geographic distribution. And the rest three clusters mixed in 6ZL and 7GS. There is a close relationship between 6ZL and 7GS. It shows that 6ZL is related to 7GS rather than to 5QJ or 8SQ, although 6ZL is close to 5QJ and 8SQ in geographic location. It seems likely that there were some populations between current populations 6ZL and 7GS, or gene flow could have been accomplished by pollen or seed dispersal.

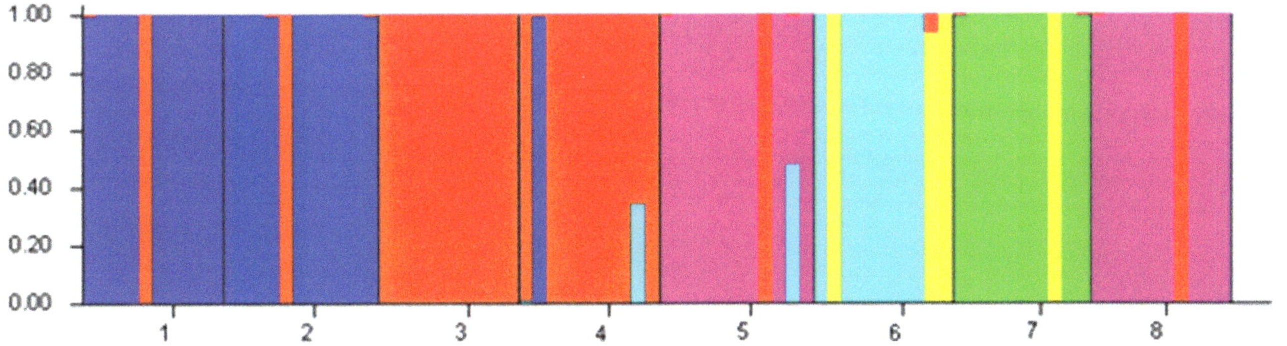

Figure 7. Diagram of results from STRUCTURE(k = 6).

Table 6. Genetic structure analysis using HICKORY ver. 1.1 [62].

	Parameter				
Model	*Dbar*	*Dhat*	*pD*	*DIC*	*f*
Full	6750.93	5395.03	1355.9	8106.82	0.980572
f=0	6780.4	5350.93	1429.47	8209.87	
Theta =0	12092.5	11716.4	376.128	12468.6	0.994715
f free	6844.98	5363.43	1481.55	8326.53	0.500214

Default values for computations were used as follows: Burn-in 5000; sample 100 000; thin 20.

The results of PCA based on AFLP data indicate a different result, which the eight populations form a triangle that has 1XN and 2DP at one angle, 5QJ and 8SQ at a second angle, but 3MS and 4NL mixed with 6ZL and 7GS in the third angle. The reasons for divergence of the East China region and the Nanling Range are not clear.

Guide for conservation measures

In the face of the species extinction generated by human beings [91,92], the urgent need for the conservation of biodiversity is global. In fact, many endemic species became endangered by the loss or fragmentation of habitats and change in natural conditions [26,29,64,93,94]. *D. c.* subsp. *longipes*, with its narrow archipelago-like distribution, habitat fragmentation, and reproductive-physiologic barriers, is typical species that can be easily affected. Although *D. c.* subsp. *longipes* is endangered in China, it has not yet been given high priority for protection. The original aim of this work is to provide some insight into the population biology of *D. c.* subsp. *longipes* in order to guide conservation measures. The guides which can be suggested by our study listed below. Firstly, Most of the genetic variation is located within populations, so we need to protect all the existing populations of *D. c.* subsp. *longipes* in order to preserve as much genetic variation as possible. Secondly, we should promote the further research and technology development to enable better reproduction of *Disanthus*. Finally, it is necessary to enforce the governmental law to forbid people stealing or purchasing wild *Disanthus*.

Conclusions

The results of the AFLP on populations of *Disanthus cercidifolius* subsp. *longipes* show a pattern of high within-population diversity and low among-population divergence. Although the distribution of *D. c.* subsp. *longipes* could be grouped as the East China region and the Nanling Range, the population divergence is clear. The estimates of genetic differentiation and gene flow suggest that the species primarily outcrosses, but can resort to mixed reproductive strategies. These patterns of genetic diversity and levels of genetic variation may be the result of *D. c.* subsp. *longipes* restricted to several isolated habitats and "excess flowers production, but little fruit set". The status of relic and their low reproductive success results in *D. c.* subsp. *longipes* being endangered.

Acknowledgments

We are grateful to the anonymous reviewers and the editor for their critical review; their comments substantially contributed to the revision and improvement of this work. We thank Dr. Sufang Chen for her valuable advice on writing.

Author Contributions

Conceived and designed the experiments: YY RS WL. Performed the experiments: YY RS. Analyzed the data: YY WG QF. Contributed reagents/materials/analysis tools: JJ DC. Wrote the paper: YY WL DC.

References

1. Mabberley DJ (1997) The Plant-book: A Portable Dictionary of the Vascular Plants. Cambridge University Press.
2. Pan K, Lu A, Wen J (1991) A systematic study on the genus *Disanthus* Maxim. (Hamamelidaceae). Cathaya 3: 1–28.
3. Cheng WC (1938) Observations on Mong-shan (Hunan). Science(Sci Soc China) 22: 400.
4. Chang HT (1948) Additions to the Hamamelidaceus flora of China. Sunyatsenia 7: 70.
5. Fu L (1992) China Plant Red Data Book: Rare and Endangered Plants. Beijing: Science Press.
6. State Forestry Administration and Ministry of Agriculture (1999) Key Wild Plants under State Protection. Beijing.
7. Wang S, Xie Y (2004) China Species Red List. Beijing: Higher Education Press.
8. Zavada M, Dilcher D (1986) Comparative pollen morphology and its relationship to phylogeny of pollen in the Hamamelidae. Ann Missouri bot Gard 73: 348–381.
9. Endress PK (1989) A suprageneric taxonomic classification of the hamamelidaceae. Taxon 38: 371–376.
10. Pan KY, Lu AM, Wen J (1991) A systematic study on the genus *Disanthus* Maxim. (Hamamelidaceae). Cathaya 3: 1–28.
11. Li J (1997) Systematics of the Hamamelidaceae based on morphological and molecular evidence. Ph.D. Thesis, University of New Hampshire.
12. Li J, Bogle AL, Donoghue MJ (1999) Phylogenetic relationships in the Hamamelidoideae inferred from sequences of trn non-coding regions of chloroplast DNA. Harvard Papers in Botany 4: 343–356.
13. Shi XH, Xu BM, Li NL, Sun YT (2002) Preliminary study on dormancy and germination of *Disanthus cercidifolius* Maxim var. *longipes* HT Chang seeds. Seed 6: 5–7.
14. Xiao YA, He P, Deng HP, Li XH (2002) Numerical analysis of population morphological differentiation of *Disanthus cercidiifolius* Maxim. var. *longipes* in Jinggangshan. Journal of Wuhan Botanical Research 20: 365–370.
15. Li K, Tang X (2003) The community characteristics and species diversity of the *Disanthus cercidifolius* var. *longipes* shrubland in the Guanshan nature reserve of Jiangxi Province. Journal of Nanjing Forestry University 27: 73–75.
16. Li X, Xiao Y, Hu W, Zeng J (2005) Isozymes analysis of genetic differentiation of the endangered plant *Disanthus cercidifolius* var. *longipes* HT Chang (Hamamelidaceae). Journal of Jianggangshan University (Natural Sciences) 26: 34–38.
17. Huang S, Fang Y, Tan X, Yan J, Fang S (2007) Effects of different concentrations of NAA on cutting regeneration of *Disanthus cercidifolius* var. *longipes*. Journal of Plant Resources and Environment 16: 74.
18. Gao P, Yang A, Yao X, Huang H (2009) Isolation and characterization of nine polymorphic microsatellite loci in the endangered shrub *Disanthus cercidifolius* var. *longipes* (Hamamelidaceae). Molecular Ecology Resources 9: 1047–1049.
19. Bogle A, Philbrick C (1980) A generic atlas of hamamelidaceous pollens. Contributions to the Gray Herbarium, Harvard University 210: 29–103.
20. Chang HT (1979) Hamaelidaceae. Flora Reipublicae Popularis Sinicae. Beijing: Science Press. pp.36–116.
21. Zhang ZY, Chang HT, Endress PK (2003) *Disanthus* Maximowicz. In: Z. Wu and P. H. Raven, editors. Flora of China. Beijing: Science Press and St. Louis: Missouri Botanical Garden Press.

22. Endress PK (1993) Hamamelidae. In: K. Kubitzki, J. G. Rohwer and V. Bittrich, editors. The Families and Fenera of Vascular Plants. Berlin: Springer-Verlag.
23. Xiao YA (2001) The study on population adaptability and genetic diversity of the endangered species *Disanthus cercidifolius* Maxim. var. *longipes* HT Chang. M.Sc. Thesis, Southwest China Normal University.
24. Soule M, Simberloff D (1986) What do genetics and ecology tell us about the design of nature reserves? Biological Conservation 35: 19–40.
25. Barrett S, Kohn J (1991) Genetic and evolutionary consequences of small population size in plants: implications for conservation. Genetics and conservation of rare plants: 3–30.
26. Ge S, Hong D, Wang H, Liu Z, Zhang C (1998) Population genetic structure and conservation of an endangered conifer, *Cathaya argyrophylla* (Pinaceae). International Journal of Plant Sciences 159: 351–357.
27. Millar C, LiBBY W (1991) Strategies for conserving clinal, ecotypic, and disjunct population diversity in widespread species. Genetics and conservation of rare plants: 149–170.
28. Francisco-Ortega J, Santos-Guerra A, Kim S, Crawford D (2000) Plant genetic diversity in the Canary Islands: a conservation perspective. American Journal of Botany 87: 909.
29. Qiu Y, Li J, Liu H, Chen Y, Fu C (2006) Population structure and genetic diversity of *Dysosma versipellis* (Berberidaceae), a rare endemic from China. Biochemical Systematics and Ecology 34: 745–752.
30. Song J, Murdoch J, Gardiner SE, Young A, Jameson PE, et al. (2008) Molecular markers and a sequence deletion in intron 2 of the putative partial homologue of LEAFY reveal geographical structure to genetic diversity in the acutely threatened legume genus *Clianthus*. Biological Conservation 141: 2041–2053.
31. Perez-Collazos E, Segarra-Moragues JG, Catalan P (2008) Two approaches for the selection of Relevant Genetic Units for Conservation in the narrow European endemic steppe plant *Boleum asperum* (Brassicaceae). Biological Journal of the Linnean Society 94: 341–354.
32. Yang J, Qian ZQ, Liu ZL, Li S, Sun GL, et al. (2007) Genetic diversity and geographical differentiation of *Dipteronia* Oliv. (Aceraceae) endemic to China as revealed by AFLP analysis. Biochemical Systematics and Ecology 35: 593–599.
33. Prohens J, Anderson GJ, Herraiz FJ, Bernardello G, Santos-Guerra A, et al. (2007) Genetic diversity and conservation of two endangered eggplant relatives (*Solanum vespertilio* Aiton and *Solanum lidii* Sunding) endemic to the Canary Islands. Genetic Resources and Crop Evolution 54: 451–464.
34. Jian SG, Zhong Y, Liu N, Gao ZZ, Wei Q, et al. (2006) Genetic variation in the endangered endemic species *Cycas fairylakea* (Cycadaceae) in China and implications for conservation. Biodiversity and Conservation 15: 1681–1694.
35. Pfosser M, Jakubowsky G, Schluter PM, Fer T, Kato H, et al. (2005) Evolution of *Dystaenia takesimana* (Apiaceae), endemic to Ullung Island, Korea. Plant Systematics and Evolution 256: 159–170.
36. Strand AE, Leebens-Mack J, Milligan BG (1997) Nuclear DNA-based markers for plant evolutionary biology. Molecular Ecology 6: 113–118.
37. Vos P, Hogers R, Bleeker M, Reijans M, van de Lee T, et al. (1995) AFLP: a new concept for DNA fingerprinting. Nucleic Acids Res 23: 4407–4414.
38. Armstrong TTJ, De Lange PJ (2005) Conservation genetics of *Hebe speciosa* (Plantaginaceae) an endangered New Zealand shrub. Botanical Journal of the Linnean Society 149: 229–239.
39. Kim SC, Lee C, Santos-Guerra A (2005) Genetic analysis and conservation of the endangered Canary Island woody sow-thistle, *Sonchus gandogeri* (Asteraceae). Journal of Plant Research 118: 147–153.
40. Andrade IM, Mayo SJ, Van den Berg C, Fay MF, Chester M, et al. (2007) A preliminary study of genetic variation in populations of *Monstera adansonii* var. *klotzschiana* (Araceae) from north-east brazil, estimated with AFLP molecular markers. Annals of Botany 100: 1143–1154.
41. Tang SQ, Dai WJ, Li MS, Zhang Y, Geng YP, et al. (2008) Genetic diversity of relictual and endangered plant *Abies ziyuanensis* (Pinaceae) revealed by AFLP and SSR markers. Genetica 133: 21–30.
42. Wolfe A, Liston A (1998) Contributions of PCR-based methods to plant systematics and evolutionary biology. Molecular systematics of plants II: 43–86.
43. Chase MW, Hills HH (1991) Silica gel: an ideal material for field preservation of leaf samples for DNA studies. Taxon 40: 215–220.
44. Doyle JJ, Doyle JL (1987) A rapid DNA isolation procedure for small quantities of fresh leaf tissue. Phytochemical bulletin 19: 11–15.
45. Yeh FC, Yang RC, Boyle TBJ, Ye ZH, Mao JX (1997) POPGENE, the user-friendly shareware for population genetic analysis. Molecular Biology and Biotechnology Centre, University of Alberta, Canada.
46. Schönswetter P, Tribsch A (2005) Vicariance and dispersal in the alpine perennial *Bupleurum stellatum* L.(Apiaceae). Taxon: 725–732.
47. Ortiz M, Tremetsberger K, Talavera S, Stuessy T, García-Castaño JL (2007) Population structure of *Hypochaeris salzmanniana* DC.(Asteraceae), an endemic species to the Atlantic coast on both sides of the Strait of Gibraltar, in relation to Quaternary sea level changes. Mol Ecol 16: 541–552.
48. Ehrich D (2006) AFLPdat: a collection of R functions for convenient handling of AFLP data. Mol Ecol Notes 6: 603–604.
49. Nei M (1978) Estimation of average heterozygosity and genetic distance from a small number of individuals. Genetics 89: 583–590.
50. Sneath PHA, Sokal RR (1973). Numerical Taxonomy, 2nd edn. San Francisco: Freeman.
51. Rohlf FJ (2000) NTSYS-pc: numerical taxonomy and multivariate analysis system, version 2.1. New York: Exeter Software.
52. Nei M, Li WH (1979) Mathematical model for studying genetic variation in terms of restriction endonucleases. Proc Natl Acad Sci USA 76: 5269–5273.
53. Van de Peer Y, De Wachter Y (1994) TREECON for Windows: a software package for the construction and drawing of phylogenetic trees for the Microsoft Windows environment. Comp Applic Biosci 10: 569–570.
54. Excoffier L, Laval G, Schneider S (2005) Arlequin (version 3.0): An integrated software package for population genetics data analysis. Evolutionary Bioinformatics 1: 47–50.
55. Excoffier L, Smouse P, Quattro J (1992) Analysis of molecular mariance inferred from metric distances among DNA haplotypes: Application to human mitochondrial DNA restriction data. Genetics 131: 479–491.
56. Wright S (1950) Genetical structure of populations. Nature 166: 247–249.
57. Nie Z-L, Wen J, SUN H (2009) AFLP Analysis of *Phryma* (Phrymaceae) Disjunct between Eastern Asia and Eastern North America. Acta Botanica Yunnanica 31: 289–295.
58. Inc S (2007) SPSS 16.0 for Windows. SPSS Inc, Chicago, IL.
59. Pritchard JK, Stephens M, Donnelly P (2000) Inference of population structure using multilocus genotype data. Genetics 155: 945–959.
60. Evanno G, Regnaut S, Goudet J (2005) Detecting the number of clusters of individuals using the software STRUCTURE: a simulation study. Mol Ecol 14: 2611–2620.
61. Zhang DL, Zhang HL, Wang MX, Sun JL, Qi YW, et al. (2009) Genetic structure and differentiation of *Oryza sativa* L. in China revealed by microsatellites. Theor Appl Genet 119: 1105–1117.
62. Holsinger K, Lewis P (2003) HICKORY: a package for analysis of population genetic data V. 1.0. University of Connecticut, Storrs, USA.
63. Cardoso M, Provan J, Powell W, Ferreira P, De Oliveira D (1998) High genetic differentiation among remnant populations of the endangered *Caesalpinia echinata* Lam.(Leguminosae-Caesalpinioideae). Molecular Ecology 7: 601–608.
64. Bouza N, Caujapé-Castells J, González-Pérez M, Batista F, Sosa P (2002) Population structure and genetic diversity of two endangered endemic species of the Canarian laurel forest: *Dorycnium spectabile* (Fabaceae) and *Isoplexis chalcantha* (Scrophulariaceae). International Journal of Plant Sciences 163: 619–630.
65. Hamrick J, Godt M (1989) Allozyme diversity in plant species. In: A. H. D. Brown, M. T. A. Clegg, L. Kahler and B. S. Weir, editors. Plant Population Genetics, Breeding and Genetic Resources. Sunderland, Mass: Sinauer. pp.43–63.
66. Ge XJ, Yu Y, Zhao NX, Chen HS, Qi WQ (2003) Genetic variation in the endangered Inner Mongolia endemic shrub *Tetraena mongolica* Maxim.(Zygophyllaceae). Biological Conservation 111: 427–434.
67. Hamrick JL, Godt MJW (1996): Conservation genetics of endemic plant species. In Conservation genetics: case histories from nature. In: Avise JC, Hamrick JL, Editors. New York: Chapman and Hall. pp.281–304.
68. Karron J (1991) Patterns of genetic variation and breeding systems in rare plant species. Genetics and Conservation of Rare Plants: 87–98.
69. Ge XJ, Yu Y, Zhao NX, Chen HS, Qi WQ (2003) Genetic variation in the endangered Inner Mongolia endemic shrub *Tetraena mongolica* Maxim. (Zygophyllaceae). Biol Conserv 111: 427–434.
70. Wu Z, Lu A, Tang Y, Chen Z, Li DZ (2003) The families and gerera of angiosperms in China. Beijing: Science Press.
71. Lewis PO, Crawford DJ (1995) Pleistocene Refugium Endemics Exhibit Greater Allozymic Diversity than Widespread Congeners in the Genus *Polygonella* (Polygonaceae). American Journal of Botany 82: 141–149.
72. Kwon J, Morden C (2002) Population genetic structure of two rare tree species (*Colubrina oppositifolia* and *Alphitonia ponderosa*, Rhamnaceae) from Hawaiian dry and mesic forests using random amplified polymorphic DNA markers. Mol Ecol 11: 991–1001.
73. Lacerda DR, Acedo MDP, Filho JPL, Lovato MB (2001) Genetic diversity and structure of natural populations of *Plathymenia reticulata* (Mimosoideae), a tropical tree from the Brazilian Cerrado. Molecular Ecology 10: 1143–1152.
74. Jian S, Zhong Y, Liu N, Gao Z, Wei Q, et al. (2006) Genetic variation in the endangered endemic species *Cycas fairylakea* (Cycadaceae) in China and implications for conservation. Biodiversity & Conservation 15: 1681–1694.
75. Schaal B, Hayworth D, Olsen K, Rauscher J, Smith W (1998) Phylogeographic studies in plants: problems and prospects. Mol Ecol 7: 465–474.
76. Slatkin M (1987) Gene flow and the geographic structure of natural populations. Science 236: 787.
77. Zawko G, Krauss S, Dixon K, Sivasithamparam K (2001) Conservation genetics of the rare and endangered *Leucopogon obtectus* (Ericaceae). Mol Ecol 10: 2389–2396.
78. Maguire T, Sedgley M (1997) Genetic diversity in *Banksia* and *Dryandra* (Proteaceae) with emphasis on Banksia cuneata, a rare and endangered species. Heredity 79: 394–401.
79. Xiao YA, Zeng JJ, Li XH, Hu WH, He P (2006) Pollen and resource limitations to lifetime seed prodution in a wild population of the endagered plant *Disanthus cercidifolius* Maxim. var. *longipes* H. T. Chang (Hamamelidaceae). Acta Ecologica Sinica 26: 496–502.
80. Huang Y, Zhang C, Li D (2009) Low genetic diversity and high genetic differentiation in the critically endangered *Omphalogramma souliei* (Primulaceae): implications for its conservation. Journal of Systematics and Evolution 47: 103–109.

81. Kothera L, Richards C, Carney S (2007) Genetic diversity and structure in the rare Colorado endemic plant *Physaria bellii* Mulligan (Brassicaceae). Conservation Genetics 8: 1043–1050.
82. Jacquemyn H, Honnay O, Galbusera P, Roldan-Ruiz I (2004) Genetic structure of the forest herb Primula elatior in a changing landscape. Molecular Ecology 13: 211–219.
83. Juan A, Crespo M, Cowan R, Lexer C, Fay M (2004) Patterns of variability and gene flow in *Medicago citrina*, an endangered endemic of islands in the western Mediterranean, as revealed by amplified fragment length polymorphism (AFLP). Mol Ecol 13: 2679–2690.
84. Morjan CL, Rieseberg LH (2004) How species evolve collectively: implications of gene flow and selection for the spread of advantageous alleles. Mol Ecol 13: 1341–1356.
85. Allard R, Jain S, Workman P (1968) The genetics of inbreeding populations. Adv Genet 14: 55–131.
86. Antonovics J (1968) Evolution in closely adjacent plant populations. V. Evolution of self-fertility. Heredity 23: 219–238.
87. Jain S (1976) The evolution of inbreeding in plants. Annual Review of Ecology and Systematics 7: 469–495.
88. Lloyd D (1980) Demographic factors and mating patterns in angiosperms.
89. Xiao YA (2005) Studies on Reproductive Ecology and photosynthetic adaptability of the Endangered Plant *Disanthus cercidifolius* var.*longipes* HT Chang. Ph.D. Thesis, Southwest China Normal University.
90. Liu R (1999) *Disanthus cercidifolius* var. *longipes*. Plants 4: 7.
91. Vitousek P, Mooney H, Lubchenco J, Melillo J (1997) Human domination of Earth's ecosystems. Science 277: 494.
92. Chapin F, Zavaleta E, Eviner V, Naylor R, Vitousek P, et al. (2000) Consequences of changing biodiversity. Nature 405: 234–242.
93. Palacios C, Gonzalez-Candelas F (1997) Lack of genetic variability in the rare and endangered *Limonium cavanillesii* (Plumbaginaceae) using RAPD markers. Molecular Ecology 6: 671–675.
94. Qiu Y, Luo Y, Comes H, Ouyang Z, Fu C (2007) Population genetic diversity and structure of *Dipteronia dyerana* (Sapindaceae), a rare endemic from Yunnan Province, China, with implications for conservation. Taxon 56: 427–437.

8

The Genetic Integrity of the *Ex Situ* Population of the European Wildcat (*Felis silvestris silvestris*) Is Seriously Threatened by Introgression from Domestic Cats (*Felis silvestris catus*)

Kathrin A. Witzenberger[1,2], Axel Hochkirch[2*]

1 Zoo Hoyerswerda, Hoyerswerda, Germany, **2** Trier University, Department of Biogeography, Trier, Germany

Abstract

Studies on the genetic diversity and relatedness of zoo populations are crucial for implementing successful breeding programmes. The European wildcat, *Felis s. silvestris*, is subject to intensive conservation measures, including captive breeding and reintroduction. We here present the first systematic genetic analysis of the captive population of *Felis s. silvestris* in comparison with a natural wild population. We used microsatellites and mtDNA sequencing to assess genetic diversity, structure and integrity of the ex situ population. Our results show that the ex situ population of the European wildcat is highly structured and that it has a higher genetic diversity than the studied wild population. Some genetic clusters matched the breeding lines of certain zoos or groups of zoos that often exchanged individuals. Two mitochondrial haplotype groups were detected in the in situ populations, one of which was closely related to the most common haplotype found in domestic cats, suggesting past introgression in the wild. Although native haplotypes were also found in the captive population, the majority (68%) of captive individuals shared a common mtDNA haplotype with the domestic cat (*Felis s. catus*). Only six captive individuals (7.7%) were assigned as wildcats in the STRUCTURE analysis (at K = 2), two of which had domestic cat mtDNA haplotypes and only two captive individuals were assigned as purebred wildcats by NewHybrids. These results suggest that the high genetic diversity of the captive population has been caused by admixture with domestic cats. Therefore, the captive population cannot be recommended for further breeding and reintroduction.

Editor: William J. Murphy, Texas A&M University, United States of America

Funding: This work is part of a PhD project that is financially supported by the Scholarship Programme of the German Federal Environmental Foundation. The funders had no role in study design, data collection and analysis, decision to publish, or preparation of the manuscript.

Competing Interests: The authors have declared that no competing interests exist.

* Email: hochkirch@uni-trier.de

Introduction

Due to the ongoing threats to biodiversity, species conservation is still a challenging task [1]. Although in situ conservation represents by far the most effective way to protect endangered species, it is evident that not all species can be preserved in their natural habitats. Therefore, ex situ conservation and reintroductions have become common measures of species conservation [2,3]. Despite the great advances in organization and logistics (studbooks, online databases) and increasing expertise in veterinary medicine, inbreeding and outbreeding still pose serious threats for ex situ conservation [4–7]. Captive breeding may be compromised by unknown founder origin and relationships [8], undetected hybridisation [9], limited representation of natural genetic diversity [10] or studbook errors [11,12]. These problems can affect both the genetic integrity of the captive population as well as the success of reintroduction projects. Therefore, data on the genetic diversity and relatedness in zoo populations can provide valuable information for improving conservation programmes [13].

The influence of captive breeding on the genetic diversity of endangered species has already been subject to a number of theoretic and genetic studies [7,14–16,17]. However, the majority of the genetic studies focus on highly endangered species, for which studbooks or international breeding programmes already exist [18–20]. This circumstance has two drawbacks: (1) Species with small population size might have a reduced genetic diversity even in their wild populations. Therefore, the genetic diversity of captive populations of such highly endangered species might not reflect the effects of captive breeding. (2) Studbooks are usually kept in a way to avoid the loss of genetic diversity or inbreeding depression. Hence, it is likely that genetic analyses of ex situ populations will confirm the success of such coordinated breeding schemes [16,21]. However, studbooks exist only for ca. 850 species (WAZA 2010), while ca. 10.000 species are currently kept in zoos and bred without any coordination (ISIS 2010). Many of these species are still rather common in the wild, but nevertheless some of them are regionally endangered and sometimes even subject to reintroductions. In order to assess the influence of uncoordinated breeding on the genetic diversity of a captive population, it is useful to study one of these species rather than well managed captive populations. Comparing the genetic variability of ex situ populations and wild populations also requires that a target species has viable wild populations and a sufficiently large captive population. Furthermore, it should have been successfully bred for many generations in captivity as genetic erosion may accumulate over time [18].

The European wildcat (*Felis silvestris silvestris*) is an ideal study object for such an analysis. It is still rather common in Europe and classified as Least Concern on the IUCN Red List of Threatened Species [22]. However, the populations of this species are fragmented and in some countries declining [22]. Wildcats are kept and bred in many European zoos and captive individuals have already been reintroduced in three regions in Germany [23]. However, to date no coordinated breeding programme or studbook exists for the European wildcat. Hence, there is no information available on the captive stock concerning founder size, founder origin, captive population size and relatedness. As it is planned to establish a studbook for the European wildcat in the near future (A. Sliwa pers. comm.), a genetic analysis of the breeding stock can provide important basic information for this studbook. Another challenge in the conservation of the European wildcat is hybridization with domestic cats, which is a serious threat in some wild populations [24–26]. Thus, an additional advantage of a genetic analysis prior to coordinated breeding is the exclusion of potential hybrids.

Here, we present the first systematic genetic analysis of the European captive population of *Felis s. silvestris*. We used both microsatellites and mitochondrial DNA (mtDNA) sequencing in order to assess the genetic variability, structure and integrity of the *ex situ* population. We also analysed samples from a wild population to evaluate if the genetic variability in the captive population is comparable to the diversity found in the wild.

Methods

Sampling and DNA extraction

We contacted 124 zoos, 64 of which confirmed keeping European wildcats. We obtained 80 samples (12 buccal swabs, 8 tissue samples from deceased individuals, 10 blood samples and 50 hair samples) from 30 zoos spanning six European countries (see Text S2 in File S1 and Tab. S4 in File S1 for an overview). All ex situ samples were taken during routine veterinary treatments (like vaccination, or the placement of transponders for juveniles) or whenever a veterinary treatment was necessary. The collection of hair samples does not pose a severe stress or hurt to the cats, the same is true for buccal swaps from young individuals. Blood samples were only taken from anaesthetized individuals.

The in situ sample consisted of 89 individuals (26 tissue samples and 63 hair samples) from a natural wild population in the Harz mountains (Germany). The hair samples were collected in the context of a radio telemetry study (M. Götz). The tissue samples were from a monitoring program for roadside casualties. Both studies were carried out with the permit of the Landesamt für Umweltschutz Sachsen-Anhalt which included the collection of samples from European Wildcats (see Text S1 in File S1 for further information). Additionally, we received six samples (3 tissue and 3 hair samples) from road kills and live traps collected in the wild populations in Rhineland-Palatinate and Saarland.

Furthermore, we sampled 33 domestic cats with the permission of their owners, in order to detect hybrids. 15 of the domestic samples came from a private animal shelter which was located near the in situ population. We chose this shelter, as it would provide a realistic insight into the potential influence local feral cats might have on the wild population. For the remaining samples we contacted private owners. The private owners tore out a small bunch of hair, so that the sample included the follicles. This bunch was then sent or given to us in an envelope.

DNA was extracted from buccal swabs, tissue and blood using the DNeasy blood and tissue kit (Qiagen) following the manufacturer's instructions (with special protocols for blood and buccal swabs as provided by the manufacturer). For the hair samples, we extracted DNA with a modified Chelex 100 protocol, using a 10% Chelex solution with addition of 7 µl Proteinase K (18 mg/ml) per sample and overnight lysis [27,28]. No DNA could be extracted from seven of the 208 samples (5 from wild populations and 2 from the captive population). The respective samples from the wild population were hair samples which had been taken several years ago and had then been stored at room temperature. The captive samples had probably been conserved in denatured ethanol.

Genotyping

The six samples from the wild caught individuals in Saarland and Rhineland-Palatinate were not genotyped, as the samples size from these populations was too small for population genetic analysis. They served as reference samples for the mtDNA analysis (see below). Therefore only 194 of the 201 samples were genotyped at ten di-nucleotide repeat microsatellite loci (FCA08, FCA23; FCA43; FCA58, FCA77, FCA78, FCA90, FCA126, FCA132 and FCA149) characterized in the domestic cat [29,30]. Amplification was performed in a Mastercycler (Eppendorf) using the 2.5×5PRIME HotMasterMix (5PRIME). For each PCR we used 5 µl reaction mix containing 1.2 µl genomic DNA, 2.2 µl HotMasterMix, 2.2 µl water and 0.1 µl forward and reverse primers. The PCR conditions were as recommended by the manufacturer, with an annealing temperature of 55°C for most primers (exceptions: FCA90: 60°C; FCA78: 50°C and addition of 2.5 mM Mg^{2+}). The 5′-end of each forward primer was labelled with a fluorescent dye, either 5-FAM, TAMRA or JOE. The products were genotyped on a MegaBACE 1000 automated DNA sequencer (GE Healthcare). Fragment lengths were determined using Fragment Profiler 1.2 (Amersham Biosciences). To minimize genotyping errors due to low DNA concentrations (especially in the hair samples), we applied a multiple tube approach as recommended by Taberlet et al. [31]. Samples were only included into further analyses, if they yielded unambiguous results in three independent replications (190 samples met this criterion, Tab. S4 in File S1). Two captive samples had identical genotypes in all three replications and it turned out that this individual had been transferred to another zoo and sampled twice. Therefore, the duplicate sample was excluded from further statistical analysis and this left 189 samples for analysis.

Sequencing

For 70 wild, 66 captive and 33 domestic individuals we sequenced the mitochondrial DNA fragment NADH dehydrogenase subunit 5 (ND5) based on the study by Driscoll et al. [32]. Most juveniles of sampled mothers and siblings were excluded and later assigned to the same mtDNA haplotype as all sequenced juveniles had the same haplotype as their dam. We finally obtained 88 sequences of wild, 77 of captive and 33 of domestic individuals (Tab. S4 in File S1). The primers CD-ND56-F1C and CD-ND56-R4 [32] were used for amplification in a Multigene Gradient Thermal Cycler (Labnet). We used the 5PRIME HotMasterMix (5PRIME) for amplification. The PCR product was purified using the High pure PCR product purification kit (Roche) according to the manufacturer's protocol. Sequencing was performed with the DYEnamic ET terminator cycle sequencing kit (GE Healthcare) for sequencing reactions run on a MEGABACE 1000 automated sequencer (GE Healthcare). Base-calling was performed in Sequence Analyzer 4.0 (Amersham Biosciences).

Data analysis

Genotyping. As the integration of hybrids and genetic subgroupings were unknown, we first searched for genetic structure in our samples using STRUCTURE 2.3.4 [33]. We assumed admixture between groups and used the correlated allele frequency model with a burn-in period of 100,000 simulations, followed by one million Markov chain Monte Carlo simulations. Tests were run for K = 1-15 with ten iterations for each K. In order to detect potential hybrids between wildcats and domestic cats, we used the STRUCTURE results at K = 2 and assigned all individuals either to the cluster "wildcats", "domestic cats" or "hybrid" [34]. The "hybrid" cluster was chosen, if Q values varied between 0.2 and 0.8 as recommended by Randi and Schulte et al. [35,36]. The optimal values for K were assessed using both the method described by Pritchard et al. [33], and the method suggested by Evanno et al. [37]. The method described by Evanno et al. [37] tends to result in low K values [38,39] and generally works better for scenarios with strong genetic differentiation [40] as it detects the highest level of differentiation. The method described by Pritchard et al. [33] might lead to inflated K-values if closely related individuals exist in the sample. Therefore, the clusters might more likely represent family groups and lineages [41]. In our case the detection of family groups and breeding lines is a desired outcome, which provides deeper insight into the genetic structure of the captive population. We, therefore, present the results obtained by the method described in Pritchard et al. [33]. However, we stopped increasing K when Q values for the next cluster dropped below 0.9 in all individuals as proposed by Schulte et al. [35]. The individuals were assigned to genetic clusters using the highest assignment probability. The dataset was checked for null alleles using MICRO-CHECKER 2.2.3 [42]. We checked all loci for potential linkage using the linkage maps provided by Menotti-Raymond et al. [30,43–45]. A test for linkage disequilibrium was performed in Fstat 2.9.3.2 using a log-likelihood ratio G-statistic with Bonferroni corrections [46]. For the latter test, we removed juveniles from the dataset, as these might erroneously suggest the presence of a linkage disequilibrium due to their close relationships [47].

Measures of genetic differentiation have lately been subject to a broad discussion [48,49]. Thus, we estimated genetic differentiation between the genetic clusters with three different estimates, D_{est}, F_{ST} and R_{ST} using GenAlEx 6.501 [50,51]. F_{ST} is still a useful measure if the population split is rather recent (e.g. for comparing differentiation among captive populations), while R_{ST} is more useful if the split is deep and if it is likely that mutations have already contributed to population differentiation (e.g. for comparing wildcats and domestic cats) [52,53]. When we tested D_{est} estimates for correlation with R_{ST} and F_{ST}, we found a strong positive correlation between D_{est} and F_{ST} ($R^2 = 0.47$, $P = 0.004$) but a negative correlation with R_{ST} ($R^2 = 0.41$, $P = 0.01$). Therefore, only the results of F_{ST} and R_{ST} are provided. For the analyses of population differentiation, we excluded all individuals which could not clearly be assigned to a cluster at K = 2 based on qi>0.8 and where thus assigned as hybrids (see above; all excluded individuals are marked with * in Table S5 in File S1). Nested AMOVAs based on F_{ST} and R_{ST} were performed in GenAlEx with the sample categories "captive", "domestic" and "wild" as regions and the six genetic clusters as populations.

To analyse the degree of hybridization between domestic cats and wildcats, we performed an analysis in NewHybrids 1.1 Beta 3 [54]. We used Jeffreys-type priors for pi and theta, a burnin of 100,000 sweeps and run 1,000,000 sweeps afterwards. Moreover, we calculated a Principal Component Analysis (PCA) using the adegenet package [55] for R 3.0.3 [56] replacing missing data by the mean frequency of the corresponding allele. For this analysis, we assigned all individuals to their most likely STRUCTURE clusters. Wild-caught individuals that clustered with domestic cats in STRUCTURE as well as captive individuals that clustered with wildcats in STRUCTURE were assigned to a unique cluster. We then simulated 100 first generation hybrids as well as 100 backcrosses with either domestic cats or wildcats and performed a second PCA to inspect the degree of overlap of captive individuals with the simulated hybrids.

For comparing genetic diversity, we calculated the mean number of alleles, allelic richness and inbreeding coefficients (F_{IS}) for the captive, domestic and wild samples as well as for the genetic clusters in Fstat. GenAlEx was used to determine the expected (H_e) and observed (H_o) heterozygosity for each locus and each population and to test for deviations from Hardy-Weinberg equilibrium (HWE). Furthermore, we tested the number and frequency of private alleles for domestic cats and wildcats (excluding potential hybrids) and analysed the number of alleles shared with captive cats as well as their mean frequencies.

Effective population sizes (N_e) for each the captive population, domestic population and the wild population (excluding the hybrids) were assessed in ONeSAMP [57], which uses a Bayesian approach. The upper and lower bounds of the prior distribution for N_e were 2 and 500, respectively. The captive and wild samples were tested for genetic signatures of population bottlenecks using the Wilcoxon signed rank statistic implemented in BOTTLENECK 1.2.02 [58]. We examined three different mutational models; the infinite alleles model (IAM), the stepwise mutational model (SMM) and the two-phase mutational model (TPM). However, we mainly considered the TPM, as this is the most likely mutation model for microsatellites [59].

The mean relatedness between individuals within the wild and captive population was calculated using Coancestry [60]. We assessed the performance of all seven relatedness estimators by comparing the results with known relationships and chose the estimator based on Wang [61], which showed the smallest deviation from known relationships and the smallest variance. Additionally, we tested the relatedness between the 19 known breeding pairs of the ex situ population. To measure individual F, four estimators are available in Coancestry. We used the TrioML estimator based on Wang [62] as it fitted our data best based on the relatedness data for the whole captive population (see above).

Sequencing. DNA sequences were corrected and aligned by eye as no indels occurred. We excluded ambiguous data from the beginnings and ends of the fragments in the analyses. The final alignment contained 828 bp (positions 12642–13469 based upon GenBank entry FCU20753). Sequences were deposited in GenBank under the accession numbers KM246612-KM246625. The identification of haplotypes was carried out in DnaSP 5 [63]. For calculating a haplotype network we included 198 sequences (Table S4 & S5 in File S1). To verify the assignment of haplotypes to wildcats and domestic cats, we also performed a second analysis including reference samples for European wildcats and domestic cats published by Driscoll et al. [32] (identical haplotypes are given in Table S5 in File S1). The program TCS 1.18 [64] was used to construct a parsimony-based network representation of the mitochondrial haplotypes with a connection limit of 95%.

Results

Genetic structure

Evidence for null alleles was detected only for single markers in single genetic clusters. All pairwise tests for linkage disequilibrium were non-significant (p>0.05). In the STRUCTURE analysis, we found a clear distinction between individuals from the captive (red cluster in Fig. 1a) and wild (green cluster in Fig. 1a) populations at K = 2. All domestic cats and all wild-caught domestic cats were assigned to the captive cluster (Fig. 1a). At K = 3 (optimal K value according to ΔK; Fig. S3 in File S1) the three groups "wild" (green cluster in Fig. 1b), "captive" (blue cluster in Fig. 1b) and "domestic" (red cluster in Fig. 1b) were clearly distinguished. The most likely number of genetic clusters (K) revealed by STRUCTURE was six (Fig. 1c), reflecting geographical groups of breeding lines (Table 1). For example wildcats assigned to cluster 6 (light blue in Fig. 1c) all originated from Scandinavian zoos or had ancestors from these, cluster 3 (yellow in Fig. 1c) comprised mainly ex situ wildcats stemming from one German zoo, and cluster 4 (dark blue in Fig. 1c) mainly included individuals from eastern European zoos. The wild population (green in Fig. 1c) represented a single cluster together with six captive samples. These six samples came mostly from zoos within the range of the wild populations and one of these individuals was a foundling. However not all of them had wildcat mtDNA haplotypes. Another cluster (red in Fig. 1c) included all domestic samples together with the wild-caught domestic cats, some of which were morphologically suspicious.

Genetic differentiation based upon F_{ST} was significant for all pairs of genetic clusters ($p = 0.001$, Table 2), but not for R_{ST}. The wild cluster (cluster 2) showed a higher differentiation to the captive clusters (mean $R_{ST} = 0.032$, mean $F_{ST} = 0.187$) than the domestic cluster (mean $R_{ST} = 0.015$, mean $F_{ST} = 0.163$). For the domestic cluster, the highest differentiation occurred to the wild cluster ($R_{ST} = 0.059$; $F_{ST} = 0.254$). The nested AMOVA based on F_{ST} revealed that most of the genetic variation occurred within individuals (80%). Yet, a statistically significant portion was explained by the difference between the six genetic clusters (12%) and between the captive, domestic and wild samples (7%). In the R_{ST}-based AMOVA, the population level (i.e. the clusters) explained no variation at all, but the three categories (i.e. domestic, wild, captive) explained 7% of the variation. With R_{ST} only 14% of the genetic variation occurred within individuals. The private alleles analysis revealed that captive cats shared 30 private alleles only with domestic cats (mean allele frequency of 21.0%±3.1% SE) and six only with wildcats (8.6%±4.2% SE).

When analysing the data in NewHybrids, only two captive individuals were assigned as purebred wildcats, whereas most captive cats were either assigned as F2 hybrids (62 individuals) or backcrosses with domestic cats (13 individuals). The projected inertia of the PCA was 15.7% for the first axis and 6.9% for the second axis. The first axis separated the domestic cats (negatively loaded, cluster 1 in Fig. 2a) from the wildcats (positively loaded, cluster 2 in Fig. 2a). The captive clusters (3–6) had intermediate positions with negative loadings on the first axis. Captive individuals that clustered with the wildcats in the STRUCTURE analysis (cluster 7 in Fig. 2a) had a strong overlap with the wildcat cluster (except for two individuals). Wild-caught individuals that had been assigned to the domestic/hybrid cluster in STRUCTURE had a strong overlap with the domestic cluster in the PCA as well (cluster 8 in Fig. 2a). When performing the same analysis including the simulated hybrids and back-crosses, there was a substantial overlap of the captive clusters with first generation hybrids (clusters 3–6 in Fig. 2b) and of the wild-caught, morphologically suspicious individuals with second generation backcrosses with domestic cats (cluster 8 in Fig 2b).

We found 14 mtDNA haplotypes in our samples (Fig. 3), two of which (WC1, WC2) were found in 31 in situ wildcats, 15 captive individuals as well as the European wildcat reference samples published by Driscoll et al. [32]. Further ten haplotypes (DC1-DC10) were found in domestic cats (including the domestic cat reference samples), six known or suspected in situ hybrids and 45 zoo samples (as well as GenBank sequences from domestic cats). One haplotype found in four captive individuals (WC3) was identical to a haplotype found in an Iberian wildcat (Fsi257) by Driscoll et al. (2007). The remaining haplotype (WC4*) was found in 36 wild individuals from the Harz population as well as two captive individuals. It was closely related to the most common domestic haplotype DC1 (two substitutions). Altogether 68% (n = 45) of the captive individuals had domestic haplotypes, whereas 32% (n = 21) had haplotypes found in wild populations of *F. s. silvestris* (Figs. 1d, 3). Only one captive cluster (STRUCTURE cluster 6) was little affected by mitochondrial introgression with only a single individual possessing a domestic haplotype. However, all individuals belonging to this cluster were assigned either as F2 hybrids or backcrosses with domestic cats in the NewHybrids analysis.

Genetic diversity in captive and wild population

The effective population sizes of the captive, wild and domestic populations were 70±1.74, 32±1.11 and 53±3.02 (mean ± SE), respectively. The captive clusters had a higher allelic richness and higher expected and observed heterozygosities than the wild cluster (Table 3). There was little variation in genetic diversity among the captive clusters.

The captive population showed significant evidence of a recent bottleneck under both IAM and TPM, but not under SMM. For the domestic population there was only significant evidence of a bottleneck under the IAM, whereas the wild population showed no signs of a bottleneck (Table 4). When the genetic clusters were analysed separately, two captive clusters showed no evidence of a bottleneck, whereas all other clusters showed evidence of a bottleneck under the IAM.

The mean relatedness between dyads was $r = -0.044 \pm 0.007$ (mean ± SE) in the wild population and $r = -0.042 \pm 0.004$ in the captive population. The mean individual inbreeding coefficients were 0.149±0.019 in the wild and 0.139±0.019 in the captive population. In both populations most individuals (61%) had an individual inbreeding coefficient <0.125. The number of individuals with an inbreeding coefficient between 0.125 and 0.25 was 18% and 14% for the wild and captive population respectively and the number of individuals with values above 0.25 was 23% and 25%.

Within the breeding pairs we found a mean r of 0.068±0.059. Most breeding pairs (47%) had a low relatedness $r \leq 0.05$. However, there were five breeding pairs which showed relatedness values above the half-sib level (0.25) with two of them almost reaching the full sib level (0.5).

Discussion

Genetic structure of the European wildcat populations

We found four distinct genetic clusters within the captive population of *F. s. silvestris* (Fig. 1c), which reflected their origin from certain zoos or regional groupings of zoos. The differentiation of the clusters is probably caused by the separation of breeding lines of groups of zoos that often exchanged individuals or zoos that were successful breeders. The existence of isolated breeding lines confirms a lack of knowledge in zoos concerning the stocks of other zoos. Therefore, an exchange of individuals is usually restricted to known partners. When testing the data in STRUCTURE, only six captive individuals were assigned to the wild cluster, some of which were foundlings from the in situ populations that had recently been integrated into the captive

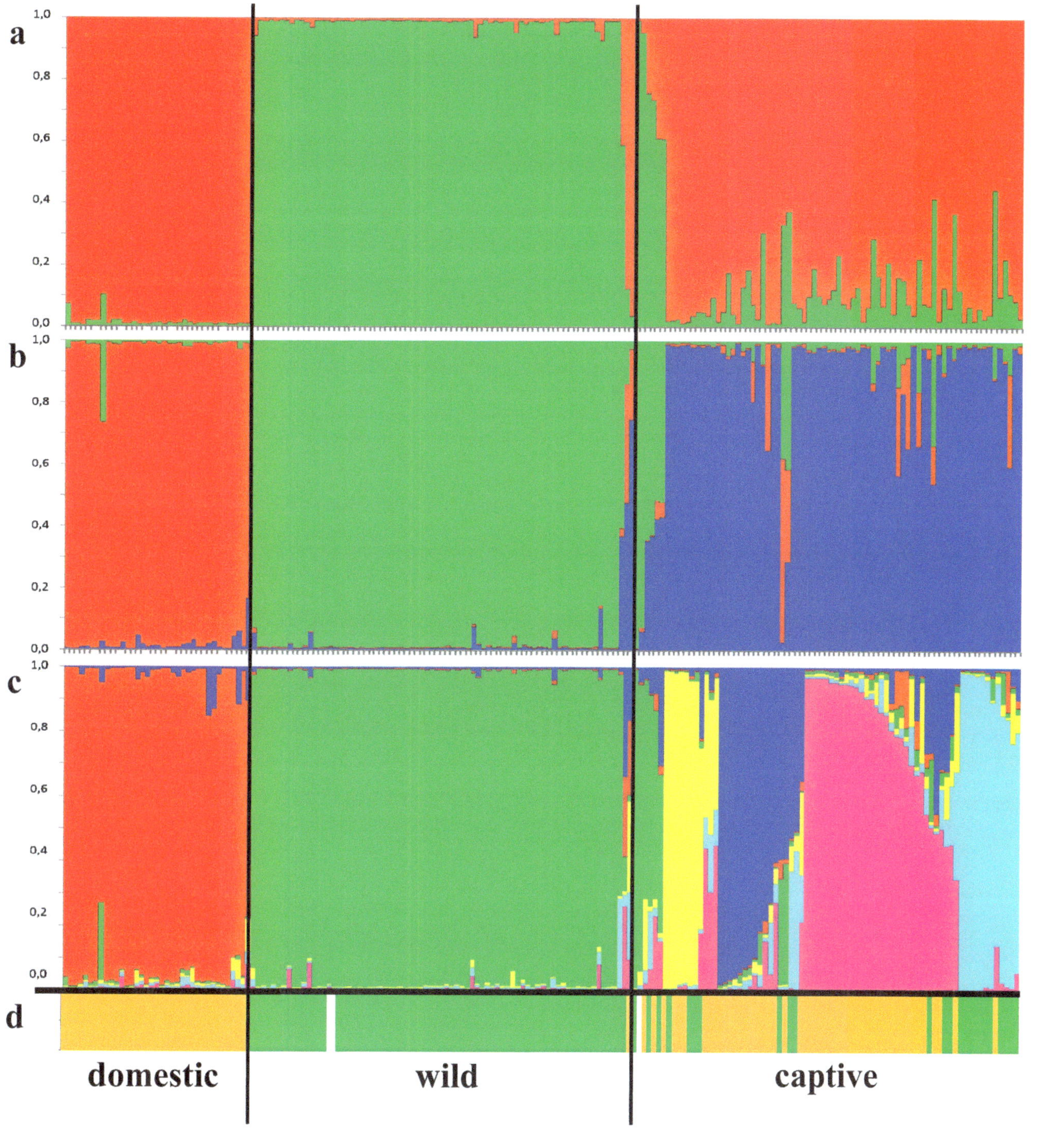

Figure 1. Genetic clusters obtained from the STRUCTURE analysis (n = 189). Each individual is represented by a single vertical line, divided into K colours. The coloured segment shows the individual's estimated proportion of membership to that genetic cluster; (a) assignment for K = 2 (red = domestic/captive, green = wild); (b) assignment for K = 3 (red = domestic, green = wild, blue = captive); (c) assignment for K = 6 (cluster 1: domestic; cluster 2: wild, clusters 3–6: captive). (d) Analogue to the cluster assignments the haplotype affiliation for each individual is indicated by a vertical bar (orange = domestic haplotype, green = wildcat haplotype, white = missing data). For the definition of clusters (geographic origin and ancestry of individuals) see Table 1.

population. However, only four of them carried wild mtDNA haplotypes. At K = 2 we found a clear distinction between wild individuals on the one hand and captive individuals together with domestic cats on the other hand. When analysing the data in NewHybrids, only two captive individuals were assigned to purebred wildcats and in the PCA there was a strong overlap of the captive clusters with simulated hybrids and backcrosses. Furthermore, the level of differentiation based upon R_{ST} values was much lower between captive clusters and domestic cats than between captive clusters and wildcats. This suggests that the

Table 1. STRUCTURE based clusters combined with available data on origin and ancestry of the included individuals.

cluster	grouping	individuals lived in or originated from zoos in
cluster 1	domestic/hybrid	all domestic samples and some wild-caught (morphologically suspicious) individuals
cluster 2	wild	wild-caught individuals from the Harz population and single ex situ samples from some German zoos
cluster 3	single German zoo	almost exclusively individuals originating from one German zoo; all four individuals with the Iberian haplotype (WC3) were assigned to this cluster
cluster 4	east-european zoos	cats originating from eastern European zoos or with ancestors from eastern European zoos
cluster 5	various zoos	various zoos
cluster 6	skand. zoos	cats originating from Scandinavian zoos or with ancestors from Scandinavian zoos

captive population is strongly influenced by gene flow with domestic cats and that the captive cats are not representative for the wild population.

In the wild population we found two distinct mtDNA clades, which is in agreement with other genetic studies [65,66]. One mtDNA lineage (consisting of the haplotypes WC1, WC2 and WC3) also included wildcat samples from Rhineland-Palatinate and the Saarland as well as wildcat reference samples of Driscoll et al. [32] (see also table S5 in File S1). This lineage ("wildcat", Fig. 3) was well separated from the other clades and probably represents the autochthonous mitochondrial lineage of the European wildcat (including the substantially differentiated Iberian clade WC3). The second in situ lineage ("wildcat introgressed", WC4*, consisting exclusively of individuals from the Harz population and two captive individuals) was closely related to a common domestic cat mtDNA haplotype (DC1). We, therefore, suspect that this lineage originated by ancient introgression from domestic cats. As this mtDNA haplotype is already derived from the domestic cat haplotype and no domestic cat shared exactly the same haplotype, we suppose that introgression probably occurred hundreds to thousands of years ago (due to the relatively small genetic distances a molecular clock is not applicable). This scenario is a reasonable explanation as the wild population in the Harz region has had a long history of geographic isolation due to the extinction of most adjacent populations [26,65,66]. During this period of isolation the population has most likely declined due to persecution and some of the remaining wild individuals might well have mated with domestic cats. On the other hand, it remains unknown if this lineage is even more widespread in eastern or south-eastern Europe as samples from this region are scarce. Due to the differences to pure domestic cat haplotypes, we treat this introgressed in situ lineage as a second wildcat lineage, which is also representative for a natural population.

In the wild population we found only few individuals (most of which were roadkills that were known or suspected to be hybrids) who shared haplotypes with domestic cats, confirming that hybridization still occurs in the wild but at a low frequency. Hybridization with domestic cats has been detected in several free-ranging populations of the European wildcat [25,26,32,65–69]. In contrast to the past introgression event in the wild population, the captive population seems to be strongly affected by recent introgression, as two thirds of the individuals shared haplotypes with domestic cats (Fig. 1 and 3). We infer, based on the data presented here, that some founders of the captive population were hybrids. Zoos and wildlife sanctuaries are often confronted with stray kittens, which are suspected to be wildcats and are integrated into the captive population (as was the case with some of our sampled individuals). However, there is a high chance that many of these individuals are of hybrid origin. Humans are more likely to encounter litters or stray kittens of domestic cats, as these are tamer and have a higher probability of living close to human settlements [70]. Hybrids are difficult to identify based on their morphology and even the discrimination of some domestic cats from pure wildcats remains difficult [24,32,67,71]. Without genetic screening there remains a high risk of integrating hybrids or backcrosses into the captive population.

Genetic integrity of the captive population

Our results show that the captive population of the European wildcat represents a hybrid swarm. Only six captive individuals (7.7%) were assigned as wildcats in the STRUCTURE analysis (at K = 2), two of which had domestic cat mtDNA haplotypes, and only two individuals were clearly assigned in the analysis with NewHybrids (one with a domestic cat mtDNA haplotype).

Table 2. F_{ST} values (above diagonal) and R_{ST} values (below diagonal) for genetic differentiation between the genetic clusters.

	Cluster 1 (domestic)	Cluster2 (*wild*)	Cluster 3	Cluster 4	Cluster 5	Cluster 6
Cluster 1	---	0.254*	0.185*	0.150*	0.147*	0.169*
Cluster 2	0.059*	---	0.234*	0.183*	0.147*	0.184*
Cluster 3	0.000	0.033	---	0.122*	0.132*	0.181*
Cluster 4	0.023	0.019	0.100*	---	0.108*	0.099*
Cluster 5	0.036*	0.050*	0.114*	0.161*	---	0.114*
Cluster 6	0.000	0.024	0.068*	0.121*	0.074*	---

The in situ samples were assigned to cluster 2, whereas the domestic samples were assigned to cluster 1. Significant values are marked with * ($P<0.05$).

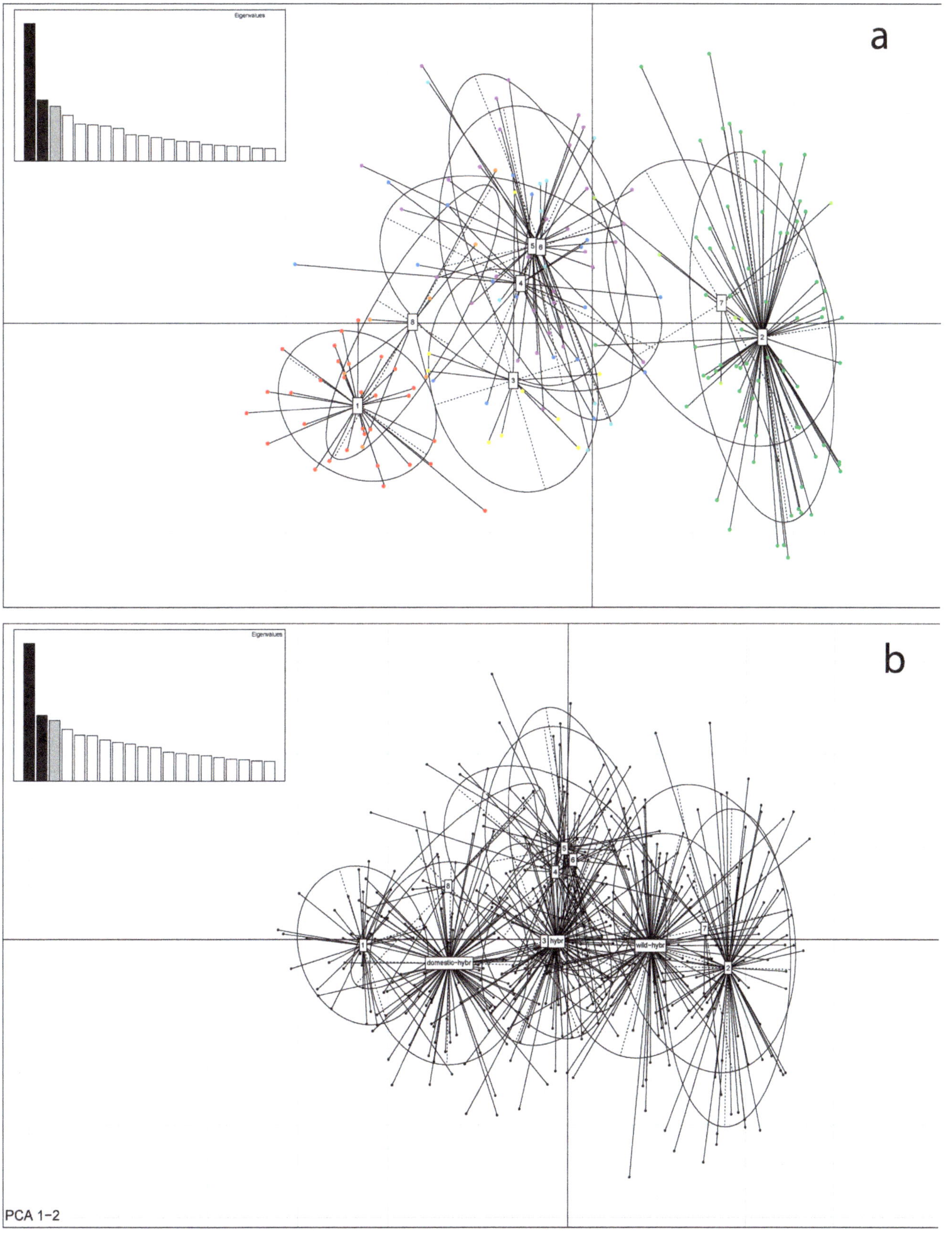

Figure 2. Plot of the first two axes of the Principal component analyses (a) including all obtained genotypes, (b) including simulated first generation hybrids as well as backcrosses with either domestic cats or wildcats. Numbers represent the cluster numbers obtained in STRUCTURE; colours in (a) correspond to the clusters in Figure 1. Captive individuals that were assigned to the wildcat cluster were assigned to a unique group (7) as well as wild-caught individuals that were assigned to the domestic/hybrid cluster (8). The inserted graph shows the distribution of Eigenvalues.

Altogether, 68% of the captive population possess domestic cat mtDNA haplotypes. Therefore, the captive population cannot be recommended for further breeding and reintroduction. As mtDNA is maternally inherited, the presence of domestic cat haplotypes in two thirds of the ex situ population does not allow any conclusions on the total amount of domestic cat nuclear introgression into the captive wildcat gene pool. Furthermore, the amount of introgression from male domestic cats remains unknown, as no Y-chromosome markers have been studied. However, the European wildcat is still widespread and it is possible to obtain new founders from the wild. Particularly the large populations in the Hunsrueck Mountains (Rhineland-Palatinate/ Saarland) currently show no indication of past introgression, but more detailed studies are required for any final conclusions. Individuals of hybrid origin cannot be recommended for breeding. This leaves many zoos with the difficult decision of what to do with the individuals which do not qualify as "pure" wildcats. One might question whether the introgression affecting the captive population occurred recently or if it reflects another trace of an ancient introgression event. Hertwig et al. [65] suggested that "pure" wildcats without any traces of past introgression do not exist in Central Europe. They hypothesized that repeated introgression from domestic cats, followed by a spread and diversification of the wildcat populations might have played a major role in the establishment of European wildcat populations. However, the domestic cat haplotypes found in the captive population were only found in few wild individuals, all of which were declared to be morphologically suspicious and most probably were wild-caught domestic cats. Although we did not sample the captive individuals ourselves, we occasionally came across some individuals, which were also morphologically suspect. This supports our hypothesis that recent hybrids were integrated into the captive population as

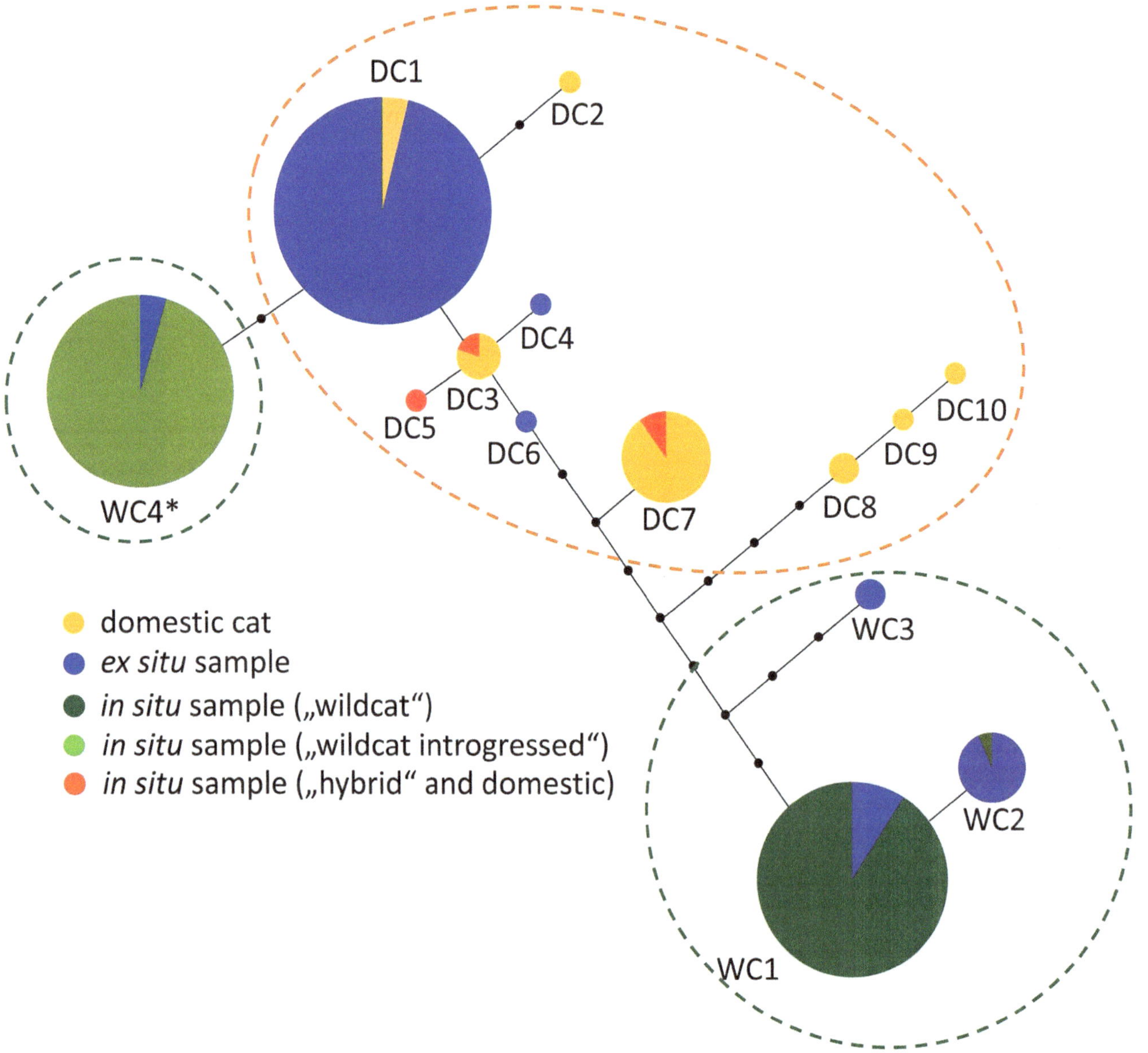

Figure 3. Unrooted parsimony network of mtDNA haplotypes based on an 828 bp sequence of ND5. The sizes of the circles roughly reflect haplotype frequencies in our samples.

Table 3. Genetic diversity and inbreeding in captive, domestic and wild populations and the STRUCTURE based clusters.

population/cluster	n	n_a	A_r	H_o	H_e	F_{IS}
cluster 1 (domestic)	32	8.2	5.9	0.731	0.741	0.005
cluster 2 (wild)	79	4.7	3.3	0.539	0.544	−0.012
cluster 3 (captive)	9	3.8	3.8	0.711	0.603	−0.170
cluster 4 (captive)	17	5.4	4.6	0.600	0.691	0.108
cluster 5 (captive)	33	6.3	4.6	0.709	0.697	−0.033
cluster 6 (captive)	12	4.1	3.9	0.717	0.664	−0.138
mean captive clusters	**71**	**4.9**	**5.7**	**0.684**	**0.649**	**−0.040**
captive	77	7.7	6.9	0.681	0.738	0.085
domestic	32	8.2	8.1	0.741	0.741	0.064
wild	73	4.1	3.7	0.532	0.533	0.021

n = sample size; n_a = number of alleles; A_r = allelic richness based on the lowest samples size (n = 9 for the cluster-based analysis; n = 32 for the three sampled populations); H_o = observed heterozygosity; H_e = expected heterozygosity; F_{IS} = inbreeding coefficient.

founders. Based upon these conclusions we recommend to abstain from any further reintroductions of captive European wildcats from the existing captive stock.

It is very likely that the reintroduced populations of the European wildcat in parts of Germany (e.g. Bavaria, Hesse) represent hybrid swarms as well as they stem from the captive population studied. Removing the reintroduced populations from the wild is not feasible as the administrative effort and costs for such a project would be far too high and it is very unlikely that all individuals could be caught and screened genetically. Moreover, it remains unknown to what extent the reintroduced wildcat populations originated from the captive population or immigrated from wild populations. A genetic survey of the reintroduced populations is therefore needed to assess their status. As the European wildcat currently expands its range (for example in Germany) [22,65,72], further reintroduction programmes are not needed and also not cost-efficient. Nevertheless, regional conservation administrations still aim at reintroduction in areas which are currently not colonized naturally.

Recommendations for a future breeding strategy

Outbreeding is a major threat for the success of *ex situ* conservation programmes [5,7,73]. Some principle decisions on the fate of the captive wildcat population have to be made by the holders and zoo associations. The first question concerns the priority of keeping the European wildcat in zoos. Space is a limited resource in zoos and it might be a wise decision to focus on breeding highly endangered species rather than non-threatened ones. However, zoos also have an educational function, which legitimates the display of native wild fauna. Therefore, if the strategic plan of zoo associations comprises a permanent husbandry of European wildcats, large parts of the current stock need to be replaced with pure wildcats. A future captive population should reflect the genetic structure and diversity found in the wild, particularly in populations which are little affected by hybridization with domestic cats. A sporadic integration of wild individuals into the captive breeding stock has been recommended by several authors [74,75]. In the case of the European wildcat, zoos are in the lucky position that they can obtain individuals from healthy wild populations and, therefore, are able to increase the genetic variability of their breeding stock. Nevertheless, each new individual should be screened genetically before it is integrated into the population.

The current plan to establish a studbook for the European wildcat also requires a coordinated breeding strategy. The most widely accepted management strategy is minimizing kinship [76],

Table 4. One-tailed p-values of the Wilcoxon test for heterozygosity excess under three mutational models for captive population, wild population and the genetic clusters.

population/cluster	IAM	TPM	SMM
captive	0.00049*	0.00244*	0.98389
domestic	0.00488*	0.53906	0.98779
wild	0.04199	0.31250	0.98779
cluster1	0.00684*	0.57715	0.99316
cluster2	0.00684*	0.24609	0.90332
cluster3	0.06543	0.24609	0.57715
cluster4	0.06543	0.34766	0.91992
cluster5	0.00146*	0.27832	0.98389
cluster6	0.00244*	0.00488*	0.11621

IAM: infinite alleles model; TPM: two-phase mutational model; SMM: stepwise mutational model. Significant values are marked with * ($P<0.005$).

which aims at minimizing the overall level of relationship in the population and maximizing the retention of genetic diversity [77,78]. However, for this approach, detailed pedigree data is needed to find optimal breeding pairs [77,79]. Genetic analysis during the establishment of a new captive population can help to minimize the risk of inbreeding and outbreeding. An artificial fragmentation of captive populations into several more or less independent subpopulations has been recommended by several authors to retain a maximum of genetic diversity [74,80]. Although such a fragmentation has been achieved in the current *ex situ* population of *F. s. silvestris* just by chance (due to the missing network of holders), and might have possibly helped to maintain the strong genetic diversity, it is likely that this diversity has mainly been caused by admixture with domestic cats. Increased genetic diversity caused by admixture of several evolutionary lineages has also been reported from invasive non-native populations [35].

Genetic monitoring of *ex situ* conservation projects

The importance of genetic assessments of captive populations has already been recognized decades ago [81]. Molecular genetic studies of captive populations have provided valuable information for the improvement of conservation breeding programmes [7]. These include the detection of hybrids [82], the detection of genetically valuable individuals that had been excluded from breeding [83], the validation of studbook data and detection of errors [11,12], and finally the detection of insufficient representation of genetic diversity found in the wild [10].

The present study shows that genetic analyses provide valuable insight into the status of wild and captive populations. Using two different marker systems (microsatellites, mtDNA) helps to unravel introgression patterns. Future studies should also integrate Y-chromosome markers to detect potential male introgression. The integration of hybrids into ex situ populations can have drastic effects on the genetic composition of captive populations. Hybridisation with closely related species or domesticated congeners represents a severe threat to the conservation of endangered species both in situ and ex situ [35,82,84–87]. Until now, many individuals from the captive population of the European wildcat have been reintroduced to the wild without any test of their genetic integrity. Such reintroductions may strongly deteriorate the genetic integrity of wild populations and promote cryptic extinctions [85]. In some parts of Europe (e.g. Scotland) the wild populations of the wildcat are severely threatened by hybridisation [22]. Therefore, measures to assess and constrict any further admixture of wild and domestic cats are of great importance [22]. Although introgression must have occurred also in nature (perhaps even repeatedly), the European wildcat can still be regarded as a distinct taxon that exists in sympatry with the domestic cat and needs to be managed as a separate unit.

Supporting Information

File S1 Text S1, Permits for sampling. **Text S2**, List of participating institutions for genetic analyses. **Figure S3**, Plots of LnP(D) and delta K (Evanno et al. 2005) for the microsatellite data obtained with STRUCTURE Harvester. **Table S4**, Sampling regime. 1) Most juveniles of sampled mothers and siblings were excluded from sequencing. Their haplotypes were inferred using their maternal relatives (number behind slash). 2) One captive individual was sampled twice and thus the duplicate sample was excluded. 3) The DNA quality of the sample from a preserved captive specimen was not sufficient for sequencing. **Table S5**, Information on the samples used in this study. The table contains all samples which were used in this study, together with their origin, the detected mtDNA haplotypes, the assignments to the STRUCTURE clusters at K = 2, K = 3 and K = 6 as well as the most probable assignment by NewHybrids. At K = 2 all individuals which qualify as hybrids based on qi <0.8 are marked with an *. These individuals were excluded for the analyses of population differentiation. For reasons of data protection the origin of the ex situ samples can only be given as encrypted information. Remarks: 1) DNA quality was too low for genotyping; 2) individual was classified as morphologically suspicious; 3) known hybrid; 4) domestic cat (this individual had a wildcat number as it was included in the wildcat samples as a control by the provider of in situ samples); 5) single samples from wild populations as reference for sequencing, but not suitable for population genetic approaches; 6) zoo located next to wild population in the Harz mountains; 7) foundling within region of the wild population Rhineland Palatinate; 8) the cat was transferred to a new zoo which also took a sample, the duplicate sample was excluded from analysis.

Acknowledgments

We would like to thank Carlos Driscoll for supplying us with unpublished data on wildcat phylogeny and for helpful advice and background information on informative sequences. Malte Götz from the Technische Universität Dresden provided the samples from the wild population. Samples from other wild populations were provided by the Centre for Biodocumentation, Katharina Diergarten, Daniel Hoffmann and Sebastian Kirchhoff. We also would like to thank all zoos (Text S2 in File S1) for contributing samples from the captive population. Domestic cat samples were obtained from Jana Deppermann, Winfried Gerlach, Jessica Hillen, Sabine Kettermann, Katrin Scharafin, Heiko Strunk and Kornelia Ociepka. Katrin Scharafin and Petra Willems helped us in the lab with sequencing mtDNA.

Author Contributions

Conceived and designed the experiments: AH KW. Performed the experiments: KW. Analyzed the data: KW AH. Contributed reagents/materials/analysis tools: KW AH. Wrote the paper: KW AH.

References

1. Butchart S, Walpole M, Collen B, van Strien A, Scharlemann J, et al. (2010) Global biodiversity: indicators of recent declines. Science 328: 1164–1168.
2. Fischer J, Lindenmayer DB (2000) An assessment of the published results of animal relocations. Biol Conserv 96: 1–11.
3. Storfer A (1999) Gene flow and endangered species translocations: a topic revisited. Biol Conserv 87: 173–180.
4. WAZA (2005) Building a future for wildlife - The world zoo and aquarium conservation strategy. In: P. . Dollinger, editor editors. Bern: WAZA.
5. Boakes EH, Wang JL (2005) A simulation study on detecting purging of inbreeding depression in captive populations. Genet Res 86: 139–148.
6. Frankham R, Ballou JD, Briscoe DA (2010) Introduction to conservation genetics. Cambridge: University Press. 618 p.
7. Witzenberger KA, Hochkirch A (2011) Ex situ conservation genetics: a review of molecular studies on the genetic consequences of captive breeding programmes for endangered animal species. Biodivers Conserv 20: 1843–1861.
8. Cuaron AD (2005) Further role of zoos in conservation: Monitoring wildlife use and the dilemma of receiving donated and confiscated animals. Zoo Biol 24: 115–124.
9. Marshall TC, Spalton JA (2000) Simultaneous inbreeding and outbreeding depression in reintroduced Arabian oryx. Anim Conserv 3: 241–248.
10. Tzika AC, Remy C, Gibson R, Milinkovitch MC (2009) Molecular genetic analysis of a captive-breeding program: the vulnerable endemic Jamaican yellow boa. Conserv Genet 10: 69–77.

11. Bowling AT, Zimmermann W, Ryder O, Penado C, Peto S, et al. (2003) Genetic variation in Przewalski's horses, with special focus on the last wild caught mare, 231 Orlitza III. Cytogenet Genome Res 102: 226–234.
12. Witzenberger KA, Hochkirch A (2013) Evaluating ex situ conservation projects: Genetic structure of the captive population of the Arabian sand cat. Mamm Biol 78: 379–382.
13. O'Brien SJ (2006) Animal conservation genetics – an overview with relevance to captive breeding programmes. EAZA News 57: 26–35.
14. Leberg PL, Firmin BD (2008) Role of inbreeding depression and purging in captive breeding and restoration programmes. Mol Ecol 17: 334–343.
15. Hedrick PW, Miller PS, Geffen E, Wayne R (1997) Genetic evaluation of the three captive Mexican wolf lineages. Zoo Biol 16: 47–69.
16. Henry P, Miquelle D, Sugimoto T, McCullough DR, Caccone A, et al. (2009) *In situ* population structure and *ex situ* representation of the endangered Amur tiger. Mol Ecol 18: 3173–3184.
17. Boakes EH, Wang JL, Amos W (2007) An investigation of inbreeding depression and purging in captive pedigreed populations. Heredity 98: 172–182.
18. Kraaijeveld-Smit FJL, Griffiths RA, Moore RD, Beebee TJC (2006) Captive breeding and the fitness of reintroduced species: a test of the responses to predators in a threatened amphibian. J Appl Ecol 43: 360–365.
19. Lau J, Alberts AC, Chemnick LG, Gerber GP, Jones KC, et al. (2009) Isolation and characterization of 23 polymorphic microsatellite loci for a West Indian iguana (*Cyclura pinguis*) from the British Virgin Islands. Mol Ecol Res 9: 1412–1414.
20. Roldán VA, Navarro JL, Gardenal CN, Martella MB (2010) May captive populations of Greater rhea (*Rhea americana*) act as genetic reservoirs in Argentina? Zoo Biol 29: 1–6.
21. Ramirez O, Altet L, Ensenat C, Vilà C, Sanchez A, et al. (2006) Genetic assessment of the Iberian wolf *Canis lupus signatus* captive breeding program. Conserv Genet 7: 861–878.
22. Driscoll C, Nowell K (2010) *Felis silvestris*. 2013 IUCN Red List of Threatened Species. IUCN
23. Grabe H, Worel G (2001) Die Wildkatze. Zurück auf leisen Pfoten. Amberg: Buch und Kunstverlag Oberpfalz.
24. Macdonald DW, Yamaguchi N, Kitchener AC, Daniels M, Kilshwa K, et al. (2010) Reversing cryptic extinction: the history, present, and future ot the Scottish wildcat. In: D. W. . Macdonald and A. . Loveridge, editors. Biology and conservation of wild felids. Oxford, UK: Oxford University Press. pp.471–491.
25. Lecis R, Pierpaoli M, Birò ZS, Szementhy L, Ragni B, et al. (2006) Bayesian analyses of admixture in wild and domestic cats (*Felis silvestris*) using linked microsatellite loci. Mol Ecol 15: 119–131.
26. Pierpaoli M, Birò ZS, Herrmann M, Hupe K, Fernandes M, et al. (2003) Genetic distinction of wildcat (*Felis silvestris*) populations in Europe, and hybridization with domestic cats in Hungary. Mol Ecol 12: 2585–2598.
27. Walsh PS, Metzger DA, Higuchi R (1991) Chelex® 100 as a medium for simple extraction of DNA for PCR-based typing from forensic material. BioTechniques 10: 506–513.
28. Estoup A, Largiader CR, Perrot E, Chourrout D (1996) Rapid one-tube DNA extraction for reliable PCR detection of fish polymorphic markers and transgenes. Mol Mar Biol Biotechnol 5: 295–298.
29. Menotti-Raymond M, O'Brien SJ (1995) Evolutionary conservation of ten microsatellites loci in four species of felidae. The Journal of Heredity 86: 319–322.
30. Menotti-Raymond M, David VA, Lyons LA, Schäffer AA, Tomlin JF, et al. (1999) A genetic linkage map of microsatellites in the domestic cat (*Felis catus*). Genomics 57: 9–23.
31. Taberlet P, Griffin S, Goossens B, Questiau S, Manceau V, et al. (1996) Reliable genotyping of samples with very low DNA quantities using PCR. Nucleic Acids Res 24: 3189–3194.
32. Driscoll CA, Menotti-Raymond M, Roca AL, Hupe K, Johnson WE, et al. (2007) The near eastern origin of cat domestication. Science 317: 519–523.
33. Pritchard JK, Stephens M, Donnelly P (2000) Inference of population structure using multilocus genotype data. Genetics 155: 945–959.
34. Vähä J-P, Primmer C (2006) Efficiency of model-based Bayesian methods for detecting hybrid individuals under different hybridization scenarios and with different numbers of loci. Mol Ecol 15: 63–72.
35. Schulte U, Veith M, Hochkirch A (2012) Rapid genetic assimilation of native wall lizard populations (Podarcis muralis) through extensive hybridization with introduced lineages. Mol Ecol 21: 4313–4326.
36. Randi E (2007) Detecting hybridization between wild species and their domesticated relatives. Mol Ecol 17: 285–293
37. Evanno G, Regnaut S, Goudet J (2005) Detecting the number of clusters of individuals using the software STRUCTURE: a simulation study. Mol Ecol 14: 2611–2620.
38. Campana MG, Hunt HV, Jones H, White J (2010) CorrSieve: software for summarizing and evaluating Structure output. Mol Ecol Res.
39. Hausdorf B, Hennig C (2010) Species delimitation using dominant and codominant multilocus markers. Syst Biol 59: 491–503.
40. Waples R, Gaggiotti O (2006) What is a population? An empirical evaluation of some genetic methods for identifying the number of gene pools and their degree of connectivity. Mol Ecol 15: 1419–1439.
41. Rodrigues ST, Wang J (2012) The effect of close relatives on unsupervised Bayesian clustering algorithms in population genetic structure analysis. Mol Ecol Res 12: 873–884.
42. Van Oosterhout C, Hutchinson WF, Wills DPM, Shipley P (2004) MICRO-CHECKER: software for identifying and correcting genotyping errors in microsatellite data. Mol Ecol 4: 535–538.
43. Menotti-Raymond M, David VA, Chen ZQ, Menotti KA, Sun S, et al. (2003) Second-generation integrated genetic linkage/radiation hybrid maps of the domestic cat (*Felis catus*). J Hered 94: 95–106.
44. Menotti-Raymond M, David VA, Schaffer AA, Tomlin JF, Eizirik E, et al. (2009) An autosomal genetic linkage map of the domestic cat, *Felis silvestris catus*. Genomics 93: 305–313.
45. Menotti-Raymond M, David VA, Agarwala R, Schaffer AA, Stephens R, et al. (2003) Radiation hybrid mapping of 304 novel microsatellites in the domestic cat genome. Cytogenet Genome Res 102: 272–276.
46. Goudet J (1995) FSTAT (Version 1.2): A computer program to calculate F-statistics. J Hered 86: 485–486.
47. Falush D, Stephens M, Pritchard JK (2003) Inference of population structure using multilocus genotype data: Linked loci and correlated allele frequencies. Genetics 164: 1567–1587.
48. Jost L (2008) G_{ST} and its relatives do not measure differentiation. Mol Ecol 17: 4015–4026.
49. Gerlach G, Jueterbock A, Kraemer P, Deppermann J, Harmand P (2010) Calculations of population differentiation based on G_{ST} and D: forget G_{ST} but not all of statistics! Mol Ecol 19: 3845–3852.
50. Peakall R, Smouse PE (2006) GENALEX 6: genetic analysis in Excel. Population genetic software for teaching and research. Mol Ecol Res 6: 288–295.
51. Peakall R, Smouse PE (2012) GenAlEx 6.5: genetic analysis in Excel. Population genetic software for teaching and research—an update. Bioinformatics 28: 2537–2539.
52. Slatkin M (1995) A measure of population subdivision based on microsatellite allele frequencies. Genetics 139: 457–462.
53. Lugon-Moulin N, Brünner H, Wyttenbach A, Hausser J, Goudet J (1999) Hierarchical analyses of genetic differentiation in a hybrid zone of *Sorex araneus* (Insectivora: Soricidae). Mol Ecol 8: 419–431.
54. Anderson EC, Thompson EA (2002) A model-based method for identifying species hybrids using multilocus genetic data. Genetics 160: 1217–1229.
55. Jombart T (2008) adegenet: a R packae fort he multivariate analysis of genetic markers. Bioinformatics 24: 1403–1405.
56. R Development Core Team (2014) R: A Language and environment for statistical computing. Vienna: R Foundation for Statistical Computing.
57. Tallmon DA, Koyuk A, Luikart G, Beaumont MA (2008) ONeSAMP: a program to estimate effective population size using approximate Bayesian computation. Mol Ecol Res 8: 299–301.
58. Cornuet JM, Luikart G (1996) Description and power analysis of two tests for detecting recent population bottlenecks from allele frequency data. Genetis 144: 2001–2014.
59. Piry S, Luikart G, Cornuet J (1999) Computer note. BOTTLENECK: a computer program for detecting recent reductions in the effective size using allele frequency data. J Hered 90: 502–503.
60. Wang J (2010) COANCESTRY: a program for simulating, estimating and analysing relatedness and inbreeding coefficients. Mol Ecol Res 11: 141–145
61. Wang JL (2002) An estimator for pairwise relatedness using molecular markers. Genetics 160: 1203–1215.
62. Wang J (2007) Triadic IBD coefficients and applications to estimating pairwise relatedness. Genet Res 89: 135–153.
63. Librado P, Rozas J (2009) DnaSP v5: a software for comprehensive analysis of DNA polymorphism data. Bioinformatics 25: 1451–1452.
64. Clement M, Posada D, Crandall KA (2000) TCS: a computer program to estimate gene genealogies. Mol Ecol 9: 1657–1659.
65. Hertwig ST, Schweizer M, Stepanow S, Jungnickel A, Böhle U-R, et al. (2009) Regionally high rates of hybridization and introgression in German wildcat populations (*Felis silvestris*, Carnivora, Felidae). J Zool Syst Evol Res 47: 283–297.
66. Eckert I, Suchentrunk F, Markov G, Hartl GB (2010) Genetic diversity and integrity of German wildcat (*Felis silvestris*) populations as revealed by microsatellites, allozymes, and mitochondrial DNA sequences. Mamm Biol 75: 160–174.
67. Randi E, Pierpaoli M, Beaumont MA, Ragni B, Sforzi A (2001) Genetic identification of wild and domestic cats (*Felis silvestris*) and their hybrids using bayesian clustering methods. Mol Biol Evol 18: 1679–1693.
68. O'Brien J, Devillard S, Say L, Vanthomme H, Leger F, et al. (2009) Preserving genetic integrity in a hybridising world: are European Wildcats (*Felis silvestris silvestris*) in eastern France distinct from sympatric feral domestic cats? Biodivers Conserv 18: 2351–2360.
69. Oliveira R, Godinho R, Randi E, Alves PC (2008) Hybridization versus conservation: are domestic cats threatening the genetic integrity of wildcats (*Felis silvestris silvestris*) in Iberian Peninsula? Philosophical Transactions of the Royal Society B-Biological Sciences 363: 2953–2961.
70. Germain E, Benhamou S, Poulle ML (2008) Spatio-temporal sharing between the European wildcat, the domestic cat and their hybrids. J Zool 276: 195–203.
71. Hubbard AL, McOrist S, Jones TW, Boid R, Scott R, et al. (1992) Is survival of European wildcats Felis silvestris in Britain threatened by interbreeding with domestic cats? Biol Conserv 61: 203–208.

72. Steyer K, Simon O, Kraus RHS, Haase P, Nowak C (2013) Hair trapping with valerian-treated lure sticks as a tool for genetic wildcat monitoring in low-density habitats. Eur J Wildl Res 59: 39–46.
73. Frankham R (2010) Challenges and opportunities of genetic approaches to biological conservation. Biol Conserv 143: 1919–1927.
74. Frankham R (2008) Genetic adaptation to captivity in species conservation programs. Mol Ecol 17: 325–333.
75. Williams SE, Hoffman EA (2009) Minimizing genetic adaptation in captive breeding programs: A review. Biol Conserv 142: 2388–2400.
76. Falconer DS, Mackay TFC (1996) Introduction to quantitative genetics. Harlow, England: Longman.
77. Ballou JD, Lacy RC (1995) Identifying genetically important individuals for management of genetic diversity in pedigreed populations. In: J. D. Ballou, M. Gilpin and T. J. Foose, editors. Population Management for Survival & Recovery Analytical Methods and Strategies in Small Population Conservation. New York: Columbia University Press. pp. 76–111.
78. Saura M, Perez-Figueroa A, Fernandez J, Toro MA, Caballero A (2008) Preserving population allele frequencies in *ex situ* conservation programs. Conserv Biol 22: 1277–1287.
79. Caballero A, Toro MA (2000) Interrelations between effective population size and other pedigree tools for the management of conserved populations. Genet Res 75: 331–343.
80. Robert A (2009) Captive breeding genetics and reintroduction success. Biol Conserv 142: 2915–2922.
81. Templeton AR, Davis SK, Read B (1987) Genetic variability in a captive herd of Speke's gazelle (*Gazella spekei*). Zoo Biol 6: 305–313.
82. Luo S-J, Johnson WE, Martenson J, Antunes A, Martelli P, et al. (2008) Subspecies genetic assignments of worldwide captive tigers increase conservation value of captive populations. Curr Biol 18: 592–596.
83. Goncalves da Silva A, Lalonde DR, Quse V, Shoemaker A and Russello MA (2010) Genetic approaches refine *ex situ* Lowland Tapir (*Tapirus terrestris*) conservation. J Hered 101: 581–590.
84. Gottelli D, Sillerozubiri C, Applebaum GD, Roy MS, Girman DJ, et al. (1994) Molecular-genetics of the most endangered canid - the Ethiopian wolf (*Canis simensis*). Mol Ecol 3: 301–312.
85. Rhymer J, Simberloff D (1996) Extinction by hybridization and introgression. Annu Rev Ecol Syst 27: 83–109.
86. Allendorf FW, Leary RF, Spruell P, Wenburg JK (2001) The problems with hybrids: setting conservation guidelines. Trends Ecol Evol 16: 613–622.
87. Ely JJ, Dye B, Frels WI, Fritz J, Gagneux P, et al. (2005) Subspecies composition and founder contribution of the captive US chimpanzee (*Pan troglodytes*) population. Am J Primatol 67: 223–241.

9

Genetic Diversity of and Differentiation among Five Populations of Blunt Snout Bream (*Megalobrama amblycephala*) Revealed by SRAP Markers: Implications for Conservation and Management

Wei Ji[1,2]☙, Gui-Rong Zhang[1,2]☙, Wei Ran[1], Jonathan P. A. Gardner[1,2,3], Kai-Jian Wei[1,2]*, Wei-Min Wang[1], Gui-Wei Zou[4]

1 Key Laboratory of Freshwater Animal Breeding, Ministry of Agriculture, College of Fisheries, Huazhong Agricultural University, Wuhan, P. R. China, **2** Freshwater Aquaculture Collaborative Innovation Centre of Hubei Province, Wuhan, P. R. China, **3** School of Biological Sciences, Victoria University of Wellington, Wellington, New Zealand, **4** Key Laboratory of Freshwater Biodiversity Conservation, Ministry of Agriculture, Yangtze River Fisheries Research Institute, Chinese Academy of Fishery Sciences, Wuhan, P. R. China

Abstract

The blunt snout bream (*Megalobrama amblycephala*) is an important freshwater aquaculture fish throughout China. Because of widespread introductions of this species to many regions, the genetic diversity of wild and natural populations is now threatened. In the present study, SRAP (sequence-related amplified polymorphism) markers were used to assess genetic diversity of blunt snout bream. Three natural populations (Liangzi Lake, Poyang Lake and Yuni Lake, one cultured population (Nanxian) and one genetic strain ('Pujiang No. 1') of blunt snout bream were screened with 88 SRAP primer combinations, of which 13 primer pairs produced stable and reproducible amplification patterns. In total, 172 bands were produced, of which 132 bands were polymorphic. Nei's gene diversity (*h*) and Shannon's information index (*I*) values provided evidence of differences in genetic diversity among the five populations (Poyang Lake>Liangzi Lake>Nanxian> 'Pujiang No. 1'>Yuni Lake). Based on cluster analysis conducted on genetic distance values, the five blunt snout bream populations were divided into three groups, Poyang Lake and Liangzi Lake (natural populations), Nanxian and 'Pujiang No. 1' (cultured population and genetically selected strain), and Yuni Lake (natural population). Significant genetic differentiation was found among the five populations using analysis of molecular variance (AMOVA), with more genetic divergence existing among populations (55.49%), than within populations (44.51%). This molecular marker technique is a simple and efficient method to quantify genetic diversity within and among fish populations, and is employed here to help manage and conserve germplasm variability of blunt snout bream and to support the ongoing selective breeding programme for this fish.

Editor: Hanping Wang, The Ohio State University, United States of America

Funding: This work was supported by the Modern Agriculture Industry Technology System Construction Projects of China entitled "Staple Freshwater Fishes Industry Technology System" (Grant No. CARS-46-05), the National R&D Infrastructure and Facility Development Program of China (Grant No. 2006DKA30470-002-03), and the Major Science and Technology Program for Water Pollution Control and Treatment (Grant No. 2012ZX07202-004-02). The funders had no role in study design, data collection and analysis, decision to publish, or preparation of the manuscript.

Competing Interests: The authors have declared that no competing interests exist.

* Email: kjwei@mail.hzau.edu.cn

☙ These authors contributed equally to this work.

Introduction

Freshwater fishes are very diverse given the relatively small global size of the freshwater habitat, but are among the most endangered organisms globally [1]. Many species are now considered to be under threat in terms of reductions in their natural distributional ranges, and also reductions in numbers and sizes of populations as a consequence of factors such as pollution, replacement by invasive or deliberately introduced species, over-fishing, the aquarium pet trade, river flow modification and habitat change. A large body of literature points to the effects of such threats in terms of both localised and global extinctions of many freshwater fish species on all human-inhabited continents [1–7]. It has been noted that freshwater habitats in general (i.e., as a broad ecosystem type) are particularly susceptible to climate change impacts, both direct and indirect, and that consideration of even the most optimistic climate change scenarios points to the likelihood of *ex-situ* management of many species for their survival [8]. At a lesser scale of threat, the deliberate human-mediated movement of fishes between previously isolated lakes or rivers as a fisheries management activity contributes to the eroding of genetic differences between intra-specific populations, and thereby threatens the genetic integrity and genetic adaptation of local fish populations [9,10]. As a consequence, there is now an increasing awareness of the need to balance the protection and genetic integrity of freshwater fish populations against the need to exploit many such populations as an important source of protein for human populations [11].

The magnitude of the threat faced by freshwater fishes of the world is highlighted by two review papers [1,3]. Both papers point out that freshwater fishes and their associated fisheries are important sources of employment, recreation and culture, and in developing regions they provide a significant contribution to the provision of local human food sources. Both papers also make the point that freshwater systems, because of their landscape positions, are subject to a number of externalities (threats not directly associated with or a function of the water body itself) that exacerbate problems of local fish endemism and non-substitutability. Cooke *et al.* note that many different groups and organizations have a role to play in conserving freshwater fishes and conclude by noting that failure to engage with the public will hinder conservation efforts and outcomes [1], while Dudgeon *et al.* had previously noted that threats to freshwater systems may constitute the greatest conservation challenge yet faced, and may require a new paradigm for biodiversity protection [3].

The Yangtze River is the world's third longest river system, with more than 3000 tributaries and 4000 lakes [2]. The system is notable for its speciose fish fauna and for high levels of endemism [2,12]. However, it is also notable for the modifications that it has undergone and for the threats now faced by its biota ([2,12] and references therein). The blunt snout bream (*Megalobrama amblycephala* Yih, 1955) (Teleostei: Cyprinidae) is an important endemic freshwater fish that has been widely favoured and is now cultured in China as a delicacy. It was originally distributed in the middle Yangtze River and a few accessory lakes, of which Yuni Lake, Liangzi Lake and Poyang Lake are the major sources [13]. Since having been recognized as a new species in the 1950s [14], this fish has become a principal species for freshwater aquaculture in China. This bream has been widely introduced all over the country because of its ease of culture, rapid growth rate, resistance to disease, high catchability and many other advantages [15]. In 2012, the total production of the bream reached 705,821 tonnes [16]. After domestication over a fifty year period, the aquaculture performance of many cultured populations has deteriorated as indicated by growth depression and the early onset of maturation. This deterioration is thought to be due to inbreeding and poor management of broodstocks [17]. Furthermore, as a consequence of environmental change and over-exploitation of resources, natural populations of this species have decreased substantially over the last few decades. As a consequence, the germplasm resources and gene pool diversity of this species in natural inland waters are now threatened by introductions of fish with unknown genetic histories, artificial propagation and poor stocking practices [13]. At the present time, because of these ongoing threats to the integrity of blunt snout bream populations, it is urgently required to clarify genetic diversity and population structure of this bream in order to effectively protect and utilize natural populations and their genetic resources.

Genetic diversity has been estimated in several natural and genetically selected populations of blunt snout bream using morphometrics and isozymes [13], RAPDs [18–20], mtDNA sequencing and RFLPs [21–23], and microsatellite markers [24–26]. For the first time however, we assess genetic diversity in natural populations of blunt snout bream and contrast this with genetic diversity estimates derived from a genetically selected strain and a cultured population. Sequence-related amplified polymorphism (SRAP) is a molecular marker system developed for selective amplification of open reading frames [27]. These polymorphisms result mainly from various promoters, introns and spacers among different species and individuals. SRAP is a highly reproducible and highly informative technique for assessing genetic diversity in comparison with other PCR-based techniques [28,29]. SRAP markers have been successfully applied to the assessment of genetic diversity, strain identification and linkage map construction in a number of species, principally plants of commercial value [30–34]. More recently, SRAPs have also been applied to several aquatic animals [35–39] and have been shown to be highly informative and reliable markers for the evaluation of genetic diversity and population differentiation.

In this study, SRAP markers are used to investigate population genetic diversity and differentiation among three natural populations, a genetically selected strain and a cultured population of blunt snout bream in China. Our aim is to better understand the genetic diversity and integrity of natural populations based upon geography and the genetic affinities of selected strain and cultured populations based upon history. This work contributes to the critical need for germplasm conservation while at the same time supporting the ongoing selective breeding and improvement of blunt snout bream.

Materials and Methods

Ethics Statement

No specific permits were required for the field studies described here. The sampling lakes and ponds are not privately-owned or protected in any way, and the field sampling activities did not involve endangered or protected species.

Sample collection and DNA extraction

In total, 234 individuals from five blunt snout bream populations were included in this study. Samples were collected from three natural populations (Liangzi Lake, Poyang Lake and Yuni Lake), from one selected strain ('Pujiang No. 1' from Shanghai), and from one cultured population (Nanxian County in Hunan Province). Details of the sampled populations are provided in Table 1 (we use the term 'population' for ease of use, although we recognise that each group tested here may not be a biologically defined population) and the sampling sites are showed in Figure 1. As far as can be ascertained, the Liangzi Lake, Poyang Lake and Yuni Lake populations represent a good geographic spread of natural populations from the species' original range of distribution and have been least impacted by the translocation of fish (between these sites, from other regions, from selected strains, etc). Because of the long history of human-mediated movement of fish the identification of additional natural populations is now increasing difficult, so we have focussed on what we believe to be the best source of natural (wild) fish. For sampling, all individuals were captured from lakes or ponds with a haul seine and a piece of pectoral fin (~ 0.5 cm^2) was clipped from each individual without anesthesia. After sampling, the individuals of the selected and cultured populations were returned to their corresponding ponds, whereas the individuals of the three natural populations were transported to the Ezhou fish breeding base of Huazhong Agricultural University (HZAU) for further breeding programme. The clipped fin tissues were preserved in 95% ethanol. Total DNA was extracted from fin tissues using the phenol-chloroform method [40]. DNA concentration and quality were assessed by electrophoresis on a 0.8% agarose gel. The animal research oversight committee (AROC) of HZAU had knowledge of the fish sampling plans prior to their approval of the present animal research protocol. This study was approved by the Institutional Animal Care and Use Committees (IACUC) of HZAU.

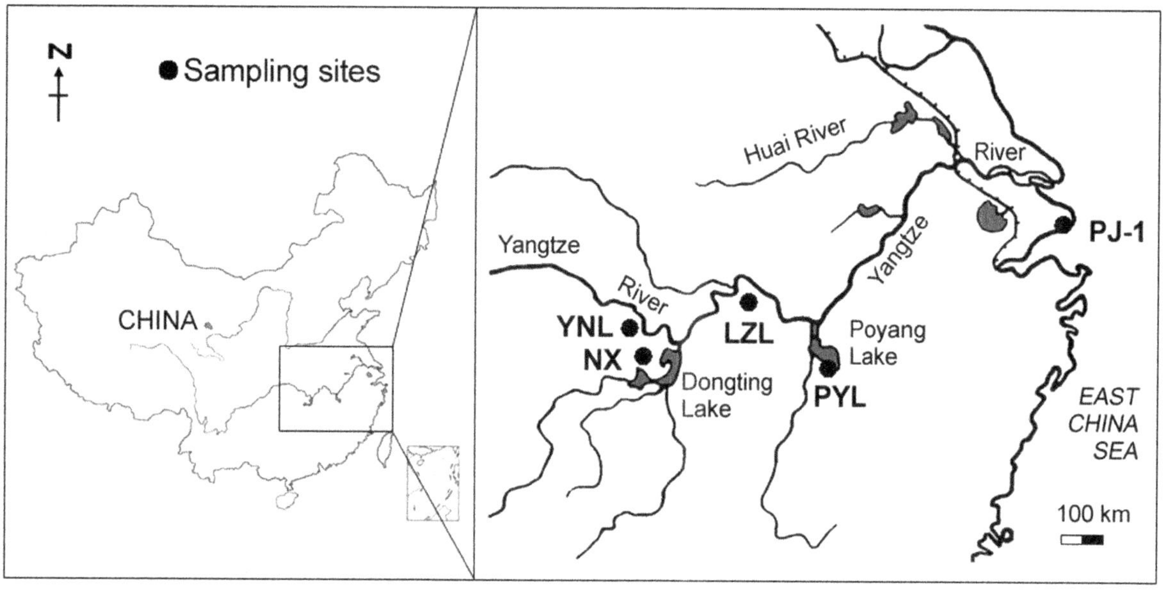

Figure 1. Sampling sites of *Megalobrama amblycephala* in China. Sample codes are given in Table 1.

SRAP reactions

A set of 19 SRAP primers (eight forward and eleven reverse, see Table 2) was obtained from Li and Quiros [27], giving a total of 88 primer pair combinations to be used to search for polymorphisms among all individuals. Thirteen polymorphic primer pairs were used for PCR amplification of fish from all five populations because, after preliminary testing, only these primers provided informative and repeatable results.

All SRAP PCR reactions were carried out in a final volume of 15 µL containing 1.5 µL 10×PCR buffer, 2.0 mM Mg^{2+}, 0.25 mM dNTP, 0.6 U *Taq* DNA polymerase (Fermentas), 0.5 µM each of primers, and 20 ng template DNA. All PCRs were conducted in a PTC-200 Thermocycler under the following conditions: an initial denaturing step was performed at 94°C for 5 min, followed by five cycles of 1 min at 94°C, 1 min at 35°C, 1 min at 72°C, then 35 cycles at 94°C for 1 min, 50°C for 1 min, 72°C for 1 min, followed by a final extension step at 72°C for 10 min. PCR products were separated on a 6% non-denaturing polyacrylamide gel via electrophoresis and silver staining was used to visualize the results.

Data analysis

Using a 100 bp DNA marker ladder as a size standard, SRAP size-dependent fragments were scored as absent (0) or present (1) to obtain a binary matrix. Only clear, strong and reproducible bands in the size range of 60 to 1124 bp were scored. POPGENE 1.32 was used to calculate number of polymorphic bands (N), percent of polymorphic bands (P), Nei's gene diversity (h), Shannon's information index (I), Nei's unbiased genetic distance (D) and genetic similarity (S) estimates [41]. Based on Nei's D, a NJ tree with bootstrap support values (10,000 replications) was constructed using POPTREE2 [42] to illustrate the relationship among populations. Genetic population structure was quantified by analysis of molecular variance (AMOVA) using ARLEQUIN 3.11 [43]. The pairwise fixation index (F_{ST}) was used to evaluate the level of genetic differentiation among populations. Sequential

Table 1. Sampling locations and sample sizes of natural, cultured and genetically selected populations of *Megalobrama amblycephala*.

Population	Code	Population type	Sampling location	Coordinates (lat., long.)	Sampling date	Sample size
Liangzi Lake	LZL	natural	Liangzi Lake, Hubei Province	30°16′N, 114°31′E	Jan 2009	47
Poyang Lake	PYL	natural	Poyang Lake, Jiangxi Province	28°36′N, 116°11′E	Jan 2009	48
Yuni Lake	YNL	natural	Yuni Lake, Hubei Province	29°50′N, 112°08′E	Jan 2009	48
'Pujiang No. 1'	PJ-1	genetically selected strain (F_7)	Songjiang fish breeding station, Shanghai	31°02′N, 121°05′E	July 2010	48
Nanxian	NX	cultured	Shuanghu fish farm, Nanxian county, Hunan Province	29°20′N, 112°18′E	July 2010	43

Table 2. Primer sequences of SRAP markers used in this study.

Forward Primer	Sequence (5′-3′)	Reverse Primer	Sequence (5′-3′)
Me1	TGAGTCCAAACCGGATA	Em1	GACTGCGTACGAATTAAT
Me2	TGAGTCCAAACCGGAGC	Em2	GACTGCGTACGAATTTGC
Me3	TGAGTCCAAACCGGAAT	Em3	GACTGCGTACGAATTGAC
Me4	TGAGTCCAAACCGGACC	Em4	GACTGCGTACGAATTTGA
Me5	TGAGTCCAAACCGGAAG	Em5	GACTGCGTACGAATTAAC
Me6	TGAGTCCAAACCGGTAA	Em6	GACTGCGTACGAATTGCA
Me7	TGAGTCCAAACCGGTCC	Em7	GACTGCGTACGAATTCAA
Me8	TGAGTCCAAACCGGTGC	Em8	GACTGCGTACGAATTCTC
		Em9	GACTGCGTACGAATTCGA
		Em10	GACTGCGTACGAATTCAG
		Em11	GACTGCGTACGAATTCCA

Bonferroni corrections [44] were used to correct for multiple testing.

To further evaluate the extent of genetic differentiation among the five populations we carried out a discriminant function analysis (DFA) to determine the population assignment success for each individual fish based on its genetic diversity. The SRAP markers had 140 variable bands, a number too large for the DFA to work with, so we summarised the genetic variation using principal components analysis (PCA). The first 40 eigen vectors, which explained 82% of the variation in the SRAP data set, were used in the DFA. The DFA was based on original group size (i.e., the number of fish per population that was actually sampled). The DFA and PCA were carried out using STATISTICA v7 software (StatSoft Ltd).

Results

SRAP analysis

Eighty-eight primer combinations were tested and of these, 13 primer pairs (Me1/Em3, Me1/Em4, Me1/Em6, Me1/Em7, Me4/Em1, Me4/Em2, Me4/Em8, Me5/Em1, Me5/Em2, Me5/Em4, Me5/Em7, Me8/Em8, and Me8/Em9) showed clear, reproducible and polymorphic banding patterns. The number of amplified bands for each primer pair ranged from 8 to 21 (average 13). In total, 172 bands were detected, of which 132 bands showed polymorphism, giving a polymorphic value of 76.7%.

Genetic diversity

For fish from the five populations (mean ± SD sample size of 46.8±2.17), the number of polymorphic bands (N) and the percentage of polymorphic bands (P) were highest in the PYL population ($N = 100$, $P = 58.14\%$), and lowest in the YNL population ($N = 50$, $P = 29.07\%$). Nei's gene diversity (h) across the five populations ranged from 0.090 (YNL) to 0.185 (PYL), and Shannon's information index (I) ranged from 0.137 (YNL) to 0.280 (PYL) (Table 3).

Genetic differentiation among populations

The SRAP markers revealed a high level of genetic differentiation among the five blunt snout bream populations. Pairwise F_{ST} values ranged from 0.351 between the geographically adjacent natural populations of Liangzi Lake (LZL) and Poyang Lake (PYL), to 0.685 between the Yuni Lake (YNL) population and the genetically selected strain of 'Pujiang No. 1' (PJ-1). The differences in genotype frequencies between all pairs of populations were highly significant ($P<0.001$) (Table 4).

Analysis of molecular variance (AMOVA) of hierarchical gene diversity indicated that 44.51% of the genetic variation was explained by within-population variation, whereas 55.49% of the variation was explained by among-population variation, with a significant fixation index ($F_{ST} = 0.555$, $P<0.001$) (Table 5).

Nei's unbiased genetic distances (D) among all populations ranged from 0.087 (LZL-PYL) to 0.270 (YNL-NX), and Nei's unbiased genetic similarities (S) ranged from 0.763 (YNL-NX) to 0.917 (LZL-PYL) (Table 6).

The NJ dendrogram constructed using Nei's unbiased genetic distance (Figure 2) showed that the populations were grouped into three clusters. The LZL (Liangzi Lake) population clustered with its geographically close population of Poyang Lake (PYL). The genetically selected population of 'Pujiang No. 1' (PJ-1) and the cultured population of Nanxian (NX) clustered together but separated from LZL and PYL samples. The YNL (Yuni Lake) population formed a distinct branch, separated from all other populations. Bootstrap support values for these clusters were all very high.

Discriminant function analysis (DFA) of the first 40 eigen values from the PCA resulted in 100% correct assignment. That is, all fishes from all populations were correctly assigned to their populations of sampling. The DFA result is consistent with the F_{ST}, AMOVA and NJ analysis, all of which reveal a high level of genetic differentiation among fish from the five populations.

Discussion

China's freshwater fish diversity is high at 920 species in 302 genera [12]. Among the recorded species, most (73%) are Cypriniformes, and across the country's 9 freshwater regions, river discharge contributes most to explaining variation in species richness [12]. Within the Yangtze River basin itself, 361 species and subspecies have been recorded, of which 177 are endemic [2]. Twenty-five fish species have been identified as being endangered [45]. It has been suggested that hydrological alterations pose the greatest threat to the basin's fish diversity and that the most immediate restoration need is for reconnection between the Yangtze River and its lakes [2]. In concluding, Fu *et al.* noted that the cluster of lakes within the middle Yangtze River region requires protection for the preservation of migratory fishes and

Table 3. Genetic diversity results for five populations of *Megalobrama amblycephala* based on SRAP markers.

	Population				
Parameter	**LZL**	**PYL**	**YNL**	**PJ-1**	**NX**
Number of polymorphic bands (*N*)	88	100	50	56	64
Percent of polymorphic bands (*P*, %)	51.16	58.14	29.07	32.56	37.21
Nei's gene diversity (*h*) (mean ± SD)	0.152±0.183	0.185±0.195	0.090±0.166	0.107±0.176	0.126±0.183
Shannon's information index (*I*) (mean ± SD)	0.235±0.265	0.280±0.280	0.137±0.240	0.162±0.256	0.191±0.266

that there is a need to identify areas of high fish biodiversity to mitigate loss of fish biodiversity in the Yangtze River basin [2]. In our study, we focus on the genetic diversity of the blunt snout bream, a cyprinid, in the cluster of lakes in the middle Yangtze River region. The focus of our research is primarily around conservation, but we recognize that there may be tension between conservation planning and fishing/aquaculture activity, and as such our ultimate aim is to contribute new knowledge to both conservation and fisheries management.

Genetic diversity

All four indices (number of polymorphic bands *N*, percent of polymorphic bands *P*, Nei's gene diversity *h*, Shannon's information index *I*) showed the same order from largest to smallest for the five populations. Both *h* and *I* indicated that the highest genetic diversity level was observed in the PYL population, followed by the LZL population, and the lowest was in the YNL population, all three being natural populations of *Megalobrama amblycephala*. As suggested by Li *et al.* [13], the genetic diversity level of fish within Niushan Lake (a subsidiary lake of Liangzi Lake) is a little higher than that within Yuni Lake. Our SRAP-based results are however, not in agreement with previous estimates of genetic diversity based on RAPD analysis that had shown that the genetic similarity of blunt snout bream from Yuni Lake (0.962) is very similar to that of fish from Liangzi Lake (0.959) [18]. However, mtDNA control region sequence analysis by Li [21] revealed a low level of population diversity within Yuni Lake compared to Liangzi Lake and Poyang Lake, with the haplotype diversity and nucleotide diversity index values detected in fish from Yuni Lake being only half of those for fish from Liangzi and Poyang Lakes. These mtDNA-based findings are similar to the results reported by us based on SRAP markers which represent the most up to date estimates of genetic diversity. The highest SRAP-based variation was observed for fish from PYL, which is the lake least affected by human activity (e.g., transfer of fish). The values reported for fish from this lake therefore give some indication of the natural level of genetic variability that may be expected to occur in wild fish, and also a measure of the impact (reduced genetic variability) associated with human-mediated activities such as breeding programmes, inbreeding, and transfer of fish from one lake to another.

For the cultured (Nanxian) and selectively bred ('Pujiang No. 1') populations, the levels of genetic diversity were much lower than within the native PYL and LZL populations. These results are in agreement with investigations on largemouth bass (*Micropterus salmoides*) and large yellow croaker (*Pseudosciaena crocea*) [46,47], and are consistent with the known impacts of hatchery-based breeding programmes in general ([9,10] and references therein). In summary, as a result of artificial selection, the chance of the effective number of individuals contributing to the population declines, with an associated decline in population genetic diversity and an increase in inbreeding depression.

An unexpected outcome of our research is that the genetic diversity level of fish from the cultivated and breeding populations was found to be greater than that of wild blunt snout bream from one native lake population (Yuni Lake, YNL). This surprising result may be due to the degradation of germplasm resources of YNL over a reasonably short period of time. Previously, Li *et al.* [13] and Zhang [18] reported that the genetic diversity of blunt snout bream from Yuni Lake was similar to that of fish from Liangzi Lake at the beginning of the early 21st century. Subsequently, Li [21] reported that the genetic diversity of fish from Yuni Lake was nearly 50% of the level of fish from Liangzi Lake, a sharp decline of genetic diversity of fish from Yuni Lake. Whilst different marker types (e.g., allozymes, mtDNA sequencing) have been used by the different studies, each study is internally consistent and points to a decrease in genetic diversity of fish from Yuni Lake over a short period of time. Our study has revealed that the genetic diversity of blunt snout bream from YNL is lower than that of a cultured population and a selectively bred population. We

Table 4. Pairwise fixation index values (F_{ST}) between pairs of blunt snout bream (*Megalobrama amblycephala*) populations.

Population	**LZL**	**PYL**	**YNL**	**PJ-1**	**NX**
LZL	—				
PYL	0.351***	—			
YNL	0.533***	0.573***	—		
PJ-1	0.573***	0.572***	0.685***	—	
NX	0.544***	0.526***	0.659***	0.434***	—

Note:
***Significant at $\alpha = 0.001$ after sequential Bonferroni correction for multiple testing.

Table 5. AMOVA analysis of genetic variation in five populations of blunt snout bream based on SRAP markers.

Source of variation	df	Sum of squares	Variance components	Percentage of variation (%)	F_{ST}-value
Among-populations	4	2376.04	12.48	55.49	0.555***
Within-populations	229	2293.57	10.02	44.51	
Total	233	4669.61	22.50		

Note:
***Significant at $\alpha = 0.001$ level.

suggest that the results obtained for fish from YNL are attributable to the rapid intensification of artificial culture, propagation, and over-fishing of blunt snout bream in Yuni Lake during the past 10 years, which have resulted in much reduced genetic diversity of bream germplasm at this location. Whilst further research is needed to determine if this is actually the case, our research highlights the potential negative impact of culture, selective breeding programmes and multiple transfer of fishes on native genetic diversity of this species of fish in lakes of the middle Yangtze River region.

Population differentiation

F_{ST} is one of the most widely used measures for evaluating the genetic differentiation between/among populations, which can provide important insights into the evolutionary processes that cause the genetic differentiation among populations [48]. Whilst the magnitude of F_{ST} values is to some extent species- or even group-specific, as a general rule F_{ST} values of 0–0.05 represent little differentiation, values of 0.05–0.25 indicate moderate differentiation, and values higher than 0.25 indicate very great differentiation among populations [49]. In the present study, all values of F_{ST} between the five populations were >0.25, indicative of a high level of genetic differentiation among these five populations. The highest F_{ST} value detected (0.685) was between PJ-1 and YNL, suggesting that pronounced genetic differentiation now exists between the selectively bred strain (PJ-1) and the natural YNL population. These findings were supported by our discriminant function analysis (DFA) which was able to assign each and every fish to its population of origin, are comparable to the results reported by [50], and indicate how rapidly genetic differentiation can be achieved between wild breeding and selectively bred fishes (the PJ-1 fish are from an F_7 strain). The five populations of blunt snout bream clustered into three groups, as revealed by NJ analysis. Two of the three wild populations, PYL and the geographically close population of LZL, clustered together, while the third wild population (YNL) was highly differentiated from the other two wild populations (in the NJ analysis YNL is identified as a distinct branch). In addition, we tried to construct a NJ tree based on pairwise F_{ST} between populations at the same time. The NJ tree based on pairwise F_{ST} was consistent with the results above (data not shown). These results confirm earlier observations that obvious genetic divergence occurred between fish from YNL and LZL or PYL, and no divergence occurred between LZL and PYL [21]. Historically, PYL, LZL and YNL (that is, Poyang Lake, Liangzi Lake, and Yuni Lake, respectively) were linked to the Yangtze River. Yuni Lake was separated from the Yangtze River about forty years ago due to human activities [13], which isolated its fish and has prevented genetic exchange among populations that were previously connected. This isolation has resulted in a build-up of genetic differentiation between fish from Yuni Lake and the two other lakes (i.e., there has been a brief but substantial period of divergence between the fish populations). LZL and PYL were still connected via the Yangtze River and closer to each other geographically, so there was still gene flow between these two lakes, therefore a closer genetic relationship was found between the populations from these two lakes. Whilst fish from these two lakes exhibit the highest levels of pairwise similarities, they are still sufficiently genetically differentiated to be distinct. These findings support the contention of Fu *et al.* [2] that lakes within the central Yangtze River region are in need of further protection and that connections between them and the river itself should be re-established. Finally, the genetically selected population of PJ-1 and the cultured population of NX clustered together to form a third, separate group that lies between the two previously described wild lake groups. The PJ-1 fish were produced through long-term artificial selective breeding since 1986, based on original wild population of blunt snout bream from Yuni Lake [26]. This selectively-bred line now has a growth rate that is increased by 30% compared to wild fish [17,51]. The distinct genetic differences among the three groups as revealed by NJ analysis indicate that genetic divergence has occurred between the selective

Table 6. Nei's unbiased genetic distance (below diagonal) and genetic similarity (above diagonal) among five populations of blunt snout bream.

Population	LZL	PYL	YNL	PJ-1	NX
LZL	——	0.917	0.831	0.837	0.844
PYL	0.087	——	0.802	0.805	0.815
YNL	0.186	0.221	——	0.776	0.763
PJ-1	0.178	0.217	0.254	——	0.916
NX	0.169	0.204	0.270	0.088	——

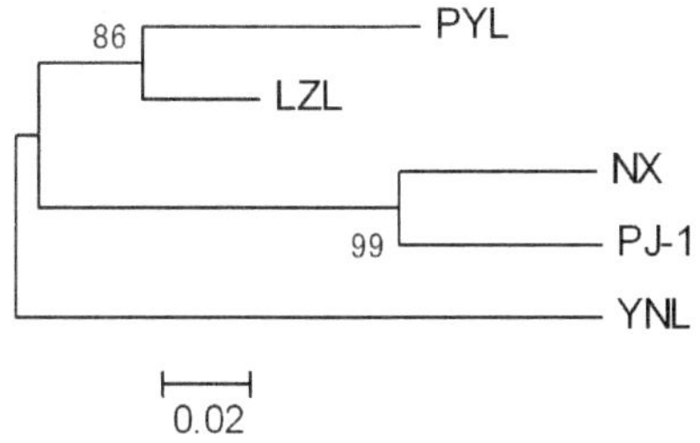

Figure 2. NJ dendrogram of relationships among blunt snout bream populations based on Nei's unbiased genetic distance. Percent bootstrap support values for 10,000 replications are indicated at nodes. See Table 1 for population abbreviation definitions.

breeding population (PJ-1), the cultured population (NX) and the wild populations (LZL, PYL and YNL). Similar results have also been reported for grass carp and mandarin fish [35,52]. This finding illustrates that restocking of lakes by fish from selectively bred programmes should be carried out with great care and a complete understanding of the genetic diversity that may exist between fish strain and recipient lake, if extensive interbreeding between donor strain and wild lake fish is expected.

In conclusion, the SRAP system has been successfully employed to assess population genetic diversity among domesticated and wild genotypes of blunt snout bream from China. Analysis revealed that cultivated, genetically selected and wild bream germplasm are highly differentiated and that wild lake populations have high levels of genetic diversity. Comparison of the genetic diversity of wild fish to that of cultured and selectively bred fish illustrates the need for greater care managing this important resource so that the aquaculture programme to support this species is not disadvantaged in the future and so that wild populations of fish are protected for both conservation and fisheries purposes.

Author Contributions

Conceived and designed the experiments: KJW GRZ WJ. Performed the experiments: GRZ WJ WR. Analyzed the data: KJW WR JPAG. Contributed reagents/materials/analysis tools: WMW GWZ. Wrote the paper: WJ GRZ KJW JPAG.

References

1. Cooke SJ, Lapointe NWR, Martins EG, Thiem JD, Raby GD, et al. (2013) Failure to engage the public in issues related to inland fishes and fisheries: strategies for building public and political will to promote meaningful conservation. J Fish Biol 83: 997–1018.
2. Fu C, Wu J, Chen J, Wu Q, Lei G (2003) Freshwater fish biodiversity in the Yangtze River basin of China: patterns, threats and conservation. Biodivers and Conserv 12: 1649–1685.
3. Dudgeon D, Arthington AH, Gessner MO, Kawabata Z-I, Knowler DJ, et al. (2006) Freshwater biodiversity: importance, threats, status and conservation challenges. Biol Rev 81: 163–182.
4. Freitas CEC, Siqueira-Souza FK, Florentino AC, Hurd LE (2013) The importance of spatial scales to analysis of fish diversity in Amazonian floodplain lakes and implications for conservation. Ecol Freshw Fish DOI: 10.1111/eff.12099.
5. Lintermans M (2013) Recovering Australia's threatened freshwater fish. Mar Freshw Res, Special Issue 64 (9): 1–133.
6. Reid GM, MacBeath TC, Csatádi K (2013) Global challenges in freshwater-fish conservation related to public aquariums and the aquarium industry. International Zoo Yearbook 47: 6–45.
7. Reis RE (2013) Conserving the freshwater fishes of South America. International Zoo Yearbook 47: 65–70.
8. Pearce-Kelly P, Khela S, Ferri C, Field D (2013) Climate-change impact considerations for freshwater-fish conservation, with special reference to the aquarium and zoo community. International Zoo Yearbook 47: 81–92.
9. Ariki H, Schmid C (2010) Is hatchery stocking a help or harm? Evidence, limitations and future directions in ecological and genetic surveys. Aquaculture 308: S2–S11.
10. Hold N, Murray LG, Kaiser MJ, Hinz H, Beaumont AR, et al. (2013) Potential effects of stock enhancement with hatchery-reared seed on genetic diversity and effective population size. Can J Fish Aquat Sci 70: 330–338.
11. Bondad-Reantaso MG (2007) Assessment of freshwater fish seed resources for sustainable aquaculture. FAO Fisheries Technical Paper 501: 1–628.
12. Kang B, Deng J, Wu Y, Chen L, Zhang J, et al. (2013) Mapping China's freshwater fishes: diversity and biogeography. Fish and Fisheries DOI: 10.1111/faf.12011.
13. Li SF, Cai WQ, Zhou BY (1991) Morphology and biochemical genetic variations among populations of blunt snout bream (*Megalobrama amblycephala*). J Fish China 15: 204–211.
14. Yih PL (1955) Notes on *Megalobrama amblycephala*, sp. nov., a distinct species from *M. terminalis* (Richardson). Acta Hydrobiol Sin 2: 115–122.
15. Ko HW (1975) An excellent fresh-water food fish, *Megalobrama amblycephala*, and its propagating and culturing. Acta Hydrobiol Sin 5: 293–312.
16. Fisheries Bureau of the Agriculture Ministry of China (2013) China fishery statistical yearbook. Beijing: Chinese Agricultural Press. 31 p.
17. Li SF, Cai WQ (2003) Genetic improvement of the herbivorous blunt snout bream (*Megalobrama amblycephala*). Naga, WorldFish Center Quarterly 26(1): 20–23.
18. Zhang DC (2001) Study on genetic diversity of bluntnose black bream from Yunihu and Liangzihu lakes. J Chin Three Gorges Univ, Nat Sci 23: 282–284.
19. Zou SM, Li SF, Cai WQ (2005) SCAR transformation of a RAPD marker in blunt snout bream "Pujiang No. 1". J Fish China 29: 296–299.
20. Li SF, Yang HY, Zou SM (2005) Effects on genetic structure from quick inbreeding and estimation of inbreeding response in cultured populations of *Megalobrama amblycephala*. J Fish China 29: 161–165.
21. Li HH (2008) mtDNA sequence variation and genetic structure of the *Megalobrama amblycephala* from Yuni Lake, Liangzi Lake and Poyang Lake. Freshw Fisheries 38(4): 63–65.
22. Tang SJ, Li SF, Cai WQ (2008) Analysis of mitochondrial DNA in blunt snout bream (*Megalobrama amblycephala*) populations with different ploidy levels. J Fish Sci China 15: 222–229.
23. Bian CY, Dong S, Tan SZ (2007) PCR-RFLP analysis of mtDNA D-loop region of three populations of bluntnose blackbream *Megalobrama amblycephala*. J Dalian Fish Univ 22: 175–179.
24. Li SW, Chang YM, Liang LQ, Sun XW (2006) Rapid isolation of microsatellites from genome of Bluntnose black bream (*Megalobrama amblycephala*). J Fish Sci China 13: 187–192.
25. Tang SJ, Li SF (2007) Preliminary analysis of genetic variation of blunt snout bream (*Megalobrama amblycephala*) populations with different ploidy levels. J Shanghai Fish Univ 16: 97–102.
26. Tang SJ, Li SF, Cai WQ (2009) Development of microsatellite markers for blunt snout bream *Megalobrama amblycephala* using 5′-anchored PCR. Mol Ecol Resour 9: 971–974.
27. Li G, Quiros CF (2001) Sequence-related amplified polymorphism (SRAP), a new marker system based on a simple PCR reaction: its application to mapping and gene tagging in *Brassica*. Theor Appl Genet, 103: 455–461.
28. Budak H, Shearman RC, Parmaksiz I, Dweikat I (2004a) Comparative analysis of seeded and vegetative biotypes buffalograsses based on phylogenetic relationship using ISSRs, SSRs, RAPDs and SRAPs. Theor Appl Genet 109: 280–288.
29. Jones N, Ougham H, Thomas H, Pašakinskienė I (2009) Markers and mapping revisited: finding your gene. New Phytol 183: 935–966.
30. Ferriol M, Picó B, Nuez F (2003) Genetic diversity of a germplasm collection of *Cucurbita pepo* using SRAP and AFLP markers. Theor Appl Genet 107: 271–282.
31. Budak HR, Shearman C, Parmaksiz I, Gaussoin RE, Riorsdan TP, et al. (2004b) Molecular characterization of buffalograss germplasm using sequence-related amplified polymorphism markers. Theor Appl Genet 108: 328–334.
32. Sun SJ, Gao W, Lin SQ, Zhu J, Xie BG, et al. (2006) Analysis of genetic diversity in *Ganoderma* population with a novel molecular marker SRAP. Appl Microbiol Biotechnol 72: 537–543.
33. Qiao LX, Liu HY, Guo BT, Weng ML, Dai JX, et al. (2007) Molecular identification of 16 *Porphyra* lines using sequence-related amplified polymorphism markers. Aquat Bot 87: 203–208.
34. Zhang F, Chen SM, Chen FD, Fang WM, Chen Y, et al. (2011) SRAP-based mapping and QTL detection for inflorescence-related traits in chrysanthemum (*Dendranthema morifolium*). Mol Breed 27: 11–23.
35. Zhang ZW, Han YP, Zhong XM, Zhang ZY, Cao ZM, et al. (2007) Genetic structure analyses of grass carp populations between wild and cultured ones. J Fish Sci China 14: 720–725.
36. Xin WT, Sun ZW, Yin HB, Sun Y (2009) Identification of sex-associated SRAP marker in *Pelteobagrus fulvidraco*. J Northeast Forestry Univ 37(5): 112–113.
37. Zhang HY, He MX (2009) Segregation pattern of SRAP maker in F1 generation of *Pinctada martensii* family. Mar Sci Bull 28(2): 50–56.

38. Ding WD, Cao ZM, Cao LP (2010) Molecular analysis of grass carp (*Ctenopharyngodon idella*) by SRAP and SCAR molecular markers. Aquacult Int 18: 575–587.
39. Hu ZH, Xu JZ, Wang YB, Chai XJ (2010) SRAP analysis of genetic variation between female and male swimming crab (*Portunus trituberculatus*). J Shanghai Ocean Univ 19: 734–738.
40. Sambrook J, Fritsch EF, Maniatis T (1989) Molecular cloning: a laboratory manual, 2nd edn. New York: Cold Spring Harbor Laboratory Press.
41. Yeh FC, Yang RC, Boyle T (1999) POPGENE version 1.32: microsoft window-based freeware for population genetic analysis. The Centre for International Forestry Research, University of Alberta, Alberta, Canada. Available: http://www.ualberta.ca/~fyeh/.
42. Takezaki N, Nei M, Tamura K (2010) POPTREE2: software for constructing population trees from allele frequency data and computing other population statistics with windows interface. Mol Biol Evol 27: 747–752.
43. Excoffier L, Laval G, Schneider S (2005) Arlequin ver. 3.0: an integrated software package for population genetics data analysis. Evol Bioinform Online 1: 47–50.
44. Rice WR (1989) Analyzing tables of statistical tests. Evolution 43: 223–225.
45. Yue P, Chen Y (1998) Pisces. In: Wang S editor. China red data book of endangered animals. Beijing: Science Press.
46. Liang SX, Sun XW, Bai JJ, Gao JS (2008) Genetic analysis for cultured largemouth bass (*Micropterus salmoides*) in China with microsatellites. Acta Hydrobiol Sin 32: 694–700.
47. Wang J, Su YQ, Quan CG, Ding SX, Zhang W (2001) Genetic diversity of the wild and reared *Pseudosciaena crocea*. Chin J Oceanol Limnol 19: 152–156.
48. Holsinger KE, Weir BS (2009) Genetics in geographically structured populations: defining, estimating and interpreting F_{ST}. Nat Rev Genet 10(9): 639–650.
49. Wright S (1978) Evolution and the genetics of populations, Vol. 4: variability within and among natural populations. Chicago: The University of Chicago Press.
50. Zhao Y, Li SF, Tang SJ (2009) Genetic variations among late selected strains and wild populations of blunt snout bream (*Megalobrama amblycephala*) by ISSR analysis. J Fish China 33: 893–900.
51. Li SF, Cai WQ (2000) Two-way selective response of *Megalobrama amblycephala*. J Fish China 24: 201–205.
52. Fang ZQ, Chen J, Zheng WB, Wu YY, Xiao Z (2005) RAPD analysis of wild population and cultivated population in *Siniperca chuatsi* Basilewsky. J Dalian Fish Univ 20: 16–19.

10

Fractured Genetic Connectivity Threatens a Southern California Puma (*Puma concolor*) Population

Holly B. Ernest[1,2*¤], T. Winston Vickers[1], Scott A. Morrison[3], Michael R. Buchalski[1,2], Walter M. Boyce[1]

1 Wildlife Health Center, School of Veterinary Medicine, University of California Davis, Davis, California, United States of America, 2 Wildlife and Ecology Unit, Veterinary Genetics Laboratory, School of Veterinary Medicine, University of California Davis, Davis, California, United States of America, 3 The Nature Conservancy, San Francisco, California, United States of America

Abstract

Pumas (*Puma concolor;* also known as mountain lions and cougars) in southern California live among a burgeoning human population of roughly 20 million people. Yet little is known of the consequences of attendant habitat loss and fragmentation, and human-caused puma mortality to puma population viability and genetic diversity. We examined genetic status of pumas in coastal mountains within the Peninsular Ranges south of Los Angeles, in San Diego, Riverside, and Orange counties. The Santa Ana Mountains are bounded by urbanization to the west, north, and east, and are separated from the eastern Peninsular Ranges to the southeast by a ten lane interstate highway (I-15). We analyzed DNA samples from 97 pumas sampled between 2001 and 2012. Genotypic data for forty-six microsatellite loci revealed that pumas sampled in the Santa Ana Mountains (n = 42) displayed lower genetic diversity than pumas from nearly every other region in California tested (n = 257), including those living in the Peninsular Ranges immediately to the east across I-15 (n = 55). Santa Ana Mountains pumas had high average pairwise relatedness, high individual internal relatedness, a low estimated effective population size, and strong evidence of a bottleneck and isolation from other populations in California. These and ecological findings provide clear evidence that Santa Ana Mountains pumas have been experiencing genetic impacts related to barriers to gene flow, and are a warning signal to wildlife managers and land use planners that mitigation efforts will be needed to stem further genetic and demographic decay in the Santa Ana Mountains puma population.

Editor: Adam Stow, Macquarie University, Australia

Funding: The work was supported by the following: California State Parks (WMB; http://www.parks.ca.gov/), California Department of Fish and Wildlife (HBE, WMB; https://www.dfg.ca.gov/), The Nature Conservancy (WMB, SAM; http://www.nature.org/); The McBeth Foundation (WMB; http://mcbethfoundation.com/); The Anza Borrego Foundation (WMB; http://theabf.org/), Nature Reserve of Orange County (WMB; http://www.naturereserveoc.org/); The National Science Foundation (WMB; http://www.nsf.gov/); and private donors (HBE, WMB). The Nature Conservancy (SAM) helped with sample and data collection and preparation of the manuscript. California Department of Fish and Wildlife, helped with sample and data collection. Otherwise, the funders had no role in study design, data collection and analysis, decision to publish, or preparation of the manuscript.

Competing Interests: The authors have declared that no competing interests exist.

* Email: hernest@uwyo.edu

¤ Current address: Department of Veterinary Sciences, University of Wyoming, Laramie, Wyoming, United States of America

Introduction

Genetic diversity, demography, and abundance – biological characteristics that influence population viability – can vary across a species' distribution. Species that are generally perceived as wide-ranging and abundant are sometimes relegated to status as "least conservation concern", in spite of indicators signaling concern and frequently, lack of data. Pumas (*Puma concolor*; also known as mountain lion, cougar, and in Florida, panther) epitomize this dilemma. Although pumas in California have not been subjected to hunting since 1972, and were designated as a Specially Protected Mammal in 1990 [1], there is minimal active management and little scientifically validated data on statewide or regional population numbers. Pumas in southern California have one of the lowest annual survival rates among any population in North America, on par with rates seen in hunted populations (unpublished data). They are under increasing threats from habitat loss and fragmentation, and mortality from vehicle strikes, depredation permits, poaching, public safety kills, wildfire, and poisoning [2,3]. Timely evaluation of potential threats to population viability is imperative in order to prioritize conservation activities to prevent collapse of some populations.

The human population of southern California is over 20 million [4] and expected to exceed 30 million by 2060 [5]. This increasing population will likely result in further loss, fragmentation, and degradation of natural habitats in the region. Habitat fragmentation south of greater Los Angeles has effectively turned the Santa Ana Mountain range in mostly Orange and Riverside counties into a 'mega-fragment' of habitat, surrounded to the west, north, and east by dense urban land uses. The only remaining montane and foothill habitat linkage connecting the Santa Ana Mountain range to other mountains of the Peninsular Range is a southeasterly swath of habitat bisected by a very heavily traveled 10-lane highway, Interstate 15 (I-15) (Figure 1).

Population viability of pumas in the Santa Ana Mountains (a geography henceforth referred to as distinct from the broader Peninsular Ranges to the east) has been of conservation concern

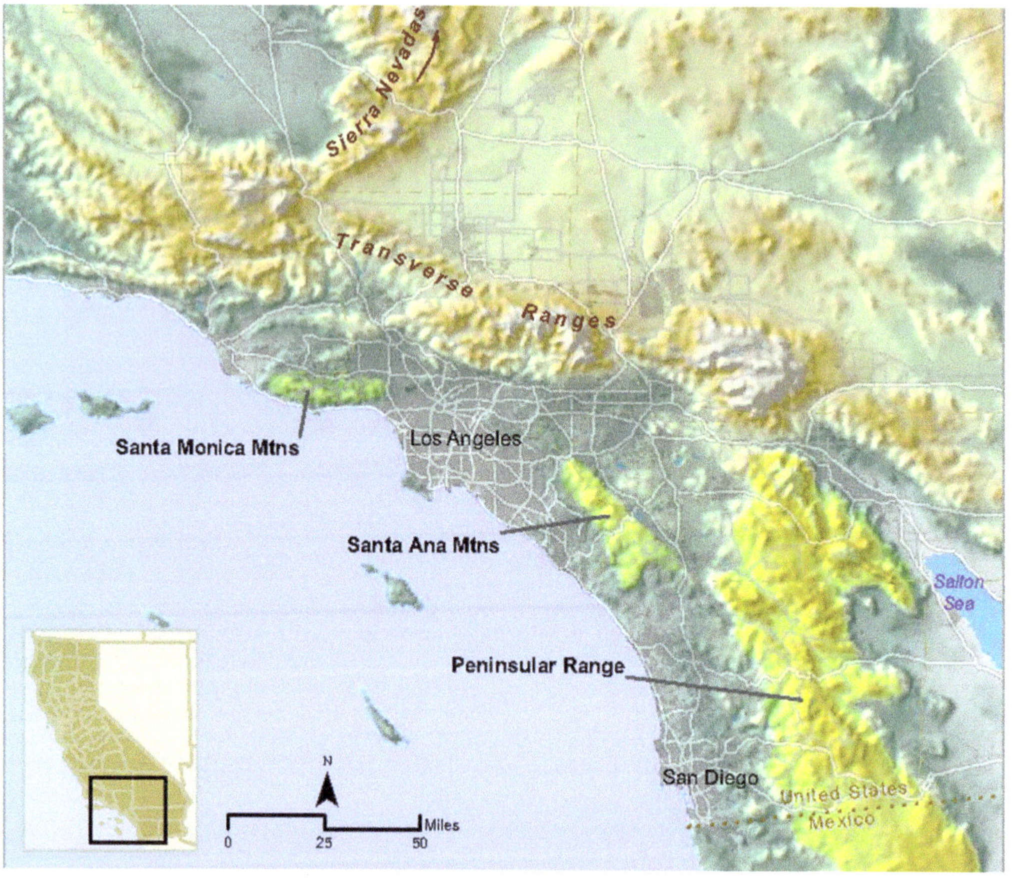

Figure 1. Topographic map depicting location of Santa Ana Mountains, eastern Peninsular Ranges in southern California, and adjacent regions. Inset shows location in the state of California.

for decades. Population monitoring and modeling in the 1980s highlighted that urbanization and highways were fragmenting puma habitat (e.g., [6]), and that in turn motivated efforts to protect habitat connectivity in the region (e.g., [7,8]). As part of a statewide assessment of puma genetic diversity and population structure, Ernest et al. [9] employed an 11-locus microsatellite panel and found that, for a limited sample size (n = 14) Santa Ana pumas had lower genetic diversity than other populations in California. Since 2001, pumas in the region have been the subject of an ongoing study by the Karen C. Drayer Wildlife Health Center of the University of California, Davis (UCD) School of Veterinary Medicine. Telemetry data from 74 pumas in the UCD study has confirmed that minimal connectivity (only one GPS-collared puma over ten years was documented to transit successfully; unpublished data) exists between the Santa Ana Mountains and the eastern Peninsular Ranges across I-15, confirming that previous connectivity concerns were warranted.

We conducted a detailed appraisal of the genetic diversity, relatedness, and population structure of southern California puma populations. Using 97 samples collected over 12 years as part of the UCD study, and a 46-locus microsatellite panel, we evaluated levels of genetic diversity, estimated effective population sizes and tested whether genetic data supported a hypothesis of recent bottleneck in the populations. We assessed whether genetics reflected our telemetry observations of infrequent puma crossings of I-15 between the Santa Ana Mountains and the Peninsular Ranges to the east. Additionally we explored inter-population gene flow at multiple time scales by employing methods that reflect recent (a few generations) and more historical (tens or more generations). Finally, we tested our hypothesis that the Santa Ana population had lower genetic diversity than those sampled from other regions in California.

Materials and Methods

Samples

We obtained blood or tissue samples for analysis of nuclear DNA from pumas captured for telemetry studies, and from those found dead or killed by state authorities for livestock depredation or public safety in San Diego, Orange, Riverside, and San Bernardino counties of southern California (n = 97) during 2001–2012 (Figure 2). Pumas captured for telemetry were captured and sampled as detailed in [10]. Forty-two samples were collected to the west of I-15 in the Santa Ana Mountains, and 55 samples were collected in the Peninsular Ranges to the east of I-15. A small number of additional samples were collected from deceased animals in San Bernardino County just to the north of the Peninsular Range across Interstate Highway 10. For population genetic comparisons with pumas sampled elsewhere throughout California, a 257 sample subset of our statewide puma DNA data archive was employed (regions and sample sizes detailed in Table 1 and depicted in Figure 1 in [9])

Ethics Statement

Animal handling was carried out in strict accordance with the recommendations and approved Protocol 10950/PHS, Animal

Table 1. Genetic diversity summary statistics for southern California pumas (n = 97) relative to other populations in California (n = 257).

Sampling Region	Abbrev.		N	Na	AR	Ho	He	I	%P
North Coast	**NC**	**Mean**	29	3.6	2.0	0.41	0.44	0.80	98%
		SE		0.2	0.1	0.03	0.03	0.05	
Modoc Plateau & Eastern Sierra Nevada	**MP-ESN**	**Mean**	51	4.2	2.4	0.52	0.54	0.98	100%
		SE		0.3	0.1	0.03	0.03	0.05	
Western Sierra Nevada	**WSN**	**Mean**	47	4.2	2.4	0.47	0.51	0.95	98%
		SE		0.2	0.2	0.03	0.03	0.06	
Central Coast: north	**CC-N**	**Mean**	83	3.2	1.9	0.41	0.41	0.70	98%
		SE		0.2	0.1	0.03	0.03	0.06	
Central Coast: central	**CC-C**	**Mean**	21	3.4	2.1	0.43	0.46	0.81	96%
		SE		0.2	0.1	0.03	0.03	0.05	
Santa Monica Mountains	**CC-S**	**Mean**		2.2	1.7	0.38	0.33	0.53	76%
Central Coast: South		**SE**	26	0.1	0.1	0.04	0.03	0.05	
Peninsular Range-East	**PR-E**	**Mean**	55	3.1	2.0	0.43	0.41	0.74	87%
		SE		0.2	0.2	0.04	0.04	0.07	
Santa Ana Mountains	**SAM**	**Mean**	42	2.3	1.6	0.33	0.32	0.54	80%
		SE		0.2	0.1	0.03	0.03	0.05	

Abbrev. = region abbreviations used in Tables and Figures. Mean with standard error (SE). N = sample size. Na = average number of different alleles per locus. AR = allelic richness, standardized to sample size. Ho = observed heterozygosity. He = expected heterozygosity. I = Shannon's information index (Sherwin et al 2006). %P = percent of polymorphic loci. Regions are detailed further in text and generally follow California Bioregions designations. (http://biodiversity.ca.gov/bioregions.html).

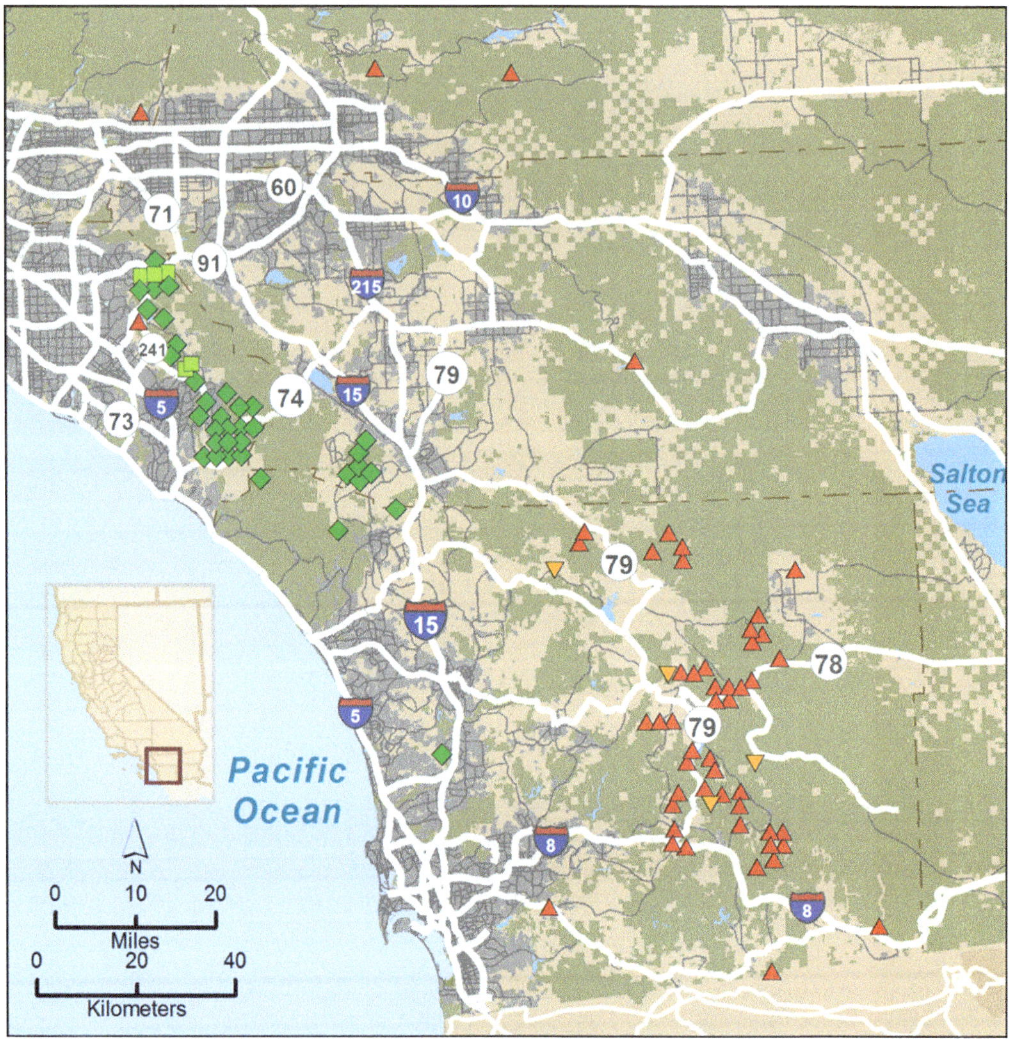

Figure 2. Map of puma capture locations in the Santa Ana Mountains and eastern Peninsular Ranges of southern California. Colors of symbols represent genetic group assignment inferred from Bayesian clustering analysis (STRUCTURE analysis, see Figure 4). Genetic group A-1 = green diamonds; A-2 = red triangles (apex at top). One male puma (M86) captured in the Santa Ana Mountains had predominant genetic assignment to the A-2 (red) genetic group. Five individuals (light green squares) captured in the Santa Ana Mountains had partial assignment to the A-2 group (M91, F92, M93, M97 and F102). Molecular kinship analysis showed that M86 and a female (F89) captured in the Santa Ana Mountains were parents of pumas M91, F92, and M93 (captured in the Santa Ana Mountains). Puma M97 assigned in parentage to M86 and F61, while F102 had unknown parentage (no parentage assignments; due possibly to her death early in project prior to collection of most of the samples). Three individuals (orange triangles, apex at bottom), had partial assignment (however, less than 20%) to A-1.

Welfare Assurance number A3433-01, with capture and sampling procedures approved by the Animal Care and Use Committee at the University of California, Davis (Protocol #17233), and Memoranda of Understanding and Scientific Collecting Permits from the California Department of Fish and Wildlife (CDFW). Permits and permissions for access to conserved lands at puma capture and sampling sites were obtained from CDFW, California Department of Parks and Recreation, The Nature Conservancy, United States (US) Fish and Wildlife Service, US Forest Service, US Bureau of Land Management, US Navy/Marine Corps, Orange County Parks Department, San Diego County Parks Department, San Diego State University, Vista Irrigation District, Rancho Mission Viejo/San Juan Company, Sweetwater Authority, California Department of Transportation (CalTrans), and the City of San Diego Water Department.

DNA Extraction and Microsatellite DNA data collection

Whole genomic DNA was extracted using the DNeasy Blood & Tissue Kit (QIAGEN, Valencia, CA, USA). Fifty microsatellite DNA primers were initially screened for this project. Forty-six loci that performed well in multiplex PCR (using the QIAGEN Multiplex PCR kit; QIAGEN) and conformed to expectations for Hardy-Weinberg and linkage equilibria were selected for ultimate analysis [11,12,13]. One sex-identification locus (Amelogenin) was used to confirm sex in samples from degraded puma carcasses [14].

PCR products were separated with an ABI PRISM 3730 DNA Analyzer (Applied Biosystems Inc., Foster City, CA, USA) with each capillary containing 1 µL of a 1:10 dilution of PCR product and deionized water, 0.05 µL GeneScan-500 LIZ Size Standard and 9.95 µL of HiDi formamide (both products Applied Biosystems Inc.) that was denatured at 95°C for 3 min. Products

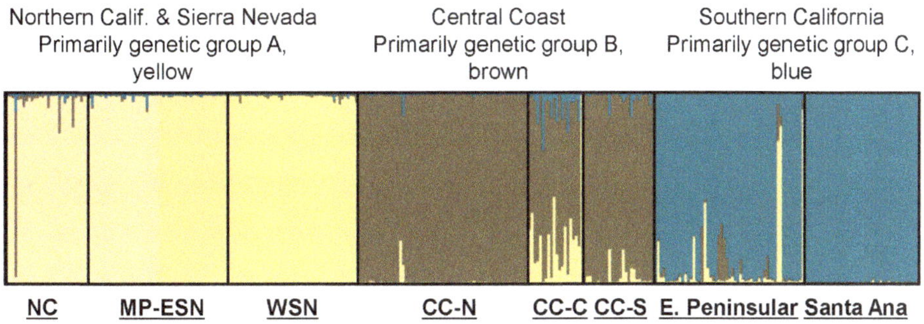

Figure 3. California puma population genetic structure. STRUCTURE bar plot displaying the genetic clustering relationship of southern California pumas relative to others in California. Three major genetic groups, A (blue, on right), B (brown, in center), and C (yellow, on left), are evident for analysis of 354 individuals sampled throughout California. Abbreviations: NC = North Coast, MP-ESN = Modoc Plateau & Eastern Sierra Nevada, WSN = Western Sierra Nevada, CC-N = Central Coast: north, CC-C = Central Coast: central, CC-S = Central Coast: South (Santa Monica Mountains), PR-E = Peninsular Range-East, SAM = Santa Ana Mountains. The plot is organized by grouping individuals in order of their geographic region sampling source. Proportional genetic assignment for each puma is represented by a vertical bar, most easily visualized for pumas that genetically assigned to a group different from most others sampled in its region (for example one individual with over 80% brown and 8% blue near far left of group A). Pumas primarily from the Sierra Nevada Range and northern California are represented by group A (yellow), group B (brown) includes primarily Central Coast pumas and group C (blue) represents primarily southern California pumas (Santa Ana Mountains and eastern Peninsular Ranges).

were visualized with STRand version 2.3.69 [15]. Negative controls (all reagents except DNA) and positive controls (well-characterized puma DNA) were included with each PCR run. Samples were run in PCR at each locus at least twice to assure accuracy of genotype reads and minimize risk of non-amplifying alleles. For >90% samples, loci that were heterozygous were run at least twice and homozygous loci were run at least three times.

Genetic diversity

The number of alleles (Na), allelic richness (AR; incorporates correction for sample size), observed heterozygosity (Ho), expected heterozygosity (He), Shannon's information index [16], and tests for deviations from Hardy-Weinberg equilibrium were calculated using software GenAlEx version 6.5 [17,18]. Shannon's information index provides an alternative method of quantifying genetic diversity and incorporates allele numbers and frequencies. Testing for deviations from expectations of linkage equilibrium was conducted using Genepop 4.2.1 [19], and we tested for the presence of null alleles using the program ML RELATE [20]. We assessed significance for calculations at alpha = 0.05 and used sequential Bonferroni corrections for multiple tests [21] in tests for Hardy-Weinberg and linkage equilibria.

The average probability of identity (PID) was calculated two ways using GenAlEx: 1) assuming random mating (PID_{RM}) without close relatives in a population [22], and 2) assuming that siblings with similar genotypes occur in a population (PID_{SIBS}) [23]. Probability of identity is the likelihood that two individuals will have the same genetic profile (genotype) for the DNA markers used. PID_{SIBS} is considered conservative since it probably conveys a higher likelihood; however, we recognized that siblings occurred in these populations.

Assessing population structure and genetic isolation

We used a Bayesian genetic clustering algorithm (STRUCTURE version 2.3.4 [24,25]) to determine the likely number of population groups (K; genetic clusters) and to probabilistically group individuals without using the known geographic location of sample collection. We used the population admixture model with a flat prior and assumed that allele frequencies were correlated among populations, and ran 50,000 Markov chain Monte Carlo repetitions following a burnin period of 10,000 repetitions. First,

Figure 4. Southern California puma population genetic structure. Bar Plot displaying results of STRUCTURE analysis focused on genotypic data from 97 southern California pumas (the blue block from Figure 3). With removal of the strong genetic signal from northern California and Central Coast samples (see Figure 3), two distinct southern California groupings were inferred, C-1 (green, on right) and C-2 (red, on left). These reflect the two regions: Santa Ana Mountains to the west of I-15 (predominantly genetic group C-1) and eastern Peninsular Ranges to the east of I-15 (predominantly genetic group C-2). Genetic clustering is dependent on genetic variance among samples included in the analysis. One male puma (M86) captured in the Santa Ana Mountains has predominant genetic assignment to the C-2 (red) genetic group (the predominant genetic cluster for PR-E), and five others had partial assignment to the C-2 group (M91, F92, M93, M97 and F102). Molecular kinship analysis showed that M86 and a female (F89) assigning to the C-1 genetic group were parents of pumas M91, F92, and M93 (all were captured in the Santa Ana Mountains).

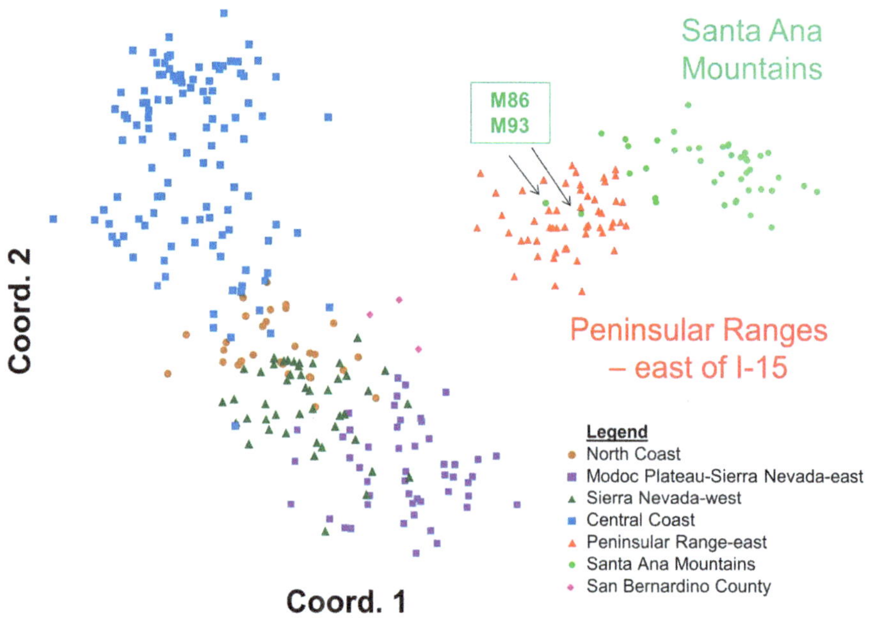

Figure 5. Principal Coordinates analyses (PCoA) constructed using genetic covariance matrices (GenAlEx) for 354 California puma genetic profiles including 97 from southern California. Patterns displayed for first two axes of variation within the genetic data set. Each point, color-coded to its sampling region, represents an individual puma. Note that colors in PCoA diagrams reflect geographic source of samples and not STRUCTURE genetic cluster assignment. Abbreviations and sample sizes per Table 1. Arrows denote pumas described in Figure 4.

an analysis including 354 statewide puma genotypes (97 from southern California and 257 from other regions) was run to estimate the probability of one through 10 genetic clusters (K), with each run iterated three times. Second, given the output of the statewide run, we ran an analysis using only the 97 southern California puma genotypes to estimate the probability of one through five K, with each run iterated three times. Employing STRUCTURE HARVESTER [26] we averaged log probability

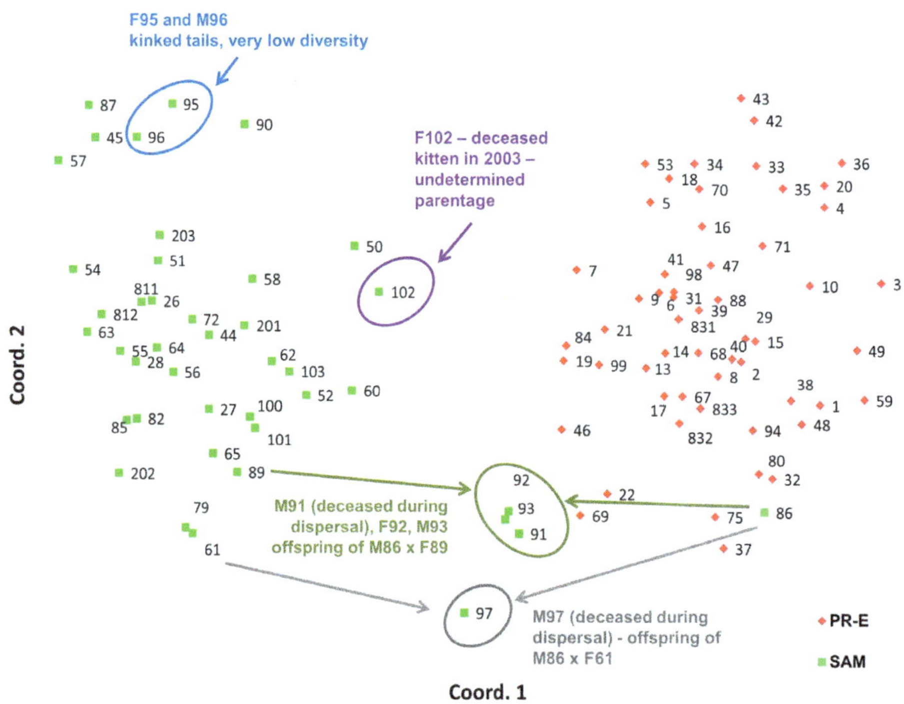

Figure 6. Principal Coordinates analyses (PCoA) via covariance matrices for 97 southern California puma genetic profiles as conducted in GenAlEx. Patterns displayed for first two axes of variation within the genetic data set. Each point represents an individual puma, and has sample identification number and color-coding to sampling region. Note that colors in PCoA diagrams reflect geographic source of samples and not STRUCTURE genetic cluster assignment. Abbreviations and sample sizes per Table 1.

Table 2. Wright's F_{ST} values indicate that southern California mountain lion populations are genetically distinct from other populations in California.

	NC	MP-ESN	WSN	CC-N	CC-C	CC-S	PR-E	SAM
North Coast (NC)	0							
Modoc Plateau & Eastern Sierra Nevada (MP-ESN)	0.09	0						
Western Sierra Nevada (WSN)	0.05	0.03	0					
Central Coast: north (CC-N)	0.13	0.14	0.12	0				
Central Coast: central (CC-C)	0.08	0.09	0.06	0.07	0			
Santa Monica Mountains (CC-S)	0.17	0.15	0.12	0.14	0.10	0		
Peninsular Range-East (PR-E)	0.13	0.08	0.09	0.12	0.09	0.16	0	
Santa Ana Mountains (SAM)	0.18	0.16	0.17	0.19	0.17	0.27	0.07	0

Note that one of the geographically closest puma populations, Santa Monica Mountains, has highest F_{ST} with the Santa Ana population, evidence of high genetic isolation for both regions. Probability, P(random> = data) based on 9999 permutations for all values are <0.001. Abbreviation definitions and sample sizes are included in Table 1.

of the data given K, log Pr(X|K), statistics across the multiple runs for each of the K estimates. In each case (statewide and southern California), we selected the K value of highest probability by identifying the set of values where the log Pr(X|K) value was maximized and subsequently selected the minimum value for K that did not sacrifice explanatory ability [27,28,29]. We defined membership to a cluster based upon the highest proportion of ancestry to each inferred cluster.

To further assess and visualize genetic relationships among regions and individuals, we performed principal coordinates analyses (PCoA) via covariance matrices with data standardization [30] using GenAlEx. This is a technique that allowed us to explore and plot the major patterns within the data sets. The PCoA process located major axes of variation within our multidimensional genotype data set. Because each successive axis explains proportionately less of the total genetic variation, the first two axes were used to reveal the major separation among individuals. Employing Genalex software, a pairwise, individual-by-individual genetic distance matrix was generated and then used to create the PCoA.

Wright's F-statistic, $F_{ST,}$ was calculated to appraise how genetic diversity was partitioned between populations. As implemented in GenAlEx, we used Nei's [31] formula, with statistical testing options offered through 9999 random permutations and bootstraps.

Detecting migrants

We used GENECLASS2 version 2.0.h [32] to identify first-generation migrants, i.e. individuals born in a population other than the one in which they were sampled. Genetic clusters identified during STRUCTURE analysis were treated as putative populations. GENECLASS2 provides different likelihood-based test statistics to identify migrant individuals, the efficacy of which depends on whether all potential source populations have been sampled. We first calculated the likelihood of finding a given individual in the population in which it was sampled, L_h, assuming all populations had not been sampled. We then calculated L_h/L_{max}, the ratio of L_h to the greatest likelihood among the populations [33], which has greater power when all potential source populations have been sampled. The critical value of the test statistic (L_h or L_h/L_{max}) was determined using the Bayesian approach of Rannala and Mountain [34] in combination with the resampling method of Paetkau et al. [33]; i.e., Monte Carlo simulations carried out on 10,000 individuals with the significance level set to 0.01.

Testing for bottlenecks and inferring effective population size

We tested for evidence of recent population size reductions in Santa Ana Mountains and eastern Peninsular Range regions with one-tailed Wilcoxon sign-rank tests for heterozygote excess in the program BOTTLENECK version 1.2.02 [35]. The program evaluates whether the reduction of allele numbers occurred at a rate faster than reduction of heterozygosity, a characteristic of populations which have experienced a recent reduction of their effective population size (Ne) [35,36]. This bottleneck genetic signature is detectable by this test for a finite time, estimated to be less than 4 times Ne generations [37]. These tests were performed using the two-phase (TPM, 70% step-wise mutation model and 30% IAM) model of microsatellite evolution and 10,000 iterations.

We then estimated contemporary Ne for each of the two regions based on gametic disequilibrium with sampling bias correction [38] using LDNE version 1.31 [39]. Ne is formally defined as the size of the ideal population that would experience the same

Table 3. Effective population size estimations and indications of recent genetic bottlenecks in southern California pumas.

	Mode	TPM	Ne (P-CI; JK-CI)
Santa Ana Mtns	Shifted mode	0.009	5.1 (3.3–6.7; 3.3–6.6)
Peninsular Range, East	Normal L	0.19	24.3 (21.7–27.3; 20.6–28.8)

Listed by column are p-values for population bottleneck tests (Wilcoxon sign-rank test; BOTTLENECK) assuming the two-phase (TPM) model of microsatellite evolution. Effective size (Ne) estimations (95% CI) based on data from 42 microsatellite loci. The Santa Ana Mountains population exhibited clear evidence of a population bottleneck. Effective population size estimate using the point estimate linkage disequilibrium method of (LDNE, Waples 2006) with 95% confidence intervals (CI) for both parametric (P) and jackknifed (JK) estimates.

amount of genetic drift as the observed population [40]. These analyses excluded alleles occurring at frequencies ≤0.05, and we used the jackknife method to determine 95% confidence intervals [38].

Relatedness analyses: pairwise coefficient and internal

Molecular kinship analysis was conducted using a number of software packages. Pairwise relatedness among individuals was evaluated using the algorithm of Lynch and Ritland [41], with reference allele frequencies calculated and relatedness values averaged within each southern California population, as implemented in GenAlEx. Partial molecular kinship reconstruction was conducted using a consensus of outputs from the GenAlEx pairwise relatedness calculator, ML Relate [20], CERVUS version 3.0.3 [42], and Colony version 2.0.3.1 [43,44]. Individual genetic diversity (also called internal relatedness) was assessed using Rhh [45] as implemented in R statistical software [46]. This is a measure of genetic diversity within each individual (an estimate of parental relatedness [47], and we averaged over individuals for each of the two regions of southern California. Significance of differences between means was evaluated using t tests.

Results

Forty-two of the 46 loci that we employed were polymorphic in southern California and selected for the subsequent analyses. The average probabilities of identity with assumptions of either random mating (PID_{RM}) or mating among sibs (PID_{SIBS}) across the 42 loci for the eastern Peninsular Ranges were (PID_{RM}) 6.3×10^{-22} and (PID_{SIBS}) 3.1×10^{-10}, and for the Santa Ana Mountains were (PID_{RM}) 2.8×10^{-15} and (PID_{SIBS}) 1.1×10^{-7} respectively. These very small values indicate that the panel of genetic markers provided very high resolution to distinguish individuals. For example, given this data the probability of seeing the same multi-locus genotype in more than one puma was less than one in nine million for Santa Ana Mountains pumas.

Genetic diversity

Measures of genetic variation including allelic diversity, heterozygosity, Shannon's information index, and polymorphism, were lower for Santa Ana pumas than most of those tested from other regions of California (Table 1). Such low genetic diversity indicators were approached only by pumas in the Santa Monica Mountains (Ventura and Los Angeles Counties), a neighboring remnant puma population in the north Los Angeles basin (Figure 1).

Population Structure

Bayesian clustering analysis (STRUCTURE; Figure 3 of statewide puma genetic profiles (n = 354), including 97 from southern California, also support genetic distinctiveness of Santa Ana Mountains and eastern Peninsular Range pumas from other populations in the state. Three main genetic groups (A, B, and C) were evident in the analysis (Figure 3) The 97 pumas sampled in southern California (right-hand set of bars in Figure 3, with samples from Santa Ana and eastern Peninsular Range pumas labeled) predominantly cluster within genetic group C. The Santa Ana pumas assign very tightly to group C (0.996 average probability assignment), while pumas of the eastern Peninsular Ranges showed more variable assignment (0.93 average probability assignment), with 9 individuals (16%) having less than 0.90 assignment. Pumas sampled in the Central Coast of California (which included Santa Monica Mountains pumas) make up the central set of bands, and those individuals predominantly assign to the genetic group B. Pumas sampled in the other regions of California (North Coast Ranges, Modoc Plateau, western Sierra

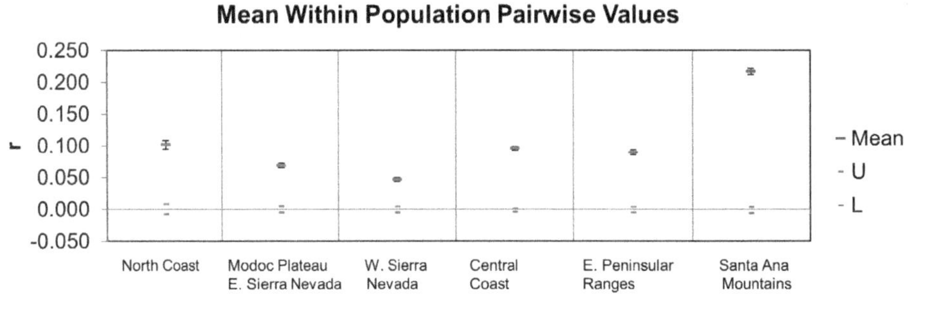

Figure 7. Average pairwise relatedness (r; blue bars with confidence intervals) for pumas sampled in southern California relative to other regions in California. Algorithm of Lynch and Ritland (1999) as implemented in GenAlEx. Expected range for "unrelated" is shown as red bars with confidence intervals. The average relatedness of Santa Ana Mountain pumas is higher than those sampled in Peninsular Ranges east of I-15 and for any other region tested in California. Relatedness in the Santa Ana Mountains pumas approaches second order family relationship (half sibs, niece-aunt, grandparent-grandchild, etc.). Abbreviations listed in Table 1.

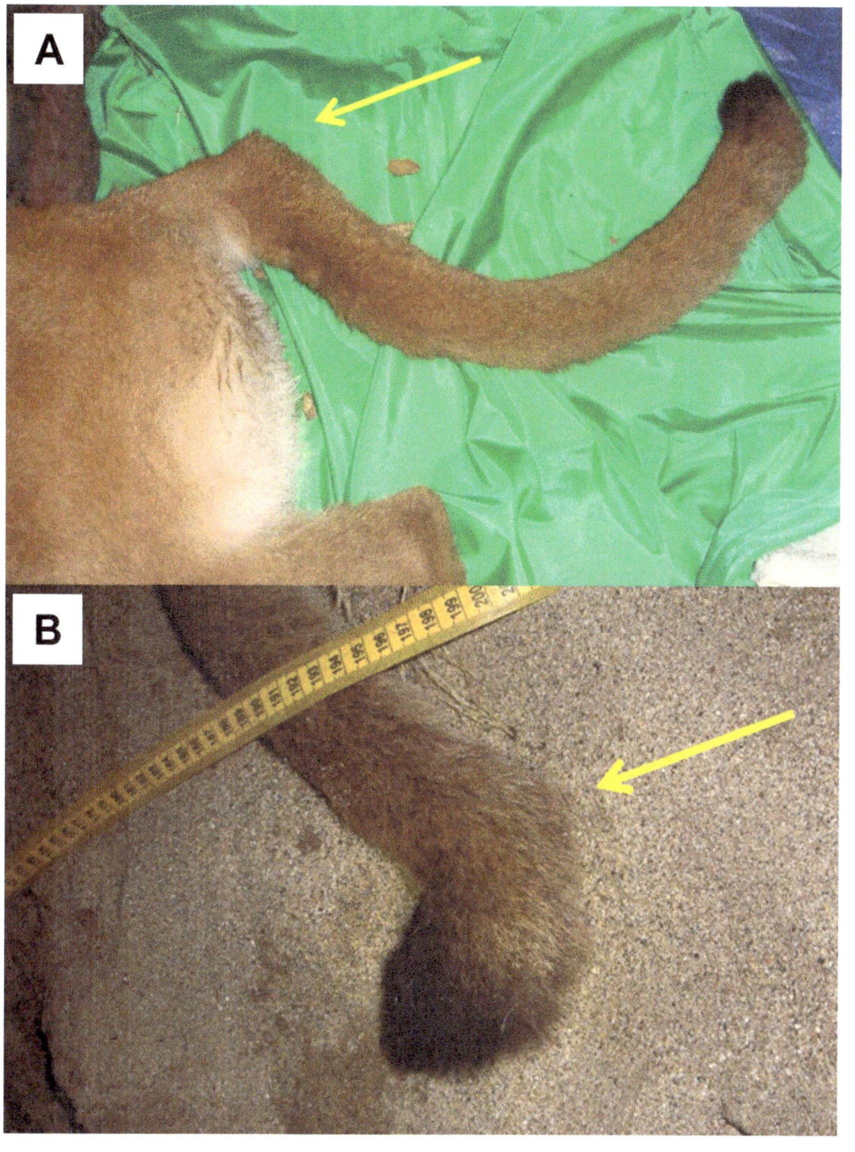

Figure 8. Photographs of kinked tails of pumas F95 (a) and M96 (b). Arrows indicate kink sites. Puma F95 had tail kink at base of tail and Puma M96 had tail kink near distal tip of tail. These two pumas had among the lowest genetic diversity measured in this study.

Nevada, and eastern Sierra Nevada) predominantly cluster with the genetic group A. Notably, there are individuals sampled in each geographic area which cluster with a genetic group that is not the dominant one in that area, suggesting dispersal events and/or genetic exchange that have occurred to varying degrees in each region.

A STRUCTURE analysis focused only on genetic data from the 97 southern California pumas indicated two distinct genetic groups (C-1 and C-2 shown in Figure 4). Pumas sampled in the eastern Peninsular Range region east of I-15 group primarily with C-2 and those of the Santa Ana Mountain region on the west side of I-15 group with C-1. An exception to the consistent genetic clustering was an adult male (M) puma (M86), that was captured in the Santa Ana Mountains but clustered with pumas from the eastern Peninsular Ranges (primarily genetic group C-2). Five other pumas captured in the Santa Ana Mountains had a 30–50% assignment to the C-2 group (M91, F92, M93, M97 and F102). Molecular kinship analysis showed that M86 and a female (F89) captured in the Santa Ana Mountains and assigned to the C-1 genetic group were the likely parents of three of these pumas (M91, F92, and M93) (results of relatedness and kinship analyses). M86 also was the likely parent of another puma in the group (M97), an offspring of another female (F61) that was sampled in Santa Ana Mountains and clustered with the C-1 genetic group. F102 was a <1 year old female killed by a vehicle in 2003 prior to collection of the majority of samples from adults in the Santa Ana Mountains.

Principal coordinates analysis of statewide puma genetic profiles (n = 354) (PCoA; Figure 5) allowed graphical examination of the first two major axes of multivariate genetic variation, and confirmed and added detail to the genetic distinctiveness of southern California pumas relative to others in California. The PCoA also reinforced the distinctiveness of pumas sampled in the Santa Ana Mountains from those sampled in the eastern Peninsular Ranges. Most pumas sampled in the Santa Ana Mountains align in a cloud of data points distinct from the eastern

Peninsular Range pumas, and were the most genetically distant from all other pumas tested in California (Figure 5). The analysis also confirms the STRUCTURE findings that M86 who was sampled in the Santa Ana Mountains genetically aligns with the pumas sampled in the Peninsular Ranges, as does one of his offspring, M93 (see Figure 6 for additional detail). The PCoA position of data points for three pumas sampled in the San Bernardino Mountains north of Peninsular Ranges (pink diamonds in Figure 5) illustrates an intermediate genetic relationship between pumas from the rest of California and pumas sampled in the eastern Peninsular Ranges and Santa Ana Mountains, and suggests that they may represent transitional gene flow signature between southern California and regions to the north and east.

PCoA analysis of only the samples collected in the Santa Ana and Peninsular Ranges (Figure 6) confirms the findings from the STRUCTURE analysis indicating genetic distinctiveness of these two populations despite geographic proximity. Siblings M91, F92, and M93 (offspring of F89 and M86 according to our kinship reconstructions) as well as M97 (likely offspring of a female puma captured in the Santa Ana Mountains, F61, and M86, according to kinship reconstructions) are located graphically midway between their parents' PCoA locations.

Genetic isolation

Wright's F_{ST} calculations (Table 2) indicate that Santa Ana Mountains pumas are the most isolated of those tested throughout California ($p = 0.0001$). Despite the short distance (as short as the distance across the I-15 Freeway) between the Santa Ana Mountains and the eastern Peninsular Range region, F_{ST} was surprisingly high (0.07) given the very close proximity of the two regions (separated only by an interstate highway). The Santa Monica Mountains pumas and Santa Ana Mountains pumas had the highest F_{ST} (0.27; lowest gene flow) of all pairwise comparisons in the state, demonstrating a high level of genetic isolation between these regions.The Santa Monica Mountains and Santa Ana Mountains are less than 100 km direct distance apart, through the center of Los Angeles. However the more likely distance for puma travel between these two mountain ranges, avoiding urban areas and maximizing upland habitat, would likely exceed 300 km (estimated using coarse measurements on Google Earth, Google, Inc.).

Detection of migrants

GENECLASS2 identified four individuals as first-generation migrants ($P<0.01$), four with the L_h method (pumas F75, M80, M86, and M99), and one with the L_h/L_{max} ratio (M86, which was detected using both likelihood methods). Pumas F75, M80, and M99 were all captured from the San Bernardino Mountains (Figure 2) at the northern extent of the study region, yet clustered with individuals from the Eastern Peninsular Range during STRUCTURE analysis. Their migrant designation may suggest immigration from populations north of Los Angeles and/or a distinct genetic population within the San Bernardino region. Puma M86 was captured in the Santa Ana Mountains, but assigned strongly to the eastern Peninsular Range genetic cluster, indicating a seemingly clear population of origin. This individual assignment is in accord with the clustering results from STRUCTURE (Figure 4).

Evidence of genetic bottlenecks

The Santa Ana Mountains population exhibited clear evidence of a population bottleneck (Table 3; Wilcoxon sign-rank test for heterozygote excess, and detection of a shift in the allele frequency distribution mode [36]; BOTTLENECK software). The eastern Peninsular Range mountain lions did not show a strong signature of a bottleneck.

Effective population size

Effective population size (Ne) estimations using the linkage disequilibrium method (LDNe program) were 5.1 for the Santa Ana Mountains population and 24.3 for mountain lions in the eastern Peninsular Ranges. Statistical confidence intervals for both regions, given the genetic data, were tight (Table 3).

Relatedness: pairwise coefficient and internal

The average pairwise coefficient of relatedness (r, Figure 7) was highest in Santa Ana Mountains pumas relative to all others tested in California (0.22; 95% confidence interval of 0.22–0.23), a level that approaches second order kinship relatedness (half-sibs, grantparent/grandchild, aunt-niece, etc). The value for the eastern Peninsular Ranges was 0.10 (confidence interval of 0.09–0.10), less than that of third order relatives (first cousins, great-grandparent/great grandchild). Other regions of California averaged similar or lower values to those of eastern Peninsular Ranges (Figure 7).

Among pumas sampled in the Santa Ana Mountains, the population average (0.14) for internal relatedness as implemented in rHH software was significantly higher (t test; $p = 5.8 \times 10^{-6}$) than for those sampled in the eastern Peninsular Ranges (0.001). Of a group of six pumas which clustered near one another in PCoA (Figure 6), five have among the lowest individual genetic diversity measured in southern California (Puma ID [Internal Relatedness value: F45 [0.37], F51 [0.37], M87 [0.28], F90 [0.21], F95 [0.38], and M96 [0.33]). Notably, pumas F95 and M96 (highest internal relatedness) were observed with kinked tails at capture in the Santa Ana Mountains (Figure 8).

Discussion

Pumas of the Santa Ana Mountains are genetically depauperate, isolated, and display signs of a recent and significant bottleneck. In general, coastal California puma populations have less genetic diversity and less gene flow from other populations than those farther inland [9] (Table 1). This study showed that two coastal populations (Santa Ana Mountains and Santa Monica Mountains) had particularly low genetic variation and gene flow from other regions. Lack of gene flow is likely due in part to natural barriers to puma movement: geography and habitat (Pacific Ocean to the west; less hospitable desert habitat bounding certain regions, etc.). However, our data suggest that anthropogenic developments on the landscape are playing a large role in genetic decay in the Santa Ana Mountains puma population. As large solitary carnivores with sizable habitat requirements, pumas are extremely sensitive to habitat loss and fragmentation [48,49].

The genetic bottleneck in the Santa Ana Mountains pumas is estimated at less than about 80 years, depending on definitions of effective population size (Ne) and puma generation time. Luikhart and Cornuet [37] state that the bottleneck signatures decay after "4 times Ne [here estimated to be 5.1] generations". Logan and Sweanor [50] estimated generation time for their New Mexico population of pumas to be 29 months (2.4 years) for females. If an allowance of 2.4–4.0 years is made for generation times (unknown) in the Santa Ana Mountains population, the maximum estimated time since a bottleneck would be about 40–80 years. This was a period of tremendous urban development and multi-lane highway construction in southern California, particularly I-15 [51]. It is likely that the potential for connectivity between the Santa Ana Mountains and the Peninsular Range-East region will continue to be eroded by ongoing increases in traffic volumes on I-15, and conversion of unconserved lands along the I-15 corridor by development and agriculture [8,48,52].

An isolated population of pumas in the Santa Monica Mountains to the north of the Santa Ana Mountains also exhibit low values relative to other western North American populations (see Table 2 in [53]. Santa Monica pumas are isolated by urbanization of a megacity and busy wide freeways (Ventura county, including greater Los Angeles region [53]. Multiple instances of intraspecific predation, multiple consanguineous matings (father to daughter, etc.), and lack of successful dispersal highlight a suite of anthropogenic processes also occurring in the Santa Ana Mountains. Our collective findings of kinked tails *and* very low genetic diversity in Santa Ana pumas F95 and M96 may portend manifestations of genetic inbreeding depression similar to those seen in Florida panthers [54,55]; however recognizing that kinked tails can have non-genetic etiologies.

Our analyses suggest that the Santa Ana Mountains puma population is highly challenged in terms of genetic connectivity and genetic diversity, a result hinted at in Ernest et al. [9] and now confirmed to be an ongoing negative process for this population. This compounds the demographic challenges of low survival rates and scant evidence of physical connectivity to the Peninsular Ranges east of I-15 (unpublished data). Beier [6] documented these same challenges during the 1990's, and data from the ongoing UCD study suggest the trends have accelerated. Substantial habitat loss and fragmentation has occurred and is continuing to occur; Burdett et al. [10] estimated that by 2030, approximately 17% of puma habitat that was still available in 1970 in southern California will have been lost to development, and fragmentation will have rendered the remainder more hazardous for pumas to utilize. Riley et al [53] document a natural "genetic rescue" event: the 2009 immigration and subsequent breeding success of a single male to the Santa Monica Mountains. This introduction of new genetic material into the population was paramount to raising the critically low level of genetic diversity, as also exemplified by the human-mediated genetic augmentation of Florida Panthers with Texas puma stock [56].

These findings raise concerns about the current status of the Santa Ana Mountains puma population, and the longer-term outlook for pumas across southern California. In particular, they highlight the urgency to maintain – and enhance – what connectivity remains for pumas (and presumably numerous other species) across I-15. Despite warnings [6,9] about potential serious impacts to the Santa Ana Mountains puma population if concerted conservation action was not taken, habitat connectivity to the Peninsular Ranges has continued to erode. We are hopeful that these new genetic results will motivate greater focus on connectivity conservation in this region. Indeed, the Santa Ana Mountains pumas may well serve as harbingers of potential consequences throughout California and the western United States if more attention is not paid to maintaining connectivity for wildlife as development progresses.

Acknowledgments

Samples, data, and expertise were provided by multiple people and agencies, including California Department of Fish and Wildlife (R. Botta, B Gonzales, M. Kenyon, P. Swift, S. Torres, D. Updike, and others), California State Parks, National Park Service, The Nature Conservancy, UC Davis Wildlife Health Center, UC Davis Veterinary Genetics Laboratory, and the US Geological Survey. We thank the following for their technical assistance: L. Dalbeck, T. Drazenovich, T. Gilliland, J. George, and M. Plancarte. GIS data management and project cartography was provided by B. Cohen and J. Sanchez. Field work assistance was provided by J. Bauer, C. Bell, P. Bryant, D. Dawn, M. Ehlbroch, D. Krucki, K. Logan, B. Martin, B. Millsap, M. Puzzo, D. Sforza, L. Sweanor, C. Wiley, E. York, and numerous volunteers. Thanks to E. Boydsen, K. Crooks, R. Fisher, and L. Lyren for assistance coordinating field projects and sample acquisition. We thank anonymous reviewers for their constructive comments.

Author Contributions

Conceived and designed the experiments: HBE TWV WMB. Performed the experiments: HBE TWV MRB WMB. Analyzed the data: HBE TWV MRB. Contributed reagents/materials/analysis tools: HBE TWV SAM WMB. Contributed to the writing of the manuscript: HBE TWV SAM MRB WMB.

References

1. Mansfield TM, Torres SG (1994) Trends in mountain lion depredation and public safety threats in California. Paper 33. Proceedings of the Sixteenth Vertebrate Pest Conference, Santa Clara, CA. Editors: WS Halverson and AC Crabb. University of California, Davis.
2. Riley SPD, Pollinger JP, Sauvajot RM, York EC, Bromley C, et al. (2006) FAST-TRACK: A southern California freeway is a physical and social barrier to gene flow in carnivores. Mol Ecol 15: 1733–1741.
3. Riley SPD, Bromley C, Poppenga RH, Uzal FA, White L, et al. (2007) Anticoagulant exposure and notoedric mange in Bobcats and Mountain Lions in urban Southern California. J Wildl Manage 71: 1874–1884.
4. U.S. Census Bureau (2012) 2010 Census of Population and Housing, Summary Population and Housing Characteristics, CPH-1-6, California, U.S. Government Printing Office, Washington, DC.
5. California Department of Finance (2013) New Population Projections: California to surpass 50 million in 2049. Press release. Available: http://www.dof.ca.gov/research/demographic/reports/projections/p-1/documents/Projections_Press_Release_2010-2060.pdf Accessed 2 December 2013.
6. Beier P (1995) Dispersal of juvenile cougars in fragmented habitat. J Wildl Manage 59: 228–237.
7. Luke C, Penrod K, Cabanero CR, Beier P, Spencer W, et al. (2004) A Linkage Design for the Santa Ana-Palomar Mountains Connection. Unpublished report. San Diego State University, Field Station Programs and South Coast Wildlands. Available at: http://www.scwildlands.org/reports/SCML_SantaAna_Palomar.pdf.
8. Morrison SA, Boyce WM (2009) Conserving connectivity: Some lessons from mountain lions in southern California. Conserv Biol 23: 275–285.
9. Ernest HB, Boyce WM, Bleich VC, May B, Stiver SJ, et al. (2003) Genetic structure of mountain lion (Puma concolor) populations in California. Conserv Genet 4: 353–366.
10. Burdett CL, Crooks KR, Theobald DM, Wilson KR, Boydston EE, et al. (2010) Interfacing models of wildlife habitat and human development to predict the future distribution of puma habitat. Ecosphere 1: 1.
11. Ernest HB, Penedo MCT, May BP, Syvanen M, Boyce WM (2000) Molecular tracking of mountain lions in the Yosemite Valley region in California: genetic analysis using microsatellites and faecal DNA. Mol Ecol 9: 433–441.
12. Kurushima JD, Collins JA, Well JA, Ernest HB (2006) Development of 21 microsatellite loci for puma (Puma concolor) ecology and forensics. Mol Ecol Notes 6: 1260–1262.
13. Menotti-Raymond M, David VA, Lyons LA, Schaffer AA, Tomlin JF, et al. (1999) A genetic linkage map of Microsatellites of the domestic cat (Felis catus). Genomics 57: 9–23.
14. Pilgrim KL, Mckelvey KS, Riddle AE, Schwartz MK (2005) Felid sex identification based on noninvasive genetic samples. Mol Ecol Notes 5: 60–61.
15. Toonen RJ, Hughes S (2001) Increased throughput for fragment analysis on an ABI Prism 377 automated sequencer using a membrane comb and STRAND software. BioTechniques 31: 1320–1324.
16. Sherwin WB, Jabot F, Rush R, Rossetto M (2006) Measurement of biological information with applications from genes to landscapes. Mol Ecol 15: 2857–2869.
17. Peakall R, Smouse PE (2006) GenAlEx 6: genetic analysis in Excel. Population genetic software for teaching and research. Mol Ecol Notes 6, 288–295.
18. Peakall R, Smouse PE (2012) GenAlEx 6.5: genetic analysis in Excel. Population genetic software for teaching and research – an update. Bioinformatics 28: 2537–2539.
19. Rousset F (2008) genepop'007: a complete re-implementation of the genepop software for Windows and Linux. Mol Ecol Resour 8: 1755–0998.
20. Kalinowski ST, Wagner AP, Taper ML (2006) ML-RELATE: a computer program for maximum likelihood estimation of relatedness and relationship. Mol Ecol 6: 576–579.
21. Rice WR 1989. Analysing tables of statistical tests. Evolution 43: 223–5.

22. Taberlet P and Luikart G (1999), Non-invasive genetic sampling and individual identification. Biol J Linn Soc 68: 41–55.
23. Waits LP, Luikart G, Taberlet P (2001) Estimating the probability of identity among genotypes in natural populations: cautions and guidelines. Mol Ecol 10: 249–56.
24. Pritchard JK, Stephens M, Donnelly P (2000). Inference of population structure using multilocus genotype data. G3 155: 945–959.
25. Hubisz MJ, Falush D, Stephens M, Pritchard J (2009) Inferring weak population structure with the assistance of sample group information. Mol Ecol Resour 9: 1322–1332.
26. Earl DA, vonHoldt BM (2012) STRUCTURE HARVESTER: a website and program for visualizing STRUCTURE output and implementing the Evanno method. Conserv Genet Resour 4: 359–361.
27. Pritchard JK, Wen W (2002) Documentation for structure version 2.3.4. As viewed June 1 2013: http://pritch.bsd.uchicago.edu.
28. Evanno G, Regnaut S, Goudet J (2005) Detecting the number of clusters of individuals using the software STRUCTURE: a simulation study. Mol Ecol 14: 2611–2620.
29. Waples RS, Gaggiotti O (2006) What is a population? An empirical evaluation of some genetic methods for identifying the number of gene pools and their degree of connectivity. Mol Ecol 15, 1419–1439.
30. Orloci L (1978) Multivariate analysis in vegetation research. 2nd ed. Dr W. Junk. The Hague. 451 p.
31. Nei M (1977) F-statistics and analysis of gene diversity in subdivided populations. Ann Hum Genet 41: 225–233.
32. Piry S, Alapetite A, Cornuet JM, Paetkau D, Baudouin L (2004) GENECLASS2: A software for genetic assignment and first-generation migrant detection. J Hered 95: 536–539.
33. Paetkau D, Slade R, Burdens M, Estoup A (2004) Genetic assignment methods for the direct, real-time estimation of migration rate: a simulation-based exploration of accuracy and power. Mol Ecol 13: 55–65.
34. Rannala B, Mountain JL (1997) Detecting immigration by using multilocus genotypes. Proc Natl Acad Sci USA 94: 9197–9201.
35. Cornuet JM, Luikart G (1997) Description and power analysis of two tests for detecting recent population bottlenecks from allele frequency data. G3 144: 2001–2014.
36. Luikart G, Allendorf FW, Cornuet JM, Sherwin WB (1998) Distortion of allele frequency distributions provides a test for recent population bottlenecks. J Hered 89: 238–247.
37. Luikart G, Cornuet JM (1998) Empirical evaluation of a test for identifying recently bottlenecked populations from allele frequency data. Conserv Biol 12: 228–237.
38. Waples RS (2006) A bias correction for estimates of effective population size based on linkage disequilibrium at unlinked gene loci. Conserv Genet 7: 167–184.
39. Waples RS, Do C (2008) LDNE: a program for estimating effective population size from data on linkage disequilibrium. Mol Ecol Resour 8: 753–756.
40. Wright S (1969) Evolution and the Genetics of Population. Vol 2. The Theory of Gene Frequencies. University of Chicago Press.
41. Lynch M, Ritland K (1999) Estimation of pairwise relatedness with molecular markers. G3 152: 1753–1766.
42. Kalinowski ST, Taper ML, Marshall TC (2007) Revising how the computer program CERVUS accommodates genotyping error increases success in paternity assignment. Mol Ecol 16: 1099–1006.
43. Jones OR, Wang J (2010) COLONY: a program for parentage and sibship inference from multilocus genotype data. Mol Ecol Resour 10: 551–555.
44. Wang J (2012) Computationally efficient sibship and parentage assignment from multilocus marker data. G3 191: 183–194.
45. Alho JS, Välimäki K, Merilä J (2010) Rhh: an R extension for estimating Multilocus heterozygosity and heterozygosity–heterozygosity correlation. Mol Ecol Resour 10: 720–72.
46. R Development Core Team (2011) R: A language and environment for statistical computing. R Foundation for Statistical Computing, Vienna, Austria.
47. Amos W, Worthington Wilmer J, Fullard K, Burg TM, Croxall JP (2001) The influence of parental relatedness on reproductive success. Proc R Soc Lond B - Biol Sci 268: 2021–2027.
48. Beier P (1993) Determining Minimum Habitat Areas and Habitat Corridors for Cougars. Conserv Biol 7: 94–108.
49. Crooks KR (2002) Relative Sensitivities of Mammalian Carnivores to Habitat Fragmentation. Conserv Biol 16: 488–502.
50. Logan K and Sweanor L (2001) Desert Puma: Evolutionary Ecology And Conservation Of An Enduring Carnivore. Island Press. 464.
51. Barbour E (2002) Metropolitan growth planning in California, 1900–2000. Public Policy Institute of California. San Francisco, California. 246.
52. Beier P (1996) Metapopulation modeling, tenacious tracking, and cougar conservation. In McCullough DR, editor, Metapopulations and wildlife management. Washington DC: Island Press. 293–323.
53. Riley SPD, Serieys LEK, Pollinger J, Sikich J, Dalbeck L, et al. Individual behaviors dominate the dynamics of an urban mountain lion population isolated by roads. Curr Biol 24: 1989–1994.
54. Roelke ME, Martenson JS, O'Brien SJ. (1993). The consequences of demographic reduction in the endangered Florida panther. Curr Biol 3: 340–350.
55. Culver M, Hedrick PW, Murphy K, O'Brien S Hornocker MG. (2008) Estimation of the bottleneck size in Florida panthers. Anim Conserv 11: 104–110.
56. Johnson WE, Onorato DP, Roelke ME, Land ED, Cunningham M, et al. (2010). Genetic restoration of the Florida panther. Science 329, 1641–1645.

11

Genetic Diversity and Structure of Dalmatian Pyrethrum (*Tanacetum cinerariifolium* Trevir. /Sch./ Bip., Asteraceae) within the Balkan Refugium

Martina Grdiša[1]*, Zlatko Liber[2], Ivan Radosavljević[2], Klaudija Carović-Stanko[1], Ivan Kolak[1], Zlatko Satovic[1]

1 Department of Seed Science and Technology, Faculty of Agriculture, University of Zagreb, Zagreb, Croatia, **2** Department of Botany, Division of Biology, Faculty of Science, University of Zagreb, Zagreb, Croatia

Abstract

Dalmatian pyrethrum (*Tanacetum cinerariifolium* Trevir. /Sch./ Bip.) is an outcrossing, perennial insecticidal plant, restricted to the eastern Adriatic coast (Mediterranean). Amplified fragment-length polymorphisms (AFLP) were used to investigate the genetic diversity and structure within and among 20 natural plant populations. The highest level of gene diversity, the number of private alleles and the frequency down-weighted marker values (*DW*) were found in northern Adriatic populations and gradually decreased towards the southern boundary of the species range. Genetic impoverishment of these southern populations is most likely the result of human-related activities. An analysis of molecular variance (AMOVA) indicated that most of the genetic diversity was attributed to differences among individuals within populations (85.78%), which are expected due to the outcrossing nature of the species. A Bayesian analysis of the population structure identified two dominant genetic clusters. A spatial analysis of the genetic diversity indicated that 5.6% of the genetic differentiation resulted from isolation by distance (IBD), while 12.3% of the genetic differentiation among populations followed the pattern of isolation by environmental distance (IBED). Knowledge of the genetic diversity patterns of the natural populations and the mechanism behind these patterns is required for the exploitation and possible conservation management of this endemic and economically important species.

Editor: Igor Mokrousov, St. Petersburg Pasteur Institute, Russian Federation

Funding: This study was supported by the Ministry of Science, Education and Sports of the Republic of Croatia (public.mzos.hr), within the framework of Project No. 178-1191193-0212 (ZS), Project No. 178-0000000-3595 (IK) and Project No. 119-1191193-1232 (ZL). The funders had no role in study design, data collection, analysis, decision to publish, or preparation of the manuscript.

Competing Interests: The authors have declared that no competing interests exist.

* Email: mgrdisa@agr.hr

Introduction

Conservation of genetic diversity is essential for any species long-term survival [1], providing the ability to respond to selective pressures caused by both anthropogenic and natural environmental changes [2]. Patterns of genetic diversity and population structure are specifically shaped in response to many factors and are result of both past history and present processes [3]. Primarily, climatic oscillations during the Quaternary causing repeated restriction-recolonization events have left distinguishing signatures in modern species genetics [4]. These oscillations have had a considerable impact on the current distribution and genetic diversity in numerous animal and plant species [5–8] in which altitudinal and latitudinal range shifts were a likely response to intense climatic changes [9–10]. Detailed phylogeographical studies have revealed that three European peninsulas of the Mediterranean area (Apennines, Balkans and Iberia) have served as refugia for many temperate species and a postglacial recolonization source for central and northern Europe [3,7,9–11]. However, this traditional interpretation of southern European peninsulas as being single glacial refugia has been modified, and the model "refugia within refugia" was developed [10,12–15]. Genetic divergence expressed as a rarity index (*DW*; frequency down-weighted marker values) as an indicator of the historical processes of a species [16] enables the distinction between old vicariance and recent dispersal. As a consequence of long-term isolation, populations in glacial refugia are characterized by high levels of *DW* values and genetic diversity [17], whereas populations that were established after the most recent glacial cycle show little genetic variation [18] as an outcome of successive founder events during postglacial colonization [6,9]. However, zones of secondary contact of divergent lineages can also exhibit high levels of genetic diversity [11,19].

The Balkan Peninsula itself represents a unique ecological and biogeographical phenomenon in Europe possessing the tremendous number of plant and animal species and pronounced richness in endemic taxa as well. This remarkable diversity is the result of the variety of its regions, complex geological history, and interactions between populations, species and ecosystems [20]. In the last decade various phylogeographic and molecular studies have been conducted in this area with the aim of elucidating the genetic relationships of its flora and fauna [10,21–25].

It is widely acknowledged that geographical and environmental factors can also play significant roles in the modeling of species-specific inter- and intra-population genetic structure. Spatially separated populations may experience isolation by distance (IBD) phenomena in which geographical barriers and physical distances cause restricted pollen and seed dispersal, i.e., gene flow [26–30]. Greater genetic divergence between populations inhabiting different environments suggests that various present and/or past ecological conditions may significantly impact the genetic differentiation of local populations [31–32], forming an isolation by environmental distance (IBED) pattern [33]. A significant positive partial correlation, after eliminating the effect of geographic distance, indicates the contribution of climatic gradients to patterns of genetic divergence [33–34].

Nevertheless, species are greatly affected by direct or indirect human activities including population fragmentation and overharvesting pressures, all resulting in loss of genetic diversity and the alteration of their population genetic structure [35]. Moreover, founder effects and conscious or unconscious human selections for high yield and quality through cultivation of wild plants likewise impose profound impacts on the genetic diversity patterns by narrowing genetic base of the plant material [36–38].

For our investigation we have chosen Dalmatian pyrethrum (*Tanacetum cinerariifolium* Trevir. /Sch./ Bip.), an insecticidal perennial plant. Dalmatian pyrethrum is an outcrossing diploid (2n = 18) [39], self-incompatible and thermophytic plant species of the family Asteraceae [40]. This species is restricted to the eastern coast of the Adriatic Sea, and its distribution range expands from the coastal region to 200 m.a.s.l., but in some cases, it is also present in higher mountainous Mediterranean zones at elevation above 500 m.a.s.l. Continuously distributed populations of Dalmatian pyrethrum can be found in extremely degraded habitats with shallow rocky soils. In Croatia, abundant populations can be found in the southern parts of the Istrian Peninsula (Premantura), the Kvarner islands (Krk, Cres and Lošinj), the Velebit and Biokovo mountains and along the Dalmatian coastal region and its islands (Brač, Hvar, Biševo, Vis, Korčula, Lastovo and Mljet) [41]. Apart from Croatia distribution range expands to the southern parts of Bosnia and Herzegovina, and coastal regions of Montenegro and Albania [42]. Plants produce the natural insecticide pyrethrin, which attacks the nervous systems of the insects causing a knock-down effect and death [43].

The cultivation and use of Dalmatian pyrethrum and its products have a documented history in Croatia. The dust from grinding dried flowers has traditionally been used in Croatian households and agriculture [44]. Originally, flowers were excessively gathered from the wild resources; however, due to the high demand for plant material, cultivation began in 1850s near Dubrovnik and rapidly expanded along the Dalmatian coastal region and the islands [45]. The first fields were established with seeds gathered from the wild. Dalmatian pyrethrum was primarily reproduced through vegetative splits and later by directly sowing the seeds [46]. It was grown in the middle and southern parts of Dalmatia, including Zadar, Biograd, Šibenik, Split, Hvar, Brač, Makarska, Korčula and Dubrovnik, as well as the northern Adriatic islands Krk and Cres [44]. In 1914, Croatia became the leading producer with high quality plant material. The highest yield was achieved in 1926 with approximately 1,359 tons of dried pyrethrum flowers [46].

The goal of this study was to analyze the genetic diversity and structure within and among 20 populations of Dalmatian pyrethrum across their geographical range in Croatia and to determine impact of (*i*) Pleistocene climatic oscillations, (*ii*) geographic (IBD) and (*iii*) environmental isolation (IBED), and (*iiii*) human activities through overexploitation and cultivation on the obtained pattern of genetic diversity. AFLP markers have found the widest application in assessing the variability and genetic structure of populations in a variety of plants [47–53] and once again have proven to be invaluable tools in describing the former. Despite its historical significance and value, the genetic diversity and structure of Dalmatian pyrethrum has never been investigated; therefore, this study using AFLP markers represents the first such attempt. Dalmatian pyrethrum is nationally strictly protected by The Croatian Ordinance on the proclamation of protected and strictly protected wild taxa (Official Gazette 99/09; http://narodne-novine.nn.hr/clanci/sluzbeni/dodatni/403046.pdf). Today, as an endemic species, Dalmatian pyrethrum is threatened by anthropogenic habitat loss and degradation related to factors such as urbanization and habitat conversion; therefore, the need and importance of these species' genetic resources protection must be recognized even more. The gathered information could be of utmost importance for future activities regarding the conservation and breeding programs of this endemic and economically important species.

Materials and Methods

Ethics statement

Collecting of all samples from natural habitats in Croatia including Nature Park Biokovo was permitted by the authority of Directorate for Nature Protection of the Ministry of Culture of the Republic of Croatia (532-08-01/6-06-02; 07.07.2006).

Sampling of plant populations

A total of 411 specimens of *T. cinerariifolium* were sampled from 20 locations along the Adriatic coast in Croatia (Figure 1). Between 13 and 25 samples per location were collected as fresh leaf tissue (Table 1). The vast majority of the species populations is distributed in Croatia, therefore we focused our study exclusively on a wide sampling of Croatian populations.

DNA isolation and AFLP analysis

The total genomic DNA was isolated from 90 to 110 mg of fresh leaves tissue with DNA-GenElute Plant Genomic DNA Miniprep Kit (Sigma Aldrich, Steinheim, Germany), according to manufacturer's instructions. Prior to isolation leaves tissue was quickly frozen in liquid nitrogen (−196°C) and ground to a fine powder with TissueLyzer homogenizer set (Qiagene, Hilden, Germany) at 30 Hz for 1 minute. The DNA concentrations were measured using a Qubit Fluorometer (Invitrogen, Carlsbad, California). The AFLP analysis was carried out following the original protocol that was described by Vos et al. [54] with some modifications [55]. Restriction of the genomic DNA was performed using the restriction enzymes *Eco*RI and *Mse*I. Specific adapters were ligated using T4 DNA Ligase (Fermentas, Thermo Scientific). The restriction digestion and adapter ligation, pre-amplification and selective amplifications were performed in a GeneAmp PCR System 9600 (Applied Biosystems, Carlsbad, SAD). Six primer combinations were selected for amplification; VIC-*Eco*RI-ACGC+*Mse*I-AGA, NED-*Eco*RI-AGA+*Mse*I-CAG, VIC-*Eco*RI-ACG+*Mse*I-CGA, FAM-*Eco*RI-ACC+*Mse*I-CGA, NED-*Eco*RI-AGA+*Mse*I-CGA, PET-*Eco*RI-ATG+*Mse*I-CGA. The amplified fragments were separated by capillary electrophoresis in an ABI3130xl Genetic Analyzer (Applied Biosystems, Carlsbad, USA). The amplified DNA fragments in the size range of 50 to 125 bp were scored with GeneMapper 4.0 software

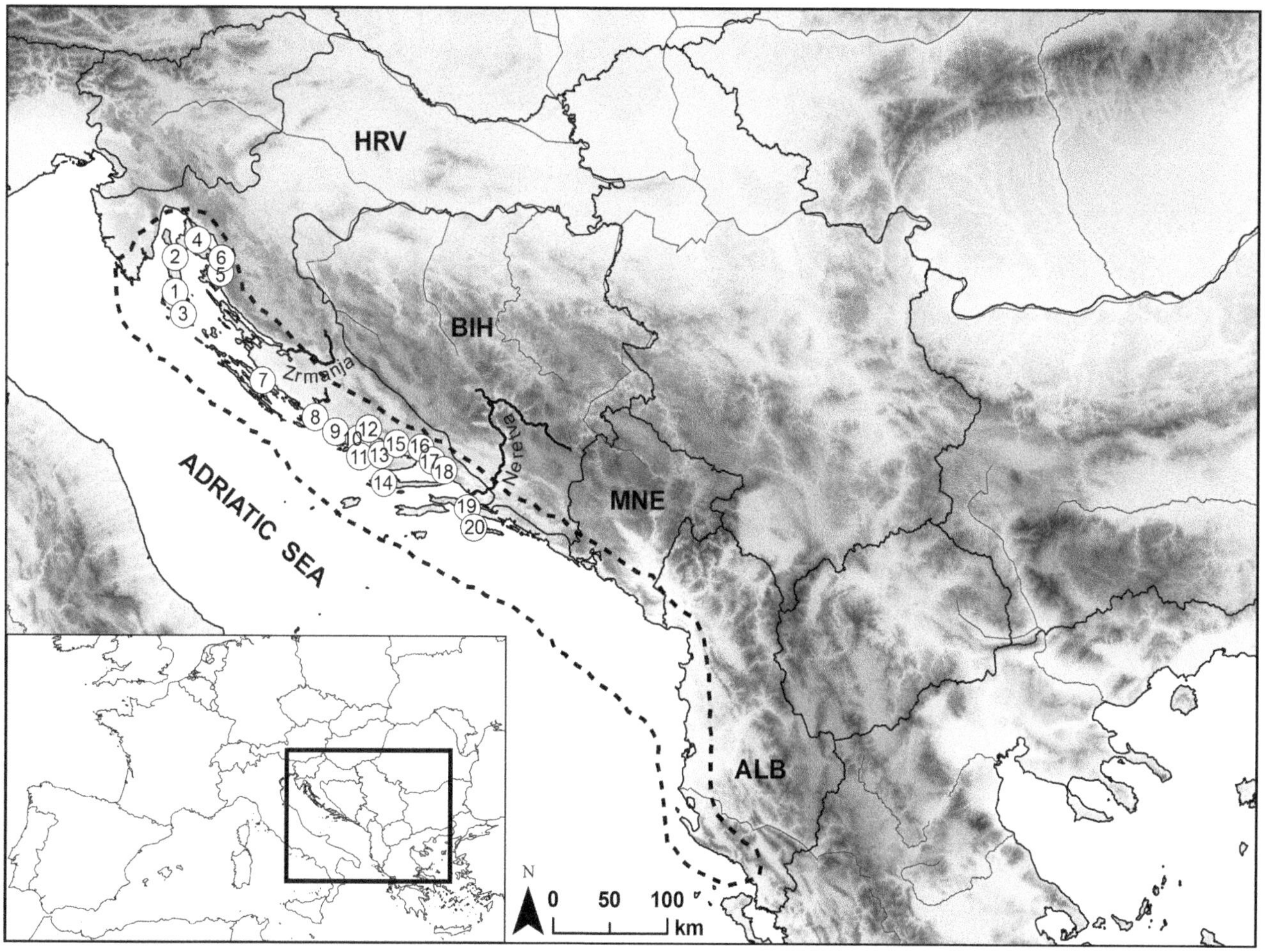

Figure 1. Map of geographic distribution and sampled populations of *Tanacetum cinerariifolium* in Croatia. Dashed line indicates distribution range of *T. cinerariifolium*. Circles numbered 1–20 represent sampled populations.

(Applied Biosystems, Carlsbad, USA) as being present (1) or absent (0) in order to create a binary matrix.

Data analysis

The selection of markers for further statistical analysis and the estimation of the error rate per primer combination were performed as suggested by Herrmann et al. [56]. Thirteen DNA samples were used as duplicated samples, two of which were chosen for multiple controls, while 14 samples represented negative controls. The output from GeneMapper (fragment size and peak heights) was imported into the marker selection algorithm scanAFLP 1.3r (www-leca.ujf-grenoble.fr/logiciels.htm).

The genetic diversity within populations was estimated by determining the percentage of polymorphic bands *(%P)*, the number of private alleles (N_{pr}) and Shannon's information index (I) [57–58]. Shannon's information index was used to measure the total diversity (H_T) as well as the average intra-population diversity (H_P) and the diversity within (H_P/H_T) and among populations $[(H_T–H_P)/H_T]$. As a measure of divergence, the frequency down-weighted marker values (DW) [23] were calculated using the software AFLPdat [59].

The marker allelic frequencies were estimated from the observed frequencies of the fragments using a Bayesian method with non-uniform prior distribution of allele frequencies proposed by Zhivotovsky [60] as implemented in AFLP-Surv ver. 1.0 [61]. This approach assumes Hardy-Weinberg equilibrium ($F_{IS} = 0$) justified by the outcrossing nature of Dalmatian pyrethrum. The obtained allelic frequencies were used in the analysis of genetic diversity within and between populations following the treatment of Lynch and Milligan [62]. The population genetic structure was described in terms of the total gene diversity (H_T), the average gene diversity within populations (H_W), the average gene diversity among populations in excess of that which was observed within populations (H_B), and Wright's F_{ST} statistics. The expected heterozygosity or Nei's gene diversity (H_E) [63] was calculated for each population as well as pairwise Nei's genetic distances (D_{NEI72}) [62,64]. The unrooted tree was constructed using the Fitch-Margoliash method in the PHYLIP ver. 3.6b software package [65]. A total of 1,000 distance matrices were computed by bootstrapping using the program AFLP-Surv, and the bootstrap values were calculated from FITCH and CONSENSE within PHYLIP.

An analysis of molecular variance (AMOVA) [66] was used to partition the total genetic variance within and among populations using Arlequin ver. 2.000 [67]. The variance components were tested statistically by non-parametric randomization tests using 10,000 permutations. Pairwise population comparisons that were examined with AMOVA resulted in φ_{ST} values that were equivalent to the proportion of the total variance that was partitioned between the two populations and could be interpreted

Table 1. Sampling sites and molecular diversity revealed by AFLP markers in 20 Dalmatian pyrethrum populations from Croatia.

No.	Accession Number[a]	Population	Latitude (N)[b]	Longitude (E)[b]	Elevation (m)	n	P%	N_{pr}	I	H_E	*DW*
1	MAP02143	Osor	44.70	14.39	2	23	44.18	2	0.242	0.112	56.76
2	MAP02158	Cres	44.96	14.41	72	22	45.70	2	0.246	0.113	58.18
3	MAP02139	Mali Lošinj	44.53	14.47	42	21	50.93	4	0.284	0.131	95.20
4	MAP02156	Krk	45.08	14.67	24	24	52.11	3	0.268	0.118	62.74
5	MAP02148	Gornja Klada	44.82	14.90	589	22	40.64	1	0.214	0.105	48.82
5	MAP02138	Senj	44.93	14.92	6	19	37.61	2	0.214	0.105	42.91
7	MAP02180	Pašman	44.00	15.29	17	22	36.26	3	0.205	0.102	41.45
8	MAP02171	Zlarin	43.70	15.84	45	25	42.33	0	0.234	0.109	45.04
9	MAP02166	Primošten	43.58	16.05	190	22	45.53	4	0.239	0.110	46.09
10	MAP02144	Čiovo	43.51	16.29	143	19	47.22	2	0.251	0.118	48.67
11	MAP02153	Šolta	43.38	16.29	114	19	40.13	0	0.226	0.105	31.59
12	MAP02145	Kozjak	43.58	16.41	516	20	41.15	1	0.220	0.106	31.25
13	MAP02155	Brač	43.38	16.52	116	22	40.81	3	0.216	0.102	47.50
14	MAP02173	Hvar	43.17	16.53	333	21	37.44	0	0.209	0.102	29.40
15	MAP02170	Omiš	43.45	16.70	2	22	38.79	2	0.205	0.102	39.09
16	MAP02142	Kotiški Stanovi	43.32	17.06	1335	13	26.14	1	0.160	0.092	18.92
17	MAP02140	Lađena	43.30	17.07	1295	18	34.40	1	0.196	0.099	47.38
18	MAP02141	Ravna Vlaška	43.29	17.08	1228	22	34.23	3	0.196	0.096	48.53
19	MAP02184	Pelješac	42.92	17.40	18	14	42.50	1	0.247	0.125	33.18
20	MAP02152	Mljet	42.77	17.45	172	21	33.39	1	0.186	0.094	40.03
		Average					40.57		0.223	0.107	45.64
		Minimum					26.14		0.160	0.092	18.92
		Maximum					52.11		0.284	0.131	95.20

[a]Accession number from The Collection of Medicinal and Aromatic Plants, Zagreb, Croatia as available at the CPGRD (http://cpgr.zsr.hr);
[b]N-North; E-East; Coordinates are in degree decimal format; n-sample size; %P-proportion of polymorphic bands; N_{pr} -number of private bands; I-Shannon's information index; H_E -gene diversity of a population assuming Hardy-Weinberg equilibrium; and *DW*-frequency down-weighted marker values.

as the inter-population distance average between any two populations [68].

The genetic structure was assessed using the model-based clustering approach implemented by the software STRUCTURE 2.3.3 [69], which simultaneously identifies clusters and assigns individuals to clusters using a Bayesian approach. Ten runs of STRUCTURE were performed by setting the number of clusters from 1 to 21 ($K = 1$–21). Each run consisted of a burn-in period of 200,000 steps followed by 10^6 MCMC (Monte Carlo Markov Chain) replicates assuming an admixture model and correlated allele frequencies. No prior information was used to define the clusters. The calculations were carried out at the Bioportal of the University of Oslo (http://www.bioportal.uio.no/). The most likely number of clusters (K) was selected by calculating an *ad hoc* statistic ΔK based on the rate of change in the log probability of the data between successive K values as described by Evanno et al. [70] and implemented in Structure-sum-2011 [71]. The runs with the maximum likelihood were selected, and by averaging the estimated membership coefficients of the individuals, the proportion of ancestry of each population in each of the clusters was calculated.

A further population mixture analysis was conducted using BAPS ver. 5.3 [72] (a) without using the geographic origin of the samples as an informative prior (*'clustering of individuals'*) and (b) with this prior (*'spatial clustering of individuals'*) [73]. BAPS was run with the maximal number of clusters (K) set to 20, and each run was replicated 10 times. The results of the mixture analysis were used as input for population admixture analysis [74] with the default settings to detect admixture between the clusters.

Isolation by distance (IBD) among the populations was tested using the method of Rousset [75]. Mantel's test [76] was used to compute and test the linear correlation between the matrix of the natural logarithm of geographic distances (in km) between pairs of populations and the matrix of pairwise $F_{ST}/(1\text{-}F_{ST})$ ratios.

The plausibility of patterns of isolation by environmental distance (IBED) was evaluated, where differences in the bioclimatic variables among the sampling sites would better explain the genetic distance patterns [33,77]. The climate data for 25 sampling sites were obtained from the WorldClim database at a spatial resolution of approximately 1 km^2 [78]. The ecological characteristics of the sampling sites were described using 19 bioclimatic variables (11 temperature- and 8 precipitation-related) representing the annual trends, seasonal variations and extremes in temperature and precipitation (Table 2). Bioclimatic variables were used for the Principal Component Analysis (PCA) and subsequently to calculate the environmental distances among populations based on principal component scores rather than original values to avoid intercorrelation among variables. PCA was performed using the PROC PRINCOMP procedure in SAS. The number of principal components was determined by checking the eigenvalues of the principal components and the cumulative proportion of the explained variance. The biplot was constructed using two principal components showing populations and bioclimatic variables (as vectors). The standardized scores of the first three principal components (eigenvalue >1) were multiplied by the root of their eigenvalues, and the pairwise Euclidean distance between populations were calculated to obtain an environmental distance matrix.

A simple Mantel's test was used to test the linear correlation between the matrix of environmental and genetic [$F_{ST}/(1\text{-}F_{ST})$] distances, and a three-way Mantel's test was conducted between the same distances matrices while considering the geographical distances among populations. Thus, the correlations between the residual environmental and genetic distances among the analyzed populations represent isolation by environmental distance (IBED) pattern. The significance levels were obtained after 10,000 permutations as implemented in NTSYS-pc ver. 2.10s [79].

Results

ScanAFLP

The initial number of polymorphic bands that were obtained with six selective primer combinations equaled 850, while after the applying scanAFLP, the remaining dataset consisted of 593 polymorphic markers (43.35% of the initial number of polymorphic markers). The estimated error rate per primer combination ranged from 0.68% to 3.50% with a mean of 1.83%, which is within the typical range found in AFLP datasets [80].

Within population genetic diversity

The percentage of polymorphic markers varied among the populations with the highest proportion in population 4 (52.11%) and the lowest in population 16 (26.14%). Shannon's information index indicated that the total diversity (H_T) was 0.297, while the average intrapopulation diversity (H_P) was 0.223. The proportion of diversity within populations (75.00%) was considerably higher than the proportion of diversity among populations. Out of 593 scored markers, 36 were private. The populations with the most private alleles (4) were pops. 3 and 9, while no such alleles were found in the pops. 8, 11 and 14. The population 3 had the highest level of expected heterozygosity (H_E), and the population 16 had the lowest (0.131 and 0.92, respectively) (Table 1, Figure 2A). The frequency down-weighted marker values (DW) ranged from 18.92 in the population 16 to 95.20 in the population 3 (Table 1, Figure 2B).

Genetic structure among populations

A moderate level of population differentiation was observed ($F_{ST} = 0.078$) ranging from 0.018 between pops. 10 and 11 to 0.161 between 6 and 18.

The AMOVA analysis showed that although the most of the genetic diversity was attributable to differences among individuals within populations (85.78%), the φ_{ST} values among populations were highly significant ($\varphi_{ST} = 0.141$ $p<0.0001$) confirming the existence of population differentiation. The φ_{ST} values ranged from 0.049 between pops. 4 and 11 to 0.278 between pops. 5 and 16.

Nei's distance (D_{NEI72}) varied from 0.002 between pops. 10 and 11 to 0.022 between pops. 6 and 18 with an average value of 0.010. The pattern of grouping in the Fitch-Margoliash tree suggests a direct relationship between the populations and their geographical origins. This pattern revealed the existence of two clades (northern vs. middle and southern Adriatics) supported by a bootstrap value of 88%. Inside the northern clade, the inland pops. 5 and 6 along with the population 2 separated from the northern island populations with a bootstrap value of 72%. The pops. 9 and 11 separated from pops. 8 and 4 with a bootstrap value of 88%. The populations from Mt. Biokovo (pops. 16, 17 and 18) grouped together and showed a clear separation from the remaining southern populations with a high bootstrap support of 99% (Figure 3).

The results of the STRUCTURE analysis are reported in Figure 2C and 2D. The average estimates of the likelihood of the data, conditional on a given number of clusters, *[ln P(X|K)]*, were obtained for each of the 10 independent runs for each K (from $K = 1$ to 21). The estimates of *[ln P(X|K]* increased progressively as K increased. Using these estimates, the highest ΔK value was observed for $K = 2$ (302.59). The second highest ΔK value was

Table 2. Correlations between 19 environmental variables (BIO01–BIO19) and the first three principal components.

No.	Environmental variable	PC1		PC2		PC3	
BIO01	Annual Mean Temperature	0.942	***	0.244	ns	0.095	ns
BIO02	Mean Diurnal Range (Mean of monthly (max temp - min temp)	0.678	**	0.546	*	−0.357	ns
BIO03	Isothermality (P2/P7) (*100)	0.733	***	0.474	*	−0.251	ns
BIO04	Temperature Seasonality (standard deviation *100)	−0.346	ns	0.545	*	−0.497	*
BIO05	Max Temperature of Warmest Month	0.935	***	0.322	ns	0.027	ns
BIO06	Min Temperature of Coldest Month	0.928	***	0.138	ns	0.218	ns
BIO07	Temperature Annual Range (P5–P6)	0.518	*	0.652	**	−0.482	*
BIO08	Mean Temperature of Wettest Quarter	0.866	***	0.285	ns	−0.179	ns
BIO09	Mean Temperature of Driest Quarter	0.935	***	0.270	ns	0.078	ns
BIO10	Mean Temperature of Warmest Quarter	0.938	***	0.263	ns	0.082	ns
BIO11	Mean Temperature of Coldest Quarter	0.950	***	0.199	ns	0.131	ns
BIO12	Annual Precipitation	−0.712	***	0.653	**	0.234	ns
BIO13	Precipitation of Wettest Month	−0.539	*	0.747	***	0.380	ns
BIO14	Precipitation of Driest Month	−0.860	***	0.363	ns	−0.275	ns
BIO15	Precipitation Seasonality (Coefficient of Variation)	0.795	***	0.061	ns	0.476	*
BIO16	Precipitation of Wettest Quarter	−0.503	*	0.789	***	0.348	ns
BIO17	Precipitation of Driest Quarter	−0.844	***	0.425	ns	−0.203	ns
BIO18	Precipitation of Warmest Quarter	−0.873	***	0.380	ns	−0.196	ns
BIO19	Precipitation of Coldest Quarter	−0.492	*	0.489	*	0.695	***
	Eigenvalue	11.55		4.00		1.97	
	% variance	60.79		21.04		10.39	

ns - non-significant;
*significant at $P<0.05$;
**significant at $P<0.01$;
***significant at $P<0.001$.

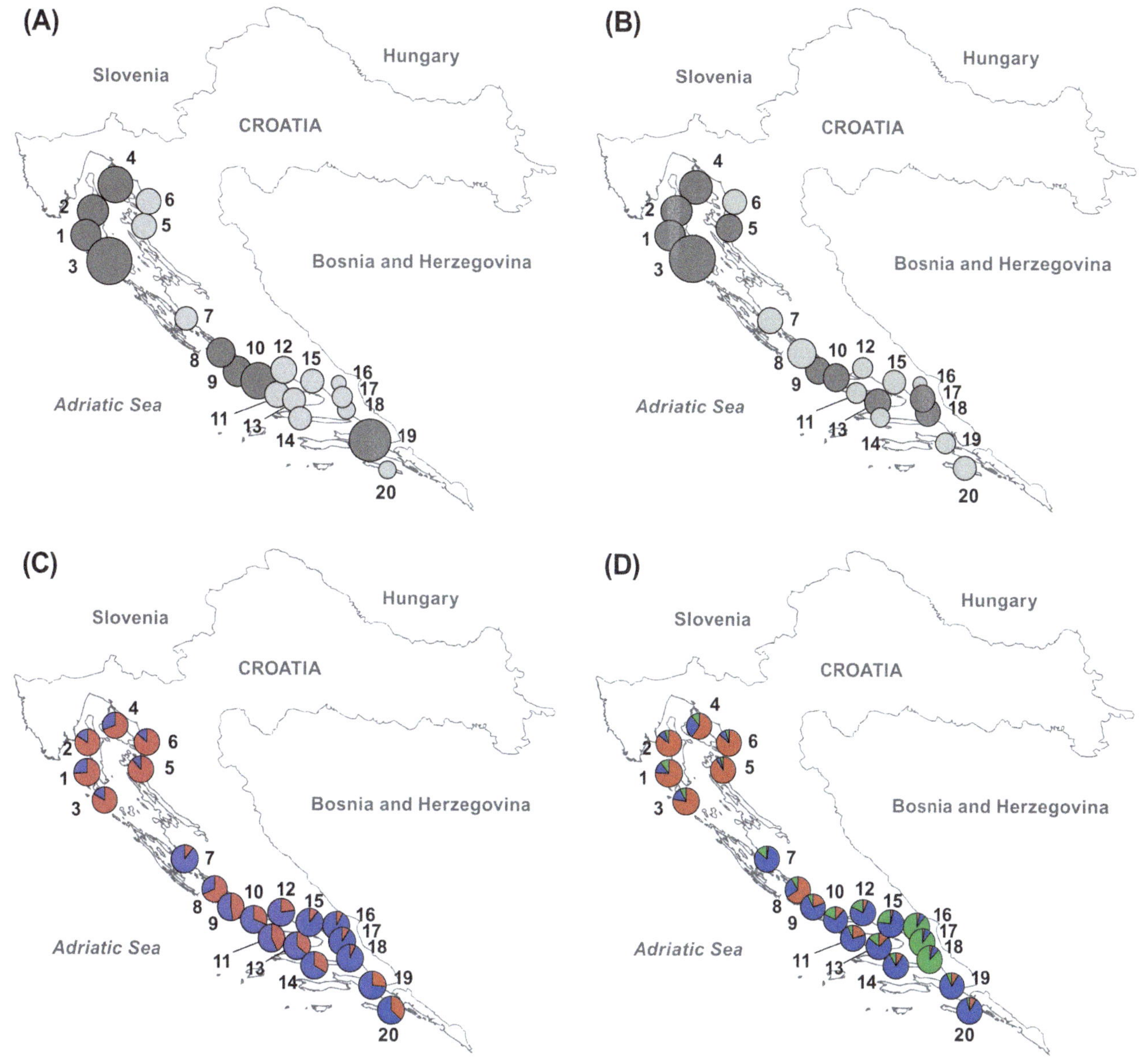

Figure 2. AFLP variation of *Tanacetum cinerariifolium*. A) Gene diversity (H_E), B) Frequency down-weighted marker values (*DW*). In (A) and (B), the size of the circles is proportional to the depicted values (dark gray color represents values above average across populations and light gray represents values below average). C) Bayesian analysis of the population structure using the software STRUCTURE assuming $K=2$, D) Bayesian analysis of the population structure using the software STRUCTURE assuming $K=3$. In (C) and (D), the proportions of the ancestry of each population in each of the defined gene pools are color-coded (gene pool A red; gene pool B-blue; and gene pool C-green).

observed for $K=3$ (180.78). At *Ks* values greater than three, the average *[ln P(X|K)]* values decreased, while the variances increased substantially and, therefore, the ΔK values were considerably lower (0.10 to 5.80) in comparison to $K=2$ and 3. Thus, the hypothesis of having more than three clusters (gene pools) was far from optimal.

The proportions of membership (ancestry) of each individual in each of the gene pools were calculated for $K=2$ to 3 based on the run with the highest *[ln P(X|K)]*, and populations were assigned to a particular gene pool (A, B or C). The results were in accordance with the results that were obtained using the distance-based method. Representative populations of each gene pool that were identified in the STRUCTURE analysis were located in proximity of each other. At $K=2$, the representative populations of gene pool A were mainly those from the northern Adriatic, while the populations from the middle and southern Adriatic were representative of a separate gene pool B. The most admixed populations were 9 and 11 with almost equal proportions of ancestry from the two gene pools (Figure 2C). At $K=3$, a newly formed gene pool C was identified and was predominant in the pops. 16, 17 and 18 from Mt. Biokovo (Figure 2D).

The BAPS mixture with or without spatially informative priors resulted in the congruent assignment of individuals to two clusters. The best partitions received log marginal likelihoods of −51646.2626 at P = 1 (without using geographical coordinates as informative priors) and −51793.7943 at P = 1 (spatial clustering). The results were consistent with the results of the STRUCTURE analysis (Figure S1).

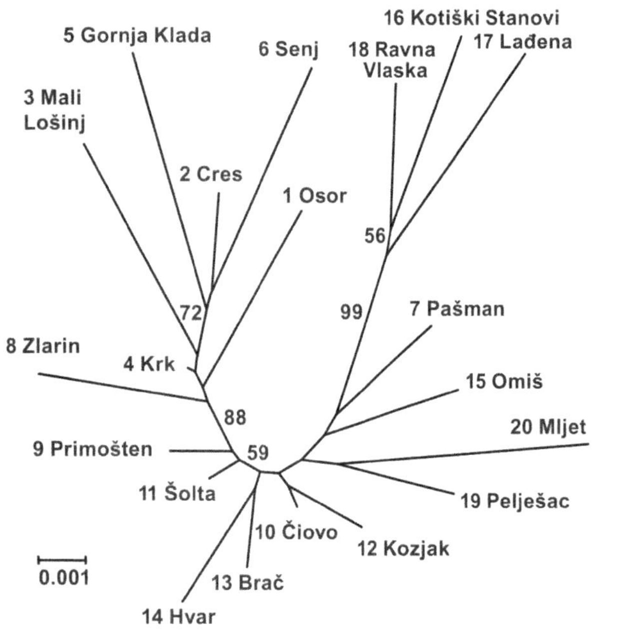

Figure 3. Unrooted Fitch-Margoliash tree of 20 *Tanacetum cinerariifolium* populations. Bootstrap values greater than 50% that were obtained after 1000 permutations are indicated on the branches.

Isolation by distance (IBD) and isolation by environmental distance (IBED)

A significant correlation was found between the genetic and geographical distances ($r = 0.236$; $P_{Mantel} = 0.012$), suggesting that 5.6% of the genetic differentiation between populations could be explained by geographical distances (Figure 4A).

As expected, 19 environmental variables were highly intercorrelated. Out of 171 pairwise correlations, 107 (62.57%) were significant at $P<0.05$ (Table S1). In 49 cases the strong positive correlations ($r>0.70$) were detected, mostly among eleven temperature-related variables (BIO01–BIO11) and among eight precipitation-related variables (BIO12–BIO19). The strong negative correlations ($r<-0.70$) were found in 10 cases, usually between temperature-related and precipitation-related variables.

A Principal Component Analysis based on the correlation matrix revealed that first three principal components had an eigenvalue >1 and explained more than 90% of the total variation (Table 2). Eight temperature-related variables (BIO01, BIO03, BIO05, BIO06, BIO08, BIO09, BIO10 and BIO11) and one precipitation-related variable (BIO15) were highly positively correlated ($r>0.70$; $P<0.001$) with the first principal component (PC1), while four precipitation-related variables (BIO12, BIO14, BIO17 and BIO18) were highly negatively correlated ($r<-0.70$; $P<0.001$). Strong positive correlations ($r>0.70$; $P<0.001$) were observed between the second principal component (PC2) and two precipitation-related variables (BIO13 and BIO16). The biplot that was constructed by the two principal components showing populations and 19 environmental factors (as vectors) is presented in Figure 5. PC1, explaining 60.79% of the total variation, separated the middle (pops. 7–11) and southern (pops. 13–15, 19 and 20) Adriatic populations occupying warmer habitats from pops. 12, 16, 17 and 18, which were sampled at higher elevation (516 to 1,335 m.a.s.l.). Along PC2, explaining 21.04% of the total variation, the northern Adriatic populations (pops. 1–6), occupying habitats that were characterized by greater rainfall, were separated from the remaining populations. The average Euclidean distance between the pairs of populations that was calculated based on the standardized scores of the first three principal components was 5.34, ranging from 0.08 (between pops. 16 and 17) to 10.12 (between pops. 8 and 16, respectively).

A greater correlation was found between the genetic and environmental distances ($r = 0.376$; $P_{Mantel}<0.001$) than between the genetic and geographical distances (Figure 4B). After considering the effect of the geographical distances in a three-way Mantel test, the correlation between the residual genetic and residual environmental distances remained significant ($r = 0.351$; $P_{Mantel} = 0.005$) suggesting that isolation by environmental distance (IBED) explains 12.3% of the observed genetic pattern (Figure 4C). The removal of the effect of environmental variation resulted in a still significant but much lower correlation between the genetic and geographic distances ($r = 0.189$; $P_{Mantel} = 0.039$).

Discussion

Identification of genetic diversity patterns is of fundamental importance for effective management and species conservation

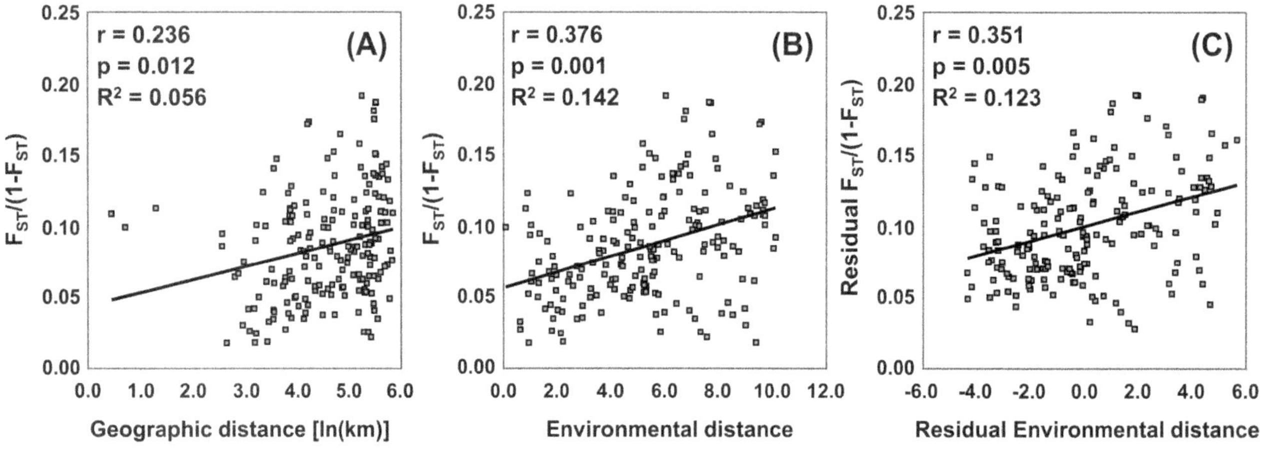

Figure 4. Isolation by distance and isolation by environmental distance. Plots of Mantel's test showing the relationships between A) the geographical and genetic distances, B) the environmental and genetic distances and C) the residual environmental and genetic distances considering the geographical distances between 20 Dalmatian pyrethrum populations.

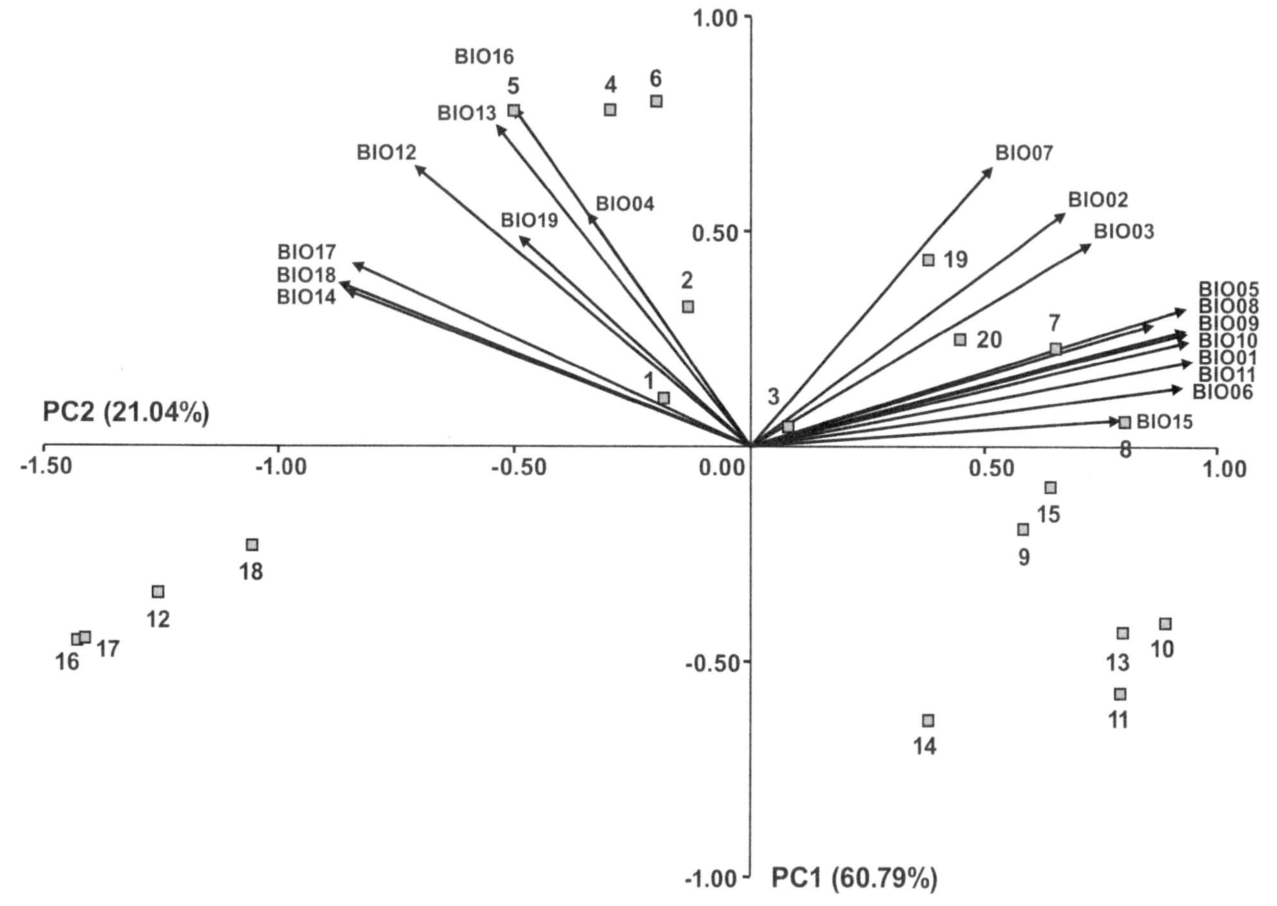

Figure 5. PCA plot based on 19 environmental variables (BIO01–BIO19) describing 20 *Tanacetum cinerariifolium* sampling sites.

[7], and as we expand our knowledge about the genetic diversity and structure of natural populations, it subsequently becomes important to understand the mechanisms behind the determined patterns. Therefore, the genetic diversity and population genetic structure of Dalmatian pyrethrum must be analyzed both in the context of demographic history and biological traits of the species [3]. Additionally, understanding the impact of geographical distances and consequently restricted gene flow among the analyzed populations as well as climatic characteristics of the sampling sites can contribute to a better comprehension of the relationships between the populations. Furthermore, the fact that human historical activities might have influenced the observed pattern must not be excluded.

The results of our study show genetic differentiation of *T. cinerariifolium* into two (Figure 2C) to three (Figure 2D) distinct groups. At $K = 2$ it is plausible that the north-south split in the study area is due to widely spaced canyon of karstic river Zrmanja in northern Dalmatia as karst rivers are often major barriers to the otherwise continuous distribution of species. The obtained results are somewhat similar to the phylogeographical patterns detected in *Edraianthus tenuifolius* [10], *Campanula pyramidalis* [25] and *Cardamine maritima* [81] showing either phylogeographical or taxonomical split in the area of river Neretva valley (central Dalmatia). However, the fact that north-south split might be the artefact of genetic impoverishment (discussed further in the text) of central and southern population must not be neglected. With additional splitting of southern gene pool at $K = 3$, a new pattern that is congruent with the survival of the species in multiple refugial areas, followed by a period (s) of isolation occurs. Three genetically and geographically distinct and well-defined groups of Dalmatian pyrethrum populations therefore support the "refugia within refugia" model that was already confirmed in numerous studies in the Balkan Peninsula [10,21,23,24]. The restricted gene flow among inter-refugial populations has led to genetic differentiation, resulting in well-defined gene pools [82]. Among the 20 populations that were included in this study, the highest levels of gene diversity, private alleles and *DW* were found in group of northern Adriatic populations from Kvarner bay. The genetic structure of these populations, which are grouped together within the same cluster as revealed by the STRUCTURE analysis, suggests their ancient origin and significant isolation through a recent period of evolution [17] Such northerly located micro-refugia for a thermophytic species like Dalmatian pyrethrum is quite unexpected but is strongly supported by the obtained results. Previous research in this region on Dalmatian wall lizard (*Podarcis melisellensis*) [21], nose-horned viper (*Vipera ammodytes*) [83] and *Edraianthus tenuifolius* [10] found no evidence of a refugium that far north. Our results could possibly be explained by numerous glaciations and interglaciation periods that are characteristic of the Quaternary and that were accompanied by vast species migrations. During the last interglacial period (approximately 130,000 yr. B. P. to 116,000 yr. B. P.), the global climate was warmer than at present [84–85], and it is possible that the populations that were initially located in the more southern and thus warmer parts of Adriatic basin expanded far to the north-west. By tracking suitable microhabitats some populations survived during the cold period, and isolation from their pool of origin eventually has led to strong genetic differentiation that can now be detected through northern Adriatic gene pool. These microhabitats in Kvarner bay served as microrefugia, and at the end of the last glacial period, these protected and long-lived populations recolonized the northern Adriatic regions. However, a major flaw of this hypothesis is that in this scenario, there should be at least some evidence of ancient populations in the southern parts of the Adriatic basin. Generally, in more southern populations, the levels of gene diversity, number of private alleles and *DW* decrease. Overall, the detected pattern is not in concordance with the general predictions that the highest levels of diversity can be detected among southern regions of species distribution range [5,82], thus suggesting either a more recent origin of these populations or some past event which resulted in abrupt changes of their genetic structure. At this point, human activity and the excessive gathering of this species during the last century, primarily in the southern and, to a lesser extent, in the northern parts of Adriatic, must be considered. In fact, until the 1850s, the only reliable source of pyrethrum material was from wild plants [45]; hence, it is very likely that overexploitation alone is responsible for such low gene diversity and *DW*. An exception is the population 19 located in the southern parts of the species' range. Because it exhibits high levels of gene diversity accompanied by low levels of private alleles and *DW*, this population is likely the zone of secondary contact between the divergent lineages and is anthropogenic in origin. Additionally, some populations show a quite admixed pattern comprising large fractions of different gene pools (pops. 8, 9, 11, etc.). The specific structure of these populations could also easily be a consequence of the admixture of divergent lineages through secondary contact after postglacial colonization, however, more likely explanation of the obtained results could lie in the cultivation practices of Dalmatian pyrethrum, which began in 1850s near Dubrovnik (Southern Dalmatia) and rapidly expanded along the Dalmatian coastal region and the islands [45]. These populations could have been founded through human activity by the exchange of seed material between farmers from different geographical regions. It is well documented that anthropogenic activities foster genetic homogenization and decrease genetic differentiation among populations [36,86–87]. Therefore, we hypnotize that for southern populations overexploitation and process of cultivation are responsible for substantial loss of private alleles, levels of gene diversity and *DW*. Moreover, extensive seed exchange between different populations caused admixture as observed in pops. 8, 9 and 11. Thus, this phylogeographical study could represent typical example of humans' shaping the genetic structure of investigated plant species. The genetic impoverishment as a result of either overexploitation of wild resources or cultivation practices has also been demonstrated for American ginseng (*Panax quinquefolius* L.) [35], *Rheum tanguticum* [88], *Scutellaria baicalensis* [37], *Eucommia ulmoides* [89], *Moringa peregrina* [90], etc.

The lowest values of gene diversity were detected in the populations from Mt. Biokovo (pops. 16, 17 and 18) ($H_E = 0.092$–0.096), which were also identified as a separate gene pool by the Bayesian cluster analysis. The Mt. Biokovo populations (pops. 16, 17 and 18) were sampled in geographically and ecologically marginal areas of the species distribution, and it is generally expected that in such peripheral populations, the amounts of diversity may be reduced [91]. This reduction can be the result of, small population size, increased isolation and consequently reduced gene flow [92–93]. At such high elevation (greater than 1200 m.a.s.l.) with atypical climatic conditions for the species, it is highly unlikely that the spontaneous formation and development of populations of such a pronouncedly thermophytic species could be expected in the present day. However, during some past periods that were warmer than present (Holocene Climatic Optimum) [94], such altitudinal shifts in species distribution are quite expected and have already been documented for the western Balkans [10]. Therefore, these populations can be treated as relict populations, which are ecologically isolated from other populations. Evidently, a sufficient period of time has passed for this group of populations to diverge and form a separate, well-defined gene pool.

Evidence that both geography and environmental heterogeneity contributed significantly to the obtained patterns of genetic divergence was found. The IBD analysis showed positive and significant correlations, which could in principal be the result of successive postglacial colonization from the refugia [28]. However, a large part of unexplained genetic variations among the populations still remains after accounting for the geographical distances. Furthermore, a significant correlation between environmental and genetic variation was found, and 12.3% of the genetic variation was explained by environmental heterogeneity. The populations from Mt. Biokovo (pops. 16, 17 and 18), located at significantly higher elevation (from 1295 to 1335 m.a.s.l.), inhabiting environmentally distinct habitat in comparison to the remaining populations, are typical cases of environmental isolation. It is well acknowledged that microclimate strongly influences plant phenology [95–96] whereas differences in temperature, photoperiod and moisture are major factors in shaping patterns of flowering and fruiting [97]. Due to cooler temperatures and higher precipitation the populations from Mt. Biokovo tend to flower later (cca. 20 days) than do the low-elevation populations (authors observations); therefore, there is no phenological overlap between them, resulting in genetic isolation and differentiation of Mt. Biokovo populations (pops. 16, 17, 18). In addition, great influence of sea surface on coastal area ecological conditions and consequently on vegetation cover [25] could have also served as a stimulus for such strong differentiation between high altitude and lowland populations. Bathymetric configuration of Adriatic Sea floor is characterized by very deep valley with steep slopes in southern region (South Adriatic Basin) and relatively shallow and plain central and northern region [98–99]. For that reason northern Adriatic part has experienced more pronounced substantial shifts in shoreline during the sea-level fluctuations, while the southern regions remained almost unchanged [100]. Therefore, in the past when the sea level was significantly lower than today [99,101], populations from middle and especially southern parts of Dalmatia were located in proximity of the sea as they are today. However, this is not the case for high altitude areas as they were never located near sea surface, which might have resulted in divergent evolution particularly observed in Mt. Biokovo populations (pops. 16, 17 and 18). The strong influence of environmental factors on genetic divergence of populations has already been documented for several plant [77,102] and animal species [31,33]. For example, Temunović et al. [77] determined genetic structuring of Croatian Continental and Mediterranean ash trees populations (*Fraxinus angustifolia* Vahl) due to contrasting ecological conditions of the sampling sites. Similarly, strong fine-scale spatial genetic structure associated with habitat factors was detected in continuous population of *Biscutella laevigata* [102]. For the Franciscana dolphins genetically isolated populations in environmentally distinct areas of Argentina were observed [33] and for the grey wolf (*Canis lupus*) genetic differentiation among local populations was correlated to contrasting climatic conditions [31].

In our study the detected IBED pattern explained more than twice the genetic divergence than did the IBD pattern (12.3 vs.

5.6%, respectively). Therefore it is clear that the revealed genetic differentiation in Dalmatian pyrethrum is more strongly influenced by the bioclimatic conditions of the sampling sites rather than by geographic distances *per se*.

Conclusion

The aim of this study was to analyze the genetic diversity and structure of natural Dalmatian pyrethrum (*Tanacetum cinerariifolium* Trevir. /Sch./ Bip.) populations from Croatia. Overall, our results show that, as a thermophytic plant, Dalmatian pyrethrum could have survived the Quaternary climatic oscillation along the eastern Adriatic coast by tracking suitable habitats, i.e., microrefugia. Northern Adriatic populations, which were identified as a separate cluster by the STRUCTURE analysis, are characterized by high levels of gene diversity and *DW* and should be treated as conservation priority sites. Human influence through overexploitation and cultivation has resulted in the low genetic variability of southern Adriatic populations. The results also provide evidence that both IBD and especially IBED contributed to the genetic differentiation among some populations.

Supporting Information

Figure S1 Genetic structure of *T. cinerariifolium* populations. (A) using software Structure and assuming $K = 2$, (B) using Structure and assuming $K = 3$; (C) using BAPS without spatially informative prior, and (D) using BAPS with spatially informative prior. Each individual plant is represented by a single vertical line divided into colors. Each color represents one cluster, and the length of the colored segment shows the individual's estimated proportion of membership in that cluster. White lines separate populations that are labelled below the figure (1–20).

Author Contributions

Conceived and designed the experiments: ZS ZL MG. Performed the experiments: MG ZL. Analyzed the data: MG ZS ZL. Contributed reagents/materials/analysis tools: ZL IR. Contributed to the writing of the manuscript: MG ZS ZL IR KCS IK.

References

1. Hamrick JL, Godt MJ (1996) Effect of life history traits on genetic diversity in plant species. Philos Trans R Soc Lond B Biol Sci 351: 1291–1298.
2. Yamagishi M, Nishioka M, Kondo T (2010) Phenetic diversity in the *Fritillaria camschatcensis* population grown on the Sapporo campus of Hokkaido University. Landsc Ecol Eng 6: 75–79.
3. Hewitt GM (1999) Post-glacial re-colonization of European biota. Biol J Linn Soc Lond 68: 87–112.
4. Pope LC, Domingo-Roura X, Erven K, Burke T (2006) Isolation by distance and gene flow in the Eurasian badger (*Meles meles*) at both a local and broad scale. Mol Ecol 15: 371–386.
5. Taberlet P, Fumagalli L, Wust-Saucy AG, Cosson JF (1998) Comparative phylogeography and postglacial colonization routes in Europe. Mol Ecol 7: 453–464.
6. Hewitt G (2000) The genetic legacy of the Quaternary ice ages. Nature 405: 907–913.
7. Schmitt T (2007) Molecular biogeography of Europe: Pleistocene cycles and postglacial trends. Front Zool 4: 11. doi:10.1186/1742-9994-4-11.
8. Hofreiter M, Stewart J (2009) Ecological change, range fluctuations and population dynamics during the Pleistocene. Curr Biol 19: R584–R594.
9. Hewitt GM (1996) Some genetic consequences of ice ages, and their role in divergence and speciation. Biol J Linn Soc Lond 58: 247–276.
10. Surina B, Schönswetter P, Schneeweiss GM (2011) Quaternary range dynamics of ecologically divergent species (*Edraianthus serpyllifolius* and *E. tenuifolius*, Campanulaceae) within the Balkan refugium. J Biogeogr 38: 1381–1393.
11. Petit RJ, Aguinagalde I, de Beaulieu JL, Bittkau C, Brewer S, et al. (2003) Glacial Refugia: Hotspots But Not Melting Pots of Genetic Diversity. Science 300: 1563–1565.
12. Gomez A, Lunt DH (2007) Refugia within refugia: patterns of phylogeographic concordance in the Iberian Peninsula. In: Weiss S, Ferrand N, editors. Phylogeography of Southern European Refugia. Dordrech: Springer. 155–188.
13. Canestrelli D, Aloise G, Cecchetti S, Nascetti G (2010) Birth of a hotspot of intraspecific genetic diversity: notes from the underground. Mol Ecol 9: 5432–5451.
14. Vega R, Amori G, Aloise G (2010) Genetic and morphological variation in a Mediterranean glacial refugium: evidence from Italian pygmy shrews, *Sorex minutus*. (Mammalia: Soricomorpha). Biol J Linn Soc Lond 100: 774–787.
15. Canestrelli D, Salvi D, Maura M, Bologna MA, Nascetti G (2012) One Species, Three Pleistocene Evolutionary Histories: Phylogeography of the Italian Crested Newt, *Triturus carnifex*. PLoS ONE 7(7): e41754. doi:10.1371/journal.pone.0041754.
16. Paun O, Schönswetter P, Winkler M, Intrabiodiv Consortium, Tribsch A (2008) Historical divergence vs. contemporary gene flow: evolutionary history of the calcicole *Ranunculus alpestris* group (Ranunculaceae) in the European Alps and the Carpathians. Mol Ecol 17: 4263–4275.
17. Schönswetter P, Tribsch A (2005) Vicariance and dispersal in the alpine perennial *Bupleurum stellatum* L. (Apiaceae). Taxon 54: 725–732.
18. Wielstra B, Crnobrnja-Isailović J, Litvinchuk SN, Reijnen BT, Andrew K, et al. (2013) Tracing glacial refugia of *Triturus* newts based on mitochondrial DNA phylogeography and species distribution modeling. Front Zool 10: 13. doi:10.1186/1742-9994-10-13.
19. Hewitt GM (2001) Speciation, hybrid zones and phylogeography - or seeing genes in space and time. Mol Ecol 10: 537–549.
20. Savić IR (2008) Diversification of the balkan fauna: its origin, historical development and present status. In: Makarov SE, Dimitrijević RN, editors. Advances in Arachnology and Developmental Biology: papers dedicated to Prof. Dr. Božidar Ćurčić. Belgrade: Institute of Zoology, Faculty of Biology; Sofia: BAS; Vienna: Faculty of Life Sciences; Belgrade: SASA; UNESCO MAB Serbia. 57–78.
21. Podnar M, Mayer W, Tvrtkovic N (2004) Mitochondrial phylogeography of the Dalmatian wall lizard, *Podarcis melisellensis* (Lacertidae). Org Divers Evol 4: 307–317.
22. Lakušić D, Novčić R, Kučera J, Marhold K (2006) *Cardamine pancicii* Hayek (Brassicaceae), a neglected species of the Balkan Peninsula: Morphological and molecular evidence. Willdenowia 36: 177–191.
23. Previšić A, Walton C, Kučinić M, Mitrikeski PT, Kerovec M (2009) Pleistocene divergence of Dinaric *Drusus* endemics (Trichoptera, Limnephilidae) in multiple microrefugia within the Balkan Peninsula. Mol Ecol 18: 634–647.
24. Kryštufek B, Buzan EV, Hutchinson WF, Hänfling B (2007) Phylogeography of the rare Balkan endemic Martino's vole, *Dinaromys bogdanovi*, reveals strong differentiation within the western Balkan Peninsula. Mol Ecol 16: 1221–1232.
25. Lakušić D, Liber Z, Nikolić T, Surina B, Kovačić S, et al. (2013) Molecular phylogeny of the *Campanula pyramidalis* species complex (Campanulaceae) inferred from chloroplast and nuclear non-coding sequences and its taxonomic implications. Taxon 62: 505–524.
26. Wright S (1943) Isolation by distance. Genetics 28: 114–138.
27. Ægisdóttir HH, Patrick Kuss P, Stöcklin J (2009) Isolated populations of a rare alpine plant show high genetic diversity and considerable population differentiation Ann Bot 104: 1313–1322.
28. Campillo S, Serra M, Carmona MJ, Gómez A (2011) Widespread Secondary Contact and New Glacial Refugia in the Halophilic Rotifer *Brachionus plicatilis* in the Iberian Peninsula. PLoS ONE 6(6): e20986. doi:10.1371/journal.pone.0020986.
29. Hou Y, Lou A (2011) Population Genetic Diversity and Structure of a Naturally Isolated Plant Species, *Rhodiola dumulosa* (Crassulaceae). PLoS ONE 6(9): e24497. doi:10.1371/journal.pone.0024497.
30. Wang IJ, Glor RE, Losos JB (2013) Quantifying the roles of ecology and geography in spatial genetic divergence. Ecol Lett 16: 175–182.
31. Pilot M, Jedrzejewski W, Branicki W, Sidorovich VE, Jedrzejewska B (2006) Ecological factors influence population genetic structure of European grey wolves. Mol Ecol 15: 4533–4553.
32. Manel S, Schwartz MK, Luikart G, Taberlet P (2003) Landscape genetics: combining landscape ecology and population genetics. Trends Ecol Evol 18: 189–197.
33. Mendez M, Rosenbaum HC, Subramaniam A, Yackulic C, Bordino P (2010) Isolation by environmental distance in mobile marine species: molecular ecology of franciscana dolphins at their southern range. Mol Ecol 19: 2212–2228.
34. Kozak KH, Wiens JJ (2006) Does niche conservatism drive speciation? A case study in North American salamanders. Evolution 60: 2604–2621.

35. Cruse-Sanders JM, Hamrick JL (2004) Genetic diversity in harvested and protected populations of wild American ginseng, *Panax quinquefolius* L. (Araliaceae). Am J Bot 91: 540–5488.
36. Doebley JF, Gaut BS, Smith BD (2006) The molecular genetics of crop domestication. Cell 127: 1309–1321.
37. Yuan QJ, Zhang ZY, Hu J, Guo LP, Shao AJ, et al. (2010) Impacts of recent cultivation on genetic diversity pattern of a medicinal plant, *Scutellaria baicalensis* (Lamiaceae). BMC Genet 11: 29.
38. Bourguiba H, Audergon JM, Krichen L, Trifi-Farah N, Mamouni A, et al. (2012) Loss of genetic diversity as a signature of apricot domestication and diffusion into the Mediterranean Basin. BMC Plant Biol 12: 49.
39. MacDonald WL (1995) Pyrethrum flowers-production in Australia. In: Casida JE, Quistad GB, editors. Pyrethrum flowers: Production, Chemistry, Toxicology, and Uses New York: Oxford University Press. 55–66.
40. Parlevliet JE (1975) Breeding pyrethrum in Kenya. Pyrethrum Post 13: 47–54.
41. Nikolić T. ed. (2014) Flora Croatica Database. Faculty of Science, University of Zagreb. Available: http://hirc.botanic.hr/fcd.
42. Heywood VH (1976) Tanacetum. In: Tutin TG, Heywood VH, Burges NA, Moore DM, Valentine DH, Walters SM, Webb DA, editors. Flora Europaea, Plantaginaceae to Compositae (and Rubiaceae). Volume 4. Cambridge: Cambridge University Press. 169–171.
43. Head SW (1973) Composition of pyrethrum extract and analysis of pyrethrins. In: Casida JE, editor. Pyrethrum, The Natural Insecticide. New York: Academic Press. 25–49.
44. Ožanić S (1930) Buhač (*Pyrethrum cinerariaefolium* D.C.). Beograd: Prosveta. Beograd.
45. Bakarić P (2005) Buhač - prirodni insekticid. Gospodarski list 17: 41–45.
46. Ožanić S (1955) Poljoprivreda Dalmacije u prošlosti. Split: Izdanje društva agronoma NRH- Podružnica Split. 230–231 p.
47. Shim S, Jørgensen RB (1999) Genetic structure in cultivated and wild carrots (*Daucus carota* L.) revealed by AFLP analysi. Theor Appl Genet 101: 227–233.
48. Schmidt K, Jensen K (2000) Genetic structure and AFLP variation of remnant populations in the rare plant *Pedicularis palustris* (Scrophulariaceae) and its relation to population size and reproductive components. Am J Bot 87: 678–689.
49. Rivera-Ocasio E, Aide TM, McMillan WO (2002) Patterns of genetic diversity and biogeographical history of the tropical wetland tree, *Pterocarpus officinalis* (Jacq.), in the Caribbean basin. Mol Ecol 11: 675–683.
50. Li Q, Xiao M, Guo L, Wang L, Tang L, et al. (2005). Genetic diversity and genetic structure of an endangered species, *Trillium tschonoskii*. Biochem Genet 43: 445–458.
51. Baldoni L, Tosti N, Ricciolini C, Belaj A, Arcioni S, et al. (2006) Genetic Structure of Wild and Cultivated Olives in the Central Mediterranean Basin. Ann Bot 98: 935–942.
52. Yang D, Ma P, Liang X, Liang Z, Zhang M, et al. (2012) Metabolic Profiles and cDNA-AFLP Analysis of Salvia miltiorrhiza and *Salvia castanea* Diel f. *tomentosa* Stib. PLoS ONE 7(1): e29678. doi:10.1371/journal.pone.0029678.
53. Wang T, Chen G, Zan Q, Wang C, Su Y-j (2012) AFLP Genome Scan to Detect Genetic Structure and Candidate Loci under Selection for Local Adaptation of the Invasive Weed *Mikania micrantha*. PLoS ONE. 7(7): e41310. doi:10.1371/journal.pone.0041310.
54. Vos P, Hogers R, Bleeker M, Reijans M, Van de Lee T, et al. (1995) AFLP: a new technique for DNA fingerprinting. Nucleic Acids Res 23: 4407–4414.
55. Carović-Stanko K, Liber Z, Politeo O, Strikić F, Kolak I, et al. (2011) Molecular and chemical characterization of the most widespread *Ocimum* species. Plant Syst Evol 294: 253–262.
56. Herrmann D, Poncet BN, Manel S, Rioux D, Gielly L, et al. (2010) Selection criteria for scoring amplified fragment length polymorphisms (AFLPs) positively affect the reliability of population genetic parameter estimates. Mol Ecol 53: 302–310.
57. Shannon CE, Weaver W (1949) The Mathematical Theory of Communication. Urbana: University of Illinois Press.
58. Lewontin RC (1972) The apportionment of human diversity. Evol Biol 6: 381–398.
59. Ehrich D (2006) AFLPdat: a collection of R functions for convenient handling of AFLP data. Mol Ecol Notes 6: 603–604.
60. Zhivotovsky LA (1999) Estimating population structure in diploids with multilocus DNA markers. Mol Ecol 8: 907–913.
61. Vekemans X, Beauwens BS, Lemaire M, Roldan-Ruiz I (2002) Data from amplified fragment length polymorphism (AFLP) markers show indication of size homoplasy and of a relationship between degree of homplasy and fragment size. Mol Ecol 11: 139–151.
62. Lynch M, Milligan BG (1994) Analysis of population genetic structure with RAPD markers. Mol Ecol 3: 91–99.
63. Nei M (1973) Analysis of gene diversity in subdivided populations. Proc Nat Acad Sci 70: 3321–3323.
64. Nei M (1972) Genetic distance between populations. Am Nat 106: 283–292.
65. Felsenstein J (1993) Phylogeny Inference Package (PHYLIP). Version 3.5. Seattle: Department of genetics, University of Washington.
66. Excoffier L, Smouse PE, Quattro JM (1992) Analysis of molecular variance inferred from metric distances among DNA haplotypes: application to human mitocondrial DNA restriction sites. Genetics 131: 479–491.
67. Schneider S, Roessli D, Excoffier L (2000) ARLEQUIN Version 2.000: A Software for Population Genetic Data Analysis. Geneva: Genetics and Biometry Laboratory University of Geneva.
68. Huff DR (1997) RAPD characterization of heterogeneous perennial ryegrass cultivars. Crop Sci 37: 557–564.
69. Pritchard JK, Stephens M, Donnelly P (2000) Inference of population structure using multilocus genotype data. Genetics 155: 945–959.
70. Evanno G, Regnaut S, Goudet J (2005) Detecting the number of clusters of individuals using the software STRUCTURE: a simulation study. Mol Ecol 14: 2611–2620.
71. Ehrich D, Gaudeul M, Assefa A, Koch M, Mummenhoff K, et al. (2007) Genetic consequences of Pleistocene range shifts: Contrast between the Arctic, the Alps and the East African mountains. Mol Ecol 16: 2542–2559.
72. Corander J, Waldmann P, Sillanpää MJ (2003) Bayesian analysis of genetic differentiation between populations. Genetics 163: 367–374.
73. Corander J, Marttinen P, Sirén J, Tang J (2008) Enhanced Bayesian modelling in BAPS software for learning genetic structures of populations. BMC Bioinformatics 9: 539.
74. Corander J, Marttinen P (2006) Bayesian identification of admixture events using multi-locus molecular markers. Mol Ecol 15: 2833–2843.
75. Rousset F (1997) Genetic differentiation and estimation of gene flow from F-statistics under isolation by distance. Genetics 145: 1219–1228.
76. Mantel N (1967) The detection of disease clustering and a generalized regression approach. Cancer Res 27: 209–220.
77. Temunović M, Fernández-Manjarrés JF, Satovic Z, Grgurev M., Frascaria-Lacoste N, et al. (2012) Environmental heterogeneity explains the genetic structure of Continental and Mediterranean populations of *Fraxinus angustifolia* Vahl. PLoS ONE 7(8): e42764. doi:10.1371/journal.pone.0042764.
78. Hijmans RJ, Cameron SE, Parra JL, Jones PG, Jarvis A (2005) Very high resolution interpolated climate surfaces for global land areas. Int J Climatol 25: 1965–1978.
79. Rohlf FJ (2000) NTSYS-pc. Numerical Taxonomy and Multivariate Analysis System, version 2.1. New York: Exeter Publications.
80. Bonin A, Bellemain E, Bronken Eidesen P, et al. (2004) How to track and assess genotyping errors in population genetics studies. Mol Ecol 13: 3261–3273.
81. Kučera J, Tremetsberger VJ, Marhold K (2008) Molecular study of the *Cardamine maritima* group (Brassicaceae) from the Balkan and Apennine Peninsulas based on amplified fragment length polymorphism. Plant Syst Evol 275: 193–207.
82. Hampe A, Petit RJ (2005) Conserving biodiversity under climate change: the rear edge matters. Ecol Lett 8: 461–467.
83. Ursenbacher S, Schweiger S, Tomović L, Crnobrnja-Isailović J, Fumagalli L, et al. (2008) Molecular phylogeography of the nose-horned viper (*Vipera ammodytes*, Linnaeus (1758)): evidence for high genetic diversity and multiple refugia in the Balkan peninsula. Mol Phylogenet Evol 46: 1116–1128.
84. Muhs DR, Ager TA, Begét JE (2001) Vegetation and paleoclimate of the last interglacial period, central Alaska. Quat Sci Rev 20: 41–61.
85. Kukla GJ, Bender ML, de Beaulieu J-L, Bond G, Broecker WS, et al. (2002) Last interglacial climates. Quat Res 58: 2–13.
86. Olden JD, Poff NL, Douglas MR, Douglas ME, Fausch KD (2004) Ecological and evolutionary consequences of biotic homogenization. Trends Ecol Evol 19: 18–24.
87. Wiersum KF, Dold AP, Husselman M, Cocks M (2006) Cultivation of medicinal plants as a tool for biodiversity conservation and poverty alleviation in the Amatola region, South Africa. In: Bogers RJ, Craker LE, Lange D, editors. Medicinal and Aromatic Plants. Dordrech: Springer. 43–57.
88. Hua Y, Wang L, Xie X, Yang J, Li Y, Zhang H (2010) Genetic diversity of wild populations of *Rheum tanguticum* endemic to China as revealed by ISSR analysis. Biochem Syst Ecol 38: 264–274.
89. Yao X, Deng J, Huang H (2012) Genetic diversity in *Eucommia ulmoides* (Eucommiaceae), an endangered traditional Chinese medicinal plant. Conserv Genet 13: 1499–1507.
90. Zaghloul MS, Hamrick JL, Moustafa AA (2012) Conservation genetics of Sinai's remnant populations of *Moringa peregrina*, an economically valuable medicinal plant. Conserv Genet 13: 9–19.
91. Eckert CG, Samis KE, Lougheed C (2008) Genetic variation across species' geographical ranges: the central–marginal hypothesis and beyond. Mol Ecol 17: 1170–1188.
92. Lesica P, Allendorf FW (2002) When Are Peripheral Populations Valuable for Conservation? Conserv Biol 9: 753–760.
93. Vucetich JA, Waite TA (2003) Spatial patterns of demography and genetic processes across the species' range: Null hypotheses for landscape conservation genetics. Conserv Genet 4: 639–645.
94. Rossignol-Strick M (1999) The Holocene climatic optimum and pollen records of sapropel 1 in the eastern Mediterranean, 9000–6000 BP. Quat Sci Rev 18: 515–530.
95. Fitter AH and Fitter RSR (2002) Rapid Changes in Flowering Time in British Plants. Science 296: 1689–1691.
96. Dahlgren JP, von Zeipel H, Ehrlen J (2007) Variation in vegetative and flowering phenology in a forest herb caused by environmental heterogeneity. Am J Bot 94: 1570–1576.
97. Warren RJ, Bahn V, Bradford MA (2011) Temperature cues phenological synchrony in ant-mediated seed dispersal. Glob Chang Bio 17: 2444–2454.

98. Ferrarese S, Cassardo C, Elmi A, Genovese R, Longhetto A, et al. (2008) Response of temperature and sea surface circulation to a Sirocco wind event in the Adriatic basin: A model simulation. J Mar Syst 74: 659–671.
99. Surić M, Richards DA, Hoffmann DL, Tibljaš D, Juračić M (2009) Sea-level change during MIS 5a based on submerged speleothems from the eastern Adriatic Sea (Croatia). Mar Geol 262: 62–67.
100. Pikelj K, Žigić V, Juračić M (2009) Origin and distribution of surface sediments in the Grgur Channel, Adriatic Sea, Croatia. Geol Croat 62: 95–105.
101. Lambeck K, Yokoyama Y, Purcell T (2002) Into and out of the Last Glacial Maximum: sea-level change during Oxygen Isotope Stages 3 and 2. Quat Sci Rev 21: 343–360.
102. Parisod C, Christin P-A (2008) Genome-wide association to fine-scale ecological heterogeneity within a continuous population of *Biscutella laevigata* (Brassicaceae). New Phytol 178: 436–447.

12

Analysis of Genetic Diversity and Population Structure of Rice Germplasm from North-Eastern Region of India and Development of a Core Germplasm Set

Debjani Roy Choudhury[1¶], Nivedita Singh[1¶], Amit Kumar Singh[1], Sundeep Kumar[1], Kalyani Srinivasan[2], R. K. Tyagi[2], Altaf Ahmad[3], N. K. Singh[4], Rakesh Singh[1*]

1 Division of Genomic Resources, National Bureau of Plant Genetic Resources, New Delhi, 110 012, India, **2** Division of Germplasm Conservation, National Bureau of Plant Genetic Resources, New Delhi, 110 012, India, **3** Department of Botany, Faculty of Science, Jamia Hamdard (Hamdard University), New Delhi, 110062, India, **4** National Research Centre on Plant Biotechnology, Indian Agricultural Research Institute, New Delhi, 110 012, India

Abstract

The North-Eastern region (NER) of India, comprising of Arunachal Pradesh, Assam, Manipur, Meghalaya, Mizoram, Nagaland and Tripura, is a hot spot for genetic diversity and the most probable origin of rice. North-east rice collections are known to possess various agronomically important traits like biotic and abiotic stress tolerance, unique grain and cooking quality. The genetic diversity and associated population structure of 6,984 rice accessions, originating from NER, were assessed using 36 genome wide unlinked single nucleotide polymorphism (SNP) markers distributed across the 12 rice chromosomes. All of the 36 SNP loci were polymorphic and bi-allelic, contained five types of base substitutions and together produced nine types of alleles. The polymorphic information content (PIC) ranged from 0.004 for Tripura to 0.375 for Manipur and major allele frequency ranged from 0.50 for Assam to 0.99 for Tripura. Heterozygosity ranged from 0.002 in Nagaland to 0.42 in Mizoram and gene diversity ranged from 0.006 in Arunachal Pradesh to 0.50 in Manipur. The genetic relatedness among the rice accessions was evaluated using an unrooted phylogenetic tree analysis, which grouped all accessions into three major clusters. For determining population structure, populations K = 1 to K = 20 were tested and population K = 3 was present in all the states, with the exception of Meghalaya and Manipur where, K = 5 and K = 4 populations were present, respectively. Principal Coordinate Analysis (PCoA) showed that accessions were distributed according to their population structure. AMOVA analysis showed that, maximum diversity was partitioned at the individual accession level (73% for Nagaland, 58% for Arunachal Pradesh and 57% for Tripura). Using POWERCORE software, a core set of 701 accessions was obtained, which accounted for approximately 10% of the total NE India collections, representing 99.9% of the allelic diversity. The rice core set developed will be a valuable resource for future genomic studies and crop improvement strategies.

Editor: Swarup Kumar Parida, National Institute of Plant Genome Research (NIPGR), India

Funding: Financial support granted by Indian Council of Agricultural Research, New Delhi, India, under NPTC project, is gratefully acknowledged. The funders had no role in study design, data collection and analysis, decision to publish, or preparation of the manuscript.

Competing Interests: The authors have declared that no competing interests exist.

* Email: rsingh@nbpgr.ernet.in

¶ These authors contributed equally to this work.

Introduction

Plant genetic resources are of paramount importance for the future and to ensure the food and nutritional security of an increasing population. A large number of genetic materials have been conserved in Genebank, but their use is being limited due to an unmanageable number of accessions and the continuous expansion of accessions numbers. Core germplasm development has been proposed for better management and use of collections available in Genebank [1]. This requires the development of a core set of accession to more precisely characterize, explore, and conserve Genebank resources, monitor the genetic drift during preservation, and identify gaps in genetic diversity [2,3]. The idea of the core set was proposed by Frankel and Brown in 1984 [2]. As proposed, a core set is a small set of accessions (usually 10 % of the population) chosen to represent the genetic spectrum of an entire collection [4].

The sampling percentage of a core set has long been under debate [5]; 20%–30% of the sampling percentage was suggested by Yonezawa *et al.* (1995) to adequately sample a given collection [6]. Mini core sets representing ~1% of total collections and have also been used to characterize very large collections [7–10]. In fact, a mini core subset from USDA rice gene bank has been developed using 26 phenotypic trait and 70 molecular markers [11]. Still the perfect ratio and fixed size for all core set does not exist, since different crops or different constructing goal needs different sampling percentages [5]. Several methods have been used to develop core set. The stratified random sampling method, in contrast to the simple random sampling method, has been successfully applied to the development of numerous core set, e.g., fodder crops, potatoes, etc. [12–16]. One common approach for constructing a core set is grouping whole collections into major ecotypes and then selecting representatives from each ecotype [5,17–18]. However, a core set formed for the purpose of capturing accessions with rare or extreme values of the desired trait(s) (e.g. high resistance to pest or high yield and yield contributing traits) should be evaluated separately with the intention of representing the (pattern of) genetic diversity in the collection. Indeed, the pattern of genetic diversity will have genetic variations among all accessions that have been accumulated as a result of natural processes, species' characteristics and historical events [19]. Until now, there is no universally accepted method for constructing a core set as many factors affect representativeness of core set, such as sampling percentage, data type, number of traits observed, genetic diversity of germplasm, grouping method and sampling method [18,20–21].

The rice genetic resource is an important source for rice breeding and makes a valuable contribution to global wealth and food security [22]. India is considered to be the origin of rice. Based on the phylo-geographical and archeological evidence it is suggested that rice was domesticated about 10,000 years ago from its wild ancestor *O. rufipogon* in the region South of Himalayan mountain range, likely in the present day eastern and North-Eastern (NE) India, extending eastward to Nepal, Myanmar and Thailand to Southern China [23–25]. The NE India is a large geographical area comprising of seven states namely Arunachal Pradesh, Assam, Manipur, Meghalaya, Mizoram, Nagaland and Tripura states, is home to a large number of indigenous rice varieties [26]. The geography of the NE states of India is unique, having snow capped peaks of the Himalayas, the ecological hot spots of the NE foothills and the Brahmaputra valley. High rainfall, humidity, varied topography and altitude, heavy natural selection pressures, environmental stresses have made the region rich both in floristic and crop diversities [27].

A large collection of the rice genetic resource from NE India is currently conserved at National Genebank (NGB), National Bureau of Plant Genetic Resources (NBPGR), New Delhi. Since these germplasm contains unique traits which can be exploited for future crop improvement programme but has not been properly characterized at molecular level. Therefore, in the present study this collection was analyzed (i) to study the genetic diversity and population structure of all 6,984 rice germplasm accessions using 36 SNP markers, which were developed and used earlier for diversity and population structure analysis of rice varieties [28] and (ii) to develop a core set representing the maximum diversity of unique NE India rice genetic resources. Taken together, the described core set can be used to effectively for genomic studies, in rice crop improvement and conservation programs in genebank.

Materials and Methods

Plant materials

A total of 6,984 accessions of NE India (Arunachal Pradesh, Assam, Manipur, Meghalaya, Mizoram, Nagaland and Tripura,) were drawn from NGB, NBPGR, New Delhi. The NE India germplasm collection also included 24 accessions with EC ID from IRRI, Philippines and two accessions from Ghana which were growing in farmer's field since their introduction in 1991. The purpose of their inclusion was to analyse their clustering in relation to Indian germplasm collection. The details of each accession along with passport information were obtained from the online database of NGB, (www.nbpgr.ernet.in) and presented in Table S1.

DNA extraction from rice seed

Seeds of each genotype (10–12 seeds) were dehusked and used for DNA isolation using QIAGEN DNeasy plant mini kit. Kernels were ground into fine powder using tissue lyser (Tissue lyser II Retsch, Germany) with a tissue lyser adapter set (QIAGENq). DNA was extracted following the procedures described by manufacturers.

Genotyping of rice accessions using SNP markers

Genomic DNA of all the 6,984 accessions was diluted to prepare working stocks of 10 ng/μl. Sequenom Mass ARRAY multiplex assays were designed for 36 SNPs (iPLEX gold chemistry), representing conserved single-copy rice genes [29], taking three genes per rice chromosome [28]. These SNP markers are located on the short arm, centromeric region and long arm of the twelve rice chromosome. The 36-plex assay(s) were designed and validated by Sequenom Corporation (San Diego). The 30-mer pre-amplification primers and variable length genotyping primers generated by the Assay Design 3.1 software were procured and used for the validation of SNPs according to the Sequenom user manual. Mass ARRAY Typer 3.4 Software was used for the visualization of SNPs and allele calling. Chromosome number, primer ID and physical position of 36 SNPs used in this study are given in Table S2.

Table 1. Estimation of gene diversity, heterozygosity, PIC and major allele frequency in NE Indian rice collection (6984)

S. No.	States	No. of accessions	Gene diversity	Heterozygosity	PIC	Major allele frequency
1	Arunachal Pradesh	663	0.006–0.49 (0.29)	0.003–0.16 (0.07)	0.006–0.37 (0.23)	0.51–0.99 (0.78)
2	Assam	2635	0.01–0.5 (0.31)	0.01–0.32 (0.10)	0.01–0.37 (0.25)	0.50–0.99 (0.76)
3	Manipur	549	0.02–0.50 (0.26)	0.004–0.26 (0.09)	0.02–0.37 (0.21)	0.50–0.98 (0.80)
4	Meghalaya	2427	0.03–0.49 (0.34)	0.01–0.29 (0.12)	0.03–0.37 (0.27)	0.50–0.98 (0.74)
5	Mizoram	107	0.01– 0.49 (0.28)	0.01–0.42 (0.16)	0.01–0.37 (0.23)	0.52–0.99 (0.79)
6	Nagaland	377	0.008–0.49 (0.31)	0.002–0.16 (0.07)	0.008–0.37 (0.25)	0.52–0.99 (0.76)
7	Tripura	226	0.004–0.49 (0.28)	0.004–0.38 (0.08)	0.004–0.37 (0.22)	0.51–0.99 (0.79)
	Total	6984				

Mean values are given in parentheses.

Table 2. State-wise Analysis of Molecular Variance (AMOVA) of NE Indian rice collections.

States	Source	df	SS	MS	Est. Var.	%
Arunachal Pradesh	Among Pops	2	1160.573	580.286	1.523	21
	Among Indiv	660	6728.744	10.195	4.305	58
	Within Indiv	663	1051.000	1.585	1.585	21
	Total	1325	8940.317		7.413	100
Assam	Among Pops	2	4963.974	2481.987	1.751	24
	Among Indiv	2632	25094.849	9.535	3.836	51
	Within Indiv	2635	4908.500	1.863	1.863	25
	Total	5269	34967.323		7.449	100
Manipur	Among Pops	2	905.987	452.994	1.422	21
	Among Indiv	546	5785.991	10.597	4.509	58
	Within Indiv	549	866.500	1.578	1.578	21
	Total	1097	7558.478		7.510	100
Meghalaya	Among Pops	4	9032.582	2258.145	2.422	30
	Among Indiv	2422	23045.557	9.515	3.745	46
	Within Indiv	2427	4914.000	2.025	2.025	35
	Total	4853	36992.138		8.192	100
Mizoram	Among Pops	2	108.378	54.189	0.747	9
	Among Indiv	104	1247.813	11.998	4.569	56
	Within Indiv	107	306.000	2.860	2.860	35
	Total	213	1662.192		8.176	100
Nagaland	Among Pops	2	40.936	20.468	0.709	8
	Among Indiv	12	168.464	14.039	6.319	75
	Within Indiv	15	21.000	1.400	1.400	17
	Total	29	230.400		8.428	100
Tripura	Among Pops	2	400.316	200.158	1.564	22
	Among Indiv	223	2084.863	9.349	3.946	57
	Within Indiv	226	329.500	1.458	1.458	21
	Total	451	2814.679		6.967	100

Statistical analyses

The major allele frequency, gene diversity, heterozygosity and Polymorphic Information Content (PIC) for each locus were calculated for SNP markers using Power Marker 3.5 [30]. In addition, Principal Component Analyses (PCA), genetic distances [31] across the genotypes and neighbor-joining (NJ) tree were calculated using Power Marker 3.5 [30]. Phylogenetic trees for all states were constructed using MEGA software version 6.0 [32]. Principle Coordinate Analysis (PCoA) and Analysis of Molecular Variance (AMOVA) were performed using software GenAlEx V6.5 [33]. SNP data were numerically coded as follows: A = 1, C = 2, G = 3, T = 4 and missing data was coded as 0 as suggested in GenAlEx V6.5 user manual [33]. The model-based program, STRUCTURE 2.3.3 [34] was used to infer the population structure. For each K, three replications were run. Each run was implemented with a burn-in period of 100,000 steps followed by 100,000 Monte Carlo Markov Chain replicates [34] derived for each K and then plotted to find the plateau of the ΔK values [35]. The "Structure harvester" program was used (http://taylor0.biology.ucla.edu) to determine the final population. POWERCORE software [36] was used for the core development. Since numbers of rice accessions from each NE states were not equal, ranging from 107 accessions from Mizoram to 2,635 accessions from Assam. Therefore, to avoid the dominance of collections from individual states within the core set, we first developed core sub-sets using the rice collections from each state. Next, all accessions of the core sub-sets were pooled together to create the final core set of NE India rice germplasm.

Results

Genetic diversity

Genetic diversity of NE India rice collections (6,984 accessions), available at NGB, was estimated using 36 SNP markers. Alleles generated with all 36 SNP markers were scored to study the genetic diversity. Since these accessions cover seven states of NE India, they were analyzed state-wise for precise estimation of the level of their genetic diversity. The accessions collected from Arunachal Pradesh (663), Assam (2635), Manipur (549), Meghalaya (2427), Mizoram (107), Nagaland (377) and Tripura (226) were analyzed for major allele frequency, heterozygosity, gene diversity and PIC (Table 1). The highest mean major allele frequency was present in Manipur accessions (0.80) and lowest was found in Meghalaya accessions (0.74). The comparison of mean heterozygosity of alleles across the seven states revealed that Mizoram accessions were highly heterozygous (0.16); whereas

Table 3. State-wise Principal Coordinate Analyses of NE Indian rice collections, with percentage of variation explained by the first 3 axes.

States	Per cent variation	1-axis	2-axis	3-axis
Arunachal Pradesh	%	26.74	9.23	7.30
	Cum %	26.74	35.97	43.27
Assam	%	20.79	8.30	6.50
	Cum %	20.79	29.09	35.58
Manipur	%	21.54	7.23	5.73
	Cum %	21.54	28.77	34.50
Meghalaya	%	22.45	11.45	5.55
	Cum %	22.23	33.54	39.05
Mizoram	%	19.10	7.96	6.62
	Cum %	19.10	27.06	33.68
Nagaland	%	24.95	11.05	7.30
	Cum %	24.95	36.01	43.31
Tripura	%	14.96	10.15	7.57
	Cum %	14.96	25.12	32.69

Cum- Cumulative.

heterozygosity in the accessions belonging to Arunachal Pradesh (0.07) and Nagaland (0.07) were recorded minimum (Table 1). The mean gene diversity across the seven states varied from 0.34 (in the collections from Meghalaya) to 0.26 in those from (Manipur) and PIC varied from 0.21 (in the collections from Manipur) to 0.27 (from Meghalaya). Out of 36 primers used for characterization of total rice collections, five SNP primers (01-608-4_C_375, 02-3029-1_C_474, 03-3478-1_C_206, 05-48-1_C_279 and 08-4218-5_C_129) exhibited the maximum PIC and gene diversity value 0.37 and 0.49, respectively (Table S2). The primer 04-19-4_C_240 was found to be the least informative with PIC and gene diversity values equaling 0.02 and 0.02, respectively, and major allele frequency of 0.99 (Table S2), SNP markers which showed low PIC values in the present study may not be considered for future studies on genetic diversity and population structure in rice. The maximum heterozygosity was observed with primer 04-1801-20_C_428 was 0.28, while the minimum heterozygosity was observed with primer 04-19-4_C_240 (0.01).

Cluster analysis was performed using the NJ method for all the collections from seven states and unrooted phylogenetic trees were constructed (Fig. S1). Rice collections of each state were grouped in to three major clusters. Genetic distance for collections of Arunachal Pradesh varied from 0.00 to 0.75, Assam from 0.00 to 0.66, Manipur from 0.00 to 0.74, Meghalaya from 0.00 to 0.77, Mizoram from 0.00 to 0.68, Nagaland from 0.00 to 0.66, and Tripura from 0.00 to 0.67.

Population structure

A model-based program, STRUCTURE, was used to determine the genetic relationship among individual rice accessions. The membership of each accession was run from K = 1 to K = 20 for all the collections from all seven states to estimate the number of populations. Structure Harvester (http://taylor0. biology.ucla.edu) was used to determine final number of populations. The number of populations in Meghalaya collections was estimated as five, in Manipur collections as four and three for collections belonging to remaining states, i.e. Arunachal Pradesh, Assam, Mizoram, Nagaland and Tripura (Fig. S2, Table S3). The collection from Arunachal Pradesh contained 265 pure and 114 admix individuals in population1, 65 pure and 16 admix individuals in population2 and 131 pure and 72 admix individuals were present in population3. The collection from Assam showed 300 pure and 329 admix in population1, 1073 pure and 568 admix in population2, 288 pure and 77 admix in population3. The Nagaland collections 61 pure and 10 admix were present in population1, 164 pure and 91 admix in population2 and 27 pure and 24 admix were present in population3. Tripura collections showed 14 pure and 1 admix in population1, 107 pure and 21 admix in population2, 54 pure and 29 admix in population3. The collections from Mizoram had 39 pure and 19 admix in population1, 4 pure and no admix in population2 and 31 pure and 14 admix in population3. Similarly, Manipur collections showed 100 pure and 63 admix in population1, 53 pure and 55 admix in population2, 22 pure and 29 admix in population3 and 100 pure and 128 admix in population4. The accessions included in the core are marked with boxes in the bar plot (Fig. S3).

Analysis of molecular variance (AMOVA)

An AMOVA study of all rice accessions, belonging to seven different NE states, was performed to analyze the distribution of genetic diversity between and within the populations. The number of populations for each state was considered the same, as revealed by the population structure. AMOVA analyses showed that 21% diversity exists among populations in the collection from Arunachal Pradesh, while the collections from Assam, Manipur, Meghalaya, Mizoram, Nagaland and Tripura showed 24, 19, 29, 9, 8 and 22% diversity, respectively (Table 2). Maximum diversity of 29% among population was observed in Meghalaya collections, while minimum diversity of 8% was observed in Nagaland collections (Fig. S4). The highest level of diversity among individuals was present in Nagaland collections (75%) and the least diversity was observed in Meghalaya collections (46%).

Principal coordinate analyses (PCoA)

PCoA revealed significant diversity in the NE India rice germplasm collections. Meghalaya collections, which exhibited

Table 4. A comparison of percent of alleles generated in total collection vs core set.

Alleles	Arunachal Pradesh		Assam		Manipur		Meghalaya		Mizoram		Nagaland		Tripura	
	Total collection (663 acc.)	Core set (60 acc.)	Total collection (2635 acc.)	Core set (259 acc.)	Total collection (549 acc.)	Core set (60 acc.)	Total collection (2427 acc.)	Core set (242 acc.)	Total collection (107 acc.)	Core set (14 acc.)	Total collection (377 acc.)	Core set (35 acc.)	Total collection (226 acc.)	Core set (31 acc.)
AA	20.08	21.36	19.43	20.63	17.73	18.02	20.66	20.93	18.27	19.07	21.23	20.98	19.33	20.31
CC	21.16	17.68	18.23	18.44	13.51	15.91	16.98	17.22	15.82	16.92	16.83	17.44	18.43	16.96
GG	16.72	18.44	21.10	18.93	24.67	21.07	20.07	18.90	19.55	19.07	20.53	19.86	22.19	20.01
TT	37.83	32.12	34.94	30.20	38.62	34.43	35.17	30.64	36.55	31.70	37.11	31.33	35.32	33.82
CT/TC	1.57	3.70	2.31	3.87	2.14	3.71	2.67	4.45	3.66	4.29	1.59	3.78	2.21	3.81
GA/AG	1.53	3.65	2.32	4.36	2.00	3.94	2.42	4.25	3.76	5.48	1.49	3.17	1.46	3.05
TA/AT	0.33	0.78	0.46	0.96	0.49	0.86	0.56	1.08	0.83	1.31	0.26	0.75	0.37	0.61
GT/TG	0.70	2.00	1.07	2.20	0.80	1.83	1.29	2.10	1.46	1.91	0.87	2.24	0.67	1.22
CA/AC	0.07	0.27	0.14	0.41	0.04	0.22	0.17	0.43	0.10	0.24	0.09	0.47	0.03	0.20

the maximum number of population (k = 5) showed a very distinct PCoA plot, population1 in Meghalaya collections concentrated only in quadrant 3 and 4, whereas population4 concentrated in 1 and 2 quadrant and unlike population3, population4 and population5 did not show intermixing with each other (Fig. S5d). PCoA of collections from Arunachal Pradesh showed that population2 was very distinct, forming a separate group (Fig. S5a). PCoA plot of Assam, Manipur and Nagaland collections showed that all three populations were intermixed (Fig. S5b, Fig. S5c & Fig. S5f). In Mizoram collection population2 and Tripura collection population1 were highly distinct and formed a tight group in quadrant 1 and 4, respectively (Fig. S5e, Fig. S5g). The maximum cumulative variation (%) was observed for Nagaland rice collections with a value of 43.31% (Table 3); whereas, the minimum variation was found in Tripura accessions (32.53%). Overall, <44% variation were observed by first three components of principal coordinates, which indicates that accessions from all seven states were very diverse from one another.

Development of Core Set

Of the 6,984 rice accessions studied, a core set of 701 accessions (i.e., 60 from Arunachal Pradesh, 259 from Assam, 60 from Manipur, 242 from Meghalaya, 14 from Mizoram, 35 from Nagaland, and 31 from Tripura) were selected using POWER-CORE (Table S4). Nine allele types were produced by 36 SNP markers, of which four alleles were homozygous and five were heterozygous. Out of six types of substitutions possible (2 transitions and 4 transversion), five types of substitutions were observed and C/G or G/C type substitutions were not observed in our study. Allele frequency was calculated for all collections from seven states and core set (Table 4). A comparison of allele frequency showed that there was no loss of alleles in the resulting core set and they remained 99.9% identical.

Genetic diversity of Core Set

The genetic diversity of the core set was estimated to know the extent of diversity captured from total collection. Major allele frequency; gene diversity, heterozygosity and PIC were separately estimated for core-sets of individual states (Table 4). Comparisons of all genetic parameters (major allele frequency, gene diversity, heterozygosity and PIC) revealed that values for core set (Table 5) were greater than total collections (Table 1). For example, the Arunachal Pradesh gene diversity was 0.29, which increased to 0.39 in the core set. Similarly, the PIC for the Arunachal Pradesh collection was 0.23, while its core set PIC was 0.31. Cluster analysis showed that, core set accessions were getting distributed in to four clusters. The nine accessions with EC ID which were also part of core set were represented in three out of four clusters. The genetic diversity for the core set varied from 00.00 to 0.68 (Fig. 1).

Population structure of Core Sets

STRUCTURE analysis grouped core set of 701 rice accessions into three populations (Fig. 2). Populations were studied for the number of pure and admix individuals. Population1 had 62 pure and 74 admix, population2 had 111 pure and 53 admix and population3 had 250 pure and 151 admix (Fig. 3).

AMOVA and PCoA of Core Set

AMOVA analyses for core set showed 20% variance among populations (Table 6, Fig. 4); whereas, 48% variance was present among individuals. The partitioning of molecular variance was similar to the total collection and maximum diversity among individuals. PCoA could not be useful to identify significant

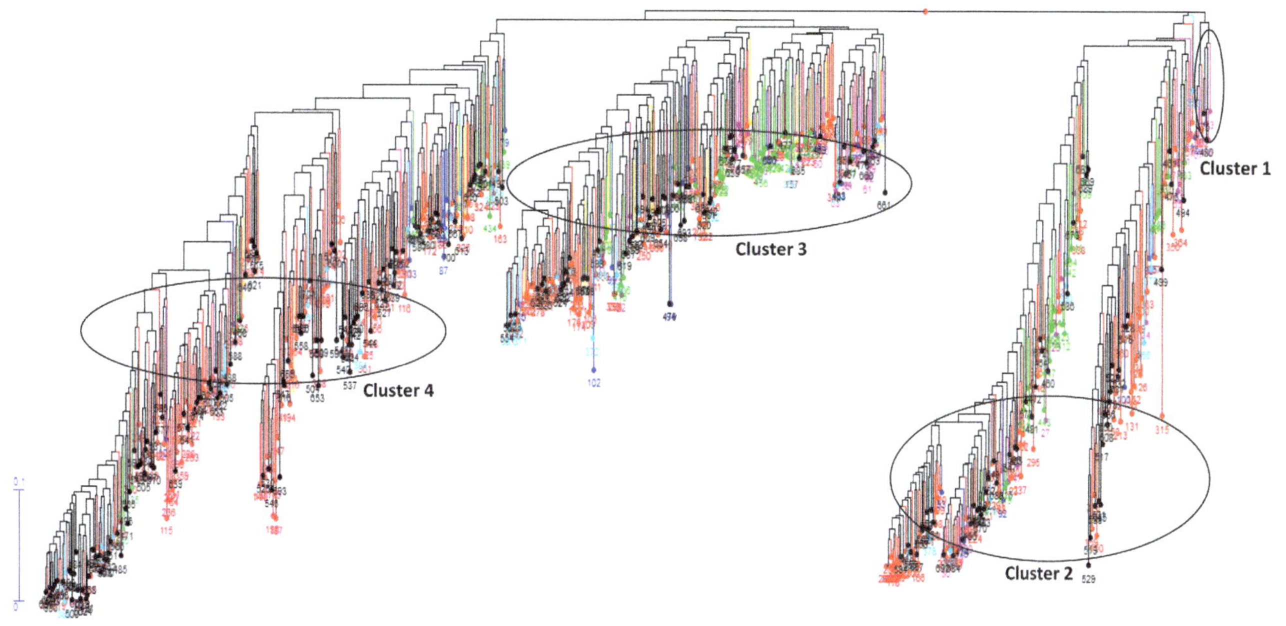

Figure 1. Phylogenetic tree of NE India rice core set (701 accessions) constructed based on SNP data. Germplasm accessions collected from the NE states are depicted with different colours Arunachal Pradesh - green, Assam - red, Manipur - purple, Meghalaya - black, Mizoram -yellow, Nagaland - light blue, Tripura - dark blue.

isolation in populations except, for population2, which was tightly clustered in quadrant 3 and 4 (Fig. 5). Maximum cumulative percent variation was explained by the first three coordinates with 43.08% variance (Table 7). A plot of percent allele frequency of total collection *versus* core set showed that nearly all alleles with similar frequency were represented in the core collection (Fig. 6) and 99.9% alleles have been retained in the collections designated as core set.

Discussion

Genome level profiling of huge germplasm collections in crop species is essential to identify small and diverse sets of accessions for their efficient use in crop improvement programmes. Despite the advancement in genomics, the Indian rice collection remained uncharacterized at the molecular level, with respect to parameters such as genetic diversity and population structure. This has been the major limiting factor in their utilization and development of improved cultivars. The present study is the first major effort to

Table 5. Estimation of gene diversity, heterozygosity, PIC and major allele frequency in NE India rice core set (701 accessions).

S. No.	States	core sample	Gene diversity	Heterozygosity	PIC	Major allele frequency
1	Arunachal Pradesh	60	0.0726–0.5000 (0.3962)	0.0377–0.2778 (0.1713)	0.0700–0.3750 (0.3124)	0.5000–0.9623 (0.6940)
2	Assam	259	0.1670–0.5000 (0.4018)	0.0843–0.3618 (0.1908)	0.1531–0.3740 (0.3172)	0.5000–0.9080 (0.69271)
3	Manipur	60	0.1372–0.5000 (0.3749)	0.0455–0.3455 (0.1745)	0.1278–0.3750 (0.2982)	0.5000–0.9259 (0.7142)
4	Meghalaya	242	0.2390–0.4995 (0.4100)	0.1013–0.3493 (0.2001)	0.2105–0.3747 (0.3231)	0.5161–0.8612 (0.6873)
5	Mizoram	14	0.0689–0.4970 (0.3506)	0.0714–0.4615 (0.2213)	0.0665–0.3735 (0.2813)	0.5385–0.9643 (0.7416)
6	Nagaland	35	0.1087–0.5000 (0.3854)	0.0385–0.3846 (0.1682)	0.1028–0.3846 (0.3049)	0.5000–0.9423 (0.7033)
7	Tripura	31	0.0351–0.5000 (0.3596)	0.0323–0.3667 (0.1524)	0.0345–0.3750 (0.2875)	0.5000–0.9821 (0.7344)

Mean values are given in parentheses.

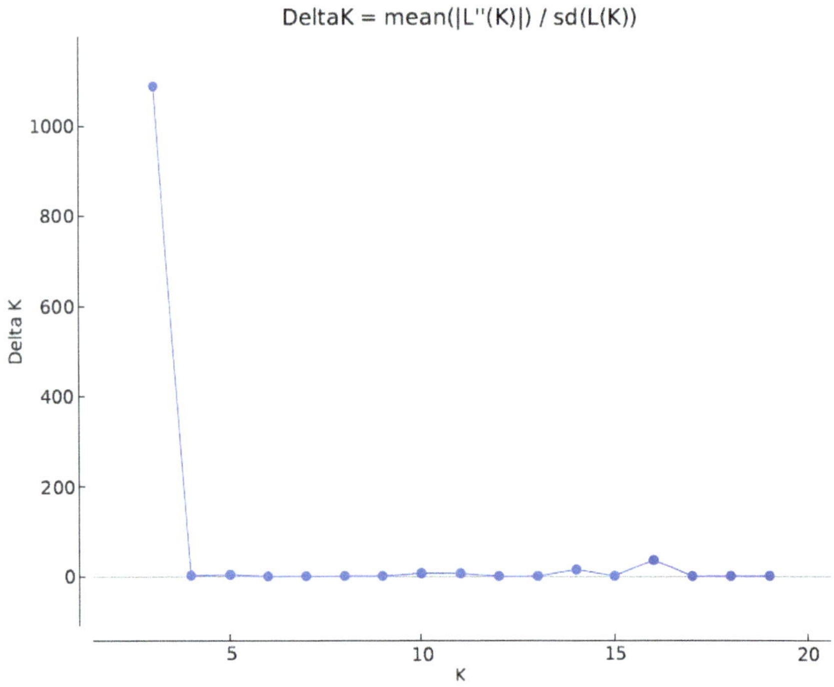

Figure 2. Estimation of populations in NE India rice core set using LnP(D) derived Δk for k from 1 to 20.

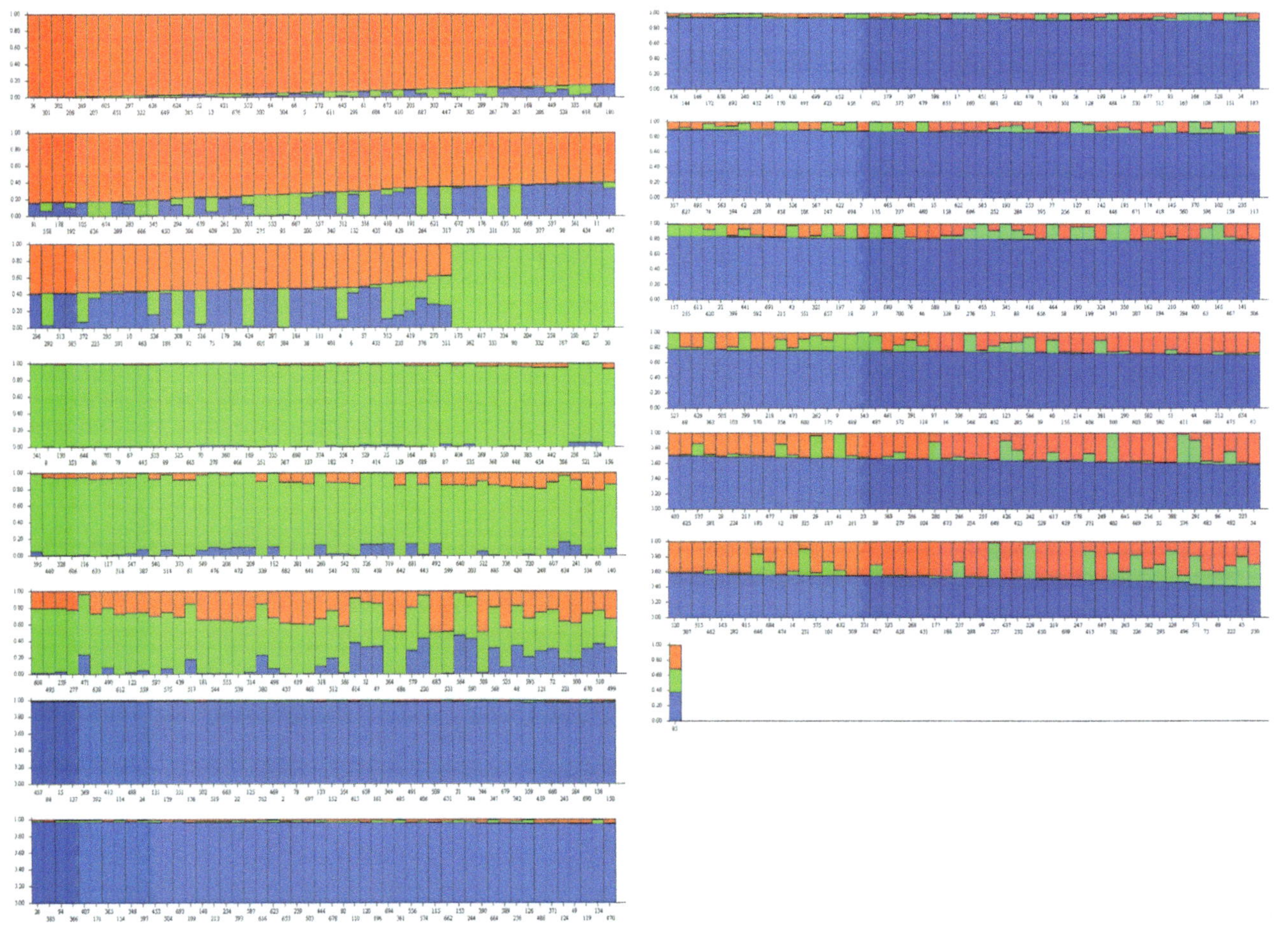

Figure 3. Model based clustering of the NE India rice core set (701 accessions).

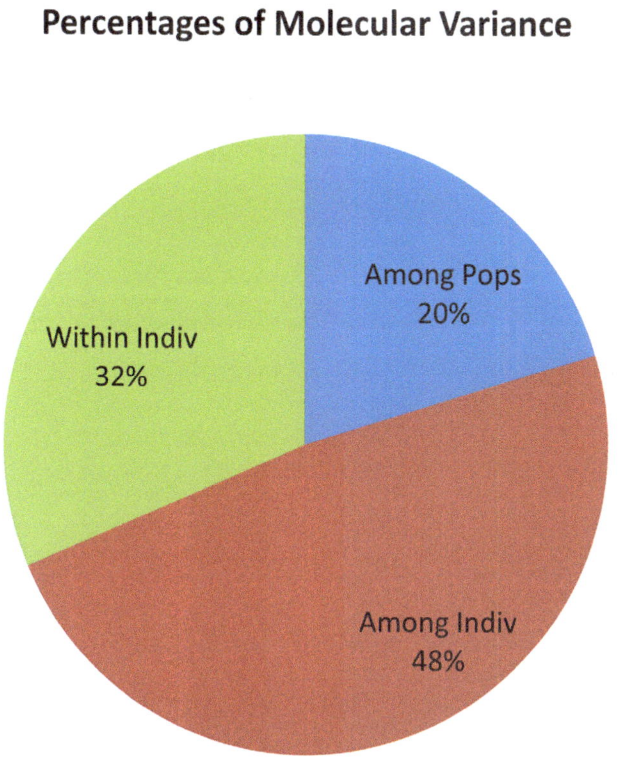

Figure 4. Analysis of Molecular Variance (AMOVA) analyses of NE India rice core set.

characterize the NE India rice collections. This study shows that highly efficient and selective SNP markers can be used to enhance genome based analysis, e.g., genome wide association mapping.

State wise analysis of the genetic diversity of the NE India collection showed interesting results. The average PIC values for these SNP markers did not vary much across the states and ranged from 0.23 (Manipur) to 0.27 (Meghalaya). The average PIC value recorded in the present study was in the range reported by researchers in previous studies. Singh *et al.*, (2013) [28] reported average PIC value 0.25 for the SNP markers in a set of 375 Indian rice varieties. The highest values for the parameters like PIC and gene diversity were found in collections from Meghalaya, which indicates the presence of a large rice genetic diversity in Meghalaya collections as compared to those from other six states. The same was also confirmed from PCoA, which displayed cluttered plot for the Meghalaya collection. The highest heterozygosity was observed in the collections from Mizoram, while Manipur had the highest major allele frequency.

Analysis of statistical genetic parameters across the collections from different states indicated a high level of variability in the NE India collection. There is relatively a high level of genetic diversity among the NE India rice collections, as reported by others using morphological [37] and molecular markers such as RAPD [38], ISSR [39] and SSR [26,40]. Comparison of our results with previous findings revealed that PIC and genetic diversity were nearly half of the SSR-based studies which were on the expected line [26,28,40]. Similar results have been reported in other species that have used SSR and SNP markers [41,42]. In the past, analysis of rice accessions from all seven states revealed three major clusters. Vairavan *et al.*, (1973) [37] reported three clusters among a set 400 accessions of NE India rice collection based on morphological analysis. Das *et al.*, (2013) [40] showed four groups were present in NE India rice varieties, while Choudhury *et al.*, (2013) [26] reported only two groups, based on SSR marker analysis.

Population structure study has been reported for the first time in NE rice collections and the number of populations estimated five for Meghalaya, four for Manipur and three for the remaining states. Admixture has been reported in all the states besides pure lines and this may be due to germplasm taken in this study which is more heterogeneous in comparison to varieties.

There have been no other studies for characterization of a large number of NE India rice accessions at the molecular level that

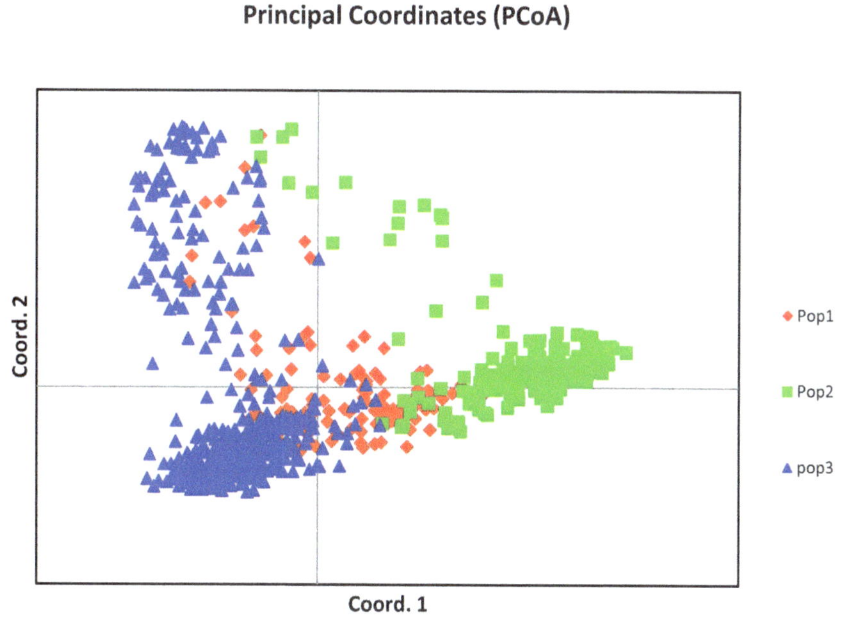

Figure 5. Principal Coordinate Analysis (PCoA) analyses of NE India rice core set.

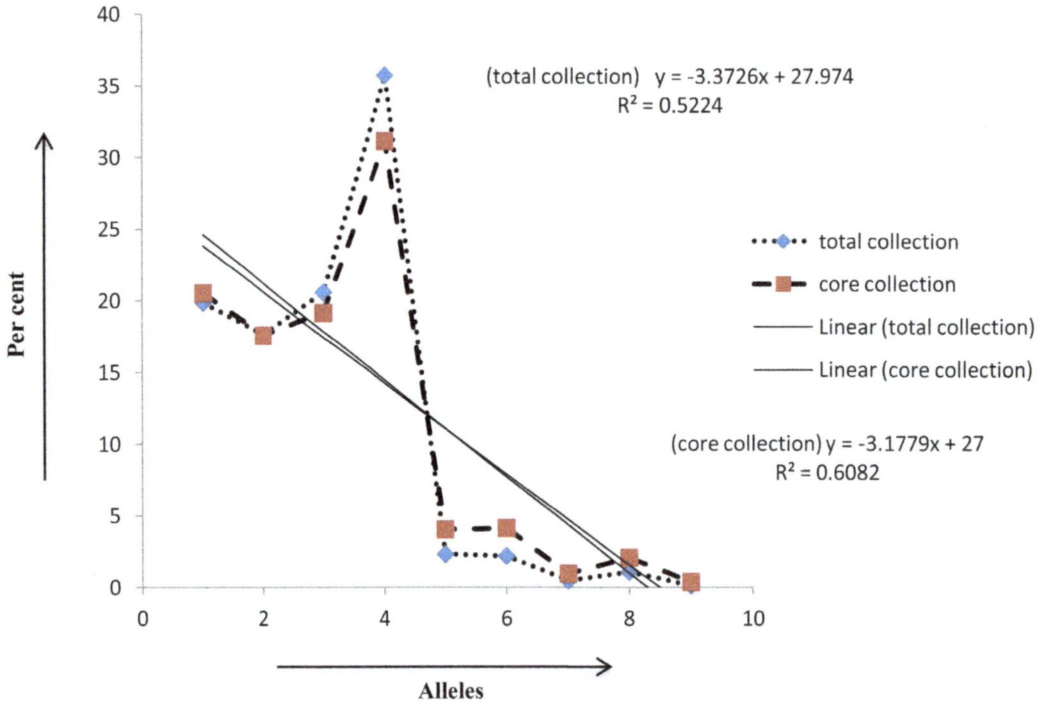

Figure 6. Comparison of allele frequency occurrence in total NE rice collections vs core set identified in the present study.

could be used for comparison of our results. AMOVA analysis showed maximum variation among individuals for accessions belonging to Nagaland (75%) and minimum for Meghalaya collection (37%). Choudhury *et al.*, (2013) [26] conducted AMOVA analysis on NE rice, which revealed 25% variation among individuals, 66% among varieties and 9% among cultivated types. The variation among individuals found in our study was higher than that reported previously, which may be attributed to the large sample size used in present study. Also partitioning of variability at population and individual levels in AMOVA among states were different because sample size taken from different states varied from 107 accessions from Mizoram to 2,635 accessions from Assam.

The genomic characterization revealed huge variability among the NE India rice collection; therefore, we decided to identify a core set for the NE India accessions to facilitate the rice breeders to effectively use the same in their crop improvement programmes and efficient conservation and management of germplasm in the genebank. In recent past, various types of molecular markers have been used to develop core sets for different annual crops. These markers include SNPs for Arabidopsis [43], SSRs for rice, wheat and common bean [44–46], AFLPs for barley [47], RFLPs for wild wheat [48] and RAPD for common bean [49]. The present study used advanced SNP markers due to their biallelic nature, co-dominance inheritance and amenability to high throughput analysis. Another crucial step for development of core set is to formulate a suitable strategy that can identify either major and minor variances or diversity, using a fully representative accession from the large collections. One of the most commonly followed approach for constructing a core set from a collection is to rely on grouping an entire collection based on growing regions and then selecting core accessions from each group followed by combining accessions from each group to form final core set [5].

In the current study, we divided the NE India rice collection into seven (on the basis of source states) subsets and later all the accessions of subsets were pooled to form a core set of NE India rice germplasm. Statistical parameters, genetic distance and cluster analysis were taken into consideration to develop this core set; these methods were followed in accordance with other studies [4,50,51,52,53,54]. We developed the NE India rice core set collection using SNP markers employing POWERCORE software (using advanced M strategy implemented through a modified heuristic algorithm). We relied on molecular data more than the passport and phenotypic data, because molecular markers can

Table 6. AMOVA of core set of NE Indian rice (701 accessions).

Source	df	SS	MS	Est. Var.	%
Among Pops	2	1649.152	824.576	1.996	20
Among Indiv.	698	8774.397	12.571	4.740	48
Within Indiv.	701	2166.500	3.091	3.091	32
Total	**1401**	**12590.049**		**9.826**	**100**

Table 7. Percentage of variation explained by the first 3 axes in Principal Coordinate Analyses of NE Indian rice core set (701 accessions).

Per cent variation	1-axis	2-axis	3-axis
%	18.41	18.04	6.63
Cum %	18.41	36.45	43.08

accurately represent the genetic diversity of the original collection and free of problems related to missing data or environmental interactions that are typically found in passport and phenotypic data. POWERCORE was used to develop a core set of 701 rice accessions (10% of total NE India rice collection). Nearly all alleles were represented in the core set; hence, there was no loss of alleles during the core set development.

In present study, we developed a core set of 10% of total collection; while there is no perfect core set constructing theory or universal methods to fit all crops, which might be due to large number of germplasm resources, absence of data for accessions, or complication and diversification of data type, and so on [50]. The core set was statistically supported with various other statistical parameters including genetic distance and cluster analysis that were taken into consideration in the final development of core set which is in accordance with previous studies [50]. The gene diversity value 0.54 and PIC value 0.48 estimated by SSR markers on rice core [22], is a higher range than the values found in our study (gene diversity value 0.39 and PIC value 0.31). This variation is obvious due to nature of marker used for development of core set. In present study, we have used SNP marker which is biallelic in nature whereas, Zhang et al., 2011, [22] have used SSR marker which is multi-allelic. A study conducted with SNP markers in tropical maize [51] reported a gene diversity value 0.29 and heterozygosity value 0.25 on core sets, and similar gene diversity was observed in present study. The gene diversity, heterozygosity, PIC values of core set were higher than those of total collection which was expected because the diversity increases with elimination of genetically similar accessions during core set development [11].

In conclusion, SNP marker based molecular characterization of NE India rice collection revealed that large variation exists among the accessions and the pattern of the genetic diversity and population structure varied across the seven states. Collections from Meghalaya were found to be the most diverse and possessed maximum number of populations. Core sets have been developed state wise as well as for the entire NE collection that can be potentially exploited by rice breeders in conventional breeding programmes and molecular breeders/researchers for genomic studies such as association mapping for gene discovery for various desirable traits and identification of suitable parents for their use in rice improvement. Further development of core set and sub-sets for NE India rice is likely to significantly reduce the task of the genebank curators as they will need to multiply and maintain small number accessions in the medium term storage as active collections to meet the requirement of the rice breeders.

Supporting Information

Figure S1 Circular phylogenetic trees of NE rice collection constructed based on SNP data using MEGA software (a) Arunachal Pradesh, (b) Assam, (c) Manipur, (d) Meghalaya, (e) Mizoram, (f) Nagaland and (g) Tripura. (Red marked accessions was selected for core by Power Core).

Figure S2 Estimation of populations in NE rice collections using LnP(D) derived Δk for k from 1 to 20 in (a) Arunachal Pradesh, (b) Assam, (c) Manipur, (d) Meghalaya, (e) Mizoram, (f) Nagaland and (g) Tripura.

Figure S3 Population structure of NE rice collection based on SNP data (a) Arunachal Pradesh, (b) Assam, (c) Manipur, (d) Meghalaya, (e) Mizoram, (f) Nagaland and (g) Tripura.

Figure S4 Analysis of Molecular Variance in NE rice collection based on SNP data (a) Arunachal Pradesh, (b) Assam, (c) Manipur, (d) Meghalaya, (e) Mizoram, (f) Nagaland and (g) Tripura.

Figure S5 Principal Coordinate Analysis in NE rice collection based on SNP data (a) Arunachal Pradesh, (b) Assam, (c) Manipur, (d) Meghalaya, (e) Mizoram, (f) Nagaland and (g) Tripura.

Table S1 List of NE rice collections studied using SNP markers.

Table S2 List of SNP primers used for genotyping of NE rice collection (6984 rice accessions) alongwith gene diversity, PIC and Major allele frequency.

Table S3 Output Table generated from Structure Harvester software for Evanno method. Yellow highlight shows the largest value in the Delta K column.

Table S4 List of rice accessions identified as NE core set based on SNP markers.

Acknowledgments

We are grateful to Dr James W. Lillard, JR., Professor and Associate Dean for Research, Morehouse School of Medicine, Atlanta, GA 30310, for manuscript editing. Thanks are due to the Director, NBPGR, New Delhi, for providing the facilities to carry out this study.

Author Contributions

Conceived and designed the experiments: RS NKS. Performed the experiments: DRC NS. Analyzed the data: RS AKS SK. Contributed reagents/materials/analysis tools: KS RKT AA. Wrote the paper: RS AKS RKT.

References

1. Frankel OH (1984) Genetic perspective of germplasm conservation. In: Arber WK. Llimensee, Peacock WL, Starlinger P, editors. Genetic manipulation: Impact on man and society. Cambridge University Press, Cambridge, England, pp. 161–170.
2. Frankel O H, Brown AHD (1984) Plant genetic resources today: A critical appraisal. In Holden JHW, Williams JT. Crop Genetic Resources: Conservation and Evaluation. London, UK: George Allen and Unwin Ltd. pp. 249–257.
3. Brown AHD (1989) The case for core collections. In: Brown AHD, Frankel OH, Marshall DR, Williams JT, editors. The Use of Plant Genetic Resources. Cambridge University Press, Cambridge, UK. pp. 136–155.
4. Ebana K, Kojima Y, Fukuoka S, Nagamine T, Kawase M (2008) Development of mini core collection of Japanese rice landrace. Breed Sci 58: 281–291.
5. Wang J, Guan Y, Wang Y, Zhu L, Wang Q (2014) A Strategy for finding the optimal scale of plant core collection based on Monte Carlo simulation. Sci. World Journal doi: 10.1155/2014/503473.
6. Yonezawa K, Nomura T, Morishima H (1995) Sampling strategies for use in stratified germplasm collections. In: Hodgkin T, Brown AHD, Hintum van Th JL, Morales EAV, editors. Core collections of plant genetic resources John Wiley and sons Chichester UK pp. 35–53.
7. Upadhyaya HD, Yadav D, Dronavalli N, Gowda CLL, Singh S (2010) Mini core germplasm collections for infusing genetic diversity in plant breeding programs. Electronic J Plant Breed 1: 1294–1309.
8. Jiang Q, Hou J, Hao C, Wang L, Ge H, et al. (2011) The wheat (*T. aestivum*) sucrose synthase 2 gene (*TaSus2*) active in endosperm development is associated with yield traits. Funct Integr Genomics 11: 49–61.
9. Upadhyaya HD, Reddy KN, Sharma S, Varshney RK, Bhattacharjee R, et al. (2011) Pigeonpea composite collection and identification of germplasm for use in crop improvement programmes. Plant Genet Resour 9: 97–108.
10. Sharma M, Rathore A, Naga Mangala U, Ghosh R, Sharma S, et al. (2012) New sources of resistance to Fusarium wilt and sterility mosaic disease in a mini-core collection of pigeonpea germplasm. Eur J Plant Pathol 133: 707–714.
11. Agrama HA, Yan WG, Lee F, Fjellstrom R, Chen MH, et al. (2009) Genetic assessment of a mini-core subset developed from USDA Rice Genebank. Crop Science 49: 1336–1346.
12. Balfourier F, Charmet G, Prosperi JM, Goulard M, Monestiez P (1998) Comparison of different spatial strategies for sampling *a* core collection of natural populations *of* fodder crops. Genet Sel Evol 30:215–235.
13. Chandra S, Huaman Z, Hari Krishna S, Ortiz R (2002) Optimal sampling strategy and core collection size of Andean tetraploid potato based on isozyme data: A simulation study. Theor Appl Genet 104: 1325–1334.
14. Hu J, Zhu J, Xu HM (2000) Methods of constructing core collections by stepwise clustering with three sampling strategies based on the genotypic values of crops. Theor Appl Genet 101: 264–268.
15. Peeters JP, Martinelli JA (1989) Hierarchical cluster analysis as a tool to manage variation in germplasm collections. Theor Appl Genet 78: 42–48.
16. Franco J, Crossa J, Warburton ML, Taba S (2006) Sampling strategies for conserving maize diversity when forming core subsets using genetic markers. Crop Sci46: 854–864.
17. Reeves PA, Panella LW, Richards CM (2012) Retention of agronomically important variation in germplasm core collections: implications for allele mining. Theor Appl Genet 124: 1155–1171.
18. Upadhaya HD, Gowda CLL, Pundir RPS, Reddy VG, Singh S (2006) Development of a core subset of finger millet germplasm using geographical origin and data on 14quantitative traits. Genet Resour Crop Evol 53: 679–685.
19. Odong TL, Jansen J, van Eeuwijk FA, van Hintum TJL (2013) Quality of core collections for effective utilization of genetic resources review, discussion and interpretation. Theor Appl Genet 126: 289–305.
20. Díez CM, Imperato A, Rallo L, Barranco D, Trujillo I (2012) Worldwide core collection of olive cultivars based on simple sequence repeat and morphological markers. Crop Sci 52:211–221.
21. Rao ES, Kadirvel P, Symonds RC, Geethanjali S, Ebert AW (2012) Using SSR markers to map genetic diversity and population structure of *Solanum pimpinellifolium* for development of a core collection. Plant Genet Resour 10:38–48.
22. Zhang P, Li J, Li X, Liu X, Zhao X (2011) Population Structure and Genetic Diversity in a Rice Core Collection (*Oryza sativa L.*) investigated with SSR markers. PloS ONE 6 (12): e27565 doi: 10. 1371/journal.pone.0027565.
23. Chang TT (1976) The origin, evolution, cultivation, dissemination, and diversification of Asian and African rice. Euphytica 25: 425–441.
24. Khush GS (1997) Origin, dispersal, cultivation and variation of rice. Plant Mol Biol 35: 25–34.
25. Londo JP, Chiang YC, Hung KH, Chiang TY, Schaal BA (2006) Phylogeography of Asian wild rice, *Oryza rufipogon*, reveals multiple independent domestications of cultivated rice, *Oryza sativa*. Proc Natl Acad Sci USA 103: 9578–9583.
26. Choudhury B, Khan ML, Dayanandan S (2013) Genetic structure and diversity of Indigenous Rice varieties (*Oryza sativa*) in Eastern Himalayan region of Northeast India. Springer Plus 2: 228–237.
27. Hore DK (2005) Rice diversity collection, conservation and management in northeastern India. Genet Resour Crop Evol 52:1129–1140.
28. Singh N, Choudhury DR, Singh AK, Kumar S, Srinivasan K, et al. (2013) Comparison of SSR and SNP markers in estimation of genetic diversity and population structure of Indian rice varieties. PLoS ONE 8(12): e84136.
29. Singh NK, Dalal V, Batra K, Singh BK, Chitra G, et al. (2007) Single copy genes define a conserved order between rice and wheat for understanding differences caused by duplication, deletion, and transposition of genes. Funct Integr Genomics 7: 17–35.
30. Liu K, Muse SV (2005) Power Marker: an integrated analysis environment for genetic marker analysis. Bioinformatics 21: 2128–2129. doi:10.1093/bioinformatics/bti282.
31. Nei M, Tajima F, Tateno Y (1983) Accuracy of estimated phylogenetic trees from molecular data. II. Gene frequency data. J Mol Evol 19: 153–170.
32. Tamura K, Stecher G, Peterson D, Fillipski A, Kumar S (2013) MEGA6 Molecular Evolutionary genetics Analyses version 6.0. Mol Biol Evol 30: 2725–2729.
33. Peakall R, Smouse P (2012) GenAlEx 6.5: Genetic analysis in Excel. Population genetic software for teaching and research - an update. Bioinformatics 28: 2537–2539.
34. Pritchard JK, Stephens M, Donnelly P (2000) Inference of population structure using multilocus genotype data. Genetics 155: 945–959.
35. Evanno G, Regnaut S, Goudet J (2005) Detecting the number of clusters of individuals using the software STRUCTURE: a simulation study. Mol Ecol 14: 2611–2620.
36. Kim KW, Chung HN, Cho GT, Ma KH1, Chandrabalan D, et al (2007) PowerCore: a program applying the advanced M strategy with a heuristic search for establishing core sets. Bioinformatics 23: 2155–2162.
37. Vairavan S, Siddiq EA, Arunachalam V, Swaminathan MS (1973) A study on the nature of genetic divergence in rice from Assam and Northeast Himalayas. Theor Appl Genet 43:213–221.
38. Sarma RN, Bahar B (2005) Genetic variation of bora rice (glutinous rice) of Assam as revealed by RAPDs. Plant Genetic Resources Newsletter 144:34–38, FAO. IPGRI
39. Bhuyan N, Borah BK, Sarma RN (2007) Genetic diversity analysis in traditional lowland rice (Oryza sativa L.) of Assam using RAPD and ISSR markers. Curr Sci 93:967–972
40. Das B, Sengupta S, Parida SK, Roy B, Ghosh M, et al. (2013) Genetic diversity and population structure of rice landraces from Eastern and North Eastern States of India. BMC Genetics 14:71–85.
41. Van Inghelandt D, Melchinger AE, Lebreton C, Stich B (2010) Population structure and genetic diversity in a commercial maize breeding program assessed with SSR and SNP markers. Theor Appl Genet 120: 1289–1299.
42. Varshney RK, Baum M, Guo P, Grando S, Ceccarelli S, et al. (2010) Features of SNP and SSR diversity in a set of ICARDA barley germplasm collection. Mol Breed 26: 229–242.
43. McKhann HI, Camillery C, Berard A, Bataillon T, David JL, et al. (2004) Nested core collections maximizing genetic diversity in *Arabidopsis thaliana*. Plant J 38: 193–202.
44. Yan WG, Agrama H, Jia M, Fjelistrom R, McClung A (2010) Geographic description of genetic diversity and relationships in the USDA rice world collection. Crop Sci 50: 2406–2417.
45. Balfourier F, Roussel V, Strelchenko P, Exbrayat-Vinson F, Sourdille P, et al. (2007) A Worldwide bread wheat core collection arrayed in a 384-well plate. Theor Appl Genet 114: 1265–1275.
46. Blair MW, Diaz LM, Buendia HF, Duque MC (2009) Genetic diversity, seed size association and population structure of a core collection of common beans (*Phaseolus vulgaris* L.). Theor Appl Genet 119: 955–972.
47. van Treuren R, Tchoudinova I, van Soest LJM, van Hintum TJL (2006) Marker- assisted acquisition and core collection formation: A case study in barley using AFLPs and pedigree data. Genet Res Crop Evol 53: 43–52.
48. Zaharieva M, Santoni S, David J (2001) Use of RFLP markers to study genetic diversity and to build a core collection of the wild wheat relative Ae-geniculata Roth (= Ae-ovarta L.). Genet Sel Evol 33S: 269–288.
49. Skroch PW, Neinhuis J, Beebe S, Tohne J, Pedraza F (1998) Comparison of Mexican common bean (*Phaseolus vulgaris* L.) core and reserve germplasm collections. Crop Sci 38: 488–496.
50. Xiao-ling L, Yong-gen L, Jin-quan L, Hai-ming X, Shahid MQ (2011) Strategies on sample size determination and qualitative and quantitative traits integration to construct core collection of rice (*Oryza sativa*). Rice Sci 18: 46–55.
51. Wen W, Franco J, Chavez-Tovar VH, Yan J, Taba S (2012) Genetic Characterization of a Core Set of a Tropical Maize Race Tuxpeño for Further Use in Maize Improvement. PLoS ONE 7(3): e32626.
52. Kuroda Y, Tomooka N, Kaga A, Vaughan D A (2009) Genetic diversity of wild soybean (*Glycine soja* Sieb. et Zucc.) and Japanese cultivated soybeans [*G. max* (L.) Merr.] based on microsatellite (SSR) analysis and the selection of a core collection. Genet Resour Crop Evol 56:1045–1055.

53. Jansen J, van Hintum TJL (2007) Genetic distance sampling: a novel sampling method for obtaining core collections using genetic distances with an application to cultivated lettuce. Theor Appl Genet 114:421–428

54. Soto-Cerda B J, Diederichsen A, Ragupathy R, Cloutier S (2013) Genetic characterization of a core collection of flax (*Linum usitatissimum* L.) suitable for association mapping studies and evidence of divergent selection between fiber and linseed types. BMC Plant Biol 13: 78–91.

13

Genetic, Epigenetic, and HPLC Fingerprint Differentiation between Natural and Ex Situ Populations of *Rhodiola sachalinensis* from Changbai Mountain, China

Wei Zhao[1,3,4], Xiaozheng Shi[2,3], Jiangnan Li[1,3], Wei Guo[1,3,4], Chengbai Liu[1,3], Xia Chen[1,3]*

1 National and Local United Engineering Laboratory for Chinese Herbal Medicine Breeding and Cultivation, School of Life Science, Jilin University, Changchun, Jilin, China, **2** Key Laboratory for Molecular Enzymology and Engineering, the Ministry of Education, Jilin University, Changchun, Jilin, China, **3** School of Life Science, Jilin University, Changchun, Jilin, China, **4** Institute of Botany, Changbai Mountain Academy of Sciences, Erdao, Jilin, China

Abstract

Rhodiola sachalinensis is an endangered species with important medicinal value. We used inter-simple sequence repeat (ISSR) and methylation-sensitive amplified polymorphism (MSAP) markers to analyze genetic and epigenetic differentiation in different populations of *R. sachalinensis*, including three natural populations and an *ex situ* population. Chromatographic fingerprint was used to reveal HPLC fingerprint differentiation. According to our results, the ex situ population of *R. sachalinensis* has higher level genetic diversity and greater HPLC fingerprint variation than natural populations, but shows lower epigenetic diversity. Most genetic variation (54.88%) was found to be distributed within populations, and epigenetic variation was primarily distributed among populations (63.87%). UPGMA cluster analysis of ISSR and MSAP data showed identical results, with individuals from each given population grouping together. The results of UPGMA cluster analysis of HPLC fingerprint patterns was significantly different from results obtained from ISSR and MSAP data. Correlation analysis revealed close relationships among altitude, genetic structure, epigenetic structure, and HPLC fingerprint patterns ($R^2 = 0.98$ for genetic and epigenetic distance; $R^2 = 0.90$ for DNA methylation level and altitude; $R^2 = -0.95$ for HPLC fingerprint and altitude). Taken together, our results indicate that ex situ population of *R. sachalinensis* show significantly different genetic and epigenetic population structures and HPLC fingerprint patterns. Along with other potential explanations, these findings suggest that the ex situ environmental factors caused by different altitude play an important role in keeping hereditary characteristic of *R. sachalinensis*.

Editor: Zhong-Jian Liu, The National Orchid Conservation Center of China; The Orchid Conservation & Research Center of Shenzhen, China

Funding: This study was supported by the Project for Wildlife Conservation and Management of the National Forestry Bureau of China (No relevent grant number was provided when we applied for this fund.), the Program for Science and Technology Support of Jilin, and the National Special Research Foundation for the Earthquake Industry of China (Grant number: 201208005). The funders had no role in study design, data collection and analysis, decision to publish, or preparation of the manuscript.

Competing Interests: The authors have declared that no competing interests exist.

* Email: chenxiajlu@163.com

Introduction

Plant is one of the most important resources that human being depend on, because of the direct benefits to humans that arise from its exploitation in new agricultural and horticultural crops. The development of medical drugs and the pivotal role played by plants are the functions of all natural ecosystems [1]. Plant diversity conservation has become a topic of common concern in the world. Ex situ conservation is the process of protecting an endangered species of plant or animal outside its natural habitat. In situ conservation is the best method of plant diversity conservation. But unfortunately, in situ conservation is not an option in many cases. The natural habitat of many species have already been completely degradation or loss caused by human over disturbance, and those of many others have been so reduced in size and so fragmented that the species are in danger of extinction [2]. Even when habitat preservation can be practical, it frequently requires the reintroduction of species. In such cases, ex situ conservation is needed to avoid species extinction in nature [1]. Ex situ conservation plays the most important role in the plant conservation especially through Botanical garden.

In recent years, some studies of population genetic structure in ex situ conservation species indicated that they may exist in a potential genetic risk, such as the loss of genetic diversity, inbreeding depression, outbreeding depression and genetic adaptation [3,4,5,6]. For ex situ conservation of endangered plants, due to environmental changes, species may generate inheritable variation in phenotype, resistance, life history etc. Those changes will lead to changes in species evolution route and a certain difficulty to future the reintroduction of species [7,8]. Consequently, the amount of genetic variation holding in the captive population is critical to assure the success of ex situ conservation and subsequent releasing. In addition, evaluation of genetic status for ex situ conservation species is also very important.

Rhodiola sachalinensis A. Boriss. is a perennial herbaceous plant in the family Crassulaceae. In China, this species is mainly distributed in alpine zones of Changbai Mountain and the southeastern Zhangguangcai Mountain, where plants are found at altitudes ranging from 1,700 to 2,500 m [9]. This species is rare with a fragment distribution. Due to heavy disturbance by human activities, it is now endangered, and has been listed as an endangered plant species in the Chinese Plant Red Book [10]. *R. sachalinensis* is a medicinal plant, containing the effective component of salidroside in roots and rhizomes, and has a reputation for stimulating the nervous system, decreasing depression, relieving fatigue, resisting anoxia and microwave radiation, preventing high altitude sickness, and improving sleep [11,12,13].

From the year 2000, Changbai Mountain Academy of Sciences (CBMAS), Changbai Mountain Natural Reserve, began to carry out ex situ conservation of *R. sachalinensis*. More than 0.2 kg seeds and 60 seedings were transplanted into the ex situ conservation area in CBMAS. The ex situ conservation site is only decades of kilometers away from the main range of distribution of *R. sachalinensis*, and its main climate conditions are similar in a certain degree (Table 1). Up to 2007, about 500 individuals survived in ex situ conservation site. Based on our investigation, although there are morphological differences among natural populations of *R. sachalinensis*, we found distinct morphological differences between ex situ and natural populations (Figure 1, Table S1 and Figure S1).

To understand these changes, we selected as study subjects *R. sachalinensis* populations i.e., three natural populations (TZ1, TZ2, TZ3) and one ex situ population (ESP), and detected the genetic, epigenetic, and HPLC fingerprint differentiation between them by using ISSR, MSAP and HPLC, with an intent to check whether ex situ population conserved representative natural populations, and to shed light on the strategy for ex situ conservation of *R. sachalinensis*.

Materials and Methods

Plant materials

We confirm that the permits were obtained from Changbai Mountain Natural Reserve where collecting took place. We confirm that the location accessed is not privately owned and all Sample collection complied with "Regulations of the People's Republic of China on Natural Reserves".

Fresh leaves and roots of *R. sachalinensis* were collected in June, 2010 from three natural populations (Population TZ1, TZ2 and TZ3) and one ex situ population (ESP) cultivated in CBMAS (Table 2). ESP was reproduced by previous ex situ population in May, 2003, and they were almost at the same age. A random sample of 10–15 individuals was obtained from each population. Leaf samples were dried directly in silica gel and held at −80°C for DNA extraction. Root samples were air-dried under a shed and stored at −20°C until chemical extraction.

DNA extraction

Total DNA was extracted from silica-dried leaves using a Plant Genomic DNA kit (Dingguo, Beijing, China). DNA concentrations were quantified using a 752 UV-Vis spectrophotometer (Shanghai Spectrum Instruments, Shanghai, China). DNA quality was verified by electrophoresis on a 1.2% agarose gel. The DNA samples were diluted to 20 ng μl^{-1} and stored at −20°C until further analysis.

ISSR

Thirty-six ISSR primers (Shenggong, Shanghai, China) designed by the University of British Columbia were screened initially using 12 individuals from the four populations. Twelve primers which produced clear and reproducible banding patterns were selected for amplification of all *R. sachalinensis* individuals in this study (Table 3). The optimized ISSR-PCR reaction system (25 µl) for *R. sachalinensis* was constructed of 2 µl template DNA, 2.5 µl of 10× *Taq* DNA polymerase buffer, 0.25 mM dNTPs, 1.5 mM $MgCl_2$, 0.4 µM primer, 1.25 U of *Taq* polymerase (Takara, Dalian, China), and double-distilled water. The amplifications were carried out in a Mastercycler Gradient thermal cycler (Eppendorf, Hamburg, Germany) under the following conditions: pre-denaturation for 5 min at 94°C, followed by 40 cycles of denaturation for 30 s at 94°C, annealing for 60 s at various temperatures (see Table 3), and elongation for 90 s at 72°C, with a final extension for 10 min at 72°C.

ISSR-PCR products were separated via electrophoresis on a 1.8% agarose gel at 100 V for 150 min in 1×TBE (Tris base, boric acid, and 0.5 M EDTA) buffer, and stained with ethidium bromide (0.5 µg ml^{-1}). Additionally, 100-bp ladders (Dingguo, Beijing, China) were loaded onto each gel. Images were identified using a gel-imaging system (Bio-Rad, CA, USA).

MSAP

The methylation-sensitive amplification polymorphism (MSAP) technique, a modification of the amplified fragment length polymorphism (AFLP) method, exploits the differential sensitivity of a pair of isoschizomers (*Hpa* II and *Msp* I) to cytosine methylation. *Hpa* II and *Msp* I recognize and excise DNA sequences at CCGG restriction sites, but with different digestion sensitivities to cytosine methylation. *Msp* I cleaves DNA when no cytosines are methylated or when the internal C is methylated (5′-CmCGG-3′) on one or both DNA strands. *Hpa* II cleaves DNA in the case of no cytosine methylation or in the case of hemimethylation of external (5′-mCCGG-3′) or internal (5′-CmCGG-3′) cytosines [14]. The MSAP protocol and PCR amplification conditions were described by Schellenbaum et al. [15] with some modifications.

Digestion and ligation reactions. DNA samples were separately digested with *Eco*R I/*Hpa* II and *Eco*R I/*Msp* I (Fermentas, Shenzhen, China). Digestion and ligation were performed simultaneously in 20-µl volumes containing 200 ng DNA template, 5 U of T_4 DNA ligase (Fermentas, Shenzhen, China), 5 U of *Eco*R I, 5 U of either *Msp* I or *Hpa* II, 5 pM of *Eco*R I adaptor (Eco-A1 and Eco-A2; Table 3), 50 pM of *Hpa* II-*Msp* I adaptor (HM-A1 and HM-A2; Table 3), 2.0 µl of 10× T_4 DNA ligase buffer, and double-distilled water. The mixture was incubated at 37°C for 12 h, inactivated at 65°C, and stored at −20°C.

Pre-amplification PCR. The optimized Pre-amplification PCR reaction system (20 µl) for *R. sachalinensis* was constructed of 2 µl digestion-ligation products, 2.5 pM of each pre-amplification primer (E00 and HM00; Table 4), 1 U of *Taq* polymerase, 2.0 µl of 10× *Taq* DNA polymerase buffer, 0.20 mM dNTPs, 1.5 mM $MgCl_2$, and double-distilled water. The pre-amplification conditions consisted of pre-denaturation for 5 min at 94°C, followed by 30 cycles of denaturation for 30 s at 94°C, annealing for 60 s at 56°C, and elongation for 80 s at 72°C, with a final extension for 10 min at 72°C and storage at −20°C.

Selective amplification PCR. Pre-amplification products were diluted 1 to 20 (v/v) with ddH_2O and selective amplification using six different primer combinations (Table 4) The optimized selective amplification PCR reaction system (25 µl) for *R.*

Figure 1. Morphological differences exhibited by *R. sachalinensis* at different altitudes. (a) Growth at 2,594 m; (b) Growth at 2,293 m; (c) Growth at 2,007 m; (d) Growth at 763 m (cultivated population).

sachalinensis was constructed of 2 μl diluted pre-amplification product, 0.2 mM dNTPs, 2 pM *Eco*R I and *Msp* I/*Hpa* II primers, 1 U of *Taq* polymerase, 2.5 μl of 10× *Taq* DNA polymerase buffer, 0.20 mM dNTPs, 1.5 mM $MgCl_2$, and double-distilled water. For selective amplification, touchdown PCR conditions were as follows: pre-denaturation for 5 min at 94°C, followed by 30 cycles of 30 s at 94°C, 30 s at 65°C decreasing by 0.7°C per cycle, and 80 s at 72°C, then followed by 23 cycles of 30 s at 94°C, 30 s at 55°C, and 80 s at 72°C, and a final extension step of 10 min at 72°C.

After selective amplification, the products were mixed 1:1 with a loading buffer (98% deionized formamide, 10 mM EDTA, 0.1% bromophenol blue, and 0.1% xylene cyanol), denatured at 95°C for 5 min, and separated on 6% denaturing polyacrylamide gels (6% polyacrylamide and 7 M urea) in 1× TBE buffer at 70 W for 4.5 h. Gels were stained according to the silver staining method [16].

HPLC fingerprint analysis

HPLC fingerprint, a comprehensive and quantifiable identification method, is able to reveal chemical information about herbal medicines using chromatograms, spectrograms, and other graphs derived by analytical and chemical techniques [17,18]. The State Food and Drug Administration of China (SFDA) require that all injections made from herbal medicines or their raw materials must be standardized by HPLC fingerprint [19].

Standards and solvents. Salidroside used as standards, were provided by Jilin Food and Drug Administration. Methanol used as solvents, were of HPLC grade and were purchased from Fisher Scientific (USA). After intensive drying salidroside was weighed and dissolved in 1 mL of methanol to achieve serial concentrations, and three injections were performed for each dilution. The standard curve was calibrated using the linear least-squares regression equation derived from the peak area. The concentration of salidroside in different samples was calculated according to the regression parameters derived from the standard curves.

Extraction of chemical composition. To extract chemical composition, root samples from the four populations were pulverized, and the powder was then screened through 180-μm sieves. Fine powder (200 mg) was accurately weighed; 10 ml of 50% methanol was then added and the mixture was weighed again. The mixture was extracted by ultrasonication for 1 h at 60°C. After cooling, 50% methanol was added to restore the initial weight. The supernatant fluid was filtered through a 0.45-μm membrane filter, and the filtrate was analyzed by HPLC.

HPLC analysis. Chromatographic separations were performed on an Agilent 1100 HPLC system (Agilent, Böblingen, Germany) equipped with a quaternary pump, autosampler, degasser, automatic thermostatic column compartment, DAD detector, and a computer with Chemstation software for analysis

Table 1. Comparison of climate and soil condition between native habitat and ex situ conservation site of *R. sachalinensis* populations.

Location	Altitude (m)	T_{mean} (°C)	T_{min} (°C)	T_{max} (°C)	*P* (mm)	*Hr* (%)	Soil
Tundra Zone	2000	−3.4	−35. 9	29.6	1060. 6	73.0	Mountain tundra soil
CBMAS	763	3.6	−33.6	33.3	695.3	72.0	Albic dark-brownforest earths

T_{mean}, annual average temperature; T_{min}, extreme low temperature; T_{max}, extreme high temperature; *P*, annual precipitation; *Hr*, relative humidity.

Table 2. Details of sampled *R. sachalinensis* populations.

Populations		Altitude (m)	Latitude (N)	Longitude (E)	Estimated population size	Number of samples		C
						Leaves	Roots	
NP	**TZ1**	2594	42°01′45.34″	128°03′56.87″	60	15	10	$A_{1\sim15}$
	TZ2	2293	42°02′40.20″	128°03′09.68″	70	15	10	$B_{1\sim15}$
	TZ3	2007	42°03′21.96″	128°03′55.29″	100	15	10	$C_{1\sim15}$
ESP	**ESP**	763	42°22′53.28″	128°01′27.42″	320	15	10	$D_{1\sim15}$

NP is natural population of *R. sachalinensis*; *ESP* is ex situ population; C is code of individuals per populations.

Table 3. Sequences of ISSR primers and amplification results.

Primer	Sequence of primer (5′ to 3′)	T_A (°C)	N_B	N_{PB}	*PPB* (%)
UBC807	AGAGAGAGAGAGAGAGT	59.2	28	25	89.29
UBC810	GAGAGAGAGAGAGAGAT	51.6	34	26	76.47
UBC825	ACACACACACACACACT	51.6	31	28	90.32
UBC840	GAGAGAGAGAGAGAGAYC	59.2	31	25	80.65
UBC841	GAGAGAGAGAGAGAGAYT	52.7	37	30	81.08
UBC842	GAGAGAGAGAGAGAGAYG	58.1	28	19	67.86
UBC856	ACACACACACACACACYA	54.0	31	23	74.19
UBC860	TGTGTGTGTGTGTGTGRA	52.7	26	23	88.46
UBC873	GACAGACAGACAGACA	50.2	31	30	96.78
UBC887	DVDTCTCTCTCTCTCTC	54.0	27	20	74.07
UBC888	BDBCACACACACACACA	52.7	22	16	72.73
UBC889	DBDACACACACACACAC	55.4	26	20	76.92
Total			349	285	81.66

B: C, G or T; *D*: A, G or T; *Y*: C or T; T_A: Annealing temperature; N_B: Number of bands; N_{PB}: Number of Polymorphic bands; *PPB:* Percentage of polymorphic bands.

of the HPLC data. An Agilent C_{18} reversed-phase column (250×4.6 mm; 5 μm) was used with the column temperature set at 25°C. The mobile phase consisted of 15:85 (v/v) methanol-ddH_2O, which was maintained for 40 min. The flow rate was 1.0 ml min^{-1} and the injection volume was 20 μl. The detection wavelength was set to 277 nm.

Data analysis

ISSR fragment analysis. ISSR-amplified DNA fragments (bands) were scored as present (1) or absent (0), and a data matrix of the ISSR banding patterns of all populations was assembled for further analysis. The software program POPGENE v1.31 [20] was used to estimate the following genetic diversity parameters: Shannon's information index (*I*) [21], percentage of polymorphic loci (*PPL*), and genetic distance. Total genetic diversity (*Ht*) and mean genetic diversity within populations (*Hs*) were tested using Nei's [22] genetic diversity statistics. The proportion of diversity among populations was calculated as the coefficient of gene differentiation, G_{ST}, according to the equation $G_{ST} = (Ht-Hs)/Ht$. Gene flow was estimated using the formula: $Nm = 0.5\times[(1-G_{ST})/G_{ST}]$ [23]. These calculations were carried out using POPGENE v1.31. A hierarchical analysis of molecular variance (AMOVA) was employed to examine population genetic differentiation (F_{ST}) within and among the four different altitudinal populations using AMOVAPREP and AMOVA 1.55 [24]. The significance of this F_{ST} was tested with 1,000 random permutations. A UPGMA tree was constructed based on genetic distance [25] among different populations to determine their genetic relationships using NTsys-pc v2.02a [26].

MSAP fragment analysis. Only clear fragment patterns were considered. *Hpa* II and *Msp* I have different sensitivities to methylation patterns: *Hpa* II is only sensitive to external hemimethylated cytosine (single-strand methylation), whereas *Msp* I only recognizes internal fully methylated cytosine (methylation of both strands). We therefore scored the differential methylation pattern based on band presence or absence [27]. MSAP banding patterns were classified into four types: Type 1/1 bands represented non-methylated loci (bands present in both profiles), Type 1/0 bands corresponded to loci hemi-methylated at the external C of the restriction site (bands present in *Eco*R I/*Hpa*

Table 4. Adaptors and primers used for MSAP analysis.

Adaptors	
Eco-A1 5′- CTCGTAGACTGCGTACC-3′ Eco-A2 5′- AATTGGTACGCAGTC -3′	HM-A1 5′- GATCATGAGTCCTGCT -3′ HM-A2 5′- CGAGCAGGACTCATGA -3′
Preamplification primers	
E00 5′- GACTGCGTACCAATTCA -3′	HM00 5′- ATCATGAGTCCTGCTCGG -3′
Selective amplification primers	
*Eco*R primer	*MspI/Hpa*II primer
E7 = 5′- GACTGCGTACCAATTCAAG -3′	HM13 = 5′- ATCATGAGTCCTGCTCGGTCG -3′
E7 = 5′- GACTGCGTACCAATTCAAG -3′	HM14 = 5′- ATCATGAGTCCTGCTCGGTCC -3′
E7 = 5′- GACTGCGTACCAATTCAAG -3′	HM15 = 5′- ATCATGAGTCCTGCTCGGTTG -3′
E7 = 5′- GACTGCGTACCAATTCAAG -3′	HM16 = 5′- ATCATGAGTCCTGCTCGGTTA -3′
E7 = 5′- GACTGCGTACCAATTCAAG -3′	HM17 = 5′- ATCATGAGTCCTGCTCGGTGA -3′
E7 = 5′- GACTGCGTACCAATTCAAG -3′	HM18 = 5′- ATCATGAGTCCTGCTCGGTAC 3′

Table 5. Genetic diversity of *R. sachalinensis* distributed among different altitudinal populations.

Populations	*PPL* (%)	*I*	G_{ST}	*Nm*
TZ1	52.15	0.2633		
TZ2	50.14	0.2351		
TZ3	55.30	0.2794		
ESP	63.90	0.3195		
Mean	55.37	0.2743		
Species level	97.71	0.4299	0.3490	0.9327

PPL: percentage of polymorphic loci; *I*: Shannon's information index; G_{ST}: genetic differentiation coefficient; *Nm*: gene flow.

II only), Type 0/1 bands indicated full methylation of the internal cytosine of both strands (bands present in *Eco*R I/*Msp* I only), and Type 0/0 bands represented non-informative loci (bands absent in both profiles). Non-informative loci were removed from our dataset because they could not discriminate between hyper-methylation and nucleotide mutation, and the number of hemi-methylated and internal fully methylated loci was then counted in each sample. Ratios of hemi-methylation, internal full methylation, and total methylation (Type 1/0+Type 0/1) were calculated against the total number of detected bands (Type 1/1+ Type 1/0+ Type 0/1). To estimate population epigenetic structure in the four different habitats, MSAP data were also converted into a methylation-susceptible matrix according to Herrera and Bazaga [28]. More specifically, band absence in either *Eco*R I/*Msp* I or *Eco*R I/*Hpa* II profiles was scored as "1" and band presence in both *Eco*R I/*Msp* I and *Eco*R I/*Hpa* II profiles was scored as "0"; bands absent in both *Eco*R I/*Msp* I and *Eco*R I/*Hpa* II profiles were treated as "missing", as they could have been produced by either genetic or epigenetic factors. The resulting data were analyzed with POPGENE v1.31 [20] to calculate diversity parameters, including genetic differentiation (G_{ST}), Nei's gene diversity, gene flow (*Nm*), and genetic distance, to search for epigenetic differentiation. AMOVAPREP and AMOVA 1.55 [24] were used to further examine population epigenetic differentiation within and among the four different altitudinal populations. The software package Ntsys-pc v2.02a [26] was used for Mantel testing and UPGMA dendrogram generation.

HPLC fingerprint analysis. HPLC fingerprints of *R. sachalinensis* root extracts were analyzed using the professional software package Similarity Evaluation System for Chromatographic Fingerprint of Traditional Chinese Medicine (2004 version), recommended by the SFDA of China for evaluating similarities of traditional Chinese medicine chromatographic profiles [29]. Hierarchical cluster analysis of samples from the four different altitudinal populations was performed based on retention time and peak area variation patterns using the UPGMA method as implemented in Ntsys-pc v2.02a [26].

Correlation analysis. We used the DPS 7.05 data processing system (Zhejiang University, Hangzhou, China) to analyze relationships among environment, heredity, and HPLC fingerprint in different populations of *R. sachalinensis*.

Results

Genetic polymorphism and population structure

To assess the genetic diversity of *R. sachalinensis*, we surveyed 12 ISSR primers in 60 individuals from the different populations sampled in 2010. A total of 349 bands (Table 3) were produced, which corresponded to an average of 29.08 bands per primer. Of these bands, 81.66% (Table 3) were polymorphic among the 60 individuals.

Genetic diversity and differentiation. At the species level, the percentage of polymorphic loci (*PPL*) and Shannon's information index (*I*) were 97.71% and 0.4299, respectively. At the population level, genetic diversity of *R. sachalinensis*, the ESP, was relatively high, while TZ1, TZ2, and TZ3 genetic diversities were more or less similar to one another. *PPL* ranged from 63.90% to 50.14%, with an average of 55.37%. *I* ranged from 0.3195 to 0.2351, with an average of 0.2743 (Table 5).

The calculated coefficient of genetic differentiation (G_{ST}) suggested that genetic differentiation of *R. sachalinensis* populations was relatively high (0.3490), indicating that about 34.90% of the genetic variation was among populations. The calculated gene flow was 0.9327 migrants (*Nm*) per generation (Table 5). AMOVA of the four populations revealed that 45.12% and 54.88% of the observed genetic variation occurred among and within populations, respectively (Table 6). The AMOVA results were consistent with Nei's genetic structure estimations (Table 6). Based on ISSR data, the mean overall F_{ST} value for individual loci was 0.451. This value was significantly different from zero ($P<0.001$).

Cluster analysis. Based on the ISSR markers data, a UPGMA dendrogram of all sampled *R. sachalinensis* individuals was reconstructed using the Nei genetic distance matrix. In this dendrogram, all the individuals in a given population formed a distinct cluster. Each population had its own node. The 60 *R. sachalinensis* individuals from natural and ex situ conservation populations clustered into three major groups, with TZ1 and TZ2 comprising single branches and TZ3 and ESP clustered together on a third branch (Figure 2a).

A principal coordinates (PCO) analysis based on genetic distances provided a spatial representation of the 60 individuals from the four sampled populations (Figure 2b). The two-dimensional PCO plot clearly differentiated populations TZ1 and TZ2. Individuals from TZ3 and ESP clustered together near the bottom of the PCO plot. There were several individuals with minimal overlap between TZ3 and ESP.

DNA Methylation Levels

DNA methylation levels in the different *R. sachalinensis* populations were evaluated by MSAP fingerprint. Based on band absence or presence, cytosine methylation patterns were classified into four types: Type 1/1, Type 1/0, Type 0/1, and Type 0/0 (Figure 3; Table 7). After removing Type 0/0 bands, which corresponded to non-informative loci, total methylation levels in populations TZ1, TZ2, TZ3, and ESP4 were 38.18%, 34.80%, 30.62%, and 25.00%, respectively; corresponding internal full

Table 6. Genetic variation within and among populations of *R. sachalinensis* revealed by AMOVA.

Source of variation	df	SSD	MSD	Variance component	%Total	F_{st}	*P* value
Among populations	3	1410.1833	470.061	28.9868	45.12	0.451	<0.0010
Within populations	56	1974.5333	35.260	35.2595	54.88		
Total	59	3384.7167	505.321	64.2463	100		

df the degree of freedom, *SSD* sums of the squared deviations, *MSD* mean squared deviations, *%Total* the percentage of the total variance.

methylation levels were 20.83%, 19.79%, 18.75%, and 15.17%, and hemi-methylation levels were 17.35%, 15.01%, 11.87%, and 9.83%, respectively (Table 7). The four *R. sachalinensis* populations thus differed to some degree with respect to DNA methylation levels. Total methylation, internal full methylation, and hemi-methylation levels gradually decreased along the altitudinal gradient, with TZ1 the most methylated and ESP the least.

Epigenetic polymorphism and population structure

To estimate population epigenetic structure, MSAP data were converted into a methylation-susceptible matrix according to Herrera and Bazaga [28]. We then used POPGENE, AMOVA, and Ntsys-pc software to calculate epigenetic polymorphism levels and population structure of the four populations.

Epigenetic diversity and differentiation. At the species level, *PPL* and *I* were 85.99% and 0.3778, respectively. At the population level, epigenetic diversity of ESP was relatively high, while epigenetic diversity of TZ1, TZ2 and TZ3 gradually decreased in that order. *PPL* ranged from 53.85% to 21.39%, with an average of 34.91%. *I* ranged from 0.2443 to 0.0971, with an average of 0.1563 (Table 8).

The coefficient of genetic differentiation ($G_{ST}=0.5892$) indicated that a certain degree of epigenetic differentiation existed among the four *R. sachalinensis* populations, corresponding to about 58.92% of the observed epigenetic variation. The calculated gene flow was 0.3486 migrants (*Nm*) per generation (Table 8). AMOVA of the four populations revealed that 63.87% of the epigenetic variation occurred among populations, while 36.13% occurred within populations (Table 9). The AMOVA results were consistent with values calculated based on Nei's genetic structure (Table 9). Based on MSAP markers, mean overall F_{ST} for individual loci was 0.639, which was significantly different from zero ($P<0.001$).

Cluster analysis. Based on the methylation-susceptible matrix, a UPGMA dendrogram was reconstructed from epigenetic distances for 40 *R. sachalinensis* individuals from the different altitudes. This dendrogram was similar to the ISSR marker-based dendrogram, with individuals clustering according to population into three distinct groups. In the dendrogram, TZ1 and TZ2 comprised two groups, and TZ3 and ESP were joined together along a third major branch (Figure 4a).

PCO analysis based on epigenetic distances among the 40 sampled individuals was used to obtain a spatial representation of their relationships (Figure 4b). The two-dimensional PCO plot clearly differentiated TZ1 and TZ2. TZ3 and ESP populations were clustered near the top of the PCO plot; their epigenetic distances were very small, but no individual was shared between them.

Chemical fingerprint

Salidroside content. The salidroside content in different populations significantly varied, from 0.25 to 11.12 mg/g, as listed in Table 10. The salidroside content of natural populations (TZ1, TZ2 and TZ3) were 1.54, 11.12, 6.96 mg/g respectively. The ex situ population (ESP) was the lowest, only 0.25 mg/g.

Similarity of HPLC fingerprint. The major HPLC fingerprint of *R. sachalinensis*, were obtained by HPLC. HPLC fingerprints of *R. sachalinensis* from the different populations revealed an abundant diversity of HPLC fingerprint. Twenty chemical components were identified based on peak retention times, with relative peak areas corresponding to concentration (Figure 5). HPLC fingerprints were compared using the Similarity

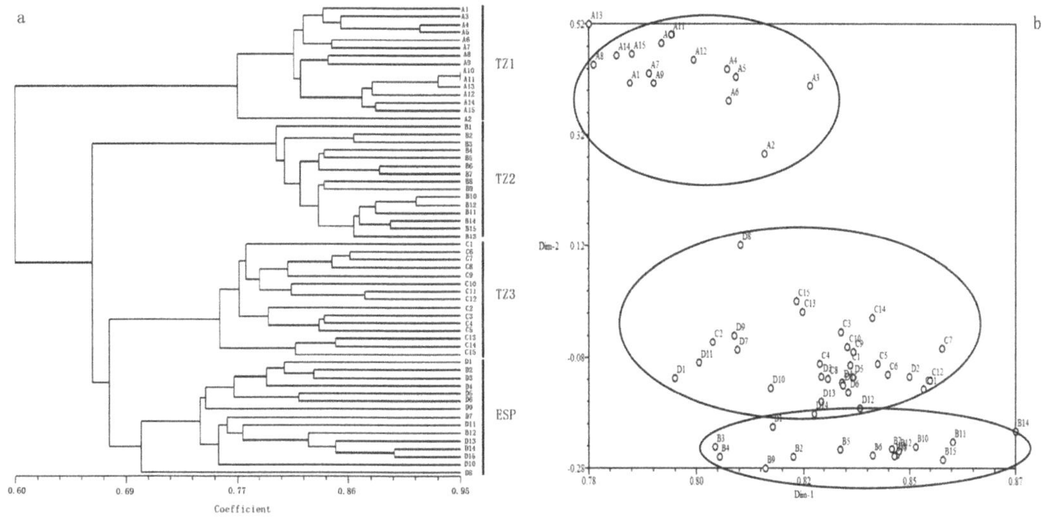

Figure 2. UPGMA dendrogram (a) and principal coordinate plot (b) based on ISSR markers for *R. sachalinensis* individuals from different populations. Black dots represent sampled individuals (for individual codes, see Table 2).

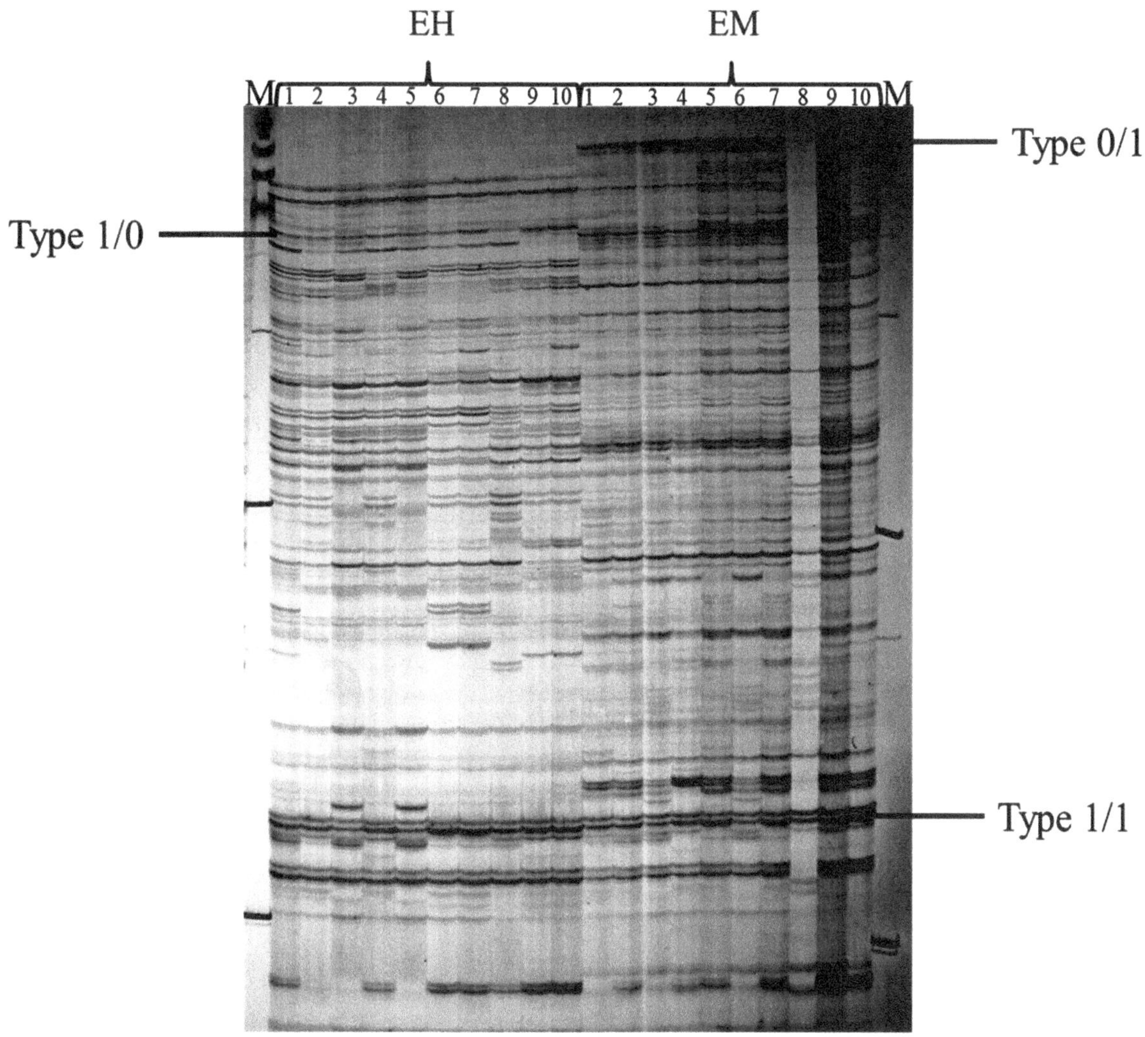

Figure 3. DNA methylation patterns in different populations of *R. sachalinensis*. *EH* and *EM* represent *Eco*R I/*Hpa* II and *Eco*R I/*Msp* I digestion combinations, respectively. Type 1/1 bands were present in both *Eco*R I/*Hpa* II and *Eco*R I/*Msp* I digestion combinations. Type 1/0 bands were present only in *Eco*R I/*Hpa* II combinations, and Type 0/1 bands were present only in *Eco*R I/*Msp* I digestions.

Table 7. DNA methylation levels of four different populations of *R. sachalinensis*.

Types	Methylation Status	Populations			
		TZ1	TZ2	TZ3	ESP
Type 1/1	CCGG CCGG GGCC GGCC	531	547	545	603
Type 1/0	CCGG GGCC	149	126	95	79
Type 0/1	CCGG GGCC	179	166	150	122
Total bands	–	859	839	800	804
Hemi-methylation ratio (%)	–	17.35	15.01	11.87	9.83
Internal full methylation Ratio (%)	–	20.83	19.79	18.75	15.17
Total methylation (%)	–	38.18	34.80	30.62	25.00

Total no. of bands of three types = I+II+III, ratio for hemi-methylation of external cytosine (%) = II/(I+II+III) × 100%, ratio for full methylation of internal cytosine (%) = III/(I+II+III) × 100%, ratio for total methylation (%) = (II+III)/(I+II+III). Each of populations contains with 10 individuals.

Evaluation System for Chromatographic Fingerprint of Traditional Chinese Medicine (2004 version) (Table 11). Based on this analysis, the higher similarity of HPLC fingerprint, over 0.89, was found among natural populations (TZ1, TZ2 and TZ3). While the similarities between ESP and natural populations were less than 0.85.

We then performed cluster analysis on *R. sachalinensis* populations using Ntsys-pc based on HPLC fingerprint similarities (Figure 6). This analysis uncovered three major clusters: TZ1, ESP, and a cluster consisting of TZ2 and TZ3. These results are obviously different from those obtained from cluster analysis of genetic and epigenetic marker data.

Correlation analysis

Correlation analysis of altitude, genetic distance, epigenetic distance, DNA methylation level, and HPLC fingerprint similarities generated interesting results (Table 12). Genetic and epigenetic distances were significantly correlated ($R^2 = 0.98$). DNA methylation level and altitude were also significantly correlated with one another ($R^2 = 0.90$), and negatively correlated with HPLC fingerprint similarity ($R^2 = -0.88$). Chemical fingerprint similarity and altitude were apparently negatively correlated ($R^2 = -0.95$) as well. Genetic and epigenetic distances were somewhat correlated with HPLC fingerprint similarity, but the correlation coefficients were not significant ($R^2 = 0.35$ and $R^2 = 0.48$).

Discussion

Population genetic structure

ISSR revealed a high of genetic diversity in endangered *R. sachalinensis*, with average of 2.53% percentage of polymorphic loci (*PPL*) and 0.2593 Shannon's information index (*I*) in three natural populations. An allozyme analysis of *R. sachalinensis* natural population genetic structure by Yan et al. [30] found a low level of genetic variation. Two factors may be responsible for these conflicting results and our observations. First, allozymes reflect only a subset of genes, i.e., those that are expressed, rather than the majority of genes including non-functional ones, whereas DNA molecular markers can more widely sample the genome. Second, breeding system, life form, and natural selection play very important roles in genetic differentiation. Hamrick and Godt [31] listed eight factors that affect genetic variation. In particular, breeding system and gene flow have particularly strong impact on genetic structure, with widespread species with high levels of gene flow and outcrossing having higher genetic variation. Generally, long-lived perennial species with a mixed breeding system have relatively high genetic diversity [32]. *R. sachalinensis* is a perennial alpine herb with predominant outcrossing, characteristics associated with high genetic diversity. In addition, *R. sachalinensis* reproduces both sexually and clonally via sprouting rhizomes. Although "pollen limitation" was invoked by Yan et al. [30] to explain the low levels of genetic variation they observed in *R. sachalinensis* populations, recent studies have demonstrated that bumblebees play a positive role in pollen dispersal of alpine plants [33,34]. These characteristics, i.e., a dual reproductive system and pollen dispersal, can be viewed as a strategy for maximal heterozygosity and reproductive success, and are common in alpine plants exposed to such drastic environments [35,36].

Ex situ population (ESP) had higher genetic diversity ($PPB = 63.9\%$ and $I = 0.3195$) than the other three natural populations. Most researches show that the populations of ex situ populations have a low level of genetic diversity, but we found a

Table 8. Epigenetic diversity of *R. sachalinensis* distributed in different altitudinal populations.

Populations	*PPL* (%)	*I*	G_{ST}	*Nm*
TZ1	53.85	0.2443		
TZ2	39.09	0.1757		
TZ3	25.29	0.1080		
ESP	21.39	0.0971		
Mean	34.91	0.1563		
Species level	85.99	0.3778	0.5892	0.3486

Table 9. Epigenetic variation within and among populations of *R. sachalinensis* revealed by AMOVA.

Source of variation	df	SSD	MSD	Variance component	%Total	F_{st}	*P* value
Among populations	3	3543.4750	1181.158	111.7928	63.87	0.639	<0.0010
Within populations	36	2276.3000	63.2310	63.2305	36.13		
Total	39	5819.7750	1244.389	175.0233	100		

different result. On the one hand, rich genotypes of species and cultivation conditions may be the major reason. More than 0.2 kg seeds and 60 seedings from natural populations were transplanted into the ex situ conservation area in CBMAS. The quantity of *R. sachalinensis* ensured variation of genotypes and genetic information, and cultivation conditions are good at establishing artificial population of *R. sachalinensis*; On the other hand, according to archived cultivation information, the artificial population of *R. sachalinensis* is derived from natural populations growing at 2,000 m, and has experienced reproduction by seed (sexual reproduction) for several generations. All these factors are better for holding a higher level of genetic diversity.

Analysis of ISSR marker data using different approaches (AMOVA and Nei's genetic diversity) found similar results. Partitioning of genetic divergence by AMOVA revealed that most of the genetic variation, 54.88%, was within-population variation. The observed level and pattern of genetic differentiation is consistent with the reproductive strategy of *R. sachalinensis*, a gravity-dispersed species with mixed mating.

High genetic differentiation (G_{ST} = 0.3490) was observed between natural *R. sachalinensis* populations and ex situ population. An effective gene flow of more than four migrants per generation is generally sufficient to counteract genetic drift and to prevent genetic differentiation between populations [37]. In this study, a gene flow of 0.9327 was estimated, indicating that some gene exchange is occurring among *R. sachalinensis* populations. However, this relatively low gene flow was not the only contributor to the high genetic differentiation observed for *R. sachalinensis* [38].

Population epigenetic structure

Genetic variation has traditionally been regarded as the fundamental source of adaptation capacity of species to respond to environmental stress [39,40]. However, in recent years, epigenetic variation has also been linked to species adaptive flexibility [41,42,43,44,45,46,47].

At the species level, high levels of epigenetic variation were uncovered in *R. sachalinensis* from different populations, with PPB = 85.99%, I = 0.3778, and G_{ST} = 0.5892 (Table 5). Most of the epigenetic variation was among population, as evidenced by a value of 63.87% revealed by AMOVA (Table 6). Because gene flow was relatively low (Nm = 0.3486), it may not be the major factor responsible for the high epigenetic differentiation of *R. sachalinensis*. During the past decade, many studies have demonstrated that epigenetic variation is usually associated with phenotypic novelties that can also contribute to the ability of plants to respond to diverse environments [28,47,48,49]. Because populations of *R. sachalinensis* distributed at different altitudes differ greatly in morphology, environmental conditions at different altitudes may be a major factor responsible for the observed high epigenetic variation. Additionally, *R. sachalinensis* DNA methylation levels, including total methylation, internal full methylation, and hemi-methylation levels, gradually decreased along an altitudinal gradient. TZ1, at the highest altitude, was the most methylated, whereas ESP as an ex situ population, at the lowest altitude, was the least. Previous studies of epigenetic variation in *Pisum sativum* L., *Nicotiana tabacum* L., and *Zea mays* L. illustrate that species can adapt to adverse environmental conditions by altering DNA methylation levels [50,51,52]. Consequently, epigenetic variation in *R. sachalinensis* was a response to environmental stress. Although these epigenetic variation are favor for survival of *R. sachalinensis* in ex situ conservation, a long-term view there is adversely for maintaining the original property of *R. sachalinensis*.

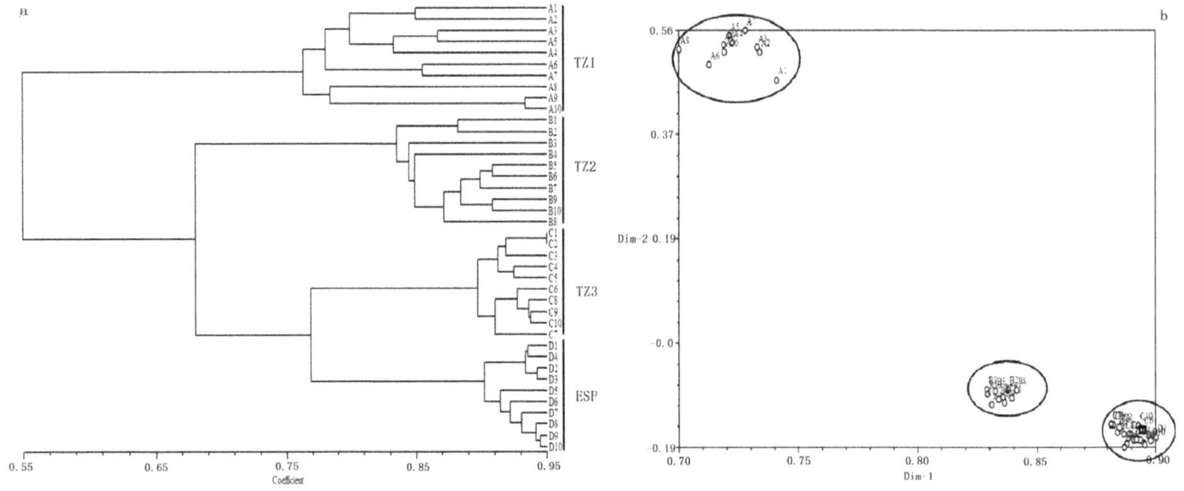

Figure 4. UPGMA dendrogram (a) and principal coordinate plot (b) based on MSAP markers for *R. sachalinensis* distributed in different altitudinal populations. Black dots represent sampled individuals (for individual codes, see Table 2).

HPLC fingerprint of *R. sachalinensis*

Plant metabolites have appeared over the course of evolution as plants adapted to their environments, and play important roles in many processes, such as physiology regulation and environmental response [53]. In additional, *R. sachalinensis* is a kind of medicinal plant, so HPLC fingerprint is self-evident importance of maintaining medicinal properties. In this study, we surveyed chromatographic fingerprints of *R. sachalinensis* individuals from different populations using HPLC fingerprint (Table 11). According to our results, there were notable differences in HPLC fingerprints among the different populations; ESP was the most distinct, as corroborated by the results of cluster analysis, and the content of salidroside is the lowest, only 0.25 mg/g. (Figure 6 and Table 10). These results are clearly different from those obtained from cluster analysis of genetic and epigenetic markers. These noticeable differences may be explained by the fact that ESP, as an ex situ population at 763 m, has experienced an environment unlike the alpine conditions of the other populations.

It is well known that HPLC fingerprints are influenced by heredity and environmental factors. For *R. sachalinensis* individuals distributed in different populations, environmental factors associated with their habitats are the major influences on HPLC fingerprint. Previous studies of HPLC fingerprint in *R. sachalinensis* have demonstrated that environmental factors such as temperature, light quality, soil, and precipitation affect secondary metabolite formation and accumulation [54,55]. Alpine environmental stress may thus be responsible for the observed differences in HPLC fingerprints among the different populations. Because ESP, at 763 m, has been subjected to the most divergent environmental stress compared with natural populations, its HPLC fingerprint would be more obviously different. These changes to the original property preservation of *R. sachalinensis* may be adversely.

Correlations among altitude, genetic and epigenetic structure, and HPLC fingerprint

Plant genetic structure, epigenetic structure, and chromatographic fingerprints are all affected by environmental factors. We performed a correlation analysis to gain a better understanding of their inter-relationships, and uncovered close relationships among these factors.

According to our results, genetic structure was significantly correlated with epigenetic structure ($R^2 = 0.98$), consistent with the similar dendrograms generated by cluster analysis of genetic and epigenetic marker data. A correlation between genetic and epigenetic changes has been previously suggested, but an equivalent relationship as revealed in this study has seldom been experimentally demonstrated [11]. Our results indicate that the pattern of epigenetic variation (cytosine methylation) in *R. sachalinensis* from different altitudes may be correlated with genomic sequence variation (genetic variation), and that epigenetic structure may be influenced by genetic structure.

Another characteristic of the epigenetic variation observed in *R. sachalinensis* was that it was correlated positively with altitude ($R^2 = 0.90$) and that HPLC fingerprint similarity and altitude were negatively correlated ($R^2 = -0.95$). Altitudinal gradients are

Table 10. Comparison of salidroside content among *R. sachalinensis* populations.

Populations	Standard equation	R^2	Salidroside content (mg/g)
TZ1	$y = 2401.4x - 11.567$	0.9998	1.54
TZ2			11.12
TZ3			6.96
ESP			0.25

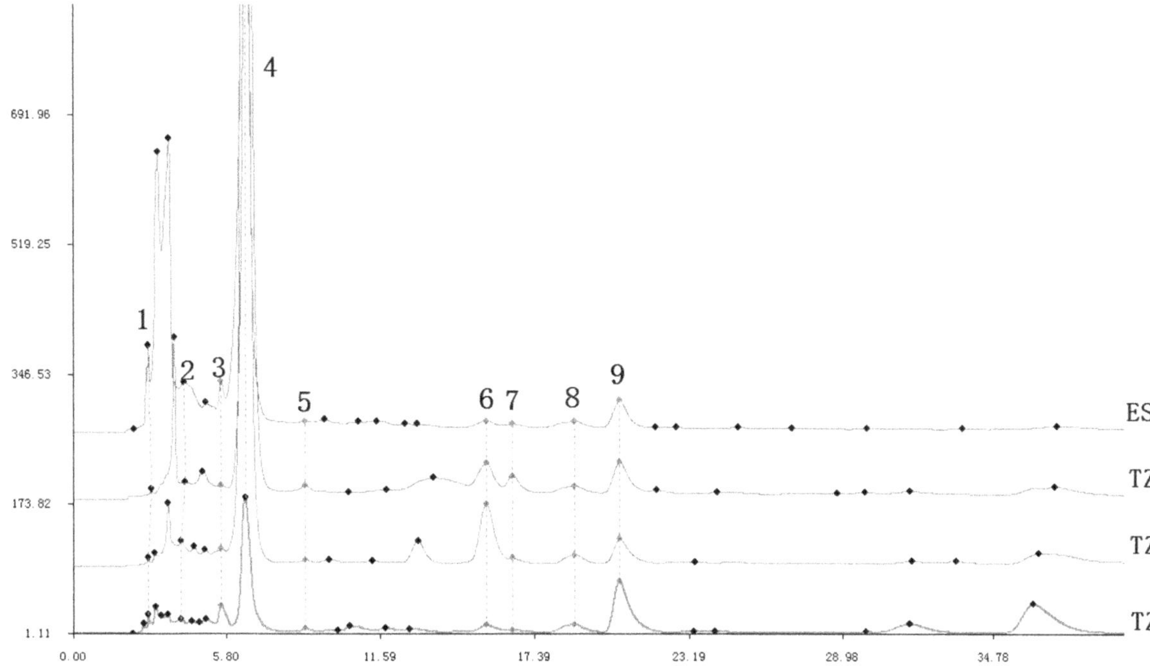

Figure 5. HPLC fingerprints of *R. sachalinensis* populations. HPLC fingerprints obtained from natural populations (TZ1, TZ2 and TZ3) and ex situ conservation population (ESP). Black dots represent peak. 1–9 correspond to 9 common peaks.

associated with substantial changes in various environmental conditions over a short distance, while epigenetic variation and HPLC fingerprints are affected by environmental factors [56,57]. Changes in epigenetic variation and HPLC fingerprint are thus inevitable along an altitudinal gradient.

There is an interesting result that epigenetic variation and HPLC fingerprints were negatively correlated ($R^2 = -0.88$). It is well known that plant metabolism and DNA methylation have strong organ and tissue specificity. So the correlate methylation patterns of leaves with HPLC fingerprint variation of roots may be unreasonable. However, this correlation may exist, because as a very important tissue of photosynthesis, DNA methylation of leaves may be an indirect influence on the formation and accumulation of secondary metabolites of roots. It may be a point of interest in the future.

Through correlation analysis of altitude, genetic structure, epigenetic structure, and HPLC fingerprint patterns, we determined that environmental conditions are a major factor influencing genetic and epigenetic structure and HPLC fingerprint. The clear genetic variation among *R. sachalinensis* populations is the result of a long process of adaptation to alpine environmental stress [58]. Environmental factors can also more directly regulate epigenetic variation to produce adaptive changes; when the environment changes, DNA methylation can regulate gene expression to allow rapid response to environmental fluctuation [28,59,60]. Recent studies of natural populations have indicated that alterations in cytosine methylation status are usually associated with changes in physiological and morphological traits of individuals, and that these phenotypic novelties are available for the action of natural section [28,48]. In this study, we also uncovered alterations in *R. sachalinensis* phenotypes among the different altitudinal populations, including changes in morphology and HPLC fingerprint.

Taken together, for natural populations of *R. sachalinensis*, although they have a higher level of genetic, their gene flow is lower, only 0.9327. So besides ex situ conservation, the most important way to conserve *R. sachalinensis* is through the protection of the habitat and reintroduction in which it lives by in situ conservation. Furthermore, for ex situ conservation and artificial cultivation of endangered plants, especially medicinal plants, not only should their survival be considered, but also the

Table 11. Similarities of chromatographic fingerprints of *R. sachalinensis* distributed in different altitudinal populations.

Populations	TZ1	TZ2	TZ3	ESP
TZ1	1	0.895	0.899	0.762
TZ2	0.895	1	0.992	0.781
TZ3	0.899	0.992	1	0.843
ESP	0.762	0.781	0.843	1

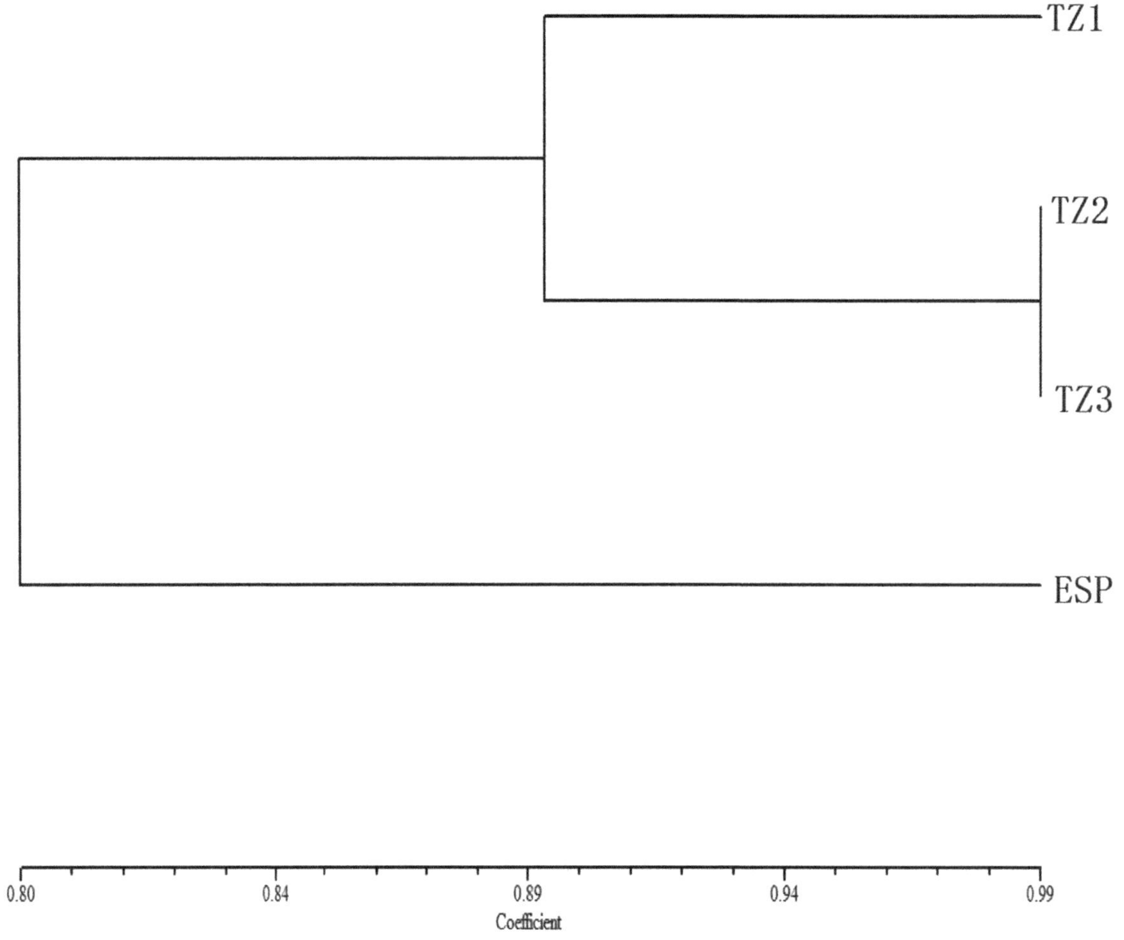

Figure 6. UPGMA dendrogram analysis derived from the Chromatographic fingerprints of *R. sachalinensis* distributed in different altitudinal populations.

impact of environmental factors on their genetic and epigenetic structures and physiological metabolism characteristics.

Conclusions

The results of this study demonstrate that environmental conditions in different habitats are major factor influenced on hereditary characteristic of species. The ex situ population of *R. sachalinensis* have been generated some variations especially epigenetics and physiological metabolism in a short period (for nearly 10 years until June, 2010). Those mean that it is a certain degree of difficult to representative natural populations from the ex situ conservation population.

In addition, *R. sachalinensis* is an important, endangered medicinal plant species. Combined knowledge of HPLC finger-

Table 12. Correlations between genetic structure, epigenetic structure, altitude, and chromatographic fingerprints.

Correlation coefficient and *P*-value	Altitude	Genetic distance	Epigenetic distance	DNA methylation level	Similarity of chromatographic fingerprints
Altitude	1	−0.17, $p=0.7405$	−0.35, $p=0.4441$	0.90**, $p=0.0054$	−0.95**, $p=0.0012$
Genetic distance	−0.17, $p=0.7405$	1	0.98**, $p=0.0002$	−0.44, $p=0.3186$	0.35, $p=0.4380$
Epigenetic distance	−0.35, $p=0.4441$	0.98**, $p=0.0002$	1	−0.61, $p=0.1441$	0.48, $p=0.2758$
DNA methylation level	0.90**, $p=0.0054$	−0.44, $p=0.3186$	−0.61, $p=0.1441$	1	−0.88**, $p=0.0099$
Similarity of chromatographic fingerprints	−0.95**, $p=0.0012$	0.35, $p=0.4922$	0.48, $p=0.2758$	−0.88**, $p=0.0099$	1

*p<0.05 **p<0.01.

print in exploited medicinal plants should provide a useful guide for their efficient utilization, breeding, and management.

Supporting Information

Figure S1 Morphological variation displayed by *R. sachalinensis* plants from different altitudes on Changbai Mountain. (a) Clone number; (b) Plant height; (c) Leaf length; (d) Leaf width; (e) Leaf thickness; (f) Stem diameter.

Table S1 Morphological summary of Changbai Mountain *R. sachalinensis* populations.

Author Contributions

Conceived and designed the experiments: WZ XC. Performed the experiments: WZ WG JNL. Analyzed the data: WZ XZS CBL. Contributed reagents/materials/analysis tools: WZ JNL WG. Contributed to the writing of the manuscript: WZ XZS XC.

References

1. Maxted N, Ford-Lloyd BV, Hawkes JG (1997) Complementary conservation strategies[M]//Plant genetic conservation. Springer Netherlands. 15–39.
2. Li QM, Xu ZF, He TH (2002) Ex situ genetic conservation of endangered *Vatica guangxiensis* (Dipterocarpaceae) in China. Biological Conservation 106: 151–156.
3. Navascues M, Emerson BC (2007) Natural recovery of genetic diversity by gene flow in reforested areas of the endemic Canary Island pine, *Pinus canariensis*. Forest Ecology And Management 244: 122–128.
4. Bijlsma R, Bundgaard J, Boerema AC (2000) Does inbreeding affect the extinction risk of small populations? predictions from Drosophila. Journal of Evolutionary Biology 13: 502–514.
5. Wolf DE, Takebayashi N, Rieseberg LH (2001) Predicting the risk of extinction through hybridization. Conservation Biology 15: 1039–1053.
6. Frankham R (2008) Genetic adaptation to captivity in species conservation programs. Molecular Ecology 17: 325–333.
7. Shepherdson D (1994) The role of environmental enrichment in the captive breeding and reintroduction of endangered species. In Creative Conservation Springer Netherlands. 167–177.
8. Huang S, Li R, Luo W, et al. (2000) Variation in characteristics of rare and threatened plants after ex-situ conservation. Chinese Biodiversity 4: 359–365.
9. Wu WC, Liu Y Yu QH (1987) Changbai Mountain rarity plant: *Rhodiola sachalinensis*. Jilin Science & Technology Press, Changchun, China.
10. Fu LG (1992) Chinese Plant Red Book. Science Press, Beijing.
11. Li TL, Xu GG, Wu LL, Sun CY (2007) Pharmacological studies on the sedative and hypnotic effect of salidroside from the Chinese medicinal plant *Rhodiola sachalinensis*. Phytomedicine 14: 601–604.
12. Zhou GB, Guan YQ, Chen HZ, Ye JN (2007) Simultaneous determination of pharmacologically active ingredients in Rhodiola by capillary chromatography with electrochemical detection. Journal Of Chromatography A 1142: 236–239.
13. Nakamura S, Li X, Matsuda H, Ninomiya K, Morikawa T, et al. (2007) Bioactive constituents from chinese natural medicines. XXVI. Chemical structures and hepatoprotective effects of constituents from roots of Rhodiola sachalinensis. Chemical & Pharmaceutical Bulletin 55: 1505–1511.
14. Vos P, Hongers R, Bleeker M (1995) AFLP: a new technique for DNA fingerprint. Nucleic Acids Res 23: 4407–4414.
15. Schellenbaum P, Mohler V, Wenzel G (2008) Variation in DNA methylation patterns of grapevine somaclones (*Vitis vinifera* L.). BMC Plant Biol 8: 78–87.
16. Bassam BJ, Caetano-Anolles G, Gresshoff PM (1991) Fast and sensitive silver staining of DNA in polyacrylamide gels. Anal Biochem 196: 80–83.
17. van Beek TA (2002) Chemical analysis of Ginkgo biloba leaves and extracts. J Chromatogr A 967: 21–55.
18. Gu M, Ouyang F, Su Z (2004) Comparison of high-speed counter-current chromatography and high-performance liquid chromatography on fingerprinting of Chinese traditional medicine. J Chromatogr A 1022: 139–144.
19. SFDA State Food and Drug Administration of China (2000) Requirements for studying fingerprint of traditional Chinese medicine injection (Draft). Chin Tradit Pat Med 22: 671–675.
20. Yeh F, Yang R, Boyle T (1997) POPGENE, the user friendly shareware for population genetic analysis. Molecular Biology and Biotechnology Center, University of Alberta, Edmonton, Canada.
21. Lewontin RC (1972) The apportionment of human diversity. Evol Biol 6: 381–398.
22. Nei M (1973) Analysis of gene diversity in subdivided populations. Proc Natl Acad Sci U S A 70: 3321–3323.
23. McDermott JM, McDonald BA (1993) Gene flow in plant pathosystems. Annual Rev Phytopathol 31: 353–373.
24. Excoffier L, Smouse PE, Quattro JM (1992) Analysis of molecular variance inferred from metric distances among DNA haplotypes: application to human mitochondrial DNA restriction data. Genetics 131: 479–491.
25. Nei M (1978) Estimation of average heterozygosity and genetic distance from a small number of individuals. Genetics 89: 583–590.
26. Rohlf FJ (1997) NTSYS-Pc: Numerical Taxonomy and Multivariate Analysis System, version 2.02a. Exeter Software, New York.
27. Sakthivel K, Girishkumar K, Ramkumar G, Shenoy VV, Kajjidoni ST, et al. (2010) Alterations in inheritance pattern and level of cytosine DNA methylation, and their relationship with heterosis in rice. Euphytica 175: 303–314.
28. Herrera CM, Bazaga P (2011) Untangling individual variation in natural populations: ecological, genetic and epigenetic correlates of long-term inequality in herbivory. Molecular Ecology 20: 1675–1688.
29. Liang YZ, Xie PS, Chan K (2004) Quality control of herbal medicines. Journal Of Chromatography B-Analytical Technologies In the Biomedical And Life Sciences 812: 53–70.
30. Yan TF, Zu YG, Yan XF, Zhou FJ (2003) Genetic structure of endangered *Rhodiola sachalinensis*. Conservation Genetics 4: 213–218.
31. Hamrick JL, Godt MJW (1989) Allozyme Diversity in Plant Species In: Plant Population Genetics, Breeding and Genetic Resources (eds. Brown AHD, Clegg MT, Kahler AL, Weir BS). Sinauer Associates, Massachusetts.
32. Nybom H, Bartish IV (2000) Effects of life history traits and sampling strategies on genetic diversity estimates obtained with RAPD markers in plants. Perspect Plant Ecol 3: 93–114.
33. Zhu XF, Wan JP, Li QJ (2010) Nectar robbers pollinate flowers with sexual organs hidden within corollas in distylous *Primula secundiflora* (Primulaceae). Biology letters 6: 785–787.
34. Zhang ZQ, Zhang YH, Sun H (2011) The reproductive biology of *Stellera chamaejasme* (Thymelaeaceae): A self-incompatible weed with specialized flowers. Flora 206: 567–574.
35. Rafii ZA, Dodd RS (1998) Genetic diversity among coastal and Andean natural populations of *Araucaria araucana* (Molina) K. Koch. Biochem Syst Ecol 26: 441–451.
36. Ma XC, Xie CX, Guan M (2014) High levels of genetic diversity within one population of *Rheum tanguticum* on the Qinghai-Tibet Plateau have implications for germplasm conservation. Pharmaceutical Crops 5: 1–8.
37. Slatkin M (1987) Gene flow and the geographic structure of natural populations. Science 236: 787–792.
38. Hou Y, Lou AR (2014) Phylogeographical patterns of an alpine plant, *Rhodiola dumulosa* (Crassulaceae), inferred from chloroplast DNA sequences. Journal of Heredity 1: 101–110.
39. Falconer DS, MacKay TFC (1996) Introduction to quantitative genetics. 4th edition. Longman, Essex, England.
40. Linhart YB, Grant MC (1996) Evolutionary significance of local genetic differentiation in plant. Annual Rev Ecol Syst 27: 237–277.
41. Cervera MT, Ruiz-Garcia L, Martinez-Zapater JM (2002) Analysis of DNA methylation in *Arabidopsis thaliana* based on methylation-sensitive AFLP markers. Mol Genet Genomics 268: 543–552.
42. Kalisz S, Purugganan MD (2004) Epialleles via DNA methylation: consequences for plant evolution. Trends Ecol Evol 19: 309–314.
43. Rapp RA, Wendel JF (2005) Epigenetics and plant evolution. New Phytol 168: 81–91.
44. Richards EJ (2006) Inherited epigenetic variation–revisiting soft inheritance. Nat Rev Genet 7: 395–401.
45. Bossdorf O, Richards CL, Pigliucci M (2008) Epigenetics for ecologists. Ecol Lett 11: 106–115.
46. Zhang X (2008) The epigenetic landscape of plants. Science 320: 489–492.
47. Jablonka E, Raz G (2009) Transgenerational epigenetic inheritance: prevalence, mechanisms, and implications for the study of heredity and evolution. Q Rev Biol 84: 131–176.
48. Paun O, Bateman RM, Fay MF, Luna JA, Moat J, et al. (2011) Altered gene expression and ecological divergence in sibling allopolyploids of Dactylorhiza (Orchidaceae). BMC Evol Biol 11: 113.
49. Mirouze M, Paszkowski J (2011) Epigenetic contribution to stress adaptation in plants. Curr Opin Plant Biol 14: 267–274.
50. Labra M, Ghiani A, Citterio S, Sgorbati S, Sala F, et al. (2002) Analysis of cytosine methylation pattern in response to water deficit in pea root tips. Plant Biology 4: 694–699.
51. Dyachenko OV, Zakharchenko NS, Shevchuk TV, Bohnert HJ, Cushman JC, et al. (2006) Effect of hypermethylation of CCWGG sequences in DNA of Mesembryanthemum crystallinum plants on their adaptation to salt stress. Biochemistry-Moscow 71: 461–465.
52. Boyko A, Kovalchuk I (2008) Epigenetic control of plant stress response. Environmental And Molecular Mutagenesis 49: 61–72.
53. Duan CR, Wang BC, Xu SR (2003) The effects of the environment stress on the plant HPLC fingerprint. J Chongqing Univ 26: 67–71.

54. Wang Y, Shang XH, Yan XF (2003) Effects of N Levels on Growth and Salidroside Content in *Rhodiola sachalinensis*. J Plant Physi Molr Biol 29: 357–359.
55. Yan XF, Wang Y, Shang XH (2004) Effects of field light intensity and quality on biomass and salidroside content in roots of *Rhodiola sachalinensis*. Acta Ecologica Sinica 24: 674–678.
56. Lin MJ, Tang LY, Reddy MN, Shen CKJ (2005) DNA methyltransferase gene dDnmt2 and longevity of Drosophila. Journal Of Biological Chemistry 280: 861–864.
57. Pichersky E, Gang DR (2000) Genetics and biochemistry of HPLC fingerprint in plants: An evolutionary perspective. Trends Plant Sci 5: 439–445.
58. Falk DA, Holsinger KE (1991) Genetics and Conservation of Rare Plants. Oxford University Press, New York.
59. Gao L, Geng Y, Li B, Chen J, Yang J (2010) Genome-wide DNA methylation alterations of Alternanthera philoxeroides in natural and manipulated habitats: implications for epigenetic regulation of rapid responses to environmental fluctuation and phenotypic variation. Plant Cell Environ 33: 1820–1827.
60. Paun O, Bateman RM, Fay MF, Hedren M, Civeyrel L, et al. (2010) Stable epigenetic effects impact adaptation in allopolyploid orchids (Dactylorhiza: Orchidaceae). Mol Biol Evol 27: 2465–2473.

14

Captive Breeding Programs Based on Family Groups in Polyploid Sturgeons

Elisa Boscari[1]*, Jose Martin Pujolar[2], Isabelle Dupanloup[3,4], Riccardo Corradin[5], Leonardo Congiu[1]

1 Department of Biology, University of Padova, Padova, Italy, **2** Department of Bioscience, Aarhus University, Aarhus C, Denmark, **3** Institute of Ecology and Evolution, University of Berne, Berne, Switzerland, **4** Swiss Institute of Bioinformatics, Lausanne, Switzerland, **5** Department of Statistics, University of Padova, Padova, Italy

Abstract

In species with long life cycles and discontinuous availability of individuals to reproduction, implementing a long-term captive breeding program can be difficult or impossible. In such cases, managing diversity among familiar groups instead of individuals could become a suitable approach to avoid inbreeding and increase the possibility to accomplish a breeding scheme. This is the case of several sturgeon species including the Adriatic sturgeon, whose recovery depends on the management of a few captive stocks directly descended from the same group of wild parents. In the present study, relatedness among 445 potential breeders was inferred with a novel software for pedigree reconstruction in tetraploids ("*BreedingSturgeons*"). This information was used to plan a breeding scheme considering familiar groups as breeding units and identifying mating priorities. A two-step strategy is proposed: a short-term breeding program, relying on the 13 remaining F0 individuals of certain wild origin; and a long-term plan based on F1 families. Simulations to evaluate the loss of alleles in the F2 generation under different pairing strategies and assess the number of individuals to breed, costs and logistical aquaculture constraints were performed. The strategy proposed is transferable to the several other tetraploid sturgeon species on the brink of extinction.

Editor: Tianzhen Zhang, Nanjing Agricultural University, China

Funding: Research was funded by the "Parco Regionale Veneto del Delta del Po" Progetto 2022EPPR12. The funder had no role in study design, data collection and analysis, decision to publish, or preparation of the manuscript.

Competing Interests: The authors have declared that no competing interests exist.

* Email: elisa.boscari@gmail.com

Introduction

Conservation management aimed at preserving the genetic variability of wild populations is essential in preventing the extinction of endangered species [1], [2]. Heritable genetic variation is a prerequisite for the long-term survival of species since the rate of evolutionary change able to occur in a group of organisms depends on the amount of genetic variation present in the gene pool; hence, organisms exhibiting low levels of genetic variability will show an impaired evolutionary potential to face future environmental changes [3]. Given the imminent risk of extinction of most critically endangered species and the low number of reproductive populations available in nature, *in situ* conservation efforts (e.g. translocation, breeding in a protected area of wild habitat, supplementary feeding) are often not sufficient to reduce or prevent the effects of anthropogenic and stochastic threats [1]. Therefore, conservation activities must frequently rely on *ex situ* conservation through breeding of captive individuals [1], [3], [4], [5], [6].

Captive breeding has long played an important role in the conservation of threatened and endangered species and generally involves two basic principles. The first one is the complete genetic characterization of all breeders available in captivity in order to quantify the amount of genetic variation and to assess the purity of the stocks. Typically, captive stocks contain only a fraction of the original genetic variability present in the larger wild population. A second goal aims at retaining and transmitting most of the residual genetic diversity to the future generations of captive breeders, thus securing the long-term future of the species [7], [8].

In this regard, pedigree-based management practices are used to identify candidate breeders, *i.e.* the "*breeders unit*", for the production of the future generations [9], [10]. Evidences from empirical studies suggest that the best strategy to retain gene diversity is to minimize average kinship when pairing individuals [11]. Therefore, an accurate pedigree of the captive breeders is critical, together with information on the relatedness among the original founder population, although in most captive programs founders are assumed to be equally unrelated (e.g. "founder assumption") [5]. Due to their high variability, microsatellites are the most adequate marker of choice for parentage assignment in animal conservation [2], [5], [12].

Among fishes, the need of establishing long-term captive programs is especially urgent for sturgeons, which according to the recent report (March 2010) of the *International Union for Conservation of Nature* (IUCN) are the most threatened group of species in the world, with 85% of sturgeons being at risk of extinction (http://www.iucnredlist.org). The decline of sturgeon populations is mostly attributed to human activities, including overexploitation for the harvesting of caviar, habitat degradation, damming of rivers and poaching [13], [14]. So far, recovery actions have been based on supplementation of natural populations with juveniles produced in captivity from wild breeders. However, for several species/populations the low availability of wild animals makes the retention of part of the fingerlings produced necessary to be used as future breeders [15]. Some particular life-characteristics of sturgeons (*i.e.* slow growth, late age at maturity, 2-years or more resting stage between reproductions) complicate breeding plans as selected individuals might not be available when needed for reproduction. Moreover, sturgeons show high juveniles mortality, which might translate into different number of retained fingerlings from each reproduction that reaches maturity. This can result in broodstocks composed by different familiar groups with unbalanced sizes. Consequently, in the absence of pedigree information, the risk of mating close relatives can be high, thus increasing relatedness and distorting founder diversity [8].

With the aim of proposing adequate approaches for the management of broodstocks composed by familiar groups that differ in size and genetic composition, we focused on the Adriatic sturgeon (*Acipenser naccarii*), an endemism of the Adriatic region. Given its tetraploid status [16], we also aim at developing adequate tools for the management of tetraploid genetic data in pedigree-based breeding programs.

The Adriatic sturgeon has been listed by IUCN since May 2010 as "critically endangered" and possibly extinct in the wild (IUCN 2012). A breeding program was initiated in 1977 with the transfer of ca. 50 immature wild individuals (F0) to the private aquaculture plant Azienda Agricola V.I.P. (Orzinuovi, Brescia, Italy). Since the late 1980s, following the first successful reproduction in captivity [17], several F1 broodstocks have been established from randomly paired F0 breeders. Over 250,000 juveniles have been reintroduced in the last 25 years, and although recaptures have been occasional, up to now there is no evidence of natural reproduction. Therefore, the F0 stock at the V.I.P. aquaculture plant, currently reduced to 13 individuals, represents the only living animals of unequivocal wild origin. All Adriatic sturgeons reared in Europe directly descend from this limited stock. The recovery of the Adriatic sturgeon depends mostly on the management of F1 captive stocks that have now reached sexual maturity.

In the present study, we used a pedigree-based approach to conduct the characterization of the stocks available at the Azienda Agricola V.I.P. Plant, including (1) all remaining individuals of the parental F0 stock, and (2) a large number of F1 individuals retained from many breedings performed in the facilities during the last 30 years. After pedigree reconstruction using a new allocation program specific for tetraploid species ("*BreedingSturgeons*"), candidate breeders were identified. A breeding plan was then designed based on family groups as breeding units. The rationale of using families rather than individuals is based on the fact that selected individuals might not be available for reproduction when needed due to the particular complex life history traits of sturgeons (i.e. they reproduce every other year), but related individuals might be available instead. Our breeding program was also supported by simulations in order to estimate the loss of genetic diversity under different scenarios as well as to evaluate the sustainability of the plan from an economic and logistic point of view. The present approach can also be applied to many other species of sturgeons whose recovery depends on the establishment of *ex situ* strategies based on captive broodstock.

Materials and Methods

Sample collection and DNA extraction

The University of Padova ethic board C.E.A.S.A. (Comitato Etico di Ateneo per la Sperimentazione Animale) exempted this study from review because it was an extra moenia activity. Nevertheless, we tried to minimize the impact of sample collection on the animals. All samples were collected in aquaculture facilities by the owners. Fish were kept into the pond water to minimize stress and were released immediately after sample collection, which consisted of painless clippings from the caudal fin. No mortality, pain or stress was observed.

We analyzed a total of 42 individuals from the parental stock F0, already genotyped at 24 microsatellite loci [10]. All analyses were conducted considering the full data set (referred as Wild; N = 42) and considering only those individuals alive at present (referred as Wild-Alive, N = 13).

We also analyzed 445 F1 individuals. In July 2003, about half of the F1 individuals died due to a poisoning incident and only 233 animals survived. All analyses were conducted considering the full F1 data set (F1, N = 445) and considering those F1 alive at present (F1-Alive, N = 233).

Our analysis also includes 133 individuals from the Ticino River Park in Italy for comparison. This individuals are part of a stock established from a very limited number of breeders [18].

Genomic DNA was extracted from fin-clip (10–100 mg) using the DNA Easy Tissue Extraction Mini Kit (Euroclone) and stored at −4°C.

Mitochondrial analysis

Prior to the analysis, species status of all individuals was assessed using mitochondrial DNA haplotypes plus species-specific SNPs at the S7 and Vimentin genes [19]. Amplification of the mitochondrial control region (d-Loop) was performed for all individuals using the primer pair PRO1F-PHE1R, following the standard experimental conditions [10]. All PCR products were purified by enzymatic reaction with ExoSAP-it (Usb) and sequenced at the external service BMR Genomics using an ABI Prism3730XL automatic sequencer. A multi-alignment was created using ClustalW in Mega5 [20] and a haplotype network was constructed using TCS v. 1.13 [21] following the statistical parsimony approach [22]. Genetic diversity was measured using haplotype diversity (H) and nucleotide diversity (Pi) estimated from number of segregating sites and from mean number of pairwise differences using DnaSP software v. 5 [23].

Finally, in order to detect the presence of other species or interspecific hybrids, the analyses of the mitochondrial control region was supplemented with the scoring of species-specific SNPs in two genes, the ribosomal Protein S7 and Vimentin, following the approach developed in [19].

Microsatellite analysis

All F1 individuals were genotyped at 7 microsatellite loci: LS-39 [24], AnacE4, AnacC11, AnacA6 [25], An20 [26], AoxD234 and AoxD64 [27], selected among the 24 loci that were originally used to genotype the F0 stock [10]. In case of multi-allocations (when more than one parent pair resulted compatible), 3 additional microsatellite loci were genotyped, An16 [26], AoxD241 and AoxD161 [27]. Microsatellite loci were amplified following the conditions reported in the original references. Microsatellite scoring was performed using GeneMapper software version 1.95 (Soft Genetics LLS^{R}). The interpretation of microsatellite patterns in tetraploids is not straightforward due to the presence of multiple alleles at each locus that can be present in more than one copy, thus the true genotype cannot be accurately resolved. As an alternative, we used the classic band-sharing approach as proposed in [10] which considers presence/absence of bands, disregarding the number of alleles present in each individual. Data from all microsatellite loci were combined by creating individual profiles where bands were coded as a string of 0 s (absent) and 1 s (present). Band sharing (BS) between individuals was used as a measure of similarity. Pairwise genetic distances were calculated as 1-BS. Number of alleles at the different loci was used to compare F1 and F0 stocks and to evaluate the efficiency of allele transmission.

Pedigree reconstruction

A new software ("*BreedingSturgeons*") was implemented to estimate an allocation compatibility index for each F1 individual to all possible parent pairs in the F0 stock, assuming a tetrasomic mode of inheritance with no dominance [28].

For all possible pairs of putative males and females in the F0 stock and the F1 individuals, we inferred all possible tetraploid genotypes compatible with the observed microsatellite profiles at all loci. We defined a single locus compatibility index, *ci*(x), of an individual to a putative parent pair as: $ci(x) = \sum[A(p1)+A(p2)]/4$, where $A(pi)$ is the number of alleles that could have been inherited from parent-*i*. $A(pi)$ can vary between 0 (if no allele present in parent i is seen in the possible tetraploid genotypes inferred for the F1 individual) and 2 (if 2 alleles could have been transmitted from that parent to the F1 individual). And ci(x) can vary between 0 and 1. Multi locus estimates of compatibility indexes were obtained as average across all microsatellites loci. In this procedure the following assumptions must be satisfied: a) all the offspring alleles should be shared with the parental pair, b) two of the offspring alleles must be shared with the mother and two with the father, c) a complete heterozygote can not transmit two copies of the same allele. The above conditions should be satisfied by at least one of the inferred genotypes in the two parents and in the offspring.

Each F1 individual was allocated to the parent pairs associated with a compatibility index above a threshold (0,857) chosen to account for possible genotyping errors at one locus.

Individuals that could not be assigned to any parent pairs (not-allocated) probably represent the progeny of wild F0 individuals died before the characterization of the F0 stock. All individuals allocated with a compatibility index higher than the threshold value were checked for mitochondrial concordance with the putative mother.

We also used the program to test the power of the set of 7 microsatellite loci used for parental allocation. To do so, we generated 1,000 virtual F1 profiles from the F0 individuals. We identified the allocation power of our microsatellite panel as the fraction of virtual F1 profiles unequivocally allocated back to the correct parental pair.

In order to identify closely related individuals (full-sibs or half-sibs) among not-allocated animals, distributions of pairwise genetic distances between full- and half-sibs were created and threshold values under which 99% of comparisons between animals with a given degree of relatedness were identified [10]. The threshold values estimated for our panel of loci (0.58 for sibs and 0.66 for half-sibs) were applied to the distance matrix between F1 individuals to identify putative groups of related individuals.

Breeding strategy

A "whole-family" approach (rather than a single-breeder approach) was used. The parent pair of each family was considered to be representative of the genetic diversity of the corresponding progeny. This allowed using a higher amount of genetic information since all the F0 individuals are genotyped at 24 loci [10]. Accordingly, the genetic distance between two families is represented by the genetic distance between the two cumulative parental profiles.

In the case of families with unknown pedigree, estimation of genetic distances from the cumulative profile of parents was not possible. In these cases, parental profiles were inferred cumulating the phenotypes of their offspring using the 7 loci genotyped for parental allocation analysis. Genetic distances estimated on these partial profiles were included in the matrix after observing a significant positive correlation ($r = 0,770$; $p < 0.001$) between distances estimated at 24 and 7 loci by using the Mantel test implemented in PopTools [29].

The strategy proposed for the breeding plan is based on the following three criteria: (a) prioritize families with higher genetic value, (b) exclude crosses between families with a shared parent and (c) avoid already-performed crosses. Priority of the different families (named "*priority families*") for reproduction was estimated based on how often the parents of each family were already represented in the F1 "*breeders unit*". This value was estimated by averaging the number of progeny of the two parents across all the families in the "*breeders unit*". In this way, the smallest families with under-represented parents were the first to be selected for reproduction. The above values were then converted into an ordinal scale ("priority"). Following the established order, each priority family was then paired with the more distant one (named "*mating family*") selected from among all of the families available in the "breeders unit".

With the aim of excluding crosses between half-sibs families, starting from the F0 profiles at 24 loci, all the possible combinations of virtual families (cumulative profiles) sharing one parent were created and the pairwise genetic distances estimated. The highest distance observed (0.35) was used as threshold value to exclude all crosses involving a shared parent. In the selection of the mating families we imposed the constraint that each *mating family* could not be crossed with more than 3 *priority families*. Consequently, the maximum number of crosses in which a given family can be involved is 4 (3 as "*mating family*" and 1 as "*priority family*"). This choice was necessary to avoid that some families with a high genetic distance from many others were used in too many crosses. This process was implemented in the R software "*BreedingPlanSturgeons*" also used to compare the different pairing strategies described in the following paragraph.

For the families whose parents were unknown, the information at 7 loci was used, which yielded a threshold value of 0.45. These families were excluded from the automated analyses and for each of them the more distant "*mating family*" was manually identified and a single cross suggested.

Simulations

A first R-script, "*CostsBreedingSturgeons*" (script available as additional material), was developed to infer the number of crosses that can be simultaneously performed in a given hatchery. It includes the estimate of costs and yields of a given breeding program based on the number of individuals generated and reared until release. It also considers that at the beginning of the third year (after tagging of all individuals), a fraction of the individuals are retained in the facilities in order to preserve the captive stock, while the rest of individuals can be released into the wild.

We took into consideration the following constraints: number of individuals that can be induced every year, availability of ponds for rearing families separately until tagging and maximum number of animals that can be hosted. We also considered biological features (*e.g.* mortalities and growth rates at different life stages) and costs (*e.g.* reproductions, food, tagging, manipulation, overheads) in order to estimate the total financial cost of the project. Additional variables are reported in Table A in File S1. Logistic factors can be modified according to the capacities of the different hatcheries and the biological features of the species under consideration. In our example, the different variables were based on the "Storioni Ticino hatchery" located in Cassolnovo (Pavia, Italy), where the F1 V.I.P. Stock will be maintained in the future. The run was performed assuming six tanks available for the first two years of rearing and a common pond of 1,200 square meters for the following years. Biological parameters such as average female fecundity (30,000 eggs), hatching rate (50%) and mortalities rates in the first (87%), second (66%) and following years (10% every year) were based on personal communication by the V.I.P. owners, who have more than 20 years of experience in the reproduction of Adriatic sturgeon in aquaculture.

A second R-script, "*BreedingPlanSturgeons*" (also available as additional material), was developed to estimate the expected fraction of alleles of the F1 stock that are successfully transmitted to the progeny under different breeding strategies. To this purpose, the program "*BreedingSturgeons*" (previously described and used for parental allocation) was applied to generate a virtual F2 progeny from the observed F1 microsatellite phenotypes. The minimum number of individuals to be crossed in order to ensure the transmission of all F1 alleles was estimated. Based on the order of the "*priority families*", four different scenarios were simulated in which the "*mating families*" were chosen by "Maximum Distance", selecting 1, 2, 3 or 4 individuals per family (variables "nInd.Prior" and "nInd.inc" in Table B in File S1). For each scenario, 100 replicates (variable "nRep" in Table B in File S1) were performed and the mean cumulative number of alleles successfully transmitted was counted.

Once the optimal number of individuals per family was fixed, we also tested three alternative strategies to select and pair families: random selection of both paired families (Random – Random) vs. priority order of the first family with "*mating family*" selected at random (Priority – Random) or with "*mating family*" selected for having the higher genetic distance (Priority – Maximum distance). The process of random selection of the "*mating family*" was repeated 100 times for each "*priority family*" and for each resulting combination breeders were randomly extracted from the two families. Also the extraction of the breeders was repeated 100 times for each family to minimize possible effects of differences due to intra-family diversity. Finally, from each virtual cross, 20 fingerlings (variable "nSons" in Table B in File S1) were generated and the total number of alleles counted. The total number of alleles successfully transmitted during the process was averaged among replicates and used to compare the different pairing strategies. The choice of generating 20 virtual individuals is consistent with the number of animals that will be retained in captivity as future breeders from each reproduction.

In all the above simulations, the sex of animals of the "*breeders unit*" was not considered since this information was not available. Moreover, all families for which less than 24 loci were genotyped were excluded from the simulations. For the above simulations RStudio was used.

Results

Genetic diversity

Prior to the analysis, species status of all individuals was assessed using mitochondrial DNA haplotypes (Table 1) plus species-specific SNPs at the S7 and Vimentin genes. In total, two hybrids were identified (one hybrid white sturgeon female × Adriatic sturgeon male and one hybrid Adriatic sturgeon female × white sturgeon male) and were consequently removed from the analysis.

The number of individuals showing each haplotype in the two stocks (F1/F1-alive) were: 0/0; 118/53, 220/106, 0/0, 75/45, 4/

Table 1. Mitochondrial diversity indices.

Sample	N	S	S_i	H	H_d	θ_w	θ_π
Wild	43*	13	3	7	0.783	0.004	0.006
Wild-Alive	13	9	0	4	0.744	0.004	0.006
F1	443	10	0	5	0.665	0.002	0.002
F1-Alive	230	10	0	4	0.681	0.002	0.002

Diversity indices for all samples including number of individuals (N), number of segregating sites (S), number of singletons (S_i), number of haplotypes (H), haplotype diversity (H_d), nucleotide diversity estimated from number of segregating sites (θ_w) and nucleotide diversity estimated from mean number of pairwise differences (θ_π). Samples include the original parental F0 stock (Wild), the current F0 stock (Wild-Alive), the full F1 data set excluding the two detected hybrids (F1) and those F1 alive at present (F1-Alive). *Includes one heteroplasmic female that was considered as two females with different haplotypes.

0, 37/23 for haplotype 1 to 7 respectively. A total of 11 individuals (one female F0 and 10 of its progeny) were heteroplasmic, presenting a double peak in the chromatogram sequence, and were counted as two different haplotypes for the analysis. When comparing genetic diversity across generations (Table 1), the F1 sample showed a decrease in haplotype diversity (F1: H = 0.67; F1-Alive: H = 0.68) compared to the parental stock of wild origin (Wild: H = 0.78; Wild-Alive: H = 0.74). Two out of the 7 original F0 haplotypes [19] were lost (hapl-1 and hapl-4) due to the random selection of mating pairs, while a third haplotype was lost after the 2013 poisoning event. Similarly, nucleotide diversity declined significantly when comparing the F0 and F1 samples, with dropping from 0.006 to 0.002.

Measures of genetic diversity at microsatellite loci are detailed in Table 2. Number of alleles ranged from 8 to 14 across loci in the wild population. A drop in the average number of alleles was observed when comparing the original Wild and the Wild-Alive (F0) stock, from 10.29 to 8.29. However, the F1 stock showed similar allele numbers in comparison with the original F0 stock (10 before poisoning; 9.86 after poisoning). The lowest average number of alleles (8.57) was observed in the F1-Ticino stock reared at the Ticino River Park. It is worth noting that some of the alleles observed in the F1 stocks were not observed in the F0 generation.

Parental Allocation

Prior to the parental allocation analysis, the power of the markers was assessed using 1,000 virtual F1 individuals. Using the 7 microsatellite set, a correct assignment of 92% was observed. Hence, we can consider the set of microsatellites to be informative enough for parental allocation procedures.

Parental allocations at 7 microsatellite loci for all the F1_V.I.P. individuals (N = 445) are summarized in Table C in File S1, including mitochondrial haplotypes used to solve ambiguities in case of multiple parental allocations. In total, 382 out of 445 individuals were successfully assigned to a single parent pair (30 families). No multi allocations were observed, which confirms the high resolution power of the markers. No discrepancies were observed between mitochondrial haplotypes and parental allocations. The remaining 63 individuals were not-allocated to any of the possible F0 parent pairs. However, four groups of full-sibs with different mitochondrial haplotypes were detected: Fam-unknownA (haplotype 3, 23 individuals), Fam-unknownB (haplotype 5, 4 individuals), Fam-unknownC (haplotype 2, 11 individuals) and Fam-unknownD (haplotype 6, 4 individuals). In total, the stock is composed of 30 families with known pedigree and at least 4 not-allocated families. Comparison of the F1 and F1-Alive stocks (Table C in File S1) showed the loss of few families, so that the F1-Alive stock is composed of 27 families of known pedigree and 3 not-allocated families (A and C were reduced to 11 and 5 individuals, respectively; B was unchanged; D was lost).

Short-term breeding scheme (F0×F0)

A short-term breeding plan including the 13 F0 individuals still alive at present (7 males, 6 females) was designed after evaluation of the genetic relatedness of the individuals (Fig. 1). We aimed at having the contribution of each remaining F0 breeder in at least three families within the F1 generation. Mating pairs were selected among the more distant males and females based on the pairwise genetic distances at 24 loci (Fig. 1), trying to maximize the number of parents represented, with the minimum number of crosses. Following this strategy, 18 crosses were selected as priority (Fig. 1), among which only one was already present within the F1 stock with a limited number of individuals.

Long-term breeding scheme (F1×F1) and selection of candidate breeders

The long-term plan was designed based on the F1 progeny, using a family-based approach. A total of 32 crosses was planned, one for each of the 32 families that composed the "*breeders unit*": 30 reared at V.I.P. (27 allocated and 3 not-allocated) plus two families from the Ticino River Park (1 allocated and 1 not-allocated). Pairwise genetic distances among families are shown in Fig. 2, together with details on recommended crosses and those crosses to be avoided (distance values <0.35 for 24 loci and <0.45 for 7 loci).

In the case of 4 not-allocated and 2 allocated families, generated by a single F0 female (reported in Fig. 2 as "Indecisa") whose sample was accidentally loss, a reliable estimation of the genetic distances with the rest of families could not be achieved since they were only genotyped at 7 loci. Thus, these families were excluded from next simulations and only one cross was considered in order to guarantee their contribution to the next generation.

Table 2. Number of alleles per locus in the different stocks and average band sharing (BS) values at 7 microsatellite loci.

Locus	Wild (N = 42)	Wild-Alive (N = 13)	F1 (N = 443)	F1-Alive (N = 231)	F1-Ticino (N = 133)
LS-39	8	8	8/1	8/0	7/0
AoxD64	9	8	9/1	9/1	9/0
AnacE4	10	7	10/3	10/3	8/0
AnacC11	11	7	9/1	9/1	8/0
AoxD234	14	12	14/0	14/0	13/0
An20	11	9	11/1	11/0	8/2
AnacA6	9	7	9/0	8/0	7/0
Mean	10.29	8.29	10/1	9.86/0.71	8.57/0.29
BS	0.40	0.39	0.45	0.45	0.50

Samples include the original parental F0 stock (Wild), the current F0 stock (Wild-Alive), the full F1 V.I.P. data set (F1) and those F1 alive at present (F1-Alive). For the F1 stocks, the values presented correspond to alleles shared with the F0 generation and newly observed alleles, respectively. Detected hybrids were excluded from the analysis.

ScoreA	ScoreB	Score A+B	Alive F0	Sex	Nac4	Nac7	Nac14	Nac20	Nac25	Nac28	Nac5	Nac6	Nac13	Nac17	Nac23	Nac30	Nac31
0	3	3	Nac4	F	0	0	0	0	0	0	0	0	0	0	0	0	0
1	2	3	Nac7	F	0,43802	0	0	0	0	0	0	0	0	0	0	0	0
0	4	4	Nac14	F	0,59259	0,4955	0	0	0	0	0	0	0	0	0	0	0
0	3	3	Nac20	F	0,53968	0,53488	0,58621	0	0	0	0	0	0	0	0	0	0
0	3	3	Nac25	F	0,57627	0,57025	0,5	0,50794	0	0	0	0	0	0	0	0	0
1	3	4	Nac28	F	0,47619	0,48837	0,62069	0,50746	0,57143	0	0	0	0	0	0	0	0
0	3	3	Nac5	M	0,48718	0,58333	0,60748	0,488	0,60684	0,472	0	0	0	0	0	0	0
1	2	3	Nac6	M	0,54955	0,5614	0,68317	0,54622	0,65766	0,29412	0,52727	0	0	0	0	0	0
1	2	3	Nac13	M	0,55	0,5122	0,6	0,46875	0,48333	0,45313	0,54622	0,48673	0	0	0	0	0
2	2	4	Nac17	M	0,54955	0,54386	0,44554	0,63025	0,54955	0,52941	0,54545	0,59615	0,61062	0	0	0	0
1	3	4	Nac23	M	0,56667	0,46341	0,49091	0,54688	0,46667	0,5	0,47899	0,61062	0,45902	0,50442	0	0	0
0	3	3	Nac30	M	0,52542	0,50413	0,44444	0,50794	0,49153	0,49206	0,4359	0,58559	0,55	0,47748	0,43333	0	0
2	3	5	Nac31	M	0,51724	0,59664	0,60377	0,5	0,53448	0,56452	0,25217	0,59633	0,61017	0,54128	0,52542	0,53448	0

Figure 1. F0 breeding plan. Distance matrix at 24 microsatellite loci between the current F0 stock (Wild-Present) individuals. The already-existing families generated by each F0 parent and represented by at least 10 animals were counted (ScoreA). ScoreB represents the number of crosses per F0 that should be planned in order to get at least three families per breeder in the F1 generation (ScoreA+B). Only crosses between males and females are selected. Cells highlighted in grey indicate crosses selected for the short-term breeding plan on the basis of ScoreA and ScoreB.

Breeding plan simulations

Results obtained from the simulations using the script "*CostsBreedingSturgeons*", including inference of how many animals can be hosted in the facilities and released in the wild, as well as total costs of the project, are shown in Fig. 3.

Considering 6 ponds available at the facilities of Cassolnovo, where the progenies will be produced and reared, three crosses per year are feasible, which corresponds to about 45.000 larvae per year (15.000 per female). Induction of fish is not straightforward but assuming to induce 5 males and 5 females per cross, at least one successful reproduction is expected. Number of individuals at the 2nd year dropped to ca. 6.000 individuals (Fig. 3a) after assuming a 86.7% mortality.

In order to save pedigree information, fingerlings produced by different family combinations should be kept separated for the first two years (until tagging). At the start of the 3rd year, after tagging, the progeny is split into a fraction that is retained in a common tank with all other families to preserve the captive broodstock and a fraction that can be released into the wild. On the basis of a 10% yearly mortality, *ca.* 140 individuals are retained (Fig. 3a) in order to have an average of 20 individuals per family at the 10th year of age. These individuals will constitute the "captive population" to be kept in the facilities as an "insurance policy" for the species. The remaining individuals (ca. 1.850) can be released into the wild (Fig. 3b). Finally, estimated costs are detailed in Fig. 3c, including costs generated by induction, maintenance and tagging of the individuals as well as total annual amount required for the breeding plan.

Results from simulations using the script "*BreedingPlanSturgeons*" to estimate the optimal number of breeders are presented in Fig. 4a, in which the cumulative number of alleles transmitted to the progeny is reported for different number of breeders. The fraction of alleles from the F0 generation that are successfully transmitted to the F2 is 100% when crossing either 2, 3 or 4 individuals per family. Only when crossing 1 single individual the transmission of alleles is not complete. The best strategy appears to be the use of 3 individuals, since the use of 2 individuals performed worst in the short-them, when the number of crosses was low.

Finally, results from simulations conducted to compare three alternative mating strategies are shown in Fig. 4b. While the rate of allele transmission was similar across mating strategies, both methods in which at least one family could be selected by chance ("Priority-Random" and "Random-Random") showed larger standard deviations than the "Priority-Maximum Distance" method, particularly when the number of crosses was low, suggesting than the latter is the most conservative mating strategy.

N	Priority		pelv.Nac18	Nac8.matto	Nac7.Nac6	Nac8.Nac31	Nac33.Nac11	O2.matto	Nac26.Nac29	Nac19.Nac17	Nac28.Nac17	Nac16.Nac23	Nac19.Nac31	Nac16.Nac30	Nac33.Nac9	Nac8.Nac17	Nac12.Nac27	Nac3.740	Nac8.740	Nac15.Nac13	Nac28.Nac23	Nac3.matto	Nac33.740	Nac12.Nac17	Nac3.Nac6	Nac16.740	Nac16.Nac29	Nac24.Nac29	Indecisa.matto	Indecisa.Nac11	Fam-UnknownB	Fam-UnknownA	Fam-UnknownC	Fam-UnknownPC
10	3	pelv.Nac18	0	0	0	0	0	0	0	0	0	0	0	0	0	0	0	0	0	0	0	0	0	0	0	0	0	0	0	0	0	0	0	0
5	16	Nac8.matto	0,34831	0	0	0	0	0	0	0	0	0	0	0	0	0	0	0	0	0	0	0	0	0	0	0	0	0	0	0	0	0	0	0
14	9	Nac7.Nac6	0,41622	0,4386	0	0	0	0	0	0	0	0	0	0	0	0	0	0	0	0	0	0	0	0	0	0	0	0	0	0	0	0	0	0
10	22	Nac8.Nac31	0,40331	0,24551	0,42529	0	0	0	0	0	0	0	0	0	0	0	0	0	0	0	0	0	0	0	0	0	0	0	0	0	0	0	0	0
4	4	Nac33.Nac11	0,35165	0,35714	0,32571	0,35673	0	0	0	0	0	0	0	0	0	0	0	0	0	0	0	0	0	0	0	0	0	0	0	0	0	0	0	0
8	5	O2.matto	0,17708	0,17978	0,37297	0,34807	0,34066	0	0	0	0	0	0	0	0	0	0	0	0	0	0	0	0	0	0	0	0	0	0	0	0	0	0	0
23	17	Nac26.Nac29	0,46961	0,47305	0,44828	0,49412	0,39181	0,43646	0	0	0	0	0	0	0	0	0	0	0	0	0	0	0	0	0	0	0	0	0	0	0	0	0	0
15	26	Nac19.Nac17	0,35789	0,30682	0,35519	0,35196	0,32222	0,32632	0,3743	0	0	0	0	0	0	0	0	0	0	0	0	0	0	0	0	0	0	0	0	0	0	0	0	0
27	25	Nac28.Nac17	0,37968	0,4104	0,26667	0,39773	0,32203	0,37968	0,3523	0,28649	0	0	0	0	0	0	0	0	0	0	0	0	0	0	0	0	0	0	0	0	0	0	0	0
13	10	Nac16.Nac23	0,42105	0,36364	0,39891	0,35196	0,43333	0,33684	0,3296	0,3617	0,35135	0	0	0	0	0	0	0	0	0	0	0	0	0	0	0	0	0	0	0	0	0	0	0
20	23	Nac19.Nac31	0,34737	0,375	0,38798	0,2514	0,38889	0,35789	0,4302	0,19149	0,41622	0,37234	0	0	0	0	0	0	0	0	0	0	0	0	0	0	0	0	0	0	0	0	0	0
2	2	Nac16.Nac30	0,41935	0,39535	0,41899	0,34857	0,40909	0,3871	0,4057	0,38043	0,35912	0,19565	0,38043	0	0	0	0	0	0	0	0	0	0	0	0	0	0	0	0	0	0	0	0	0
12	11	Nac33.Nac9	0,41436	0,42515	0,33333	0,36471	0,18129	0,39227	0,3294	0,32961	0,32955	0,35196	0,32961	0,3714	0	0	0	0	0	0	0	0	0	0	0	0	0	0	0	0	0	0	0	0
8	24	Nac8.Nac17	0,36667	0,20482	0,37572	0,21893	0,28235	0,3	0,4201	0,19101	0,24571	0,34831	0,35955	0,3678	0,32544	0	0	0	0	0	0	0	0	0	0	0	0	0	0	0	0	0	0	0
2	1	Nac12.Nac27	0,31148	0,34911	0,38636	0,33721	0,38728	0,3224	0,3721	0,33702	0,35955	0,33702	0,29282	0,3559	0,32558	0,39181	0	0	0	0	0	0	0	0	0	0	0	0	0	0	0	0	0	0
7	12	Nac3.740	0,38624	0,37143	0,37363	0,34831	0,27374	0,39683	0,3708	0,29412	0,3587	0,39037	0,34759	0,3989	0,29213	0,35593	0,34444	0	0	0	0	0	0	0	0	0	0	0	0	0	0	0	0	0
14	19	Nac8.740	0,3587	0,2	0,41243	0,24855	0,29885	0,30435	0,4798	0,30769	0,38547	0,37363	0,37363	0,3933	0,34104	0,22093	0,36	0,20442	0	0	0	0	0	0	0	0	0	0	0	0	0	0	0	0
26	18	Nac15.Nac13	0,35829	0,3526	0,35556	0,39773	0,32203	0,31551	0,3977	0,34054	0,35165	0,37297	0,39459	0,4144	0,31818	0,34857	0,32584	0,3587	0,34078	0	0	0	0	0	0	0	0	0	0	0	0	0	0	0
3	20	Nac28.Nac23	0,38542	0,41573	0,26486	0,42541	0,3956	0,35417	0,326	0,34737	0,14439	0,24211	0,38947	0,3871	0,33702	0,33333	0,36612	0,40741	0,41304	0,3262	0	0	0	0	0	0	0	0	0	0	0	0	0	0
1	7	Nac3.matto	0,33679	0,19553	0,34409	0,30769	0,28962	0,20207	0,3516	0,3089	0,3617	0,29843	0,36126	0,3476	0,34066	0,30387	0,32609	0,21053	0,2973	0,35106	0,36788	0	0	0	0	0	0	0	0	0	0	0	0	0
2	14	Nac33.740	0,3587	0,31765	0,34463	0,3526	0,18391	0,34783	0,3988	0,2967	0,34078	0,37363	0,36264	0,382	0,20231	0,31395	0,36	0,16022	0,15909	0,32961	0,38043	0,2973	0	0	0	0	0	0	0	0	0	0	0	0
2	21	Nac12.Nac17	0,37705	0,37278	0,36364	0,37209	0,31792	0,37705	0,3488	0,21547	0,22472	0,33702	0,37017	0,3446	0,34884	0,23977	0,24138	0,3	0,37143	0,34831	0,33333	0,31522	0,37571	0	0	0	0	0	0	0	0	0	0	0
1	6	Nac3.Nac6	0,42391	0,43529	0,22034	0,39884	0,29885	0,40217	0,3642	0,35165	0,27374	0,3956	0,37363	0,427	0,2948	0,37209	0,34857	0,22652	0,38636	0,35196	0,30435	0,24324	0,34091	0,34857	0	0	0	0	0	0	0	0	0	0
1	13	Nac16.740	0,44385	0,3526	0,46667	0,36364	0,36723	0,37968	0,4886	0,35135	0,40659	0,25405	0,36216	0,2376	0,35227	0,38286	0,35955	0,23913	0,19553	0,42857	0,44385	0,34043	0,2067	0,35955	0,44134	0	0	0	0	0	0	0	0	0
1	15	Nac16.Nac29	0,41053	0,34091	0,40984	0,34078	0,35556	0,32632	0,2291	0,34043	0,32973	0,1383	0,34043	0,1957	0,30726	0,32584	0,29282	0,36898	0,36264	0,38378	0,29474	0,27749	0,35165	0,31492	0,37363	0,25405	0	0	0	0	0	0	0	0
1	8	Nac24.Nac29	0,37634	0,38372	0,35196	0,41714	0,35227	0,33333	0,2343	0,33696	0,30387	0,30435	0,36957	0,3778	0,30286	0,35632	0,32203	0,34426	0,38202	0,29282	0,22581	0,3262	0,38202	0,35593	0,33708	0,42541	0,23913	0	0	0	0	0	0	0
2	V.I.	Indecisa.matto	0,47541	0,2	0,55932	0,41379	0,35714	0,25424	0,5636	0,38983	0,54098	0,48387	0,45763	0,5636	0,38983	0,44444	0,40351	0,4	0,35714	0,45763	0,48485	0,2459	0,35714	0,44828	0,41935	0,44828	0,44828	0,39286	0	0	0	0	0	0
1	V.I.	Indecisa.Nac11	0,5	0,37037	0,41379	0,40351	0,2	0,44828	0,4444	0,41379	0,43333	0,5082	0,51724	0,5556	0,34483	0,39623	0,42857	0,38983	0,38182	0,37931	0,44615	0,4	0,41818	0,36842	0,34426	0,50877	0,4386	0,38182	0,51724	0	0	0	0	0
4	V.I.	Fam-UnknownB	0,51724	0,57692	0,53571	0,49091	0,43396	0,5	0,1538	0,42857	0,44828	0,45763	0,42857	0,5769	0,35714	0,56863	0,44444	0,4986	0,58491	0,57143	0,30683	0,41379	0,54717	0,49091	0,45763	0,6	0,38182	0,35849	0,51613	0,42857	0	0	0	0
11	V.I.	Fam-UnknownA	0,42424	0,36667	0,4375	0,36508	0,34476	0,28125	0,3047	0,3125	0,33333	0,49254	0,375	0,5333	0,3125	0,45763	0,35484	0,26154	0,30984	0,5	0,35211	0,38182	0,37705	0,42857	0,22388	0,49206	0,39683	0,37705	0,47541	0,27273	0,42373	0	0	0
5	V.I.	Fam-UnknownC	0,41538	0,52542	0,42857	0,48387	0,4	0,36508	0,3559	0,36508	0,38462	0,42424	0,46032	0,4915	0,20635	0,48276	0,27869	0,40625	0,5	0,36508	0,34286	0,41538	0,43333	0,35484	0,33333	0,45161	0,32258	0,33333	0,35593	0,32075	0,36842	0,28571	0	0
62	V.I.	Fam-UnknownPC	0,39394	0,5	0,40625	0,46032	0,37705	0,375	0,3667	0,34375	0,36364	0,43284	0,4375	0,4667	0,21875	0,45763	0,25806	0,38462	0,47541	0,34375	0,32394	0,39394	0,44262	0,33333	0,31343	0,46032	0,33333	0,34426	0,51613	0,25	0,56667	0,42373	0,40351	0

Figure 2. F1 breeding plan. Pairwise distances among the families of the *"Breeders Unit"* estimated by comparing the cumulative profiles of the parent pairs. The black frame includes distances estimated at 24 loci and excludes the ones based on 7 loci only. Light grey and dark grey cells represent crosses to be avoided based on threshold values of 0,35 (at 24 loci) and 0,45 (at 7 loci), respectively. Black cells represent selected crosses identified by the software *"BreedingPlanSturgeons"* (within the black frame) or by visual inspection (for the families marked with V.I.). N = number of individuals per family.

a) Matrix of crosses

		N. crosses per year	Cohorts									
			[1]	[2]	[3]	[4]	[5]	[6]	[7]	[8]	[9]	[10]
Years	[1]	3	44610	0	0	0	0	0	0	0	0	0
	[2]	3	44652	5979	0	0	0	0	0	0	0	0
	[3]	3	44874	5846	141	0	0	0	0	0	0	0
	[4]	3	44682	5993	141	123	0	0	0	0	0	0
	[5]	3	45528	5966	141	124	107	0	0	0	0	0
	[6]	3	44712	5981	141	122	113	100	0	0	0	0
	[7]	3	45453	5980	141	126	112	102	92	0	0	0
	[8]	3	44838	6061	141	121	114	102	94	84	0	0
	[9]	3	44754	6043	141	132	112	101	93	82	75	0
	[10]	3	44487	5859	141	123	120	102	95	85	77	69

b) Matrix of releases

		N. of individuals three years old to release per year									
		[1]	[2]	[3]	[4]	[5]	[6]	[7]	[8]	[9]	[10]
Years	[1]	0	0	0	0	0	0	0	0	0	0
	[2]	0	0	0	0	0	0	0	0	0	0
	[3]	0	0	1840	0	0	0	0	0	0	0
	[4]	0	0	0	1844		0	0	0	0	0
	[5]	0	0	0	0	1860	0	0	0	0	0
	[6]	0	0	0	0	0	1874	0	0	0	0
	[7]	0	0	0	0	0	0	1870	0	0	0
	[8]	0	0	0	0	0	0	0	1880	0	0
	[9]	0	0	0	0	0	0	0	0	1825	0
	[10]	0	0	0	0	0	0	0	0	0	1896

c) Matrix of costs

		Costs					
		Induction	1st year	2nd year	Chip	Shared tank	Financial support
Years	[1]	5000	18000	0	0	0	23000
	[2]	5000	18000	12000	0	0	35000
	[3]	5000	18000	12000	9905	0	44905
	[4]	5000	18000	12000	9925	0	44925
	[5]	5000	18000	12000	10005	0	45005
	[6]	5000	18000	12000	10075	0	45075
	[7]	5000	18000	12000	10055	0	45055
	[8]	5000	18000	12000	10105	0	45105
	[9]	5000	18000	12000	9830	0	44830
	[10]	5000	18000	12000	10185	0	45185

Figure 3. Example of output of the R-script "*CostsBreedingSturgeons*". The run was performed using six tanks available for the first two years of rearing and one common pond of 1200 mq for the following years. (a) Number of individuals for each cohort (columns) maintained in captivity in the different years (lines). (b) Individuals that can be released per year. (c) Annual costs.

Discussion

We present the first breeding plan for the Adriatic sturgeon, a species which is believed to be possibly extinct in nature and whose future entirely relies on the implementation of an *ex situ* captive breeding plan based on genetic and demographic information upon which to establish management decisions. The program here proposed meets the two principals generally required in captive breeding: (i) the complete genetic characterization of all the individuals and (ii) the planning of a breeding scheme that minimizes genetic relatedness in order to maximize genetic variability. One particular characteristic of our breeding plan is that is "family-based" in contrast with most breeding plans that are "individual-based". One of the difficulties when working with sturgeons is that adults are only reproductively active every other year or more rarely. This represents a serious problem since a selected individual might not be available for reproduction when needed. However, this problem can be avoided by using a family rather than a single individual approach, so that related individuals might be available instead.

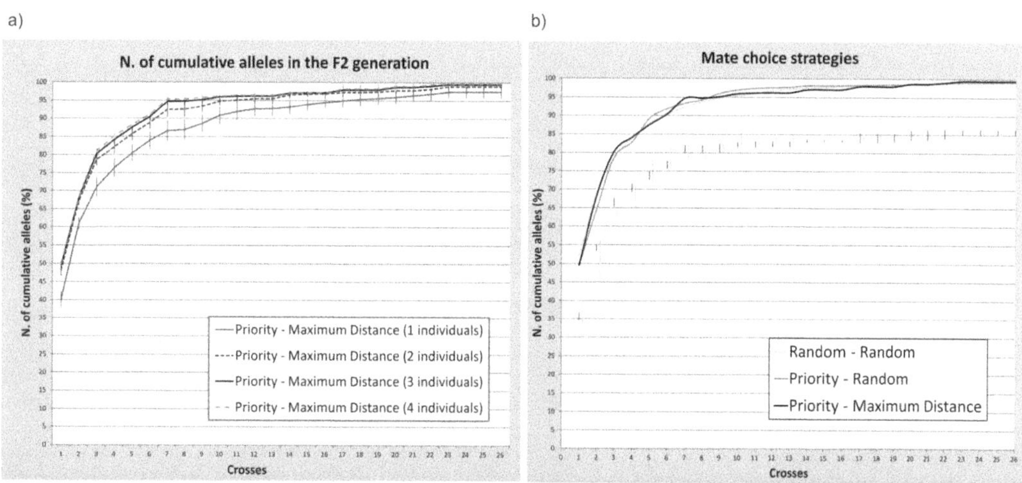

Figure 4. Simulations of allele inheritance under different strategies. Results of the simulations performed using the R-script "*BreedingPlanSturgeon*" to test: (a) the optimal number of breeders per family and (b) alternative strategies for the mating choice. The cumulative percentages of transmitted alleles are averaged on 100 replicates and reported with the corresponding standard deviation. In figure b, 3 individuals per family were crossed and standard deviation bars are shifted downwards for better visibility.

Characterization of the F1 V.I.P. stock

Pedigree reconstruction was applied to characterize the F1 stock retained at the V.I.P. plant and produced by crossing F0 individuals in the last 20 years. The approach implemented by the new software ("*BreedingSturgeons*") and based on the inference of all possible genotypes compatible with observed band profiles, showed a higher allocation efficiency compared with the simple band sharing approach [10]. Accordingly, all ambiguities were completely solved with no F1 individual being multi-allocated. Comparison of genetic variability between the original F0 (Wild) and the F1 stock showed a relatively important loss of genetic diversity at the mitochondrial level, with 2 of the 7 original haplotypes being absent in the F1 stock before poisoning and a third one lost in the F1-alive group. However, at microsatellites, genetic diversity appears to be almost intact as no significant differences in number of alleles were observed between F0 and F1, with both samples presenting an average number of alleles per locus of about 10. By contrast, a lower value of 8.57 was reported for a different F1 stock of Adriatic sturgeon retained as possible future broodstock at the Ticino River Park [18]. The Ticino F1 stock was established without any genetic input and the animals were collected from a limited number of reproductions, causing a significant erosion of genetic diversity in a single generation, as exemplified by a 16.7% loss in average allele number in comparison with the F0 stock.

In synthesis, the V.I.P. Stock resulted to be the major source of variability for this species. Several families were identified and a great part of the F0 parents are represented within the stock. In regards to the relationships among animals, the probability to select for reproduction two related individuals by chance is very low (5,6%), compared to the 35% estimated for the stock reared by the Ticino River Park [18]. However, even if the use of Ticino F1 stock as *ex situ* broodstock is not advisable, this stock includes some families that are absent or under-represented in the V.I.P. plant, so exchange of individuals between plants should be encouraged.

Finally, our results also show how important is to test for the presence of hybrids or alien species. In many aquaculture facilities, different species or interspecific hybrids are reared, so that mixing of individuals can accidentally happen, and those can go unnoticed unless properly tested. In our case, two reciprocal hybrids between Adriatic and White sturgeon were found within the F1 progeny. As expected, hybrids showed microsatellite alleles not present in the *A. naccarii* F0 gene pool but these are not the only new alleles detected. Additional alleles not present in the parental generation were also found among not-allocated F1 individuals. The presence of new alleles in the F1 might be explained by inheritance from F0 breeders that died several years ago and are not present in our sample. Alternatively, F1 individuals carrying new alleles might be hybrids produced in the hatchery and accidentally mixed into the stock. Accordingly, accepting the risk of losing rare variants we chose to exclude from the broodstock all the animals carrying extra alleles.

Is a short-term breeding plan still possible?

In the case of the Adriatic sturgeon, recruitment of new breeders from the wild is not possible since the species is possibly extinct in nature, so a breeding plan using wild breeders can only be achieved using the original F0 brought to the V.I.P. facilities in 1977. However, the F0 stock at the V.I.P. plant has been greatly reduced in size in recent years, probably due to senescence. While only 13 individuals remain at present, we propose a priority short-term breeding program to be attempted with these remaining F0 individuals. The plan includes 6 females and 7 males for a total of 18 selected crosses on the basis of pairwise genetic distance (Fig. 1). It should be noted that most of the remaining individuals had never been mated before, so that their successful reproduction would represent a relevant contribution to the future diversity of the captive stock of this species. Nevertheless, the fact that some animals have never been crossed in over 20 years of reproduction events might indicate a low reproductive potential. Hence, there is a chance that the proposed crosses between the surviving F0 adults might not be successful, and although we believe that a short-term plan using F0 should be attempted, most efforts should be conducted to the establishment of the long-term plan involving the F1 generation.

A long-term breeding plan to safeguard the Adriatic sturgeon

We present a complete long-term breeding plan based on familiar groups that aims at generating enough animals to ultimately allow for the re-introduction of the species back into its natural environment. Considering the low chances of successful reproduction of these animals and their limited number, most efforts were devoted to developing a long-term breeding strategy involving families of F1 generation, assuming that at least some of the individuals are reproducible every year.

The program starts with the exclusion of combinations between families sharing one parent, together with families that do not share any parent but show low genetic distances, which might be indicative of relatives. This allows minimizing possible errors due to the "*founder assumption*", in which wild founders of a captive population are assumed to be equally unrelated [30], although it might not always be the case.

Our choice of keeping separated for two years the animals produced by different family combinations, until the moment of tagging, limits the number of crosses that can be simultaneously performed with a finite number of ponds. However, it allows tracking the pedigree of every single animal providing useful information for monitoring the release program or for managing the following generation of captive broodstock. Alternatively, instead of crossing one individual per family, it could be considered to cross more individuals of the same family that can be pooled and do not require additional ponds. This would also increase the efficiency of allele transmission and guarantee that all F0 alleles of each family are inherited by the F2 generation, as shown by simulations that suggest the use of at least 3 breeders from the same family when possible. However, these numbers might be not easy to reach given the small size of some families and the discontinuous availability of breeders for reproduction (mostly true for females). Moreover, if breeders of the "*priority family*" are not available, one possibility is to postpone the cross until the following reproductive season and to slightly modify the priority order. If the unavailable individuals belong to the "*mating family*", one possibility is to save the priority order and to select an alternative "*mating family*", choosing the second more distant family. As shown by simulations, selection of family combinations based on priority and genetic distance were more efficient in transmitting the higher possible number of alleles to the next generation relative to the random strategies ("Priority – Random" and "Random – Random"). This was particularly true when number of crosses was low, which suggests that random approaches are not recommended for short-term programs. In fact, the higher standard deviation of the two strategies in which at least one family is randomly selected may result in a significant loss of allele diversity after few generation of captive breeding. Accordingly, the breeding plan here proposed was based on the Priority – Maximum distance approach. On one side, the establishment of a priority score allows to first cross those families whose genetic traits have the higher risk to be lost, since they are shared by a lower number of individuals in the population. On the other side, the use of maximum distance minimizes inbreeding and tends to increase the number of crosses involving families that are genetically peculiar, and consequently, are more often selected as "mating families".

Management implications

The present work represents the first breeding plan for the Adriatic sturgeon based on its detailed genetic characterization and can be considered as a reference guideline for all conservation actions based on controlled reproductions of this species. As in the case of Adriatic sturgeon, many other sturgeon species are critically endangered with greatly reduced wild populations and a consequent difficulty in collecting breeders from the wild. For such species the establishment of *ex situ* conservation plans through the retention of part of the fingerlings as future breeders is the only way forward. In this context, the approach here developed based on the management of familiar groups instead of individuals can become of major interest. Moreover, for those species that share with the Adriatic sturgeon the tetraploid condition, such as the Russian sturgeon (*A. gueldenstaedtii*), the Persian sturgeon (*A. persicus*), the Chinese sturgeon (*A. sinensis*) and many others, the methods here optimized are directly transferable. This would allow a careful identification of familiar groups thus providing the basis for the establishment of breeding programs that takes into account the genetic composition of the available broodstock together with logistic and economic constraints.

Our study also showcases the importance of genetic approaches to *ex situ* management for meeting conservation goals. The low cost of microsatellite markers facilitates the collection of genetic data that can inform breeding programs about the best pairing individuals so as to minimize mean relatedness. In our study, as few as 7 microsatellite markers were sufficient to unambiguously allocate all F1 individuals. In the event of higher number of putative parents, which might be the case in other sturgeon species, the number of microsatellites should be increased to ensure complete allocation.

Since genetic diversity is an essential prerequisite for the long-term adaptive potential, the availability of *ex situ* genetically heterogeneous broodstocks is the starting point for the rehabilitation of self-sustaining natural populations. However, in the short run, the survival rate of the released fingerlings depends on their capability do adapt to the target environment. In this sense, the establishment of adequate rearing protocols aimed to produce individuals with a high probability of survival is strongly encouraged. As clearly reported in FAO guidelines for sturgeon management and release [31], hatcheries involved in juvenile production for restocking purpose should raise fry and juveniles through training and adaptation to natural conditions, in order to maximize their fitness for survival after release.

As important as it is to use genetic studies into conservation practices, those need to be coordinated and not be carried out separately and restrictively. In the case of the Adriatic sturgeon, at present several local administrations are active in the attempt of recovering the species in its historical natural range. Unfortunately, these actions are often represented by short-term programs and are completely isolated and uncoordinated. Therefore, providing a common program like the one presented in our study that can connect and integrate the different actions would represent the first step towards a responsible concerted management of the Adriatic sturgeon. Finally, in order to implement successful conservation strategies, important actions against habitat degradation are urgently needed including the restoration of natural spawning sites and the re-establishment of river connectivity (i.e. fish passages at dams). A relevant challenge for conservation genetics is to go beyond the scientific since reintroduction is not merely a biological issue but a social, cultural and political one as well. An important part of the framework required for the successful incorporation of genetic data into conservation decision-making of culturally-important species is involvement where possible of local people in the program, together with public education about the project.

Supporting Information

File S1 Tables with additional details about the two R-scripts and samples analyzed are reported. Tables A and B include a brief description of variables used in the two R-scripts. Table C reports details about results of parental allocations.

File S2 "*CostsBreedingSturgeons*" and "*BreedingPlan-Sturgeons*" scripts. The zip file includes all system and executive files to run programs. As example, input files used to run "*BreedingPlanSturgeons*" script are also included.

Acknowledgments

We thank Giacinto Giovannini and sons, owners of the "Azienda Agricola V.I.P." and of the "Storioni Ticino hatchery". We are grateful to Giorgio Bertorelle, Andrea Benazzo, Paolo Bronzi and Lara Bellini for their help and suggestions.

Author Contributions

Conceived and designed the experiments: EB LC. Performed the experiments: EB. Analyzed the data: EB LC ID RC. Contributed reagents/materials/analysis tools: EB LC ID RC. Contributed to the writing of the manuscript: EB LC JMP.

References

1. Leus K (2011) Captive breeding and conservation. Zool Middle East 54: 151–158.
2. Witzenberger KA, Hochkirch A (2011) Ex situ conservation genetics: a review of molecular studies on the genetic consequences of captive breeding programs for endangered animal species. Biodivers Conserv 20: 1843–1861.
3. Frankham R, Ballou JD, Briscoe DA (2002) Introduction to conservation genetics. Cambridge University Press, Cambridge, UK.
4. Philippart JC (1995) Is captive breeding an effective solution for the preservation of endemic species? Biol Conserv 72: 281–295.
5. Russello MA, Amato G (2004) Ex situ population management in the absence of pedigree information. Mol ecol 13: 2829–2840.
6. Araki H, Cooper B, Blouin MS (2007) Genetic effects of captive breeding cause a rapid, cumulative fitness decline in the wild. Science 318: 100–103.
7. Fraser DJ (2008) How well can captive breeding programs conserve biodiversity? A review of salmonids. Evol App 1: 535–586.
8. Kozfkay CC, Campbell MR, Heindel JA, Baker DJ, Kline P, et al. (2008) A genetic evaluation of relatedness for broodstock management of captive, endangered Snake River sockeye salmon, *Oncorhynchus nerka*. Conserv Gens 9: 1421–1430.
9. Goldstein PZ, DeSalle R, Amato G, Vogler AP (2000) Conservation genetics at the species boundary. Conserv Biol 14(1): 120–131.
10. Congiu L, Pujolar JM, Forlani A, Cenadelli S, Dupanloup I, et al. (2011) Managing polyploidy in ex situ conservation genetics: the case of the critically endangered Adriatic sturgeon (*Acipenser naccarii*). PLoS One 6: e18249.
11. Jamie AI, Lacy RC (2012) A comparison of strategies for selecting breeding pairs to maximize genetic diversity retention in managed populations. J Hered 103: 186–196.
12. Hayes B, He J, Moen T, Bennewitz J (2006) Use of molecular markers to maximize diversity of founder populations for aquaculture breeding programs. Aquaculture 255: 573–578.

13. Ludwig A (2006) A sturgeon view on conservation genetics. Eur J Wildlife Res 52: 3–8.
14. Anders PJ, Drauch-Schreier A, Rodzen J, Powell MS, Narum S, et al. (2011) A review of genetic evaluation tools for conservation and management of North American sturgeons: roles, benefits, and limitations. J App Ichthyol 27(Suppl.2): 3–11.
15. Wuertz S, Gaillard S, Barbisan F, Carle S, Congiu L, et al. (2006) Extensive screening of sturgeon genomes by techniques revealed no sex-specific random screening marker. Aquaculture 258: 685–688.
16. Fontana F, Lanfredi M, Chicca M, Congiu L, Tagliavini J, et al. (1999) Fluorescent *in situ* hybridization with rDNA probes on chromosomes of *Acipenser ruthenus* and *Acipenser naccarii* (Osteichthyes Acipenseriformes). Genome 42: 1008–1012.
17. Bronzi P, Rosenthal H, Arlati G, Williot P (1999) A brief overview on the status and prospects of sturgeon farming in Western and Central Europe. J App Ichthyol 15: 224–227.
18. Boscari E, Congiu L (2014) The need for genetic support in restocking activities and *ex situ* conservation programmes: the case of the Adriatic sturgeon (*Acipenser naccarii* Bonaparte, 1836) in the Ticino River Park. J App Ichthyol DOI:10.1111/jai.12545.
19. Boscari E, Barmintseva A, Pujolar JM, Doukakis P, Mugue N, et al. (2014) Species and hybrid identification of sturgeon caviar: a new molecular approach to detect illegal trade. Mol Ecol Res doi:10.1111/1755-0998.12203.
20. Tamura K, Peterson D, Peterson N, Stecher G, Nei M, et al. (2011) MEGA5: Molecular Evolutionary Genetics Analysis using Maximum Likelihood, Evolutionary Distance, and Maximum Parsimony Methods. Mol Biol Evol 28: 2731–2739.
21. Clement M, Posada D, Candrall KA (2000) TCS: a computer program to estimate gene genealogies. Mol Ecol 9: 1657–1659.
22. Templeton AR, Crandall KA, Sing CF (1992) A cladistics analysis of phenotypic associations with haplotypes inferred from restriction endonuclease mapping and DNA sequence data. III. Cladogram estimation. Genetics 132: 619–633.
23. Librado NP, Rozas J (2009) DnaSP v5: a software for comprehensive analysis of DNA polymorphism data. Bioinformatics 25: 1451–1452.
24. May B, Krueger CC, Kincaid HL (1997) Genetic variation at microsatellite loci in sturgeon: primer sequence homology in *Acipenser* and *Scaphirynchus*. Canadian Journal of Fisheries and Aquatic Science 5: 1542–1547.
25. Forlani A, Fontana F, Congiu L (2007) Isolation of microsatellite loci from the endemic and endangered sturgeon *Acipenser naccarii*. Conser Gen 9: 461–463.
26. Zane L, Patarnello T, Ludwig A, Fontana F, Congiu L (2002) Isolation and characterization of microsatellite in the Adriatic sturgeon (*Acipenser naccarii*). Mol Ecol Notes 2: 586–589.
27. Henderson-Arzapalo A, King TL (2002) Novel microsatellite markers for Atlantic sturgeon (*Acipenser oxyrinchus*) population delineation and broodstock management. Mol Ecol Notes 2: 437–439.
28. Boscari E, Barbisan F, Congiu L (2011) Inheritance pattern of microsatellite loci in the polyploidy Adriatic sturgeon (*Acipenser naccarii*). Aquaculture 321: 223–229.
29. Hood GM (2010) PopTools version 3.2.5. Available: http://www.poptools.org. Accessed 2014 October 10.
30. Van Dyke F (2008) Conservation biology: foundations, concepts, applications. Wheaton College, Illinois, USA. Second Editions, Springer Science and Business Media.
31. Chebanov M, Rosenthal H, Gessner J, Van Anrooy R, Doukakis P, et al. (2011) Sturgeon hatchery practices and management for release-Guidelines FAO Fisheries and Aquaculture Technical Paper No 570. Ankara, FAO. 2011. 110.

15

Increased Extinction Potential of Insular Fish Populations with Reduced Life History Variation and Low Genetic Diversity

Michael Hellmair[1,2]*, Andrew P. Kinziger[1]

1 Department of Fisheries Biology, Humboldt State University, Arcata, California, United States of America, **2** FISHBIO, Chico, California, United States of America

Abstract

Theoretical work has shown that reduced phenotypic heterogeneity leads to population instability and can increase extinction potential, yet few examples exist of natural populations that illustrate how varying levels expressed diversity may influence population persistence, particularly during periods of stochastic environmental fluctuation. In this study, we assess levels of expressed variation and genetic diversity among demographically independent populations of tidewater goby (*Eucyclogobius newberryi*), show that reductions in both factors typically coincide, and describe how low levels of diversity contribute to the extinction risk of these isolated populations. We illustrate that, for this annual species, continuous reproduction is a safeguard against reproductive failure by any one population segment, as natural, stochastically driven salinity increases frequently result in high mortality among juvenile individuals. Several study populations deviated from the natural pattern of year-round reproduction typical for the species, rendering those with severely truncated reproductive periods vulnerable to extinction in the event of environmental fluctuation. In contrast, demographically diverse populations are more likely to persist through such periods through the continuous presence of adults with broader physiological tolerance to abrupt salinity changes. Notably, we found a significant correlation between genetic diversity and demographic variation in the study populations, which could be the result of population stressors that restrict both of these diversity measures simultaneously, or suggestive of a causative relationship between these population characteristics. These findings demonstrate the importance of biocomplexity at the population level, and assert that the maintenance of diversity contributes to population resilience and conservation of this endangered species.

Editor: Sofia Consuegra, Swansea University, United Kingdom

Funding: Funding for this study was provided by the United States Fish and Wildlife Service (www.fws.gov) to Andrew P. Kinziger, administered through the Humboldt State University Sponsored Programs Foundation. The funder had no active role in study design, data collection and analysis, or preparation of the manuscript, but consents to publication.

Competing Interests: The authors have declared that no competing interests exist.

* Email: michael.hellmair@humboldt.edu

Introduction

Life-history diversity and genetic variation are two forms of biodiversity deemed worthy of conservation by the World Conservation Union (IUCN), and are considered essential to retaining fitness and long-term adaptive potential of populations [1]–[3]. Further, it is often invoked that expressed forms of diversity contribute to the stability of biological systems: high species diversity often stabilizes ecosystem processes [4]–[7], and variation among populations of individual species buffers fluctuations in the species' cumulative abundance and the ecosystem services it provides [8]–[11]. However, the role of diversity within individual populations of a species has remained largely unexplored, though there is some experimental evidence that post-disturbance variance in abundance is stabilized by diversity, and extinction risk becomes more predictable as a consequence [12].

In this study, we investigated age composition and genetic diversity among isolated populations of tidewater goby (*Eucyclogobius newberryi*), a short-lived, endangered fish that inhabits dynamic coastal habitats subject to frequent environmental changes [13]–[15]. We discuss the importance of life history variability *within* geographically isolated populations in the context of the species' biology, and suggest that reduced levels of variation, particularly in the temporal extent of the reproductive period, render affected populations more susceptible to localized extirpation during periods of environmental disturbance.

The tidewater goby is a small (<60 mm total length), annual fish endemic to California, that occurs in brackish, isolated, and often very small (less than 10 hectares) estuaries along the coast, from just south of the Oregon border to San Diego, USA [13]–[15]. Among insular populations of tidewater goby, levels of genetic differentiation are often very high, despite geographic proximity [16]. Further, the level of genetic diversity within populations varies widely (as measured by allelic richness and heterozygosity; see [16] for a discussion on population structure of the species' northern range; Figure 1; Table 1). Migration between populations is unlikely, as suitable habitats are typically separated from the Pacific Ocean by sandbars most of the year [13], [14], [16], [17], and requires coincident breaching events between sites. Breaching generally occurs 1–2 times annually following periods of high freshwater input and large surf, but may not happen for several successive years [18]. These stochastic events commonly cause rapid draining of the estuary and influx of ocean water over subsequent tidal cycles. Breaching events are physiologically challenging for fishes due to rapid changes in water level, temperature and salinity. Abrupt changes in salinity in particular are known to be lethal to many fish species, including gobies, especially during early life history stages [19]–[21].

A number of life history adaptations allow tidewater goby populations to persist in these environmentally challenging habitats. Most notably, reproduction occurs throughout the year, though increased spawning activity is generally observed during summer and fall [13], [14], [22], [23]. Individuals can become reproductively mature at lengths under 30 mm, and can actively spawn over prolonged time periods (several months) [22], [23]. As a result, tidewater goby populations are typically characterized by a broad size and age structure through the continuous presence of both juvenile and adult individuals. This reproductive strategy may be considered a bet-hedging trait, an adaptation to unpredictable, though frequent, environmental variance in the habitat of this species that maximizes reproductive output during time periods that are generally favorable (summer and fall, when the likelihood of breaching is lowest) [18], [24], yet safeguards populations from extirpation through the continuous presence of resilient adult individuals in the event of stochastic reproductive failure by any one population segment.

In this study, (1) we used a controlled experiment to confirm that salinity increases associated with breaching causes mortality to the early life stages, and (2) show how continuous reproduction (as measured by population age composition) can buffer isolated populations of tidewater goby against adverse impacts (high mortality) resulting from these events. Further, (3) we assessed the genetic diversity of 17 populations of this species in northern California, and (4) investigated relationship between genetic diversity measures and demographic heterogeneity after accounting for habitat size (as a proxy of habitat heterogeneity). We discuss the results in the context of the biology of this endangered fish species, and argue that resilience to environmental stochasticity increases when life history variability and healthy levels of genetic variation are retained in a population.

Materials and Methods

Ethics statement

All work was conducted under California Scientific Collecting Permit SC-10527 and in concordance with institutional protocol at Humboldt State University (Institutional Animal Care and Use Committee 08/09.F.44.A & 10/11.F.78-C), where all samples have been deposited in the scientific fish collection (HSU Catalog #4890-4913).

Salinity trials

To evaluate size-specific mortality associated with salinity increases experienced by tidewater goby populations during a breaching event, we exposed two size groups of tidewater goby (small and large) captured in BL (14–35 mm TL: 11 replicates, 10

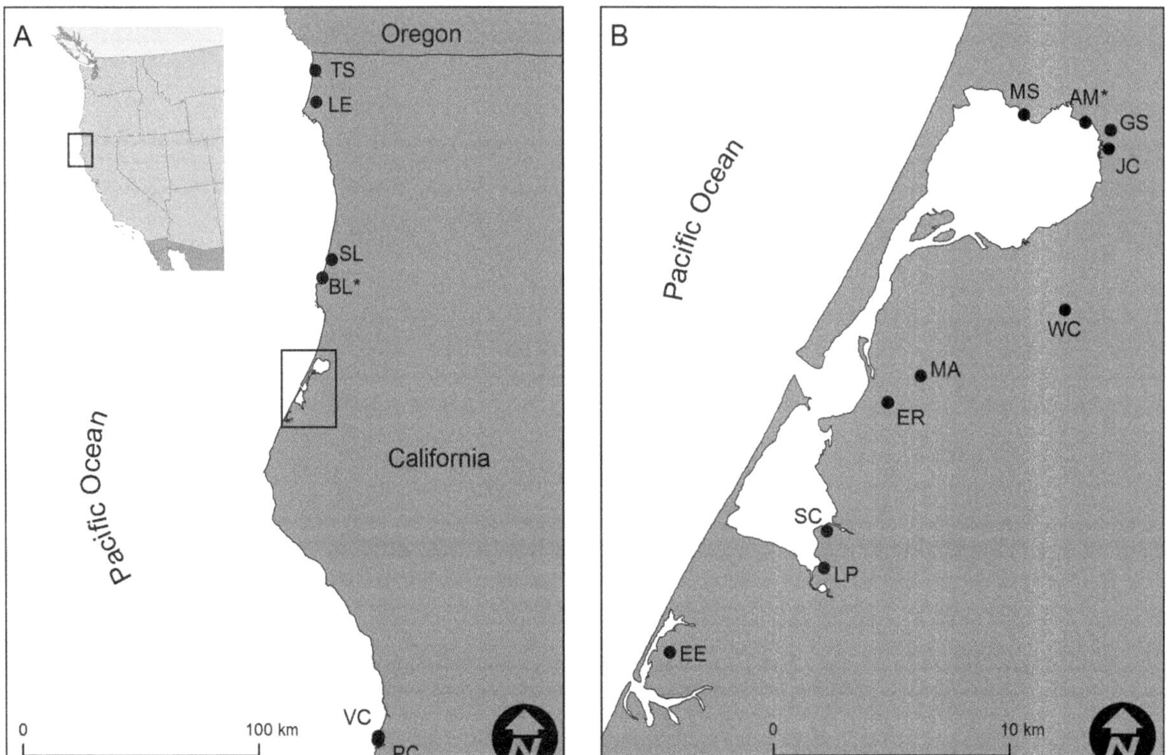

Figure 1. Map depicting locations of tidewater goby populations on the northern California coast. Panel A shows populations directly adjacent to the Pacific Ocean, including Tillas Slough (TS), Lake Earl (LE), Stone Lagoon (SL), Big Lagoon (BL*), Virgin Creek (VC) and Pudding Creek (PC). (B) McDaniel Slough (MS), Arcata Marsh (AM*), Gannon Slough (GS), Jacoby Creek (JC). Panel B illustrates isolated populations found around the margin of Humboldt Bay and the Eel River estuary, including Wood Creek (WC), Elk River (ER), Martin Slough (MA), Long Pond (LP), Salmon Creek (SC), Connick Ranch (CR) and Eel River (EE). Populations noted with an asterisk represent focal populations for otolith age analysis.

Table 1. Habitat size, location, demographic parameters and genetic diversity measures (across nine microsatellite loci) of 17 tidewater goby populations in northern California.

Population	Coordinates		Area (hectares)	n	Mean size[a] (mm)	SD	Size range[a] (mm)	H_E[b]	A_r[c]
	N	W							
Virgin Creek (VC)	39°28′11.29″	123°48′10.49″	4.5	60	28.85	3.84	18	0.57[d]	3.69
Big Lagoon (BL)	41°09′51.48″	124°07′50.16″	612.5	60	32.47	6.46	25	0.56[d]	4.92
Stone Lagoon (SL)	41°13′58.44″	124°05′03.48″	236.7	60	28.95	8.39	32	0.52[d]	4.89
Pudding Creek (PC)	39°27′10.81″	123°48′18.62″	9.5	60	35.28	5.16	20	0.45[d]	2.94
Eel River (EE)	40°39′09.45″	124°17′34.77″	108.5	60	40.78	4.38	20	0.28[d]	2.67
Elk River (ER)	40°44′51.17″	124°11′14.20″	35.1	60	32.93	6.55	27	0.28[d]	1.83
Lake Earl (LE)	41°48′59.18″	124°10′54.45″	1085.4	60	22.10	3.86	17	0.26	3.18
Tillas Slough (TS)	41°55′59.90″	124°11′25.74″	6.6	60	25.72	2.00	8	0.26	2.40
Connick Ranch (CR)	40°36′30.36″	124°19′18.02″	6.4	60	41.40	4.35	17	0.25	2.48
Long Pond (LP)	40°41′08.45″	124°13′02.22″	2.3	60	38.1	4.05	17	0.24	2.20
Salmon Creek (SC)	40°41′57.72″	124°12′58.71″	396.9	60	28.90	4.21	18	0.23[d]	2.21
Gannon Slough (GS)	40°51′02.31″	124°04′43.17″	18.2	60	41.15	2.79	13	0.22[d]	1.87
Martin Slough (MA)	40°45′27.18″	124°10′16.93″	0.2	60	30.22	3.30	14	0.21	1.78
McDaniel Slough (MS)	40°51′20.14″	124°07′20.89″	34.8	31	27.23	2.65	12	0.18[d]	1.75
Jacoby Creek (JC)	40°50′37.59″	124°04′46.21″	6.2	58	20.00	2.54	11	0.16[d]	1.66
Wood Creek (WC)	40°47′00.95″	124°05′57.38″	0.4	59	31.65	3.75	13	0.10[d]	1.36
Arcata Marsh (AM)	40°51′11.86″	124°05′28.71″	0.2	165	40.69	1.98	11	0.08	1.36

[a]Size is defined as total length (TL), and size range the difference between the largest and smallest individual from *n* individuals sampled.
[b]Mean expected Hardy-Weinberg heterozygosity.
[c]Rarefied allelic richness.
[d]Heterozygosity estimates obtained from McCraney et al. (2010).

individuals each; 36–55 mm TL: 5 replicates, 10 individuals each; Figure 1) to a 20‰ increase in salinity over ambient lagoon salinity (we mixed water from the adjacent Pacific Ocean [35‰] with lagoon water to bring the salinity to 26‰). Though the magnitude of salinity increases during natural breaching events are highly variable, depending on tidal phase, surf levels and freshwater input, salinity levels approaching that of seawater have been regularly observed. We considered an intermediate increase to 26‰ sufficient to identify breaching events as a cause of mortality and evaluate differences in size-specific survival. Two control groups (5 fish 31–55 mm TL, 16 fish 14–30 mm TL) were kept in lagoon water. These groups were not intended for inclusion in statistical analysis, but to indicate shortcomings of the experimental design in case of mortality to those individuals. All replicates were held in 10 liter tanks, aerated consistently and monitored for 16 hours. To limit the number of overall mortalities, we discontinued the experiment after only 16 hours as clear differences in survivorship between the groups had become apparent.

Sample size for this experiment was limited by numbers permitted for research take under collecting permits of this protected fish and subject to capture success in the wild. Following the observation period, we classified all individuals as dead or alive and measured them to the nearest millimeter (TL). We used Welch's two sample T-test to test for significant differences in survival between size groups.

Population age structure and genetic diversity

We captured tidewater goby using a 3.3 m by 1.3 m seine with a 1.58 mm mesh in water no deeper than the height of the seine and no shallower than 0.2 m. As neither our field observation nor any literature report indicate age- or size-specific segregation of tidewater goby, we considered collections representative subsamples of the respective goby populations.

We obtained estimates of genetic diversity and corresponding individual length measurements (total length, TL, in millimeters, mm) for eleven populations from a previous study (sampled in 2006) [16], and non-lethally collected length measurements and genetic tissue samples (fin clips) for six additional populations (in 2011), to compile a dataset representing 17 populations of tidewater goby in northern California (Table 1; Figure 1). We estimated population genetic diversity for the six additional populations using the identical set of nine microsatellite loci developed specifically for tidewater goby and according to methods outlined in detail in the literature [16], [25], [26]. We estimated rarified allelic richness using the program HP-RARE [27] to control for the correlation between observed allelic diversity and sample size [28]. Alleles were rarified to a sample size of 31, the smallest sample size of our population groups.

In organisms with indeterminate growth, particularly those with a short lifespan, the size distribution of individuals is conservatively reflective of the population's age composition. In other words, while individuals of the same age may obtain different sizes at time t (therefore inflating the measure of actual temporal spread), it is improbable that many individuals of different ages reach identical, or similar, sizes at time t (deflating temporal spread). Following this reasoning, we consider the length range of samples an approximation of the variation in ages within populations of this short-lived fish. However, to further validate that size measurements provide a nonlethal alternative to direct (lethal) age estimation, we conducted a more detailed age and growth study on two large populations using otolith analysis (described in the following section).

We evaluated the relationship between the length ranges observed in our study populations (difference, in mm, between the smallest and largest individual) and their genetic diversity (mean Hardy-Weinberg expected heterozygosity [H_E] and allelic richness [A_r]) through linear regression analysis. Alternatively, we used the standard deviation (SD) in lengths observed within populations to investigate the relationship between demographic variation and genetic diversity, as the observed length range may be strongly affected by presence of rare individuals with atypical (extreme) sizes.

As the size of the habitat to which a population is constrained may influence the duration of the reproductive period and limit populations size (at any time, larger habitats have a broader range of depths, temperatures and other environmental variation that may facilitate prolonged reproductive periods), we used area ($\log_{10}$ *hectares*) to control for effects of environmental differences among the isolated habitats in a general linear model fit.

Additional sample collection, age determination and estimation of growth parameters

Unlike most animal taxa, fishes can often be aged directly through analysis of periodically deposited increments in bony structures, most notably their otoliths (earstones). We used otolith microstructural analysis to estimate the ages (in days) for large numbers of individuals from two populations in northern California, USA, for age determination (Big Lagoon [BL] and Arcata Marsh [AM], $n = 413$ and 88, respectively; Figure 1). Due to concerns regarding periodic lethal sampling of this endangered species, we chose two populations with high abundance (several thousand individuals) and considered representative of stocks with high and low genetic diversity ($H_E = 0.56$ and 0.08, $A_r = 4.92$ and 1.36, respectively) for comparison. In Big Lagoon (BL), we collected individuals from April 2009 until August 2010, and sampled the second location, Arcata Marsh (AM), from April 2009 until October 2011. We measured all individuals in the field to the nearest millimeter (total length, TL), and sacrificed and preserved a subsample of 20–25 fish spanning the length range of goby encountered (by subjecting them to an overdose of MS222, Tricaine methanesulfonate; preservative: 90% ethanol). Concurrent with field sample collection, we obtained salinity measurements (parts per thousand, ‰) using a VEE GEE STX-3 refractometer.

We extracted the sagittal otoliths, rinsed them in water and ethanol and mounted them on a microscope slide using wet'n'wild "Wild Shine" Clear Nail Protector (type 401A). After curing, we polished the otoliths by hand using waterproof sandpaper (GatorGrit, type 1500-b) and 6 μm lapping film (Allied High Tech Products, Inc., Diamond Lapping Film #50-30265) to reveal daily increments. We counted daily increments under a compound microscope under 500X total magnification, using an oil-immersion lens (MEIJI Model #10824), and captured images captured with a Lumenera Infinity 1 camera (ImagePro Plus software, version 7.0). We began enumeration of daily increments at the first continuous increment outside the core region, and counted increments continuously to the otolith margin. We discarded the sample when continuous enumeration not possible on either of the two otoliths.

To validate daily growth increment deposition, we induced chemical marks in otoliths by immersing tidewater goby in a 5% calcein solution. We marked a total of 1441 tidewater goby and released them into their natal environment (BL, Figure 1), and used an SE-MARK detector (Western Chemical, Co.) to identify marked individuals during recapture events five, ten, twenty and twenty-eight days after marking. We retained five marked individuals during each recapture event, prepared their sagittal otoliths as indicated above and examined them under a fluorescence microscope. We estimated annual survivorship for each population using daily age estimates [29] and fit a growth model [30] to daily length at age estimates for each population (n = 413 and 88 for BL and AM, respectively):

$$L_t = \left[l_1^b + (l_2^b - l_1^b) * \frac{1 - e^{-a(t-\tau_1)}}{1 - e^{-a(\tau_2 - \tau_1)}} \right]^{\frac{1}{b}}$$

where τ_1 and τ_2 are the ages of the youngest and oldest fish in the sample (see results), l_1 and l_2 the corresponding lengths at that age, t is age (in days), a is the inverse of time, and b is a dimensionless constant. We then reconstructed birthdate distributions for these two populations by tabulating, for each fish, the estimated daily age subtracted from its collection date.

Results

Salinity trials

Mean mortality across 11 replicates of the "small" category was 0.50 (SD = 0.10), and 0.0 across the 5 replicates of the "large" category (Figure 2), illustrating that the chance of surviving a salinity increase is significantly lower for small (young) fish (Welch's 2-sample t-test; $p<0.001$). No mortality was observed in control groups of either size class. Surviving fish were released into their natal environment.

Population age structure and genetic diversity

Length-frequency distributions were highly variable among populations, with size ranges and mean sizes ranging from 8 mm to 32 mm and 20.0 to 41.4 mm, respectively. Considering that the maximum size of tidewater goby encountered during our surveys was 55 mm TL, these differences among population are substantial. Populations separated by a distance of less than four kilometers and collected within a one-week time span exhibited different mean lengths ($\hat{\mu} = 20$, 27.23 and 41.15 mm; $p<0.001$ ANOVA, Table 1), corresponding to mean ages of 42, 107, and 341 days, respectively (using growth parameters estimated from length at age data for the nearby AM population) (Table 1, Table 2, Figure S1). Monthly collections from BL over a period of 18 months confirm that an observed broad size range is not a seasonal occurrence, but a pattern that can be observed any month of the year (Figure S2).

Genetic diversity was also highly variable among populations, and expected heterozygosity (H_E) ranged from 0.08 to 0.57, while rarefied allelic richness (A_r) ranged from 1.36 to 4.92 (Table 1). No loci or populations deviated significantly from Hardy-Weinberg Equilibrium (genotypes of individuals included in this study can be found in File S1).

We detected a correlation between the temporal extent of the reproductive season and genetic diversity. A significant proportion of the variation in demography (as measured by observed size ranges and SD) was explained by expected heterozygosity ($p = 0.018$ and 0.022, respectively) after accounting for the effects of habitat area in a general liner model analysis (model: range(mm)/SD = $\log_{10}$(hectares)+H_E). Similarly, allelic richness explained a significant proportion of age/size variation (model:

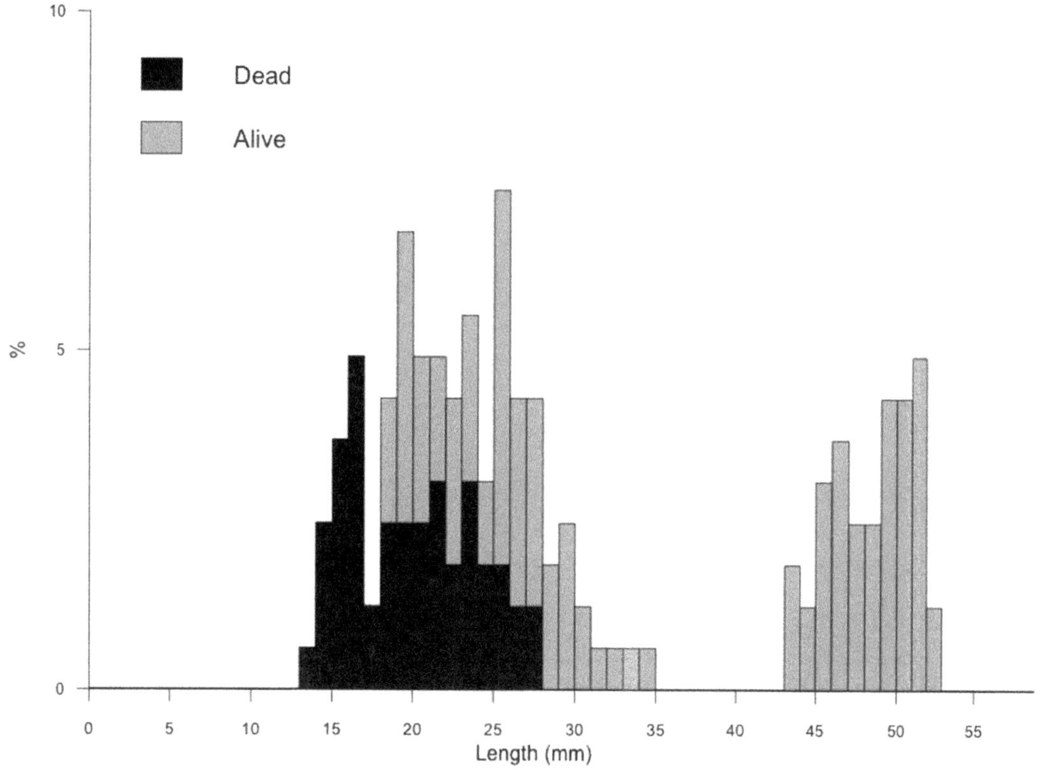

Figure 2. Size-specific mortality of tidewater goby (total n = 163) subjected to a 20‰ increase in salinity for 16 hours, illustrating high mortality of juvenile stages and resilience of adults. Mean mortality across all 11 replicates of small individuals (14–35 mm) was 50.4%, and no mortality occurred in any replicates of adult fish (43–53 mm) or the control groups.

Table 2. Estimates of growth parameters (Schnute 1981) for two northern California populations of tidewater goby, where *n* is the number of samples, l_1 and l_2 are the lengths of the youngest and oldest fish in the sample, *a* is the inverse of time, and *b* is a dimensionless constant.

	Big Lagoon	**Arcata Marsh**
n	413	88
a (SE)	−0.004(0.002)	−0.007(0.006)
b (SE)*	2.50(0.62)	5.53(2.33)
l_1 (SE)	12.85(0.96)	15.00(2.41)
l_2 (SE)*	55.94(1.24)	42.40(0.86)

range(mm)/SD = $\log_{10}$(hectares)+A_r; $p = 0.026$ and 0.020). Though explaining some of the variation, the effect of habitat area was not significant in any of the four models ($p>0.2$), and this variable was excluded from the model fits for illustration purposes. The univariate relationships between these two measures of genetic diversity and two indicators of demographic diversity in goby populations are shown in Figure 3.

Additional sample collection, age determination and estimation of growth parameters

The minimum collection goal of 150 individuals per month was met for the BL population every month from April 2009 to August 2010, except June 2009 (n = 28, Figure S2). For the AM population, the collection goal was met from April 2009 to July 2009. In August 2009, extensive sampling yielded only five adult tidewater goby (TL>35 mm), and rigorous sampling using a variety of collection methods the following 24 months suggests that tidewater goby went extinct at this site. Salinities ranged from 2‰ to 14‰ in BL in all months except May 2010 (27‰) following a natural breaching event. In AM, salinity measurements ranged from 9‰ to 12‰ between April and July 2009, increased to 34‰ in August 2009, and decreased to 15‰ in September. Both populations experienced a stochastic salinity increase during the study period. In the AM population with (low genetic diversity, little variation in individual ages/size) no individuals smaller than 30 mm TL (comprising over 94% of the population shortly prior to the salinity increase; Figure S3) were found after this event, and despite extensive repeated surveys over a period of two years, tidewater goby could no longer be documented at this location, indicating localized extinction. The BL population with (high

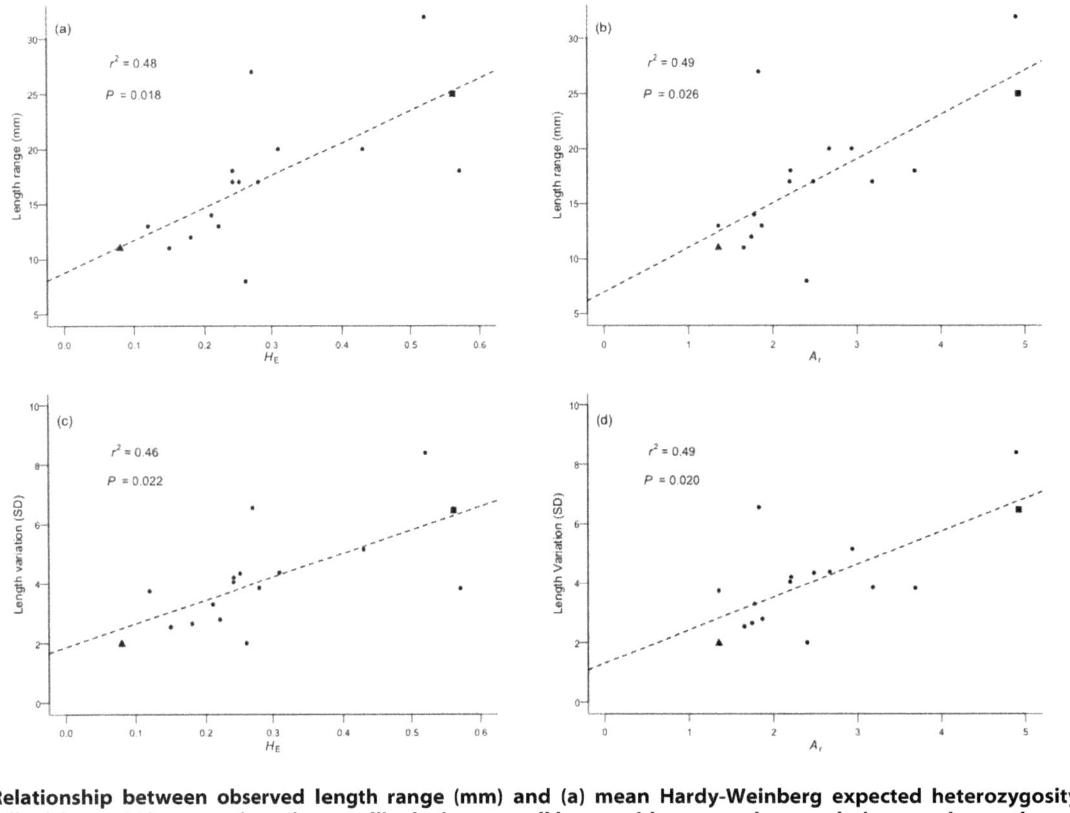

Figure 3. Relationship between observed length range (mm) and (a) mean Hardy-Weinberg expected heterozygosity (H_E), (b) rarefied allelic richness (A_r) across nine microsatellite loci among all known tidewater goby populations on the northern California coast. Populations indicated with a triangle and square indicate focal populations with low and high H_E, respectively, for otolith microstructural analysis.

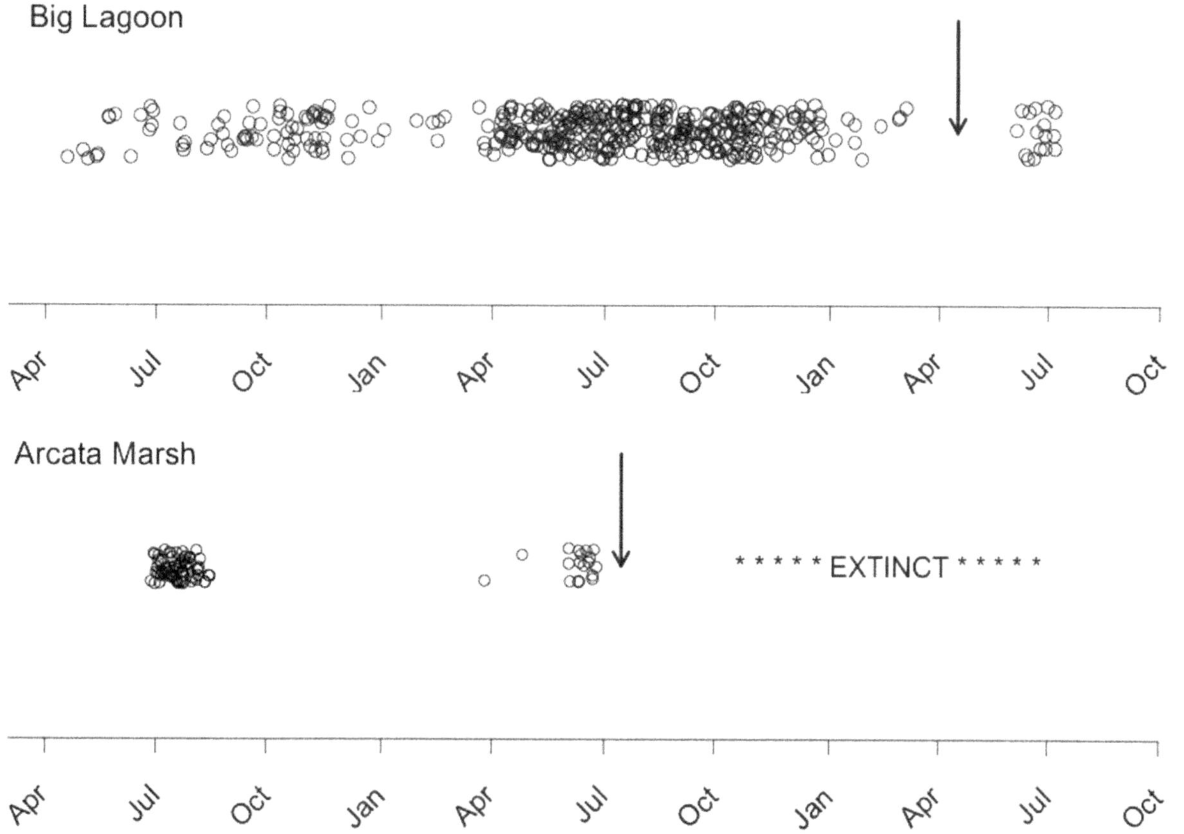

Figure 4. Individual birthdates of tidewater goby estimated from daily otolith increment counts in Big Lagoon (BL), CA, and Arcata Marsh (AM), CA. The Big Lagoon population is characterized by nearly continuous reproduction throughout the year, while the Arcata Marsh population exhibits a severely truncated reproductive period. Salinities in BL and AM ranged from 2‰ to 14‰ and 9‰ to 15‰, respectively, for the duration of the study period, with two exceptions, indicated by vertical arrows (BL: from 13‰ [April 2010] to 27‰ [May 2010]; AM: from 12‰ [July 2009] to 34‰ [August 2009]). High salinities attributed to stochastic salinity increases were lethal to young individuals and resulted in subsequent extinction of the AM population. The timeline illustrated by the horizontal axis begins in April 2008.

genetic diversity, broad range of ages/sizes) experienced a period of no births associated with the salinity increase (Figure 4), but persisted owing to the presence of large individuals tolerant of abrupt salinity changes.

Counts of increments from the fluorescent mark to the otolith margin matched the number of days between marking and sacrifice for all samples with a detectable calcein mark (ten days: n = 1, 20 days: n = 1, 28 days: n = 2). Daily age estimates ranged from 26 to 363 days for gobies sampled from the AM and from 48 to 421 days for the BL population. Estimates of instantaneous mortality rate of tidewater goby from the BL and AM populations correspond to 2.4% and 1.32% annual survivorship [30], respectively, confirming that few individuals survive longer than one year. Estimated growth curves for tidewater goby from BL and AM are illustrated in Figure S1, and parameter estimates are listed in Table 2.

Back calculation of birth dates for individuals from the AM population indicated a truncated reproductive period that manifests itself as a narrow size distribution of individuals (Figure 4; Table 1). In contrast, nearly year-round reproduction the genetically diverse BL population resulted in the presence of a broad range of individual sizes in this population (Figure 4; Table 1).

Discussion

The buffering effects of biological diversity on ecosystem processes, likened to the portfolio effect from economic theory, are frequently invoked as an important argument for the conservation of biodiversity [31]–[33]. These principles have been extended to the metapopulation level, illustrating that variability and biocomplexity among fish stocks will increase the resilience of metapopulations [8], [9]. The importance of among-population life history variability to the stability of meta-populations is increasingly recognized and has recently been quantitatively illustrated for a commercially exploited species for which long-term abundance estimates exist [9], [33]. Diversity among populations belonging to a stock complex buffers variability in overall abundance by spreading the risk of stochastic declines over many population segments, increasing the stability of cumulative meta-population abundance. This resilience is of great importance to the temporal stability of commercial exploitation, and improves the viability and sustainability of the stock complex in the long term, similar to the increased likelihood of stable returns attributed to diverse economic portfolios [8], [9], [33].

The results presented herein suggest that this portfolio effect not only applies to species which occur as distinct segments in a stock complex, but also to the resilience of individual, isolated populations. For tidewater goby, continuous reproduction independent of season (and the resulting diverse population age composition) appears to act as a natural safeguard against localized extinction by spreading the risk of high juvenile mortality across many different population segments. The implication of a truncated single spawning period for long-term population persistence was illustrated during a temporary increase in salinity at the Arcata Marsh (AM), a natural yet stochastic process in coastal habitats of tidewater goby. The timing of this disturbance coincided with a period of population turnover, shortly after the brief reproductive period but before juveniles could grow to sufficient size to better tolerate elevated salinity levels. Juvenile individuals, comprising over 94% of the population, vanished as a consequence of this disturbance (Figure S3). As length frequency distributions and results from otolith analysis suggest, the few surviving adults were nearing the maximum of longevity and their emaciated state indicated that they were post-reproductive. Extensive field surveys over the two subsequent years confirmed extinction of this previously large population. Such increased susceptibility of early life history stages, including eggs, to salinity fluctuations is not unusual and has been shown for a number of estuarine goby species [19]–[21]. A controlled experiment confirmed that increases in salinity are lethal to juvenile tidewater goby, supporting salinity change associated with breaching as the cause of extinction.

A similar salinity increase, but at a different location (Big Lagoon, BL), resulted in a stark decrease in abundance of smaller individuals (<35 mm TL, Figure S3) and an absence of new recruits to the population from the time period surrounding the breach (Figure 4). This population persisted, owing to the presence of large numbers of adult fish that can withstand physiological challenges associated with salinity increases. Though the salinity increase experienced by individuals in the Arcata Marsh (34‰) was greater in magnitude than in Big Lagoon (27‰) and during our salinity challenge experiment, adult tidewater goby have been documented in the wild at salinities of up to 42‰ [14]. These findings suggest that naturally occurring salinity increases are a significant population stressor whose effects can be buffered by demographic heterogeneity [34].

It should be emphasized that breaching events are an important mechanism in establishing connectivity among habitats typically subject to physical separation, and thus provide seasonal opportunity for migration and colonization of suitable habitats. As such, this natural process to which tidewater goby are well adapted (through broad physiological tolerances by adult individuals) is an important component of the species' biology. Consequently, the occurrence of breaching should not cause excessive stress on populations are characterized by a diverse age composition despite high juvenile mortality; rather, breaching threatens population persistence only if the population is composed of individuals of similar age, and if it occurs at a time when the great majority of individuals are of small size (shortly after peak reproduction). Further, the importance of breaching events as a population stressor may differ across the species' range: previous work suggests that populations in the southern portion of the species' range are subject to different risk factors related to a much drier local climate, such as hypersaline conditions or drying up of the estuary during the summer months [35], [36].

Our surveys of other tidewater goby in northern California not only indicate that heterogeneity in reproductive timing appears to be severely constrained in some of the remaining populations, but also suggest that such reductions in biocomplexity are often observed in populations with low levels of genetic variation (Figure 3). It should ne noted that narrow size ranges have been detected in populations from the southern portion of the species range, with increased diversification of size composition observed in subsequent surveys [14], but it is unknown how these populations compare to our study populations in terms of genetic diversity and other characteristics. Most populations with restricted size ranges of individuals (and low genetic diversity) included in this study are found along the perimeter of Humboldt Bay, in habitat patches fragmented by draining of wetlands and other human development. Though these populations persist and can reach high abundances, genetic drift following isolation is considered to have eroded genetic diversity, in absence of migration leading to significant degrees of genetic differentiation among populations [16]. The lack of heterogeneity in size composition within these populations indicates that sampled individuals were born within a constrained temporal window, despite the species' noted ability to reproduce under a wide range of environmental conditions and independent of season [14], [22]. Though respective growth rates may vary among populations, variation in individual size at age is expected to increase (individual fish of the same age may obtain different size at time t, Figure S1). Therefore, observed length range can be considered a conservative measure of the extent of the reproductive period (assuming no size selective mortality in the months preceding sampling). Large variation in sizes of tidewater goby is observed in small, isolated habitats when heterozygosity levels are high (for example, VG and PC; Table 1), yet the reproductive period appears constrained when genetic diversity is low, even in large, heterogeneous habitats (SC and LE; Table 1). Observation of this pattern across populations that are reproductively isolated yet in geographic proximity to one another, and therefore subject to nearly identical temperature and precipitation regimes, suggests a correlation between life history variation and genetic diversity in tidewater goby populations.

If reproductive timing in tidewater goby is under genetic influence, the striking asynchrony in reproductive timing in adjacent populations may be explained by the random loss of genetic variation through genetic drift. Though we interpret the results presented herein as suggestive of genetic impoverishment being a causative factor leading to decreased within-population heterogeneity, we recognize that alternative explanations for the observed patter should be considered. Specifically, a reduction in birthdate variation owing to periodic failure in recruitment or size selective mortality due to unsuitable environmental conditions may explain a limited range in the size of individuals and have caused erosion in genetic variation within those populations. However, as recruitment failures occur stochastically and not periodically, we believe that the observed pattern of demographic homogeneity in populations with low genetic diversity, considered in the context of the species' life history characteristics and their applicability to the dynamic environments in which the tidewater goby occurs, indicates that limited variation is reproductive timing is attributable to reductions in genetic diversity within isolated tidewater goby populations. Irrespective of whether a causative effect exists between genetic- and demographic diversity, limited age variation within these populations renders them particularly susceptible to environmental disturbance during certain times of the year through elimination of stabilizing effects provided by within-population variability.

We do not propose that near-term extinction of extant tidewater goby populations is accurately predicted by measured heterozygosity levels, and while stochastic events, likely in the form of natural salinity increases, will probably trigger future localized extinctions of tidewater goby in the species' northern range, we believe that this study clearly illustrates that demographically complex populations of this annual species are more resilient to such stressors. Likewise, though environmental factors doubtlessly have some influence on observed length ranges within isolated populations, the highly significant relationship between this measure of life history- and genetic diversity implies that the level of genetic variation is related to demographic heterogeneity. Though a causative relationship may be subject to debate, at a minimum, our results illustrate that a lack of genetic diversity works synergistically with environmental disturbance to influence population extirpation. As such, these results not only affirm the notion that genotypic and phenotypic diversity are tightly linked and deserve equal consideration in the quest for species conservation, but also illustrate the applicability of buffering portfolio effects to isolated populations, which contribute to their resilience against stochastic disturbances.

Our findings suggest that environmental instability during and shortly after periods of peak reproductive activity is of particular concern to the conservation of tidewater goby populations, especially those characterized by low levels of age- and genetic diversity. As a consequence, artificial breaching of estuaries during spring and summer (occasionally practiced for reasons such as draining pasture land inundated by high water levels in the estuary) should be avoided when possible. In addition, cohort assessment (analysis of size/age structure and relative abundance) should be incorporated in future surveys of tidewater goby populations, as this may permit identification of goby populations at elevated risk of near-term extirpation. Lastly, the relationship between genetic diversity and observed age variation should be investigated further, for example by experimental supplementation of genetic diversity in isolated populations through transfers of individuals from nearby, genetically diverse populations. Subsequently, frequent demographic assessments could be used then to evaluate whether reproductive variability has increased as a result of genetic supplementation, which may then receive consideration as a management tool in the quest for conservation of this endangered species.

Supporting Information

Figure S1 Schnute's (1981) four-parameter growth curves for populations of tidewater goby from Big Lagoon (BL), CA, and the Arcata Marsh (AM), CA.

Figure S2 Length frequency composition of tidewater goby from Big Lagoon, CA, sampled monthly between April 2009 and August 2010.

Figure S3 Length frequency histogram of the Big Lagoon, CA, and Arcata Marsh, CA, tidewater goby populations prior and subsequent to a stochastic, temporary increase in salinity.

File S1 Multilocus genotypes for tidewater goby, *Eucyclogobius newberryi*, from 17 populations in northern California. Information on relevant primers and amplification protocol can be found in the main text.

Acknowledgments

We thank W. McCraney for providing tidewater goby genotypes for some of our study populations, and are grateful to him, F. Allendorf, R. Waples, W. McCraney, B. Harvey, D. Hankin,T. Mulligan and two anonymous reviewers for providing insightful comments that greatly improved the quality of this paper. We also thank J. Montgomery, R. Records, G. Goldsmith, T. Scheiff, T. Laidig, M. Koenig, G. Hendrickson and J. Varkey for technical assistance. Samples were collected either by the USFWS or under California Scientific Collecting Permit SC-10527. All work was conducted according to institutional protocol at Humboldt State University (IACUC 08/09.F.44.A & 10/11.F.78-C), where all samples have been deposited in the scientific fish collection (HSU Catalog #4890-4913).

Author Contributions

Conceived and designed the experiments: MH AK. Performed the experiments: MH AK. Analyzed the data: MH. Contributed reagents/materials/analysis tools: MH AK. Contributed to the writing of the manuscript: MH AK.

References

1. Lande R, Barrowclough GF (1987) Viable Populations for Conservation. Cambridge Univ. Press, New York.
2. McNeely JA, Miller RK, Reid WV, Mittermeier RA, Werner TB (1990) Conserving the world's biological diversity. International Union for the Conservation of Nature and Natural Resources, World Resources Institute, Conservation International, WWF – US and World Banks, Washington, D.C.
3. Reed D, Frankham R (2003) Correlation between fitness and genetic diversity. Conserv Biol 17: 230–237. doi:10.1046/j.1523-1739.2003.01236.x.
4. Naeem S, Li S (1997) Biodiversity enhances ecosystem reliability. Nature 390: 507–509. doi:doi:10.1038/37348.
5. Tilman D (1996) Biodiversity: population versus ecosystem stability. Ecology 77: 350–363. doi:10.2307/2265614.
6. Yachi S, Loreau M (1999) Biodiversity and ecosystem functioning in a fluctuating environment: the insurance hypothesis. Proc Natl Acad Sci USA 96: 1463–1468. doi:10.1073/pnas.96.4.1463.
7. McCann KS (2000) The diversity-stability debate. Nature 405: 228–233. doi:10.1038/35012234.
8. Hilborn R, Quinn TP, Schindler DE, Rogers DE (2003) Biocomplexity and fisheries sustainability. Proc Natl Acad Sci USA 100: 6564–6568. doi:10.1073/pnas.1037274100.
9. Schindler DE, Hilborn R, Chasco B, Boatright CP, Quinn TP, et al. (2010) Population diversity and the portfolio effect in an exploited species. Nature 465: 609–612. doi:10.1038/nature09060.
10. Luck GW, Daily GC, Ehrlich PR (2003) Population diversity and ecosystem services. Trends Ecol Evol 18: 331–336. doi:10.1016/s0169-5347(03)00100-9.
11. Hsieh C, Yamauchi A, Nakazawa T, Wang WF (2010) Fishing effects on age and spatial structures undermine population stability of fishes. Aquat. Sci. 72: 165–178. doi:10.1007/s00027-009-0122-2.
12. Gamfeldt L, Källström B (2007) Increasing intraspecific diversity increases predictability in population survival in the face of perturbations. Oikos 116: 700–705. doi:10.1111/j.0030-1299.2007.15382.x.
13. Lafferty KD, Swenson RO, Swift CC (1996) Threatened fishes of the world: Eucyclogobius newberryi Girard, 1857 (Gobiidae). Environ Biol Fishes 46: 254.
14. Swift CC, Nelson JL, Maslow C, Stein T (1989) Biology and distribution of the tidewater goby, *Eucyclogobius newberryi* (Pisces: Gobiidae) of California. Contributions in Science 404. Natural History Museum of Los Angeles County, Los Angeles.
15. U.S. Fish and Wildlife Service (1994) Federal Register 59: 5494.
16. McCraney WT, Goldsmith G, Jacobs DK, Kinziger AP (2010) Rampant drift in artificially fragmented populations of the endangered tidewater goby (*Eucyclogobius newberryi*). Mol Ecol 19: 3315–3327. doi:10.1111/j.1365-294X.2010.04755.x.
17. Swenson RO (1999) The ecology, behavior, and conservation of the tidewater goby, *Eucyclogobius newberryi*. Environ Biol Fishes 55: 99–114.
18. Krauss NC, Militello A, Todoroff G (2002) Barrier breaching process and barrier spit breach, Stone Lagoon, California. Shore and Beach 70: 21–28.
19. Healey MC (1971) The distribution and abundance of sand gobies, *Gobius minutus*, in the Ythan estuary. J Zool 163: 177–228. doi:10.1111/j.1469-7998.1971.tb04531.x.
20. Gill HS, Potter IC (1993) Spatial segregation amongst goby species within an Australian estuary, with a comparison of the diets and salinity tolerance of the two most abundant species. J Mar Biol 117: 515–526. doi:10.1007/BF00349327.
21. Matern AS (2001) Using temperature and salinity tolerances to predict the success of the Shimofuri goby, a recent invader into California. Trans Am Fish

Soc 130: 592–599. (doi:10.1577/1548-8659(2001)130<0592:UTASTT> 2.0.CO;2).
22. Goldberg SR (1977) Seasonal ovarian cycle of the tidewater goby, *Eucyclogobius newberryi* (Gobiidae). The Southwest Nat 22: 557–559.
23. Swenson RO (1995) The reproductive behavior and ecology of the tidewater goby, *Eucyclogobius newberryi* (Teleostei: Gobiidae). Ph. D. Dissertation, University of California at Berkeley, Berkeley.
24. Simons AM (2011) Modes of response to environmental change and the elusive empirical evidence for bet hedging. Proc Biol Sci 278: 1601–1609.
25. Earl DA, Louie KD, Bardeleben C, Swift CC, Jacobs DK (2010) Rangewide microsatellite phylogeography of the endangered tidewater goby, (*Eucyclogobius newberryi*) (Teleostei: Gobionellidae), a genetically subdivided coastal fish with marine dispersal. Conserv Genet 11: 103–114.
26. Mendonca H, Smith J, Brinegar C (2001) Isolation and characterization of four microsatellite loci in the tidewater goby (*Eucyclogobius newberryi*). J Mar Biotechnol 3: 91–95. pmid: 14961369.
27. Kalinowski ST (2005) HP-RARE 1.0: a computer program for performing rarefaction on measures of allelic richness. ***Mol Ecol Notes*** 5: 187–189.
28. Leberg PL (2002) Estimating allelic richness: Effects of sample size and bottlenecks. ***Mol Ecol*** 11: 2445–2449.
29. Hoenig JM (1983) Empirical use of longevity data to estimate mortality rates. Fishery Bulletin 82: 898–903. doi:10.1890/04-0594.
30. Schnute J (1981) A versatile growth model with statistically stable parameters. Can J Fish Aquat Sci 38: 1128–1140.
31. Yachi S, Loreau M (1999) Biodiversity and ecosystem productivity in a fluctuating environment: the insurance hypothesis. Proc Natl Acad Sci USA 96: 1463–1468. doi:10.1073/pnas.96.4.1463.
32. Hughes AR, Inouye BD, Johnson MTJ, Underwood N, Vollend M (2008) Ecological consequences of genetic diversity. Ecol Lett 11: 609–623. doi:10.1111/j.1461-0248.2008.01179.x.
33. Greene CM, Hall JE, Guilbault KR, Quinn TP (2010) Improved viability of populations with diverse stock portfolios. Biol Lett 6: 382–386. doi:10.01098/rsbl.2009.0780.
34. Fox G (2005) Extinction risk of heterogeneous populations. Ecology 86: 1191–1198.
35. Lafferty KD, Swift CC, Ambrose RF (1999a). Extirpation and recolonization in a metapopulation of an endangered fish, the tidewater goby. Conserv Biol 13: 1447–1453. doi:10.1046/j.1523-1739.1999.98016.x.
36. Lafferty KD, Swift CC, Ambrose RF (1999b) Postflood persistence and recolonization of endangered tidewater goby populations. N Am J of Fish Manag 19: 618–622. doi:10.1577/1548-8675(1999)019<0618:PPAROE> 2.0.CO;2.

16

Alpha, Beta and Gamma Diversity Differ in Response to Precipitation in the Inner Mongolia Grassland

Qing Zhang[1,3], Xiangyang Hou[2], Frank Yonghong Li[4], Jianming Niu[1,3]*, Yanlin Zhou[1], Yong Ding[2], Liqing Zhao[1], Xin Li[3], Wenjing Ma[1], Sarula Kang[1]

1 School of Life Sciences, Inner Mongolia University, Hohhot, China, **2** Grassland Research Institute of Chinese Academic of Agricultural Science, Hohhot, China, **3** Sino-US Center for Conservation, Energy and Sustainability Science, Inner Mongolia University, Hohhot, China, **4** AgResearch Grasslands Research Centre, Palmerston North, New Zealand

Abstract

Understanding the distribution pattern and maintenance mechanism of species diversity along environmental gradients is essential for developing biodiversity conservation strategies under environmental change. We have surveyed the species diversity at 192 vegetation sites across different steppe zones in Inner Mongolia, China. We analysed the total species diversity (γ diversity) and its composition (α diversity and β diversity) of different steppe types, and their changes along a precipitation gradient. Our results showed that (i) β diversity contributed more than α diversity to the total (γ) diversity in the Inner Mongolia grassland; the contribution of β diversity increased with precipitation, thus the species-rich (meadow steppe) grassland had greater contribution of β diversity than species-poor (desert steppe) grassland. (ii) All α, β and γ species diversity increased significantly ($P<0.05$) with precipitation, but their sensitivity to precipitation (diversity change per mm precipitation increase) was different between the steppe types. The sensitivity of α diversity of different steppe community types was negatively ($P<0.05$) correlated with mean annual precipitation, whereas the sensitivity of β and γ diversity showed no trend along the precipitation gradient ($P>0.10$). (iii) The α diversity increased logarithmically, while β diversity increased exponentially, with γ diversity. Our results suggest that for local species diversity patterns, the site species pool is more important in lower precipitation areas, while local ecological processes are more important in high precipitation areas. In addition, for β diversity maintenance niche processes and diffusion processes are more important in low and high precipitation areas, respectively. Our results imply that a policy of "multiple small reserves" is better than one of a "single large reserve" for conserving species diversity of a steppe ecosystem, and indicate an urgent need to develop management strategies for climate-sensitive desert steppe ecosystem.

Editor: Dafeng Hui, Tennessee State University, United States of America

Funding: This study was supported by the State Key Basic Research Development Programme of China (Grant nos. 2012CB722201 and 2014CB138805), The National Basic Research Programme of China (Grant no. 31200414), The Specialised Research Fund for the Doctoral Programme of Higher Education of China (Grant no. 20121501120006), and the Start Research Funding Project of Inner Mongolia University (Grant no. 125106). The funders had no role in study design, data collection and analysis, decision to publish, or preparation of the manuscript.

Competing Interests: The authors have declared that no competing interests exist.

* E-mail: jmniu2005@163.com

Introduction

Half a century ago, Whittaker first defined species diversity from three different perspectives: alpha (α) diversity, beta (β) diversity, and gamma (γ) diversity [1]. Alpha and γ diversity are grouped as inventory diversity [2], sharing the same characteristics and differentiated only by scale. Beta diversity is defined as the difference in species composition between communities and is closely related to many facets of ecology and evolutionary biology [3,4,5]. Most research on species diversity has focused on inventory diversity, but research on β diversity has recently increased [3,6,7]. The distribution patterns and maintenance mechanisms of species diversity along environmental gradients have been core topics in ecological research [4,8,9]. It is essential to understand such patterns and mechanisms for the development of strategies and measures for conserving species diversity under environmental change. Whether or not a single large or several small (SLOSS) reserves are superior means of conserving biodiversity [10] depends on the dominant type of diversity present. For example, a high β species diversity within a community type may theoretically imply that the community occupies a heterogeneous environment. In that situation, the use of the 'several small' strategy will be superior to the "single large" strategy in reserve design for species diversity conservation [11].

Studies on the geographical patterns of species diversity generally showed a decreasing trend of α and γ diversity with increasing latitude from tropics to poles, which was primarily driven by temperature [8,12]. However, the correlations between β diversity and latitude were inconsistent; they could be positively [13,14], negatively [15,16], or not correlated [9,17]. Numerous studies have also reported species diversity patterns along other environmental gradients [18,19,20], and in many cases the species diversity-precipitation relationships have been studied [21,22,23]. Precipitation is the most important factor affecting species diversity and ecosystem functioning in arid and semiarid grasslands, such as in the Eurasian steppe, North American prairie and African savanna [23,24,25]. However, reported relationships between species diversity and precipitation have been inconsistent for

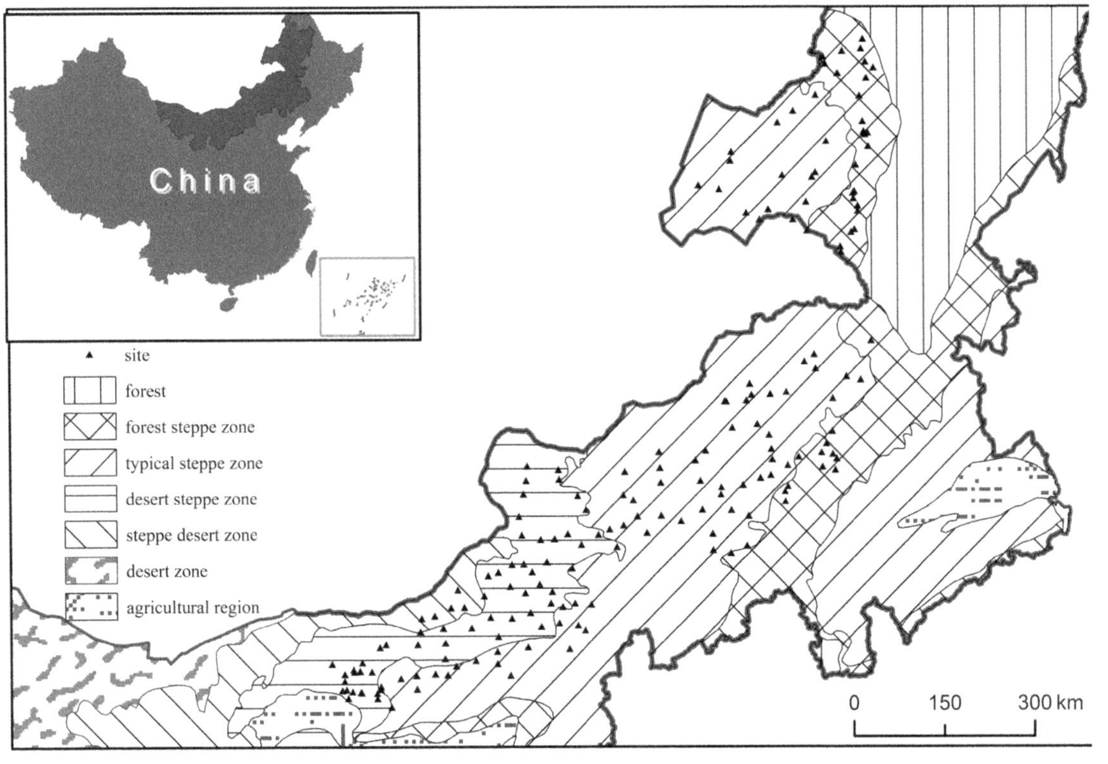

Figure 1. Map of the study region, showing the vegetation zones and sampling sites.

grassland biomes and scales; in some cases increasing precipitation promoted species diversity [21,22,26], but in other cases it had little effect [21,25].

Species evolution and diffusion, inter-specific competition, and environmental changes commonly influence the α, β and γ diversity of plant communities, but the response of species diversity pattern to these biological and environmental changes and the mechanisms for the response differ among the three species diversity measures [27,28,29,30]. Climate, geological history and ecological randomness are considered to be the main factors affecting species diversity pattern at large scales [31,32], whereas the local ecological processes and regional species pool are considered to be important at a small scale [33,34]. Environmental heterogeneity and species diffusion are closely related to β diversity [9,35]. The relative importance of different components of species diversity in maintaining the total species diversity differs with the spatial and temporal scales or across different regions [15,36].

The Inner Mongolia steppe grassland is an essential part of the Eurasian steppe [37]. Owing to the heterogeneous environment and the particular geological and evolutionary history, the grassland is rich in plant species diversity. It has more than 2,300 vascular plant species [38], and is recognised as the second most species-rich grassland biome in the world in terms of indigenous plant biodiversity after the African savanna [39]. The Inner Mongolia grassland is one of the key areas for biodiversity conservation in the world. The grassland types in Inner Mongolia show clear zones along a climatic gradient, from the desert steppe in dry areas, to the typical steppe, and through to the meadow steppe [40,41]. There have been several studies on the effects of precipitation on the species diversity and grassland productivity in the Inner Mongolia [24,42], but the composition of species diversity (α, β, γ) and their changes across different steppe grassland types on environmental gradients has not been studied.

In the present study, we measured the components of species diversity (α, β, γ) in each of the major steppe community types. The aim of the study was to present the composition of, and the relationships between α, β, and γ diversity of main steppes types in Inner Mongolia, and analyse the pattern of variation along the precipitation gradient. Then, based on these diversity patterns and relationships, we discuss the mechanism of species diversity maintenance in the Inner Mongolia grassland, and suggest the best strategies for species diversity conservation under environmental change.

Methods

Ethics statement

All the survey sites were owned and/or managed by pastoral farmers, who gave permission to the survey. The field studies did not involve any endangered or protected species.

Study area

We surveyed the grassland species diversity in the whole region of the Inner Mongolia grassland in northern China. The region stretches from latitudes 41.31°N to 50.78°N and longitudes 108.16°E to 120.39°E with elevation ranging from 532 to 1725 m above sea level (Fig. 1). The typical landforms in this region include gently rolling plains, tablelands, and hills. Mean annual temperature (MAT) ranges from −3.0 to 6.7°C, and mean annual precipitation (MAP) varies from about 150 to 450 mm, decreasing from southeast to northwest [43]. Distributed along the climate gradient from the relatively humid southeast to relatively dry northewest are soil types that very from chernozems, to chestnut- and then calcic brown soils; and grassland biomes that vary from meadow steppe, to typical steppe and then desert steppe grassland (Fig. 1).

Data collection

We investigated species diversity on 192 sites in the grassland region from late July to mid-August in 2012, when the grassland community biomass was at its peak. Seven dominant steppe community types were investigated: *Leymus chinensis* meadow steppe, *Stipa baicalensis* meadow steppe, *S. grandis* typical steppe, *L. chinensis* typical steppe, *S. krylovii* typical steppe, *S. breviflora* desert steppe, and *S. klemenzii* desert steppe. Nineteen to 39 sites were surveyed for each community type; the number of sites for each type was approximately determined by the relative size of its distribution area. The position of each site was located using GPS (Fig. 1). To focus on the relationships between species diversity and climate, and to minimize the influence of domestic animal grazing, all the surveyed sites were selected either in fenced grassland under protection, or in mowed grassland (surveyed before haymaking at the end of August). Grassland sites with obvious grazing effects (recognised by species composition) were excluded from the survey. At each site an area of 10×10 m was delineated, and ten 1×1 m plots were randomly placed in the delineated area to record all the plant species.

Meteorological records from the 156 meteorological stations in the study area were used in the analysis of the diversity-precipitation relations [44].

Data analyses

Calculation of species diversity. Since the concept of β diversity was introduced in 1972 [45], more than 40 different methods have been proposed for its calculation [6,46]. Jurasinski [2] suggested that these calculation methods could be split into two groups: the first group, designated differentiation diversity, includes a similarity coefficient, similarity attenuation slope with distance, gradient length in ordination space, and total variance of community composition [9,47]. The second group, designated proportional diversity, includes additive partition diversity and multiplicative partitioning diversity [28,45]. Additive diversity partition expresses α and β diversity in the same unit so that their relative importance can be easily quantified and interpreted, and can be directly compared across spatial and temporal scales or land-use practices [28,34,48]. We used the additive partition approach to calculate total species diversity of the studied grassland and its components. The methods were described in [27,28], and are briefly described below:

- α diversity, also called community diversity, is defined as the mean of the species richness (number) in the surveyed ten plots at a site:

$$\alpha = \frac{1}{10}\sum_{i=1}^{10} S_i \qquad \text{(Eq.1)}$$

where S_i represents the species richness in each plot.

- γ diversity, also called site diversity, is the total species richness at a site.
- β diversity, defined as the difference in species composition among the ten plots at a site, is calculated by subtracting α diversity from γ diversity:

$$\beta = \gamma - \alpha \qquad \text{(Eq.2)}$$

The β diversity has two components, one quantifying the degree of nestedness of species richness (β_N), *i.e.* the degree to which species richness differs between plots within one site from the most species-rich plot, and the other component reflecting the difference in species composition among the plots (β_R). With S_{max} representing the species number in the richest plot, β_N and β_R are calculated as follows:

$$\beta_N = \frac{1}{10}\sum_{i=1}^{10} (S_{\max} = S_i) \qquad \text{(Eq.3)}$$

$$\beta_R = \beta - \beta_N \qquad \text{(Eq.4)}$$

Consequently, the total species richness at a site (γ) is the sum of the mean species richness of all the plots at the site (α), the differences in species richness due to the nestedness (β_N) and species composition (β_R) among the plots:

$$\gamma = \alpha + \beta_R + \beta_N \qquad \text{(Eq.5)}$$

The components of species diversity (α, β and γ) were assessed for each grassland site. The species diversity at all of the sites for each steppe community type was averaged to represent the species diversity of the community type. The species diversity composition of three steppe vegetation types (i.e., the desert steppe, typical steppe and meadow steppe) were also aggregated in the same way to represent the species diversity at all grassland sites of a steppe vegetation type.

Calculation of precipitation at each site and for each grassland type. We calculated mean annual precipitation (MAP) at each vegetation site using the approach of Thornthwaite [49] based on precipitation records of the 156 meteorological stations in the region. Each site MAP was derived according to the latitude (LAT), longitude (LNG) and altitude (ALT) of the site, using a previously developed model on the relationship between MAP and the geographical coordinates of each meteorological station [44]:

$$\begin{aligned} MAP = {} & 13872.1241 - 0.8941ALT - 198.2731LAT \\ & + 2.2360LAT^2 - 0.0313LAT \times ALT \\ & - 176.0567LNG + 0.8296LNG^2 + 0.0203LNG \\ & \times ALT (r^2 = 0.84) \end{aligned} \qquad \text{(Eq.6)}$$

The MAP of each community type was calculated as the mean of the MAP at all the sites that belongs to the community type.

Relationships among species diversity components. First, the proportions of α and β to γ diversity were used to quantify the relative contribution of these two kinds of diversity to γ diversity. The composition of γ diversity (i.e., the proportions of α, β, β_R and β_N) of the seven steppe community types, and of the three steppe vegetation types, was ordered according to the precipitation of their distribution areas in order to examine the species diversity composition changes across these steppe types in relation to precipitation.

Second, we used regression analysis to model relationships between γ diversity and its components (α and β) across all 192 sites. The relationships between γ diversity and the 'occasional species' (species recorded in only one or two of the ten plots at a site) were also analysed. These relationships were used to interpret the mechanism of species diversity maintenance in the grasslands.

Table 1. The environmental characteristics (mean annual precipitation MAP, mean annual temperature MAT, and major soil types) and species diversity composition (α, β and γ) of the seven steppe community types and of the three vegetation types in the Inner Mongolia grassland.

Type	No. of sites	MAP	MAT	Major soil types	α	β	γ	β_N	β_R	α%	β%
Stipa klemenzii desert steppe	36	196	3.52	calcic brown soil	9.5±0.3	9.0±0.4	18±0.7	3.6±0.2	5.4±0.4	51.9±0.9	48.1±0.9
Stipa breviflora desert steppe	39	212	3.18	Calcic brown and light chestnut soil	10.3±0.3	10.4±0.5	20.7±0.6	3.9±0.2	6.5±0.4	50.4±1.2	49.6±1.2
Stipa krylovii typical steppe	31	263	0.85	chestnut and light chestnut soil	12.8±0.4	13.7±0.6	26.4±0.9	4.7±0.2	8.9±0.5	48.7±0.6	51.3±0.6
Leymus chinensis typical steppe	28	305	0.15	chestnut and dark chestnut soil	12.6±0.4	13.5±0.7	26.1±1.0	3.7±0.2	9.8±0.6	48.9±0.9	51.1±0.9
Stipa grandis typical steppe	21	329	−0.40	chestnut soil	15.1±0.5	17.8±0.8	32.8±1.3	5.3±0.2	12.4±0.7	46.5±0.8	53.5±0.8
Stipa baicalensis meadow steppe	18	348	−2.02	chernozem and light chernozem soil	20.7±0.8	32.3±1.7	53.0±2.5	7.8±0.4	24.5±1.5	39.4±0.9	60.6±0.9
Leymus chinensis meadow steppe	19	354	−1.85	chernozem soil	18.2±0.7	22.6±0.9	40.8±1.3	6.0±0.4	16.5±0.8	44.8±1.1	55.2±1.1
Desert steppe	75	205	3.34	mainly on calcic brown soil	9.9±0.2	9.7±0.3	19.6±0.5	3.7±0.2	5.97±0.3	51.1±0.7	48.9±0.7
Typical steppe	80	302	0.27	mainly on chestnut soil	13.3±0.3	14.7±0.5	28.0±0.7	4.5±0.2	10.2±0.4	48.2±0.5	51.8±0.5
Meadow steppe	37	351	−1.93	mainly on chernozem soil	19.4±0.5	27.3±0.2	46.7±1.5	6.9±0.4	20.4±1.0	42.2±0.8	57.9±0.8
Inner Mongolia grassland	192	273	1.05	all the soils above	13.1±0.3	15.2±0.6	28.3±0.8	4.7±0.1	10.5±0.5	48.2±0.4	51.8±0.4

The β diversity has two components of β nestedness diversity (β_N) and β replacement diversity (β_R). The diversity values are the mean ± s.e.m of species number recorded in ten plots at each site. The percentage of α and β diversity in γ diversity are also shown as α% and β%.

Patterns of species diversity along a precipitation gradient. With α, β and γ diversity and the precipitation (MAP) calculated for all the vegetation sites, we plotted each diversity measure against precipitation, and tested for a significant correlation between species diversity and MAP using linear regression analysis. We also did a linear regression analysis between the species diversity and precipitation for each of the seven steppe community types, and used the slope of the regression (i.e. diversity change per mm precipitation change) to represent the sensitivity of the species diversity of each community type to precipitation. A steep slope or high sensitivity means a greater effect of precipitation on community species diversity. The changes in sensitivity across seven community types along a precipitation gradient were tested by examining if the sensitivity and precipitation of these community types were significantly correlated.

All statistical analyses were performed using Excel 2010 and SPSS 17.0.

Results

Species diversity and diversity composition of the Inner Mongolia grassland

The species diversity (γ) and its components (α, β_R and β_N) were consistently low in desert steppe and high in meadow steppe, and in-between in the typical steppe (Table 1). The β diversity had a slightly greater contribution (51.83%) to γ diversity than α diversity (48.17%) in the Inner Mongolia grassland (Table 1 and Fig. 2A). The contribution of α and β diversity to γ diversity differs among the seven steppe community types and among the three steppe vegetation types (Fig. 2A); the contribution of α diversity was higher in desert steppe than in meadow steppe, and conversely the contribution of β diversity was lower in desert steppe than in meadow steppe. The change in the contribution of β diversity among the steppe community types was mainly due to changes in β replacement diversity (β_R), while β nestedness diversity (β_N) showed little change. When these steppe types were ordered according to the MAP of their distribution areas, a trend of decreasing α and increasing β with increasing MAP was shown (Fig. 2B).

Species diversity pattern and their components along a precipitation gradient

The α diversity increased significantly ($P<0.001$) with increasing precipitation (Fig. 3A), but the sensitivity of α diversity to precipitation across the seven steppe community types significantly decreased ($P=0.028$) with increasing precipitation (Fig. 3B).

Both β_N diversity and β_R diversity increased with precipitation, and consequently total β diversity also increased (Fig. 3C, E, G). However, there was no correlation between the sensitivity of β, β_N or β_R diversity to precipitation across the seven community types along the precipitation gradient ($P>0.05$), though the sensitivity of β_R diversity showed a non-significant increasing trend ($P=0.10$) (Fig. 3D, F, H).

The γ diversity also significantly increased with precipitation (Fig. 3I), but the sensitivity of γ diversity to precipitation showed no correlation with MAP across the seven community types (Fig. 3J).

Relations among the γ diversity and its components

The α diversity increased with γ diversity, and the increase was gradually saturates and could be described by a logarithmic curve ($P<0.001$) (Fig. 4A). The β diversity also increased with γ diversity, but the increase was, complementarily to α diversity, accelerated with the increase of γ diversity (Fig. 4B). There was a significant linear correlation between γ diversity and occasional species

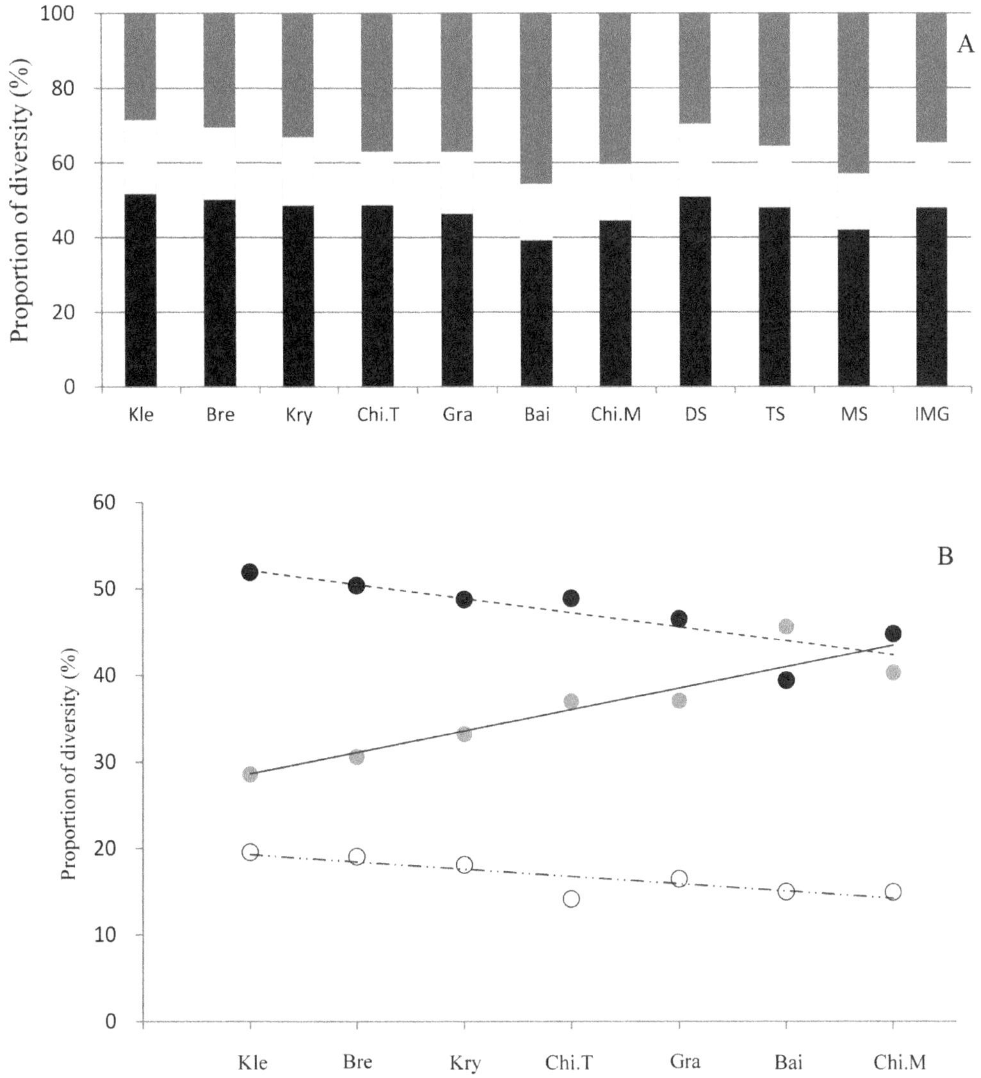

Figure 2. Species diversity of the steppe grassland in Inner Mongolia. The data are presented for seven major steppe community types: Kle (*Stipa klemenzii* desert steppe), Bre (*Stipa breviflora* desert steppe), Kry (*Stipa krylovii* typical steppe), Chi.T (*Leymus chinensis* typical steppe), Gra (*Stipa grandis* typical steppe), Bai (*Stipa baicalensis* meadow steppe), Chi.M (*Leymus chinensis* meadow steppe); three steppe vegetation types: DS (desert steppe), TS (typical steppe), MS (meadow steppe); and the Inner Mongolia grassland as a whole (IMG). A: Proportion of α diversity (black), β nestedness diversity (β_N) (white) and β replacement diversity (β_R) (grey) in γ diversity; B: The trend of the proportion of α diversity (black dot), β_N diversity (white dot) and β_R diversity (grey dot) in γ diversity in the seven grassland types with the types ordered according to the annual mean precipitation of their distribution areas, with precipitation increase from left to right.

diversity (Fig. 4C). The occasional species diversity also increased significantly ($P<0.001$) with increasing precipitation (Fig. 4D).

Discussion

Contribution of α and β to γ diversity changes across different steppe types along a precipitation gradient

The contributions of α and β diversity to γ diversity form the basis for understanding the biodiversity components [4,46]. Controversial opinions exist on the relative importance of α and β diversity to γ diversity: some believe that α diversity is more important, while the others contend that β diversity is more important. A third group suggest that α and β diversity work together [4,27,46]. We have found that β diversity contributes slightly more (51.83%) than α diversity (48.17%) to γ diversity in the Inner Mongolia grassland (Fig. 2), and that contribution of β diversity is greater in species-rich grassland (meadow steppe) in high precipitation areas than in species-poor grassland (desert steppe) in low precipitation areas. In other words, the contribution of β diversity has an increasing trend with precipitation (Fig. 2B).

The relative contribution of α and β to γ diversity in biological communities depends on the ecological heterogeneity and capability of species diffusion [27,28]. The α diversity is more important in communities with an homogeneous environment and strong-diffusion species, whereas β diversity is, on the contrary, more important in communities with an heterogeneous environment and weak-diffusion species. The increase of ecological heterogeneity with increasing precipitation in the studied grassland region [22,42] may be attributable to the increase in the contribution of β diversity in γ diversity with precipitation increase. This is supported by the high occasional species diversity in the species-rich grasslands (Fig. 4C). High occasional species

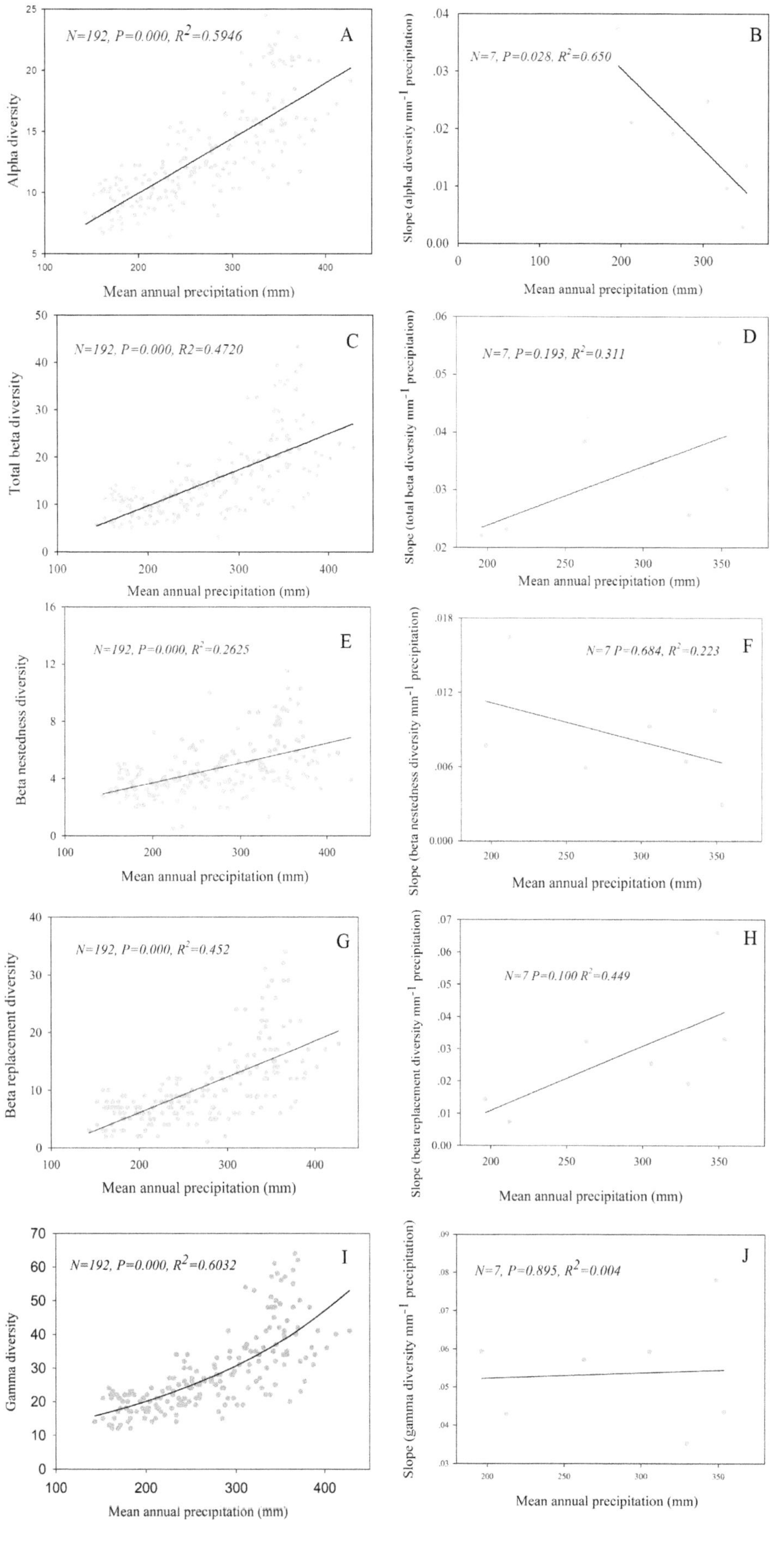
N=192, P=0.000, R2=0.5946
A
Alpha diversity
Mean annual precipitation (mm)
N=7, P=0.028, R2=0.650
B
Slope (alpha diversity mm-1 precipitation)
Mean annual precipitation (mm)
N=192, P=0.000, R2=0.4720
C
Total beta diversity
Mean annual precipitation (mm)
N=7, P=0.193, R2=0.311
D
Slope (total beta diversity mm-1 precipitation)
Mean annual precipitation (mm)
N=192, P=0.000, R2=0.2625
E
Beta nestedness diversity
Mean annual precipitation (mm)
N=7 P=0.684, R2=0.223
F
Slope (beta nestedness diversity mm-1 precipitation)
Mean annual precipitation (mm)
N=192, P=0.000, R2=0.452
G
Beta replacement diversity
Mean annual precipitation (mm)
N=7 P=0.100 R2=0.449
H
Slope (beta replacement diversity mm-1 precipitation)
Mean annual precipitation (mm)
N=192, P=0.000, R2=0.6032
I
Gamma diversity
Mean annual precipitation (mm)
N=7, P=0.895, R2=0.004
J
Slope (gamma diversity mm-1 precipitation)
Mean annual precipitation (mm)

Figure 3. Response of species diversity to precipitation (left column), and the sensitivity of species diversity of seven steppe community types to precipitation change (right column) in the Inner Mongolia grassland. The sensitivity is represented by the regression slope of the linear regression of species diversity to precipitation within each steppe type. A: response of α diversity; B: sensitivity of α diversity; C: response of β diversity; D: sensitivity of β diversity; E: response of β_N diversity; F: sensitivity of β_N diversity; G: response of β_R diversity; H: sensitivity of β_R diversity; I: response of γ diversity; J: sensitivity of γ diversity.

diversity is also in accordance with high species replacement diversity (β_R) (Fig. 2). Thus, the increase of the contribution of β to γ diversity along the gradient of increasing precipitation is most likely associated with the increase in ecological heterogeneity and the decrease in species diffusion.

Maintenance mechanism of species diversity along a precipitation gradient changes from regional species pool to the effect of local ecological processes

The maintenance mechanism of local species diversity has always been an important topic in ecology. Local ecological processes (such as predation, competition, resource supply and diffusion) [50,51] and the regional species pool [33,34] have been considered as the mechanisms for diversity maintenance. However, there is insufficient understanding on which of these two mechanisms is more important for local species diversity patterns [36]. Examining the relationship between regional and local species diversity can help quantify the importance of these two mechanisms [34]. A linear correlation between γ diversity and α diversity would suggest that the regional species pool was the main limiting factor. Alternatively, a saturated curve between α and γ diversity would suggest ecological processes are more important in maintaining diversity [34].

In the studied grassland, both α and γ diversity increased as precipitation increased (Fig. 3), but the increase of α diversity was gradually saturated (a logarithmical increase with γ diversity) (Fig. 4A). That is, the increase of α diversity with increasing γ increasing was rapid and approximately linear when γ diversity was low, but the increase slowed down when γ diversity was high in high precipitation areas. In species-poor grasslands (low γ) with low precipitation, the inter-specific competition is relatively weak resulting in much spare niche capacity. Niche theory indicates that

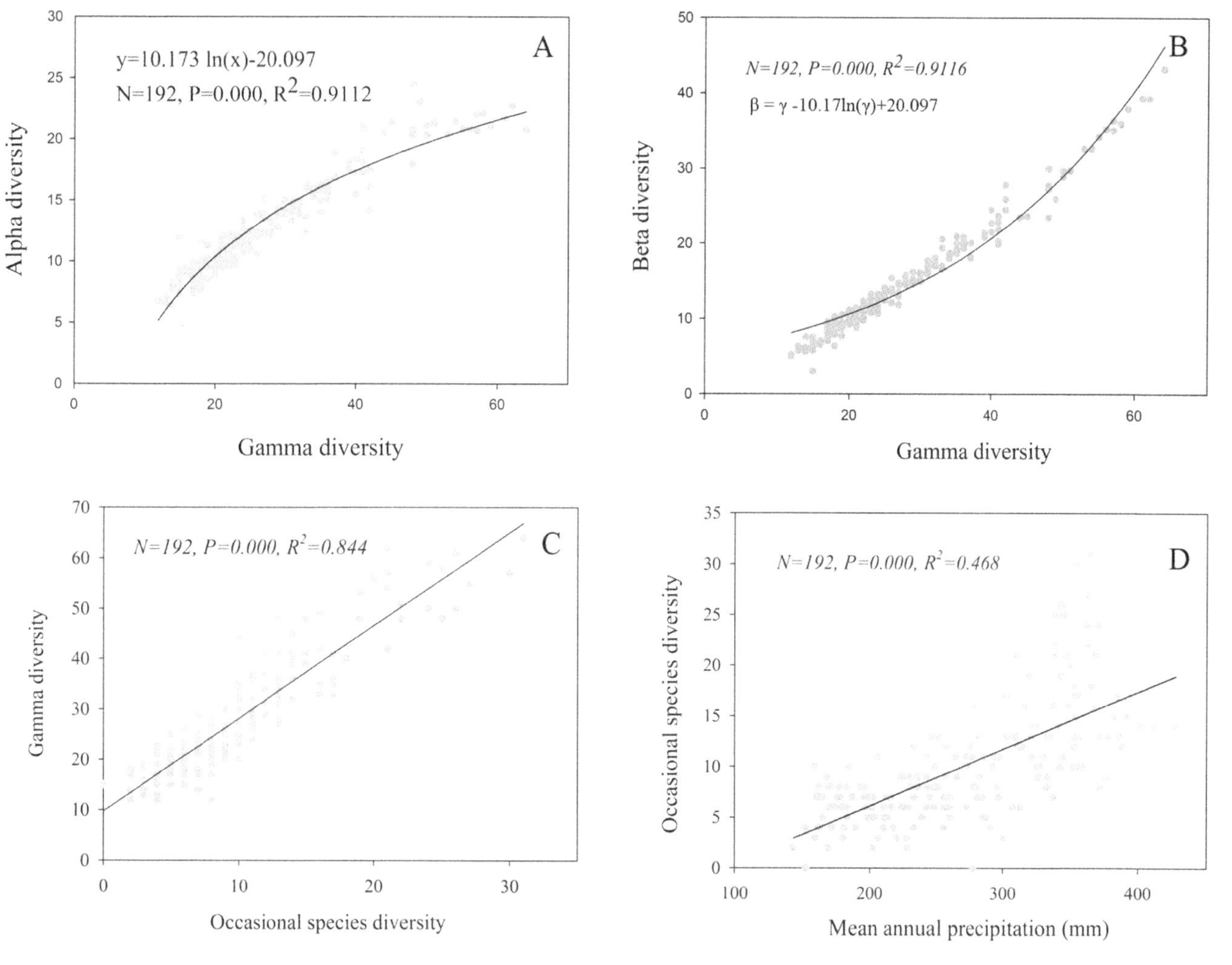

Figure 4. Relations of α diversity (A), β diversity (B) and occasional diversity (C) with γ diversity in the Inner Mongolia grassland. The β diversity is complementary to α diversity in γ diversity, and the curve in (B) is derived as $\beta = \gamma - \alpha$. D: response of occasional species diversity to mean annual precipitation.

every species occupies its unique corresponding niche [52], thus an increasing regional species pool provides the possibility for species to occupy more niches in the community. The near linear increase of γ diversity with α diversity (Fig. 4A) in low precipitation areas may indicate the importance of the regional species pool in determining local species diversity. In the high precipitation grassland region, more species appeared in the community, and inter-specific competition was relatively strong. In the case of full use of resources, the ecological niche was occupied to its fullest extent [53]. This means there was very little spare niche capacity, and no new species could be present in the community. Therefore, the regional species pool was still increasing, but local species diversity was saturated. In other words, with an increase in mean annual precipitation, the dominant maintenance mechanism of community diversity changed from the regional species pool to the effect of local ecological processes.

The maintenance mechanism of β diversity changes from niche processes to diffusion processes with increasing precipitation

The β diversity reflects different degrees of species composition. The environment heterogeneity (or the niche process [1,45]) and species diffusion processes [54] have been recognised as the main mechanisms that combine to maintain β diversity [9,29]. The relative importance of the two processes varies across regions and scales [15,35]. The nestedness species diversity (β_N) represents the degree to which species richness in each plot differs from the richest plot at a site, and reflects the extent of the variation in species number within each community. It is closely related to resource heterogeneity or the niche processes [27]. In low precipitation areas (*i.e.*, desert steppe), strong winds may erode and move the soil from grass-dominant areas to areas occupied by shrub clamps to create fertile "islands", thus increasing the environmental heterogeneity [55]. In high precipitation areas (*i.e.*, meadow steppe), β_N diversity increased with an increasing species pool. However, since α diversity is almost saturated with respect to γ diversity (Fig. 4A), β_N diversity does not increase at the same rate as γ diversity, resulting in a slight reduction in the proportion of β_N in species-rich grasslands in high precipitation areas (Fig. 2B).

On the contrary, the strong increase of β_R with increasing precipitation (Fig. 4G) indicates a strong species-replacement effect in high precipitation areas [27,28]. Communities with more weak-diffusing species, such as those recorded as occasional species, tend to form high β_R diversity [27,35]. Considering the increase of occasional species diversity with precipitation (Fig. 4D) and with γ diversity (Fig. 4C), β_R is much more important in the diversity of species-rich than species-poor grassland communities. In summary, the dominant mechanism for β diversity maintenance changes from niche processes to diffusion processes across steppe community types from species-poor desert steppe to species-rich meadow steppe along precipitation gradient.

Management implications

Changes in regional precipitation patterns under global climate change will undoubtedly affect species diversity and ecosystem function and stability [56,57], and the effects will be especially profound in arid and semiarid grassland regions [24,58]. Our results have important implications for understanding the potential effects of climate change on the semiarid grassland, and for developing biodiversity conservation strategies. First, the generally greater contribution of β diversity than α diversity to γ diversity (Fig. 2) implies it is better to construct several small reserves than a single large reserve for protecting species diversity of a steppe community in the region. The much greater contribution of β diversity in species-rich (meadow steppe) than in species-poor (desert steppe) communities suggests it more useful to apply multiple small reserves for protecting the meadow steppe grassland in relatively humid areas. Second, since species diversity provides a mechanism for maintaining ecosystem stability through compensatory interactions among species [59], the greater sensitivity of species diversity (mainly α diversity) to precipitation in desert steppe than in meadow steppe (Fig. 3B) suggests that more efforts are urgently needed to understand the effects of climate change on desert steppe for developing adaptive ecosystem management strategies. Our study has focused on species diversity composition changes along climatic gradients by excluding the grassland sites under heavy animal grazing. Human activities, mainly through animal grazing, have profound impacts on grassland species diversity [60,61,62]. The effects of animal grazing on species diversity along precipitation gradients in the Inner Mongolia grassland also need future studies.

Acknowledgments

The authors are grateful to Dr Dafeng Hui, Dr Rafael B. de Andrade and an anonymous reviewer for the comments and suggestions on the early version of this manuscript. We also thank Dr Adrian Walcroft for improving the English readability of this manuscript.

Author Contributions

Conceived and designed the experiments: QZ JN XH. Performed the experiments: QZ JN YZ LZ YD WM SK. Analyzed the data: QZ FL JN. Wrote the paper: QZ FL XL.

References

1. Whittaker RH (1960) Vegetation of the siskiyou mountains, oregon and california. Ecological Monographs 30: 280–338.
2. Jurasinski G, Retzer V, Beierkuhnlein C (2009) Inventory, differentiation, and proportional diversity: a consistent terminology for quantifying species diversity. Oecologia 159: 15–26. doi:10.1007/s00442-008-1190-z.
3. Sfenthourakis S, Panitsa M (2012) From plots to islands: species diversity at different scales. Journal of Biogeography 39: 750–759. doi:10.1111/j.1365-2699.2011.02639.x.
4. Meynard CN, Devictor V, Mouillot D, Thuiller W, Jiguet F, et al. (2011) Beyond taxonomic diversity patterns: how do alpha, beta and gamma components of bird functional and phylogenetic diversity respond to environmental gradients across France? Global Ecology And Biogeography 20: 893–903. doi:10.1111/j.1466-8238.2010.00647.x.
5. Leprieur F, Albouy C, De Bortoli J, Cowman PF, Bellwood DR, et al. (2012) Quantifying Phylogenetic Beta Diversity: Distinguishing between 'True' Turnover of Lineages and Phylogenetic Diversity Gradients. Plos One 7. doi:e4276010.1371/journal.pone.0042760.
6. Anderson MJ, Crist TO, Chase JM, Vellend M, Inouye BD, et al. (2011) Navigating the multiple meanings of beta diversity: a roadmap for the practicing ecologist. Ecology Letters 14: 19–28. doi:10.1111/j.1461-0248.2010.01552.x.
7. de Juan S, Thrush SF, Hewitt JE (2013) Counting on beta-Diversity to Safeguard the Resilience of Estuaries. Plos One 8. doi:e6557510.1371/journal.pone.0065575.
8. Wang Z, Brown JH, Tang Z, Fang J (2009) Temperature dependence, spatial scale, and tree species diversity in eastern Asia and North America. Proceedings of the National Academy of Sciences 106: 13388. doi:10.1073/pnas.0905030106.
9. Melo AS, Rangel TFLVB, Diniz-Filho JAF (2009) Environmental drivers of beta-diversity patterns in New-World birds and mammals. Ecography 32: 226–236. doi:10.1111/j.1600-0587.2008.05502.x.
10. Patterson BD, Atmar W (1986) Nested subsets and the structure of insular mammalian faunas and archipelagos. Biological Journal of the Linnean Society 28: 65–82.
11. Wiersma YF, Urban DL (2005) Beta diversity and nature reserve system design in the Yukon, Canada. Conservation Biology 19: 1262–1272.

12. Hillebrand H (2004) On the generality of the latitudinal diversity gradient. American Naturalist 163: 192–211. doi:10.1086/381004.
13. Novotny V, Miller SE, Hulcr J, Drew RAI, Basset Y, et al. (2007) Low beta diversity of herbivorous insects in tropical forests. Nature 448: 692–U698. doi:10.1038/nature06021.
14. Koleff P, Lennon JJ, Gaston KJ (2003) Are there latitudinal gradients in species turnover? Global Ecology And Biogeography 12: 483–498. doi:10.1046/j.1466-822X.2003.00056.x.
15. Qian H, Ricklefs RE (2007) A latitudinal gradient in large-scale beta diversity for vascular plants in North America. Ecology Letters 10: 737–744. doi:10.1111/j.1461-0248.2007.01066.x.
16. Dyer LA, Singer MS, Lill JT, Stireman JO, Gentry GL, et al. (2007) Host specificity of Lepidoptera in tropical and temperate forests. Nature 448: 696–U699. doi:10.1038/nature05884.
17. McKnight MW, White PS, McDonald RI, Lamoreux JF, Sechrest W, et al. (2007) Putting beta-diversity on the map: Broad-scale congruence and coincidence in the extremes. Plos Biology 5: 2424–2432. doi:10.1371/journal.-pbio.0050272.
18. Tang Z, Fang J, Chi X, Feng J, Liu Y, et al. (2012) Patterns of plant beta-diversity along elevational and latitudinal gradients in mountain forests of China. Ecography 35: 1083–1091. doi:10.1111/j.1600-0587.2012.06882.x.
19. Kluge J, Kessler M, Dunn RR (2006) What drives elevational patterns of diversity? A test of geometric constraints, climate and species pool effects for pteridophytes on an elevational gradient in Costa Rica. Global Ecology And Biogeography 15: 358–371. doi:10.1111/j.1466-822x.2006.00223.x.
20. de Andrade RB, Barlow J, Louzada J, Vaz-de-Mello FZ, Souza M, et al. (2011) Quantifying Responses of Dung Beetles to Fire Disturbance in Tropical Forests: The Importance of Trapping Method and Seasonality. Plos One 6. doi:e2620810.1371/journal.pone.0026208.
21. Adler PB, Levine JM (2007) Contrasting relationships between precipitation and species richness in space and time. Oikos 116: 221–232. doi:10.1111/j.2006.0030-1299.15327.x.
22. Bai YF, Wu JG, Pan QM, Huang JH, Wang QB, et al. (2007) Positive linear relationship between productivity and diversity: evidence from the Eurasian Steppe. Journal of Applied Ecology 44: 1023–1034. doi:10.1111/j.1365-2664.2007.01351.x.
23. Volder A, Briske DD, Tjoelker MG (2013) Climate warming and precipitation redistribution modify tree-grass interactions and tree species establishment in a warm-temperate savanna. Global Change Biology 19: 843–857. doi:10.1111/gcb.12068.
24. Bai YF, Wu JG, Xing Q, Pan QM, Huang JH, et al. (2008) Primary production and rain use efficiency across a precipitation gradient on the Mongolia plateau. Ecology 89: 2140–2153.
25. Collins SL, Koerner SE, Plaut JA, Okie JG, Brese D, et al. (2012) Stability of tallgrass prairie during a 19-year increase in growing season precipitation. Functional Ecology 26: 1450–1459. doi:10.1111/j.1365-2435.2012.01995.x.
26. Cleland EE, Collins SL, Dickson TL, Farrer EC, Gross KL, et al. (2013) Sensitivity of grassland plant community composition to spatial vs. temporal variation in precipitation. Ecology 94: 1687–1696. doi:10.1890/12-1006.1.
27. Chiarucci A, Bacaro G, Arevalo JR, Delgado JD, Fernandez-Palacios JM (2010) Additive partitioning as a tool for investigating the flora diversity in oceanic archipelagos. Perspectives in Plant Ecology Evolution and Systematics 12: 83–91. doi:10.1016/j.ppees.2010.01.001.
28. Crist TO, Veech JA (2006) Additive partitioning of rarefaction curves and species-area relationships: unifying alpha-, beta- and gamma-diversity with sample size and habitat area. Ecology Letters 9: 923-932. doi:10.1111/j.1461-0248.2006.00941.x.
29. Legendre P, Borcard D, Peres-Neto PR (2005) Analyzing beta diversity: Partitioning the spatial variation of community composition data. Ecological Monographs 75: 435–450. doi:10.1890/05-0549.
30. Boieiro M, Carvalho JC, Cardoso P, Aguiar CAS, Rego C, et al. (2013) Spatial Factors Play a Major Role as Determinants of Endemic Ground Beetle Beta Diversity of Madeira Island Laurisilva. Plos One 8. doi:e6459110.1371/journal.pone.0064591.
31. Field R, Hawkins BA, Cornell HV, Currie DJ, Diniz-Filho JAF, et al. (2009) Spatial species-richness gradients across scales: a meta-analysis. Journal of Biogeography 36: 132–147. doi:10.1111/j.1365-2699.2008.01963.x.
32. Colwell RK, Rahbek C, Gotelli NJ (2004) The mid-domain effect and species richness patterns: what have we learned so far. The American Naturalist 163: E1–E23. doi:10.1086/382056.
33. Zobel M (1992) Plant species coexistence: the role of historical, evolutionary and ecological factors. Oikos: 314–320.
34. Gering JC, Crist TO (2002) The alpha-beta-regional relationship: providing new insights into local-regional patterns of species richness and scale dependence of diversity components. Ecology Letters 5: 433–444. doi:10.1046/j.1461-0248.2002.00335.x.
35. Green JL, Ostling A (2003) Endemics-area relationships: The influence of species dominance and spatial aggregation. Ecology 84: 3090–3097. doi:10.1890/02-3096.
36. Russell R, Wood SA, Allison G, Menge BA (2006) Scale, environment, and trophic status: The context dependency of community saturation in rocky intertidal communities. American Naturalist 167: E158–E170. doi:10.1086/504603.
37. Suttie JM, Reynolds SG, Batello C (2005) Grasslands of the world. Rome: Food and Agriculture Organization of the United Nations.
38. Ma YQ (1995–1998) Flora of Inner Mongolia. Hohhot: Inner Mongolia People's Publishing House.
39. World Conservation Monitoring Centre (1992) Global biodiversity: status of the earth's living resources: Chapman & Hall.
40. Li B (1962) Basic types and eco-geographic distribution of zonal vegetation in Inner Mongolia. Acta Scientiarum Naturalium Universitatis Neimongol 4: 42–72.
41. Li Y (1996) Ecological vicariance of steppe species on Mongolian Plateau and its indication to vegetation change under climate change. Acta Phytoecologica Sinica 22: 1–12.
42. Zhang Q, Niu J, Buyantuyev A, Zhang J, Ding Y, et al. (2011) Productivity-species richness relationship changes from unimodal to positive linear with increasing spatial scale in the Inner Mongolia steppe. Ecological Research 26: 649–658. doi:10.1007/s11284-011-0825-4.
43. Inner Mongolia-Ningxia Joint Inspection Group of Chinese Sciences of Academy (1985) Vegetation of Inner Mongolia. Beijing: Science Publishing House.
44. Niu JM (2001) Climate-based digital simulation on spatial distribution of vegetation-A case study in Inner Mongolia. Acta Ecologica Sinica 21: 1064–1071.
45. Whittaker RH (1972) Evolution and measurement of species diversity. Taxon 21: 213–251.
46. Jost L (2007) Partitioning diversity into independent alpha and beta components. Ecology 88: 2427–2439. doi:10.1890/06-1736.1.
47. Chao A, Chazdon RL, Colwell RK, Shen TJ (2005) A new statistical approach for assessing similarity of species composition with incidence and abundance data. Ecology Letters 8: 148–159. doi:10.1111/j.1461-0248.2004.00707.x.
48. Veech JA, Crist TO (2010) Diversity partitioning without statistical independence of alpha and beta. Ecology 91: 1964–U1998. doi:10.1890/08-1727.1.
49. Fang JY, Yoda K (1990) Climate and Vegetation in China III water balance and distribution of vegetation. Ecological Research 5: 9–23.
50. Huston M (1979) A general hypothesis of species diversity. American Naturalist: 81–101.
51. Wimp GM, Whitham TG (2001) Biodiversity consequences of predation and host plant hybridization on an aphid-ant mutualism. Ecology 82: 440–452. doi:10.1890/0012-9658(2001)082[0440:bcopah]2.0.co;2.
52. Case TJ, Gilpin ME (1974) Interference competition and niche theory. Proceedings of the National Academy of Sciences of the United States of America 71: 3073–3077. doi:10.1073/pnas.71.8.3073.
53. Loreau M (2000) Are communities saturated? On the relationship between alpha, beta and gamma diversity. Ecology Letters 3: 73–76. doi:10.1046/j.1461-0248.2000.00127.x.
54. Cody ML (1970) Chilean bird distribution. Ecology: 455–464.
55. Zhang P, Yang J, Zhao L, Bao S, Song B (2011) Effect of Caragana tibetica nebkhas on sand entrapment and fertile islands in steppe-desert ecotones on the Inner Mongolia Plateau, China. Plant and Soil 347: 79–90. doi:10.1007/s11104-011-0813-z.
56. Hooper DU, Chapin FS, Ewel JJ, Hector A, Inchausti P, et al. (2005) Effects of biodiversity on ecosystem functioning: A consensus of current knowledge. Ecological Monographs 75: 3–35. doi:10.1890/04-0922.
57. Turnbull LA, Levine JM, Loreau M, Hector A (2013) Coexistence, niches and biodiversity effects on ecosystem functioning. Ecology Letters 16: 116–127. doi:10.1111/ele.12056.
58. Sala OE, Parton W, Joyce L, Lauenroth W (1988) Primary production of the central grassland region of the United States. Ecology 69: 40–45.
59. Bai YF, Han XG, Wu JG, Chen ZZ, Li LH (2004) Ecosystem stability and compensatory effects in the Inner Mongolia grassland. Nature 431: 181–184. doi:10.1038/nature02850.
60. Chapin III FS, Zavaleta ES, Eviner VT, Naylor RL, Vitousek PM, et al. (2000) Consequences of changing biodiversity. Nature 405: 234–242. doi:10.1038/35012241.
61. La Sorte FA, McKinney ML, Pyšek P, Klotz S, Rapson G, et al. (2008) Distance decay of similarity among European urban floras: the impact of anthropogenic activities on β diversity. Global Ecology And Biogeography 17: 363–371. doi:10.1111/j.1466-8238.2007.00369.x.
62. Li Y, Wang W, Liu Z, Jiang S (2008) Grazing gradient versus restoration succession of Leymus chinensis (Trin.) Tzvel. grassland in Inner Mongolia. Restoration Ecology 16: 572–583.

17

DNA Barcodes and Species Distribution Models Evaluate Threats of Global Climate Changes to Genetic Diversity: A Case Study from *Nanorana parkeri* (Anura: Dicroglossidae)

Wei-wei Zhou[1], Bao-lin Zhang[1,2], Hong-man Chen[1], Jie-qiong Jin[1], Jun-xiao Yang[1], Yun-yu Wang[1], Ke Jiang[1], Robert W. Murphy[1,3], Ya-ping Zhang[2]*, Jing Che[1]*

1 State Key Laboratory of Genetic Resources and Evolution, and Yunnan Laboratory of Molecular Biology of Domestic Animals, Kunming Institute of Zoology, Chinese Academy of Sciences, Kunming, China, **2** Laboratory for Conservation and Utilization of Bio-resources, Yunnan University, Kunming, China, **3** Centre for Biodiversity and Conservation Biology, Royal Ontario Museum, Toronto, Ontario, Canada

Abstract

Anthropogenic global climate changes are one of the greatest threats to biodiversity. Distribution modeling can predict the effects of climate changes and potentially their effects on genetic diversity. DNA barcoding quickly identifies patterns of genetic diversity. As a case study, we use DNA barcodes and distribution models to predict threats under climate changes in the frog *Nanorana parkeri*, which is endemic to the Qinghai-Tibetan Plateau. Barcoding identifies major lineages W and E. Lineage W has a single origin in a refugium and Lineage E derives from three refugia. All refugia locate in river valleys and each greatly contributes to the current level of intraspecific genetic diversity. Species distribution models suggest that global climate changes will greatly influence *N. parkeri*, especially in the level of genetic diversity, because two former refugia will fail to provide suitable habitat. Our pipeline provides a novel application of DNA barcoding and has important implications for the conservation of biodiversity in southern areas of the Qinghai-Tibetan Plateau.

Editor: Dirk Steinke, Biodiversity Insitute of Ontario - University of Guelph, Canada

Funding: This work was supported by grants of the Strategic Priority Research Program (B) (XDB03030113), the Ministry of Science and Technology of China (MOST) (2014FY210200), and the "Western Light" Talents Training Program of the Chinese Academy of Sciences (CAS) to ZWW; the Key Research Program (KJZD-EW-L07) of CAS, the National Natural Science Foundation of China (31090250), MOST (2011FY120200), and the Bureau of Science and Technology of Yunnan, China (2010CI045) to JC; and the Major Innovation Program of CAS (KSCX2-EW-Z-2) to ZYP. Research also was supported by Discovery Grant 3148 from the Natural Sciences and Engineering Research Council of Canada to RWM. The funders had no role in study design, data collection and analysis, decision to publish, or preparation of the manuscript.

Competing Interests: The authors have declared that no competing interests exist.

* Email: chej@mail.kiz.ac.cn (JC); zhangyp@mail.kiz.ac.cn (YPZ)

Introduction

Climatic changes influence organisms and an understanding of how this occurs is important for conservation. More than one line of evidence documents the impact anthropogenic global climate change (GCC) exerts on organisms [1]. Explorations into how past climate changes influenced organisms may serve to predict future impacts of GCC. Genetic diversity is important in conservation because higher levels maintain the evolutionary potential of species. However, the distribution of genetic diversity is often uneven across the range of a species and many factors may contribute to this. Environmental changes during glacial-interglacial cycling in the Quaternary is one of the most important historical drivers of genetic patterns [2,3]. For example, by retaining suitable habitat over several glacial cycles, refugia hold higher levels of genetic diversity compared with recently occupied areas [2,3]. Refugia also drive genetic distinctiveness within species owing to providing long-term geographic isolation. A clear understanding a species' evolutionary history and its drivers is important for planning conservation [4].

Genetic analyses form the cornerstone of conservation planning, especially in defining objective evolutionarily significant units (ESUs) and management units (MUs) [5]. DNA barcoding [6], which uses a short, universal genetic marker (*COI* in eukaryotes) to identify matrilines and species, may serve to efficiently identify ESUs and MUs.

Species distribution models (SDMs) also provide information useful for conservation planning [7–9]. Comparisons between SDMs for current and the Last Glacial Maximum (LGM) may provide a chance to explore the impacts of past climate changes to organisms. As an extension of this application, SDMs can compare the distributions of current and future habitats. This allows assessments of the risk of local extirpation and extinction caused by future habitat degradation. Whereas both DNA barcoding and SDMs provide valuable insights, their synthesis may serve to evaluate threats of GCC to organisms.

The Qinghai-Tibetan Plateau (QTP), which covers more than 2.5 million km^2 and has an average elevation of about 4000 m above sea level [10], is the largest and highest plateau on Earth. Unlike North America and Europe, no unified ice sheet formed on the QTP during the LGM, yet its environment changed substantially. The average temperature was from 6°C to 9°C lower than today and precipitation decreased by from 30% to 70% [11].

In addition to climatic drivers, geographic features also play a role in the formation of patterns of genetic diversity. In the southern QTP, two mountains stretch from east to west and the Yarlung Zangpo River (YZR; Brahmaputra River) (Fig. S2) occurs between them. The Nianqingtanggula Mountains (NM; Nyainqentanglha Mountains) in the east and Gangdisi Mountains (GM; Kailas Range) in the west form the northern mountains and the Himalayas form the southern boundary. Several rivers flow southward across the latter. The complex geography may have offered refugia during dramatic Quaternary climatic changes. Analyses of matrilineal dispersal are likely to reveal such history. Driven by the GCC, future severe temperature changes of the QTP may exceed those of lower elevations [12]. Thus, the GCC may impose enormous impacts on organisms living in the QTP. We assess this possibility by synthesizing DNA barcoding and SDMs analyses.

Because of their relatively low mobility and physiological requirements, amphibians are sensitive to environmental changes, especially temperature and precipitation. Further, amphibians retain high-resolution signals of historical responses to environmental perturbations [13]. The environment on the QTP is harsh for amphibians; suitable environment is rare except in the eastern and southern edges. A previous study on *Rana kukunoris*, a common frog on eastern QTP, revealed a unique genetic pattern compared to other taxa in the region [14]. Comparative amphibian studies from the southern QTP will yield insights into how environmental changes affect the biota fauna of the QTP.

Nanorana parkeri, a median-sized frog belonging to the family Dicroglossidae, is endemic to southern and southeastern Tibet. It occurs at elevations ranging from 2800 to 5000 m in valleys of the YZR and north of the NM [15]. Climate changes during LGM greatly affected this area. Our study evaluates samples from across the entire range of *N. parkeri*. Using the species as model system, we test the effectiveness of a pipeline that combines DNA barcoding and SDMs to identify future threats to genetic diversity in the face of GCC.

Materials and Methods

Ethics Statement

All the species included in our study (Table S1) are not endangered or protected species according to the "Law of the People's Republic of China on the Protection of Wildlife" and "Regulations for the Implementation of the People's Republic of China on the Protection of terrestrial Wildlife" (State Council Decree [1992] No. 13). The permission for field work in four major management areas (39 localities) including Xigaze, Lhasa, Nyingchi and Qamdo was issued by the Forestry Department of Xizang [Tibet] Autonomous Region, China. Local people of the Forestry Bureaus involved and helped during the whole survey. Totally, 549 tissue samples including toe tips, muscle, livers, tadpoles and egg masses were obtained, following the Animal Use Protocols approved by the Animal Care and Ethics Committee of the Kunming Institute of Zoology, Chinese Academy of Sciences. For egg masses, we separated 5–10 eggs from egg mass and stored them in 95% alcohol after removing egg jelly. For adult, within each locality only five individuals were euthanized using clove oil firstly. Following euthanization, tissues dissected from adult specimens were preserved in 95% ethanol. More other adults were just cut two toe tips and then released. Tadpoles were also euthanized using clove oil firstly, then stored in 95% alcohol after removing gut. Table S1 lists our samplings information, including species name, locality, GPS coordinates, and accession nos. in Genbank etc.

Population sampling and laboratory techniques

Our samplings covered the entire documented distribution range of *N. parkeri* (Fig. 1, Table S1). One individual of *N. pleskei*, *N. ventripunctata* and *N. liebigii* was used as an outgroup representative [16,17].

We extracted genomic DNA using standard phenol–chloroform extraction protocol [18]. Partial sequences of cytochrome c oxidase subunit I (*COI*) were sequenced for all individuals using universal primers [19]. PCR products were purified and used as the template DNA for cycle sequencing reactions performed using BigDye Terminator Cycle Sequencing Kit (v.2.0, Applied Biosystems, Foster City, USA) and an ABI PRISM 3730 DNA Analyzer.

Sequence alignment and phylogenetic analyses

Nucleotide sequences were checked by eye using LASERGENE 7.0 and aligned using CLUSTALX 1.81 [20] with default parameters. Subsequently, the aligned sequences were checked and optimized in MEGA 4.0 [21]. Identical haplotypes for mtDNA were collapsed using DNASP 5.10 [22]. The overall value of nucleotide diversity (π) and haplotype diversity (H) were also estimated using DNASP.

Phylogenetic analyses of the *COI* data were conducted using Bayesian inference (BI), maximum likelihood (ML) and maximum parsimony (MP). BI analyses were performed using MrBayes 3.1.2 [23]. We tested three different partition strategies based on codon positions (no partition; 1+2, 3; and 1, 2, 3). The best strategy was chosen based on the Bayes factor test [24] (Table 1), as it represented a robust method testing partitioning strategies [25]. Nucleotide substitution models were selected for each data partition using the Akaike information criterion in MrModeltest v2.3 [26]. BI analyses for each partition strategy were run 3 million generations while sampling trees every 1000 generations. The first 50% of the sampled trees were discarded as burn-in. The final analyses employing the best partition strategy were run for 10 million generations. AWTY [27] was used to confirm satisfactory convergence of the topological split frequencies. MP analyses were implemented using PAUP* 4.0b10 [28]. Heuristic searches with TBR were executed for 1000 random pseudoreplicates with all characters treated as unordered and equally weighted. Bootstrap analyses were conducted using 1000 replicates to assess nodal reliabilities. ML analyses were conducted using RaxML 7.0.4 [29] and the GTR+I+G model was implemented for each data partition.

We built a median-joining network (MJN) using NETWORK 4.5 [30] to visualize the frequencies of the haplotypes. To remove excessive links and median vectors, we used the MP option [31].

Divergence time estimation

Different demographic and molecular clock models were compared using path sampling and stepping-stone sampling (Table 2) [32]. Employing the best models, time to most recent common ancestor was estimated using BEAST 1.7.5 [33]. Due to the absence of a reliable fossil record and an established substitution rate of *COI* for *N. parkeri*, we used the divergence time between *N. parkeri* and *N. pleskei* (8.9±2.7 Ma) [16] as a secondary calibration point. Model selection and partition scheme were the same as used in the BI analyses. The final analyses involved two independent runs of 30 million generations each, while sampling trees every 1000th generation. Burn-in and convergence of the chains were determined with Tracer 1.5

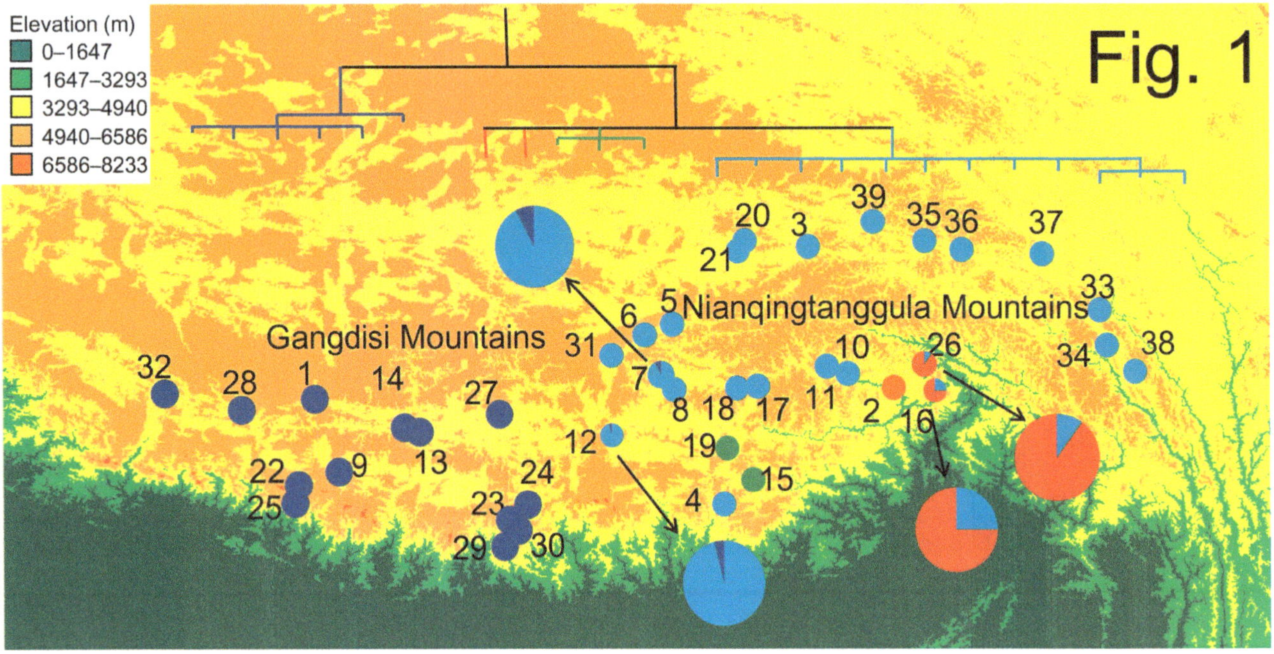

Figure 1. Sampling sites and the maternal genealogy based on mtDNA sequence data. Site numbers refer to Table S1. Colors of the matrilines match those of samplings localities on the map. Four localities containing two matrilines each are show in larger pie figures.

[34]. The measures of effective sample sizes were used to determine the Bayesian statistical significance of each parameter.

Population structure and demographic analyses

We explored population structure and genetic diversity landscape using SPADS 1.0 [35]. Groups of populations were defined as for SAMOVA [36]. We explored values of K (number of groups) ranging from 2 to 10 with 100 simulated annealing processes. The optimum value of K was identified by exploring the behavior of the indices F_{CT} and F_{SC}. Spatial patterns of genetic diversity based on allelic richness (Ar) and π across the landscape were explored using the GDivPAL function in SPADS.

To test for the influences of mountains and rivers on the genetic structure of *N. parkeri*, we measured the population structure by three independent analyses of molecular variance (AMOVA) [37]: populations north and south of the YZR; populations north of NM, west of the boundary of GM and NM, and the remaining ones; and four groups according the results of the SAMOVA. Analyses were performed using Arlequin 3.5 [38] and significance was assessed by 10000 permutations.

We investigated past changes of effective population size using the neutral test and mismatch distributions. Tajima's *D* [39] and Fu's *Fs* statistics [40] with 10000 coalescent simulations were calculated using Arlequin. Pairwise mismatch distributions [41] were calculated for each group in Arlequin. The expected distribution under a model of sudden demographical expansion was generated using 10000 permutations. The significance of deviations from this model was tested using the sum of squared deviation (SSD) and raggedness index (Rag). All analyses were performed for each lineage separately because population subdivisions could have masked the effects of expansions.

Species distribution modeling

We inferred the potential geographic range of *N. parkeri* using the maximum entropy model implemented in MAXENT 3.3.3 [42,43]. Environmental variables from the WorldClim database with resolutions of 2.5 arc-minutes [44] were downloaded as environmental layers. Because of the controversy about whether correlated variables should be removed or not, we used all of the 19 bioclimatic layers [45]. All layers were cropped to span from 83°E to 99°E and from 26°N to 33°N.

Random null distributions were built to test for the significance of our SDM. For this test, we built a new SDM using 39 random points (as same number as our sampling localities). This process was repeated 100 times and the areas under the curves (AUCs) were used as null distributions [46].

Assuming niche conservatism over time [47–49], we predicted the former distribution of *N. parkeri* by projecting our model on

Table 1. Bayes factor of each partition strategy.

Partition Strategy	Ln(BF)	Standard Deviation
no partition	−1885.375	0.134
1+2, 3	−1844.969	0.14
1, 2, 3	−1885.375	0.12

2lnBF (H_1–H_0) >10 was treated as decisive support for each hypothesis.

Table 2. Results of demographic model selection.

	strick		lognormal relaxed clock	
	PS	SS	PS	SS
constant	−1553.29	−1553.31	−1553.33	−1553.34
exponential	−1556.44	−1556.53	−1556.24	−1556.27
logistic	−1557.29	−1557.36	−1555.73	−1555.77
expansion	−1554.42	−1554.45	−1555.05	−1555.07
bayesian skyline	−1544.51	−1544.56	−1545.42	−1545.46

PS: path sampling; SS: stepping-stone sampling.

LGM climatic layers. Predicted distributions during the LGM were generated by downloading both the community climate system model (CCSM) [50] and model for interdisciplinary research on climate (MIROC) [51] from the WorldClim database. To predict the influences of GCC to this species, we also projected the model to climate data of the 2080 s based on MIROC model under the A1b scenario. In this way, we predicted the distribution changes in future.

Results

Sequence information

We obtained 549 partial *COI* sequences from 39 localities of *N. parkeri*, plus one sequence each from *N. pleskei*, *N. ventripunctata* and *N. liebigii*. The fragment consisted of 539 base pairs, of which 35 positions exhibited variation and 27 were potentially parsimony informative, resulting in 23 haplotypes in *N. parkeri*. All sequences were deposited in GenBank (Table S1). The overall value of π was 0.01746±0.00055 and H was 0.678±0.016.

Genealogical analyses of *COI*

Bayes factor test showed a preference for the partition strategy of 1+2, 3 (Table 1). The best fit substitution model for the first and second codon was HKY and GTR+G model was the best fit for third codon.

Matrilineal genealogies obtained from BI, ML and MP analyses were nearly identical (Fig. 1 & Fig. S1) and all methods recovered lineages East (E) and West (W). The boundary of GM and NM separated these two lineages. Lineage W was comprised of 6 haplotypes, which had no clear structure. Lineage E was comprised 17 haplotypes of which 12 formed highly supported sublineage E1. The MJN depicted patterns similar to those of the gene tree (Fig. 2). Haplotypes H1 and H11, which were more common than the other haplotypes, occupied central positions.

Divergence time estimates

Results of model comparisons were shown in Table 2. Both path sampling and stepping-stone sampling suggested that the BSP demographic model under strict clock outperformed the alternative models. The average divergence time estimations were shown in Fig. 3. Lineages E and W diverged about 1.4 Ma (95% CI: 0.6–2.4 Ma). Lineage E diverged about 0.41 Ma (95% CI: 0.15–0.81 Ma) and the radiation of Lineage W happened about 0.15 Ma (95% CI: 0.05–0.32 Ma).

Genetic structure and demographic history

Results of population clustering were illustrated in Figure 4. F_{CT} plateaued when K was 4 (Table S2). Samples from Lineage W grouped together (Fig. 4, dark blue). Lineage E contained 3 groups, of which populations in the southern edges of the QTP formed two small groups (localities 2, 16 and 26; and locality 15 and 19); the remaining localities comprised the third group (Fig. 4, light blue). Spatial patterns of genetic diversity based on Ar and π were shown in Figure 5 and listed in Table S3. Several peaks indicated high levels of genetic diversity across the distribution of *N. parkeri*. This was congruent with the pattern of lineage divergence.

When forming two groups based on the YZR, among-group diversity accounted for 16.04% of the overall variation and among-populations within groups accounted for 80.76%. Dividing populations according to the mountains, among-group diversity accounted for 87.90% of the overall variation and among-populations within

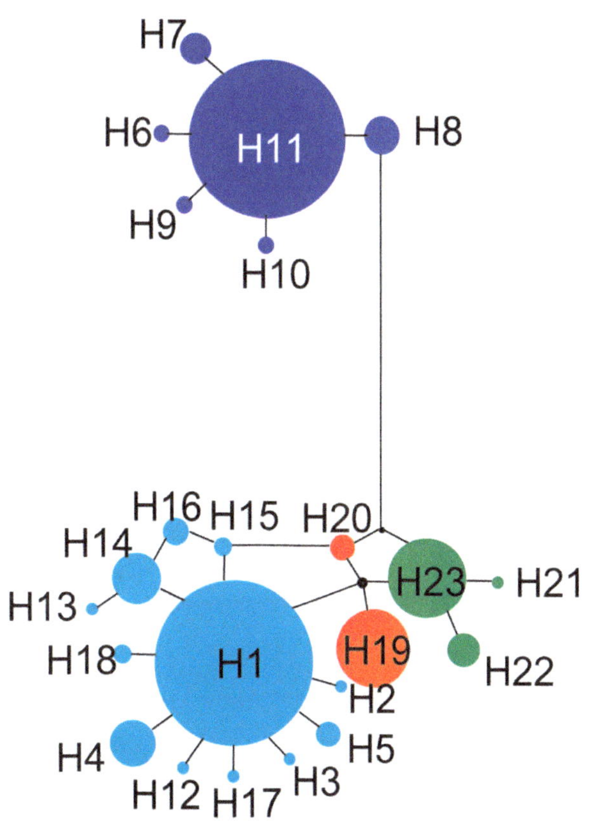

Figure 2. The median-joining network of *COI* haplotypes. Colors correspond to maternal lineages in Figure 1.

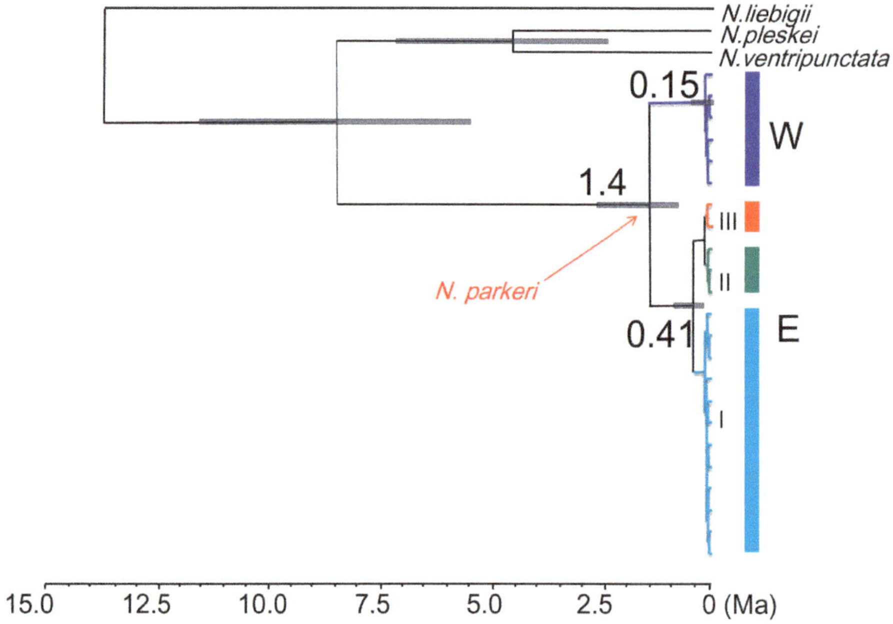

Figure 3. Estimates of divergence times obtained using BEAST 1.7.5.

groups accounted for 9.67%. Finally, after dividing populations based on results of the SAMOVA, among-group diversity accounted for 96.88% of the overall variation and among-populations within groups accounted for 0.97% (Table 3).

For Lineage W, the values of Tajima's *D* and Fu's *Fs* were significantly negative (Table 4), which indicated population expansions. The hypothesis of sudden expansion was not rejected by mismatch distribution analyses (Fig. 6) as the SSD and Rag were insignificant ($P_{SSD} = 0.381$ and $P_{RAG} = 0.658$). In sublineage E1, significantly negative values of both Tajima's *D* and Fu's *Fs* supported population expansions. The mismatch distribution analyses did not reject the sudden expansion model ($P_{SSD} = 0.545$ and $P_{RAG} = 0.609$). In sublineages E2, Tajima's *D* and Fu's *Fs* were negative, but insignificant. In sublineage E3, Tajima's *D* was negative but Fu's *Fs* was positive, yet all values were not significant. The null hypothesis of sudden expansion model was not rejected by mismatch distribution analyses.

Species distribution modeling

The AUCs from random data ranged from 0.605 to 0.800 (0.733±0.033; Table S4). For the SDMs of *N. parkeri*, the AUC of our data was 0.904 and this was significant better than that of a random model ($P<0.001$).

The MIROC model predicted that the distribution of *N. parkeri* at the LGM was much smaller than that of today (Fig. 7). The results based on CCSM model suggested a similar pattern, but with larger areas than those from MIROC model. Both models suggested range fluctuations after the LGM in northern, high altitude areas. The boundary area between GM and NM, where lineages W and E divided, has been unsuitable for *N. parkeri* since the LGM. The predicted distribution under GCC assuming the A1b scenario was shown in Figure 7. *Nanorana parkeri* is predicted to have a larger distribution in the northwestern QTP. However, the species' distribution will contract in the southeastern QTP.

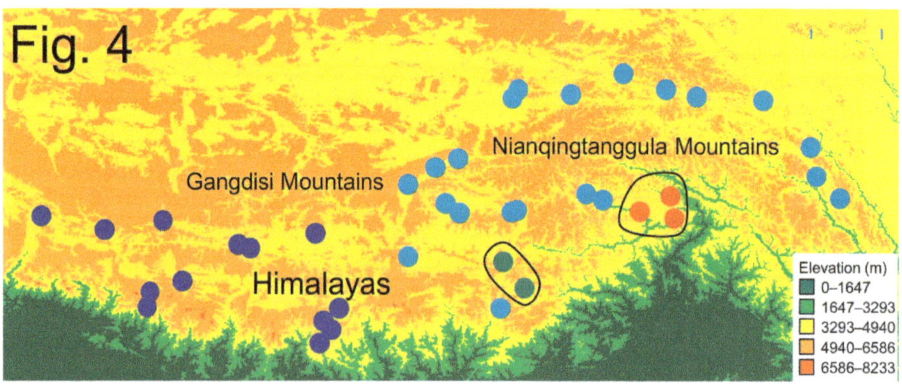

Figure 4. Results of SAMOVA. Different colors represent different groups. A solid line encircles two groups of Lineage E from microrefugia.

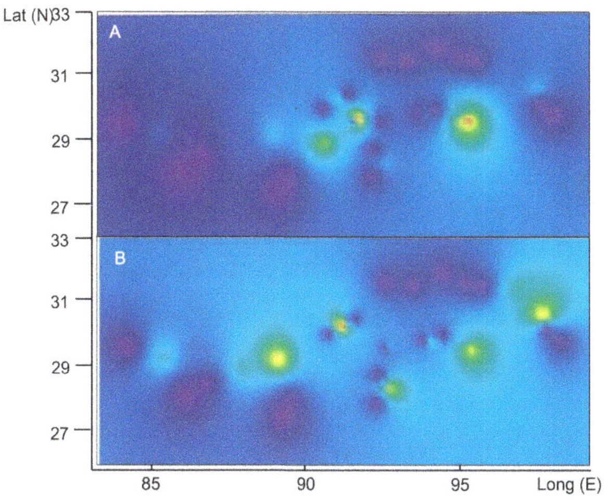

Figure 5. Genetic diversity across the range of *Nanorana parkeri*. X axis represents longitude and the Y axis latitude. A: π, pattern of nucleotide diversity; B: Ar, pattern of allelic richness.

Discussion

Lineage divergence and ESUs

ESUs, which can be determined by genetic differentiation at neutral markers caused by isolation, are important when considering conservation actions. Our study builds on the framework for delineating ESUs of *N. parkeri* by reconstructing the patterns of lineage divergence and its drivers. Our analyses suggest the recognition of ESUs that correspond to lineages W, E1, E2 and E3. Our research demonstrates that *COI* can quickly delineate ESUs.

The gene trees and MJN depict a clear east-west split for *N. parkeri*. SDMs suggest that unsuitable habitat between the GM and NM, both now and during the LGM, drives this pattern. The boundary is the demarcation line of two QTP climatic zones that are largely congruent with the 400 mm precipitation line. East of this line, the region is semi-humid, and to the west semi-arid [52]. Similar pattern was also found in *Hippophae tibetana* [53,54].

Geographic features, such as rivers or mountains, appear to contribute little to the patterns of genetic divergence. The YZR, which would impose a north-south split, does not drive population divergence. The AMOVA explains only 16.04% of the genetic variation in this scenario. Elevations of NM range from 5000 m to 6000 m, but no substantial genetic divergence distinguishes frogs in south and north of NM. Lineages E and W are separated by the boundary of GM and NM. However, habitat barriers and environmental differences more likely generate this pattern than geographical features. Two observations support this hypothesis. First, populations of lineages W and E both occur south of the GM and NM. Second, *N. parkeri* occurs along the YZR valley, which flows across the boundary and connects lineages W and E.

The pattern of genetic divergence within each major lineage differs. The 6 haplotypes in Lineage W do not yield a clear pattern (Fig. 1). The network depicts a star-like haplogroup (Fig. 2). SAMOVA also suggests populations from W comprise a single group. In contrast, Lineage E differs in having three groups (Figs. 1, 2). Results from SAMOVA present the same pattern. Most northern populations form a group as do frogs from localities 15, 19 (E2) and localities 2, 16, 26 (E3). The difference in patterns between E and W may owe to topography. The intricate topography of the East Himalayas, which is a global biodiversity hotspot, provides suitable habitats and generates barriers that limit dispersal for Lineage E. Other species exhibit a similar pattern; populations in the fringes or southern slope of Himalayas are genetically different from those on the QTP [54,55]. Expansion and contraction of distribution ranges caused by climate changes provided opportunity for populations divergence in QTP [54–57].

The genetic pattern within *N. parkeri* suggests recognition of four ESUs, W, E1, E2 and E3. Certainly gene flow occurs within and between lineages E and W, as lineages or sublineages mix at localities 8, 12, 16 and 26. However, matrilineal uniqueness occurs in populations far from the boundary.

Glacial refugia and genetic diversity of *N. parkeri*

Refugia in the QTP prevented extinction during the LGM. In doing so, they conserved high levels of genetic diversity. In contrast, genetic diversity in recently occupied areas is much lower because of founder effects [2,3]. The identification of refugia is an important part of conservation because these areas preserve genetic diversity. Synthesizing DNA barcoding and SDMs, we quickly and effectively identify refugia for *N. parkeri*. Populations in refugia retain higher levels of diversity than population in newly occupied places (Fig. 5 and Table S3).

The multiple refugia hypothesis corresponds with patterns of mtDNA divergence. The matrilineal genealogy, MJN and SAMOVA analyses indicate the absence of genetic structure within Lineage W. Thus, Lineage W likely originates from a single refugium. The SDMs indicate the presence of suitable habitat in the river valley near locality 27 (Fig. 7) and the area retains a much higher level of genetic diversity (Fig. 5) than other sites. The neutral test and mismatch distribution analyses clearly detect a population expansion in Lineage W.

Our analyses identify several refugia in Lineage E. Sublineage E1 may originate from a northern refugium near localities 7 and 8, which occur in the valley at the confluence of the Lhasa River and the YZR (Fig. 7). During the LGM the suitable environment harbored a high level of genetic diversity. The absence of genealogical structure in sublineage E1 is congruent with a sudden population expansion, as are significantly negative values of Tajima's *D* and Fu's *Fs*, and the mismatch distribution analyses. Further, our analyses identify two microrefugia in the river valleys along southern edges of the QTP. Localities 2, 16 and 26, which

Table 3. Results of AMOVA.

Group compositions	Among group	Among populations	Within populations	ΦSC	ΦST	ΦCT
By River	16.04%	80.76%	3.20%	0.96183	0.96795	0.16039
By Mountain	87.90%	9.67%	2.42%	0.79973	0.97578	0.87905
By SAMOVA	96.88%	0.97%	2.16%	0.30944	0.97844	0.96877

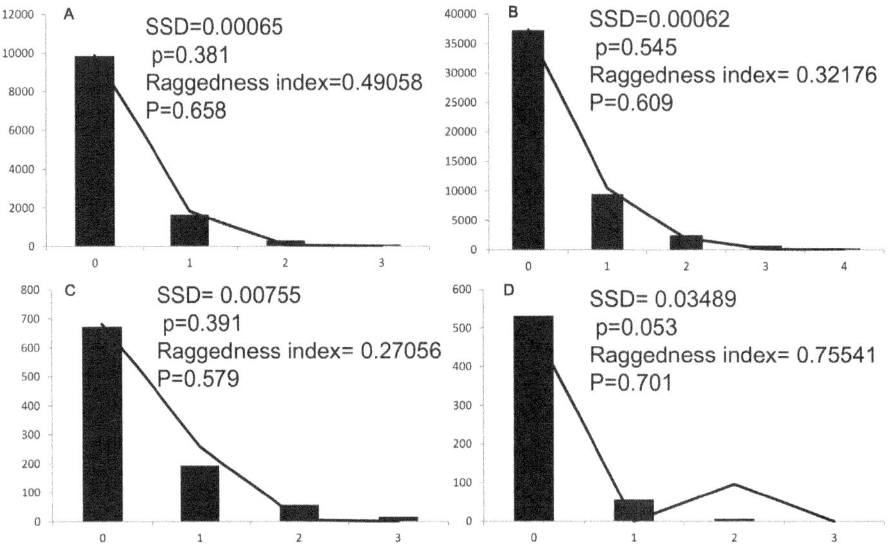

Figure 6. Results of the mismatch distribution for *Nanorana parkeri*. Bars indicate simulated mismatch distributions and lines denote the observed mismatch distributions. A: lineage W; B: sublineage E1; C: sublineage E2; D: sublineage E3.

are in the YZR Valley near Nyingchi and Medog, seemingly constitute a microrefugium. The other microrefugium consists of localities 15 and 19, which are also in a river valley. This river flows southwards across the Himalayas. The lower elevations of the river valleys make them less susceptible to the influences montane glaciers. River valleys in the Himalayas, especially those connected with southern slopes, appear to offer microrefugia.

Populations from sampling localities north of the NM also retain a high level of genetic diversity (Fig. 5). Private haplotypes H4 (locality 38), H2 and H3 (locality 33), and H5 (locality 37) occur here (Table S1). The SDMs suggest that suitable habitat existed in northern areas of NM. This area appears to harbor another microrefugium. However, these private haplotypes fail to cluster together and the most common haplotype in the area is H1 (Fig. 2 & Table S1). Thus, we cannot reject the hypothesis that these private haplotypes originated during population expansions after the LGM.

Potential threats of GCC to *N. parkeri*

Our study suggests that the combination of DNA barcoding and SDMs can detect threats of GCC to the survival of species. This pipeline facilitates conservation. SDMs predict that suitable habitat for *N. parkeri* may experience great shifts in the near future. Whereas the Northwest QTP will offer suitable habitat for *N. parkeri*, the Southeast QTP will become unsuitable. Although suitable habitat will experience an overall expansion, populations in the Southeast QTP may experience sharp decreases in population size or become extirpated. Given that amphibians have poor dispersal abilities, the latter scenario may be an unfortunate consequence of GCC.

Our pipeline suggests that many populations of Lineage E may suffer from developing unsuitable habitats. The microrefugium near Nyingchi and Medog will become unsuitable under this prediction (Fig. 7). Fortunately, suitable habitat may persist in the two other refugia. Thus, genetic variation will decrease yet all may not be lost. For Lineage W, suitable habitat will disappear near the refugium at locality 27. Loss of this refugium will greatly decrease the amount of genetic diversity.

Our analyses suggest that genetic diversity of *N. parkeri* may greatly decrease. Although this facilitates effective conservation planning, we urge caution. Our SDMs for the future assume A1b scenarios, which involve rapid global economic growth with a balance of fossil and non-fossil energy sources. Each of possible scenarios yields different predictions about the future climate. All scenarios require constant adjustment according to global economic conditions. The complex landscape in the southern QTP may supply suitable microhabitats for *N. parkeri* which cannot be detected by our SDMs. Accordingly, we suggest the urgent development of an effective monitoring program, especially for populations in refugia that may lose suitable habitats.

Conclusion

DNA barcoding detects the genetic structure of *N. parkeri* and serves to define ESUs. Our analyses recover major lineages E and

Table 4. Statistics of neutrality tests for each lineage and sublineage.

	Tajima's D	P value	Fu's Fs	P value
W	−1.62115	0.0044	−6.41652	0.0005
E1	−1.80395	0.0033	−14.10103	0
E2	−0.69425	0.2176	−0.77698	0.1858
E3	−0.65715	0.2279	0.92064	0.5177

Fig. 7

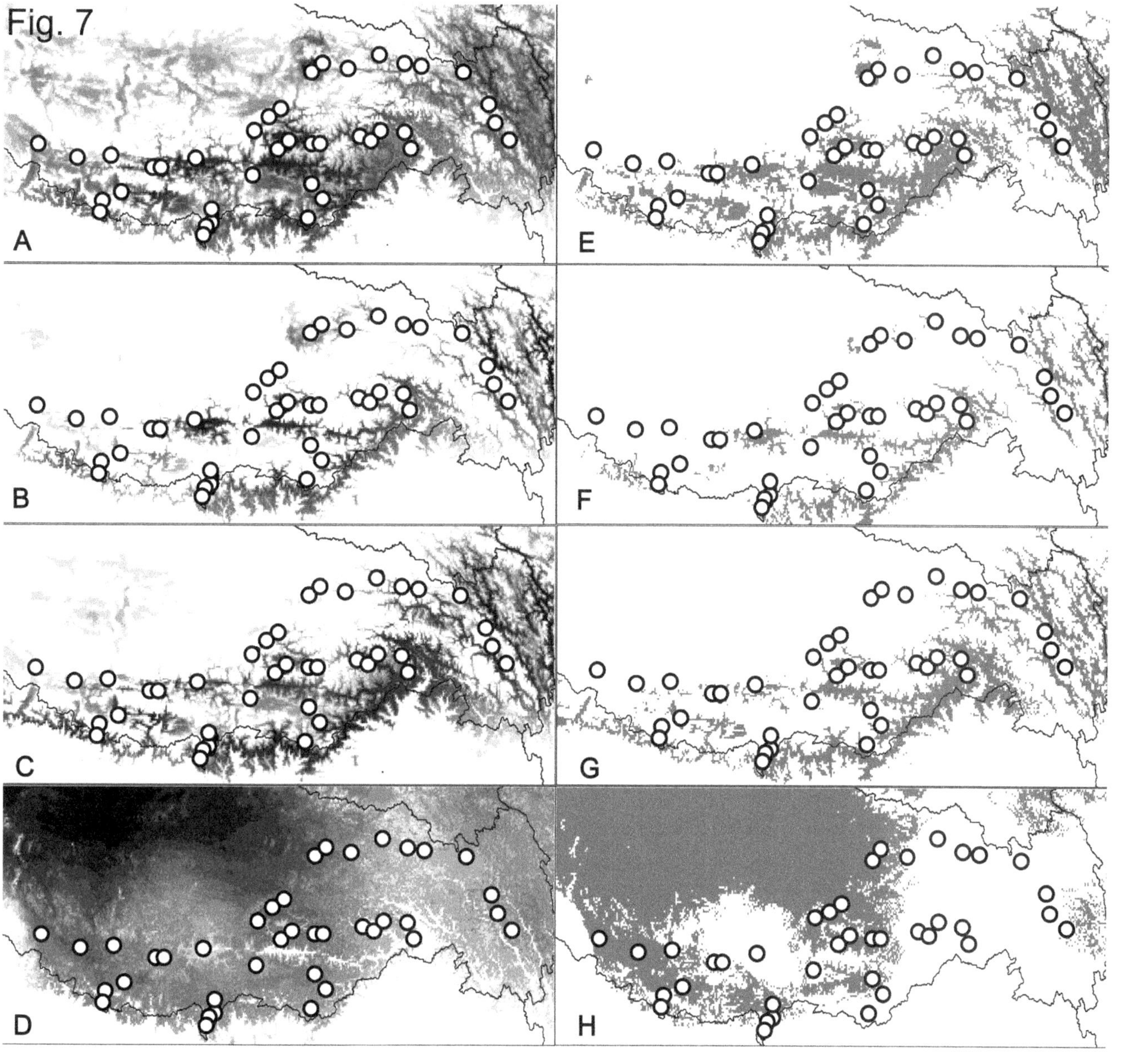

Figure 7. The predicted distribution of *Nanorana parkeri*. A: predicted distribution based on current data; B: distribution during the LGM based on MIROC model; C: distribution during the LGM based on CCSM model; D: the predicted distribution in the 2080 s under GCC; E–H: predicted distributions above the threshold for models A–D, respectively.

W, which separate at the boundary of the GM and NM. Habitat barriers and environmental differences, combined with geographic features, are the drivers of genetic divergences. Lineage E contains three parts: E1, E2 and E3. Thus our analyses define four ESUs that correspond to matrilines W, E1, E2 and E3. Genetic and environmental data identify four historical refugia, each of which corresponds to a lineage or sublineage. Lineage W originates from a refugium in river valley near locality 27. Most populations of Lineage E originate from a refugium in the river valley near Lhasa. Two microrefugia occur in river valleys along the Himalayas. Our analyses detect a population mixture after the LGM in some localities but these occurrences do not influence our designations of ESUs. Our study highlights the importance of valleys along the Himalayas for biodiversity conservation. Based on climate models under GCC, we predict the potential distribution changes and threats to genetic diversity of *N. parkeri*. Our pipeline, which combines DNA barcoding and SDMs, is an effective approach in conservation.

Supporting Information

Figure S1 Matrilineal genealogy of *Nanorana parkeri* based on BI analyses of *COI* sequence data. Bootstrap proportions ≥70% and Bayesian posterior probabilities ≥95% were treated as strongly supported (▼) and bootstrap proportions ≥70% and Bayesian posterior probabilities ≥90% were treated as being moderately supported. Bootstrap proportions<70% and Bayesian posterior probabilities <90% were treated as being unsupported (*).

Figure S2 River systems in the southern QTP.

Table S1 Detailed information for specimens included in this study.

Table S2 Values of F_{CT}, F_{ST} and F_{SC} based on population groups suggested by SAMOVA.

Table S3 Values of π and Ar for populations from each locality.

Table S4 Detailed results of AUC for random and real data.

Acknowledgments

Many thanks are owed to help from Mian Hou, Peng Guo, Kai Wang, Duan You and Yong-biao Xu in the field. We thank the local conservation departments for their help in field work. We also thank for the technical support from the laboratory of YPZ. DNA barcoding experiments were partly performed in the Southern China DNA Barcoding Center (SCDBC) in KIZ, CAS.

Author Contributions

Conceived and designed the experiments: WWZ JC YPZ. Performed the experiments: JQJ HMC BLZ. Analyzed the data: WWZ BLZ. Contributed reagents/materials/analysis tools: KJ JXY YYW. Wrote the paper: WWZ JC RWM.

References

1. Rosenzweig C, Karoly D, Vicarelli M, Neofotis P, Wu Q, et al. (2008) Attributing physical and biological impacts to anthropogenic climate change. Nature 453: 353–357.
2. Hewitt GM (2000) The genetic legacy of the Quaternary ice ages. Nature 405: 907–913.
3. Hewitt GM (2004) Genetic consequences of climatic oscillations in the Quaternary. Phil Trans R Soc B 359: 183–195.
4. Crandall KA, Bininda-Emonds OR, Mace GM, Wayne RK (2000) Considering evolutionary processes in conservation biology. Trends Ecol Evol 15: 290–295.
5. Moritz C (1994) Defining 'evolutionarily significant units' for conservation. Trends Ecol Evol 9: 373–375.
6. Hebert PD, Cywinska A, Ball SL (2003) Biological identifications through DNA barcodes. Proc R Soc Lond B Biol Sci 270: 313–321.
7. Guisan A, Thuiller W (2005) Predicting species distribution: offering more than simple habitat models. Ecol Lett 8: 993–1009.
8. Thomas CD, Cameron A, Green RE, Bakkenes M, Beaumont LJ, et al. (2004) Extinction risk from climate change. Nature 427: 145–148.
9. Guisan A, Tingley R, Baumgartner JB, Naujokaitis-Lewis I, Sutcliffe PR, et al. (2013) Predicting species distributions for conservation decisions. Ecol Lett 16: 1424–1435.
10. Zheng D (1996) The system of physico-geographical regions of the Qinghai-Xizang (Tibet) Plateau. Science in China (series D) 39: 410–417.
11. Zhao J, Shi Y, Wang J (2011) Comparison between Quaternary Glaciations in China and the Marine Oxygen Isotope Stage (MIS): An Improved Schema. Acta Geographica Sinica 66: 867–884.
12. Yao T, Liu X, Wang N (2000) The question of the range of climatic change on the Qinghai-Tibetan Plateau. Chinese Science Bulletin 45: 98–106.
13. Beebee T (2005) Conservation genetics of amphibians. Heredity 95: 423–427.
14. Zhou W, Yan F, Fu J, Wu S, Murphy RW, et al. (2013) River islands, refugia and genetic structuring in the endemic brown frog *Rana kukunoris* (Anura, Ranidae) of the Qinghai-Tibetan Plateau. Mol Ecol 22: 130–142.
15. Fei L, Ye C, Jiang J, Xie F, Huang Y (2005) An Illustrated Key to Chinese Amphibians. Sichuan Publishing House of Science and Technology, Chengdu.
16. Che J, Zhou W, Hu J, Yan F, Papenfuss TJ, et al. (2010) Spiny frogs (Paini) illuminate the history of the Himalayan region and Southeast Asia. Proc Natl Acad Sci USA 107: 13765–13770.
17. Che J, Hu J, Zhou W, Murphy RW, Papenfuss TJ, et al. (2009) Phylogeny of the Asian spiny frog tribe Paini (Family Dicroglossidae) sensu Dubois. Mol Phylogenet Evol 50: 59–73.
18. Sambrook J, Russell DW (2001) Molecular cloning: a laboratory manual. Cold Spring Harbor Laboratory Press, New York.
19. Che J, Chen H, Yang J, Jin J, Jiang K, et al. (2012) Universal *COI* primers for DNA barcoding amphibians. Mol Ecol Resour 12: 247–258.
20. Thompson JD, Gibson TJ, Plewniak F, Jeanmougin F, Higgins DG (1997) The CLUSTAL_X windows interface: flexible strategies for multiple sequence alignment aided by quality analysis tools. Nucleic Acids Res 25: 4876–4882.
21. Tamura K, Dudley J, Nei M, Kumar S (2007) MEGA4: molecular evolutionary genetics analysis (MEGA) software version 4.0. Mol Biol Evol 24: 1596–1599.
22. Librado P, Rozas J (2009) DnaSP v5: a software for comprehensive analysis of DNA polymorphism data. Bioinformatics 25: 1451–1452.
23. Ronquist F, Huelsenbeck JP (2003) MrBayes 3: Bayesian phylogenetic inference under mixed models. Bioinformatics 19: 1572–1574.
24. Brandley MC, Schmitz A, Reeder TW (2005) Partitioned Bayesian Analyses, Partition Choice, and the Phylogenetic Relationships of Scincid Lizards. Syst Biol 54: 373–390.
25. Brown JM, Lemmon AR (2007) The Importance of Data Partitioning and the Utility of Bayes Factors in Bayesian Phylogenetics. Syst Biol 56: 643–655.
26. Nylander JAA (2004) MrModeltest v2. Program distributed by the author. Evolutionary Biology Centre, Uppsala University.
27. Nylander JAA, Wilgenbusch JC, Warren DL, Swofford DL (2008) AWTY (are we there yet?): a system for graphical exploration of MCMC convergence in Bayesian phylogenetics. Bioinformatics 24: 581–583.
28. Swofford DL (2003) PAUP*: Phylogenetic Analysis Using Parsimony (* and Other Methods) Version 40b10. Sunderland, MA.: Sinauer Associates.
29. Stamatakis A, Hoover P, Rougemont J (2008) A rapid bootstrap algorithm for the RAxML Web servers. Syst Biol 57: 758–771.
30. Bandelt HJ, Forster P, Rohl A (1999) Median-joining networks for inferring intraspecific phylogenies. Mol Biol Evol 16: 37–48.
31. Polzin T, Daneshmand SV (2003) On Steiner trees and minimum spanning trees in hypergraphs. Operations Research Letters 31: 12–20.
32. Baele G, Lemey P, Bedford T, Rambaut A, Suchard MA, et al. (2012) Improving the Accuracy of Demographic and Molecular Clock Model Comparison While Accommodating Phylogenetic Uncertainty. Mol Biol Evol 29: 2157–2167.
33. Drummond AJ, Suchard MA, Xie D, Rambaut A (2012) Bayesian phylogenetics with BEAUti and the BEAST 1.7. Mol Biol Evol 29: 1969–1973.
34. Rambaut A, Drummond A (2007) Tracer v1.4. Available from http://beast.bio.ed.ac.uk/Tracer.
35. Dellicour S, Mardulyn P (2013) SPADS 1.0: a toolbox to perform spatial analyses on DNA sequence data sets. Mol Ecol Resour: in press.
36. Dupanloup I, Schneider S, Excoffier L (2002) A simulated annealing approach to define the genetic structure of populations. Mol Ecol 11: 2571–2581.
37. Excoffier L, Smouse PE, Quattro JM (1992) Analysis of molecular variance inferred from metric distances among DNA haplotypes: application to human mitochondrial DNA restriction data. Genetics 131: 479–491.
38. Excoffier L, Lischer H (2010) Arlequin suite ver 3.5: a new series of programs to perform population genetics analyses under Linux and Windows. Mol Ecol Resour 10: 564–567.
39. Tajima F (1989) Statistical method for testing the neutral mutation hypothesis by DNA polymorphism. Genetics 123: 585–595.
40. Fu Y (1997) Statistical tests of neutrality of mutations against population growth, hitchhiking and background selection. Genetics 147: 915–925.
41. Rogers AR, Harpending H (1992) Population growth makes waves in the distribution of pairwise genetic differences. Mol Biol Evol 9: 552–569.
42. Phillips SJ, Anderson RP, Schapire RE (2006) Maximum entropy modeling of species geographic distributions. Ecol Model 190: 231–259.
43. Phillips SJ, Dudík M (2008) Modeling of species distributions with Maxent: new extensions and a comprehensive evaluation. Ecography 31: 161–175.
44. Hijmans RJ, Cameron SE, Parra JL, Jones PG, Jarvis A (2005) Very high resolution interpolated climate surfaces for global land areas. Int J Climatol 25: 1965–1978.
45. Merow C, Smith MJ, Silander JA (2013) A practical guide to MaxEnt for modeling species' distributions: what it does, and why inputs and settings matter. Ecography 36: 1058–1069.
46. Smith AB (2013) On evaluating species distribution models with random background sites in place of absences when test presences disproportionately sample suitable habitat. Divers Distrib 19: 867–872.
47. Holt RD (2003) On the evolutionary ecology of species' ranges. Evol Ecol Res 5: 159–178.
48. Peterson A, Soberón J, Sánchez-Cordero V (1999) Conservatism of ecological niches in evolutionary time. Science 285: 1265–1267.
49. Wiens JJ, Graham CH (2005) Niche conservatism: integrating evolution, ecology, and conservation biology. Annu Rev Ecol Evol Syst 36: 519–539.
50. Collins WD, Bitz CM, Blackmon ML, Bonan GB, Bretherton CS, et al. (2006) The community climate system model version 3 (CCSM3). J Clim 19: 2122–2143.
51. Hasumi H, Emori S (2004) K-1 coupled GCM (MIROC) description. Center for Climate System Research, University of Tokyo, National Institute for

Environmental Studies, Frontier Research Center for Global Change, Tokyo, Japan.

52. Li S, Xu L, Guo Y, Qian W, Zhang G, et al. (2007) Change of annual precipitation over Qinghai-Xizang Plateau and sub-regions in recent 34 Years. Journal of Desert Research 27: 307–314.
53. Wang H, Laqiong,Sun K, Lu L, Wang Y, et al. (2010) Phylogeographic structure of *Hippophae tibetana* (Elaeagnaceae) highlights the highest microrefugia and the rapid uplift of the Qinghai-Tibetan Plateau. Mol Ecol 19: 2964–2979.
54. Jia D, Liu T, Wang L, Zhou D, Liu J (2011) Evolutionary history of an alpine shrub *Hippophae tibetana* (Elaeagnaceae): allopatric divergence and regional expansion. Biol J Linnean Soc 102: 37–50.
55. Wang L, Wu Z, Bystriakova N, Ansell SW, Xiang Q, et al. (2011) Phylogeography of the Sino-Himalayan Fern *Lepisorus clathratus* on the Roof of the World? PloS one 6: e25896.
56. Wang L, Abbott R, Zheng W, Chen P, Wang Y, et al. (2009) History and evolution of alpine plants endemic to the Qinghai-Tibetan Plateau: *Aconitum gymnandrum* (Ranunculaceae). Mol Ecol 18: 707–721.
57. Yang F, Li Y, Ding X, Wang X (2008) Extensive population expansion of *Pedicularis longiflora* (Orobanchaceae) on the Qinghai-Tibetan Plateau and its correlation with the Quaternary climate change. Mol Ecol 17: 5135–5145.

18

Conserving Plants in Gene Banks and Nature: Investigating Complementarity with *Trifolium thompsonii* Morton

Stephanie L. Greene[1]*, Theodore J. Kisha[2], Long-Xi Yu[2], Mauricio Parra-Quijano[3]

1 National Center for Genetic Resource Preservation, United States Department of Agriculture-Agricultural Research Service, Fort Collins, Colorado, United States of America, **2** Western Regional Plant Introduction Station, United States Department of Agriculture-Agricultural Research Service, Pullman, Washington, United States of America, **3** Departamento de Biologıa Vegetal, Universidad Politecnica de Madrid, Madrid, Spain

Abstract

A standard conservation strategy for plant genetic resources integrates *in situ* (on-farm or wild) and *ex situ* (gene or field bank) approaches. Gene bank managers collect *ex situ* accessions that represent a comprehensive snap shot of the genetic diversity of *in situ* populations at a given time and place. Although simple in theory, achieving complementary *in situ* and *ex situ* holdings is challenging. Using *Trifolium thompsonii* as a model insect-pollinated herbaceous perennial species, we used AFLP markers to compare genetic diversity and structure of *ex situ* accessions collected at two time periods (1995, 2004) from four locations, with their corresponding *in situ* populations sampled in 2009. Our goal was to assess the complementarity of the two approaches. We examined how gene flow, selection and genetic drift contributed to population change. Across locations, we found no difference in diversity between *ex situ* and *in situ* samples. One population showed a decline in genetic diversity over the 15 years studied. Population genetic differentiation among the four locations was significant, but weak. Association tests suggested infrequent, long distance gene flow. Selection and drift occurred, but differences due to spatial effects were three times as strong as differences attributed to temporal effects, and suggested recollection efforts could occur at intervals greater than fifteen years. An effective collecting strategy for insect pollinated herbaceous perennial species was to sample >150 plants, equalize maternal contribution, and sample along random transects with sufficient space between plants to minimize intrafamilial sampling. Quantifying genetic change between *ex situ* and *in situ* accessions allows genetic resource managers to validate *ex situ* collecting and maintenance protocols, develop appropriate recollection intervals, and provide an early detection mechanism for identifying problematic conditions that can be addressed to prevent further decline in vulnerable *in situ* populations.

Editor: Erin Espeland, USDA- ARS, United States of America

Funding: Sources of funding included USDA Agricultural Research Service National Program 301 and 215 and the International Treaty on Plant Genetic Resources for Food and Agriculture - FAO. The funders had no role in study design, data collection and analysis, decision to publish, or preparation of the manuscript.

Competing Interests: The authors have declared that no competing interests exist.

* Email: stephanie.greene@ars.usda.gov

Introduction

Since utilization of plant genetic resource (PGR) conservation is an important aim, it's long been recognized that effective strategies need to integrate *in situ* (on-farm or in the wild) and *ex situ* (gene or field bank) approaches. *In situ* conservation allows the natural trajectory of evolution to continue to mold genetic variation, and ensures sustained access to genetically adapted populations. An important strength of *ex situ* conservation is that it readily provides the diverse germplasm needed by plant breeders, natural research managers and basic and applied researchers. In contrast to *in situ* populations, *ex situ* accessions provide a genetic snapshot, reflecting a wild (or on-farm) population's adaptation to the biotic and abiotic conditions it was collected in. The static nature of *ex situ* conservation is frequently cast as a disadvantage; however, there is evidence that these snapshots conserve alleles that can be lost from *in situ* populations. For example, in a comparison of barley (*Hordeum vulgare* L.) landraces conserved *in situ* (on farm) in Morocco, and *ex situ* (since 1985), for resistance to current strains of powdery mildew (*Blumeria graminis* f.sp. *hordei*), *ex situ* accessions had more qualitative resistance than *in situ* accessions, for some pathotypes of powdery mildew, due to rare resistance genes being preserved *ex situ*, but lost *in situ* [1]. Other studies have reported the conservation of molecular marker bands in *ex situ* samples that have disappeared from current *in situ* populations [2,3]. These studies reaffirm the general consensus that the most effective PGR conservation strategies need to integrate both an *in situ* and *ex situ* approach [4–6].

Gene bank curators and managers generally aim to have *ex situ* samples that represent the inherent diversity of *in situ* populations at the time they are sampled. The literature suggests this might not be a straight forward proposition. *In situ* and *ex situ* accessions were compared in maize (*Zea mays* L.) landraces using morphological [7] and SSR markers [8]. *Ex situ* accessions of American Indian Hopi landraces differed significantly from their *in situ* counterparts that had been maintained on farm in Arizona and the differences were attributed to original collecting bottlenecks and seed multiplication in a different environment (Iowa) [7]. In contrast, similar levels of genetic diversity and insignificant differentiation between *in situ* and *ex situ* samples of Jala, a

Mexican maize race, were attributed to good initial sampling and effective regeneration practices [8]. Bean (*Phaseolus vulgare* L.) landraces conserved *ex situ* and on farm were compared using SSR markers [3] and phenotypic and development traits [9]. Both studies found that gene diversity was significantly less in *ex situ* subpopulations. SSR data indicated significant genetic differentiation in *ex situ* subpopulations as well as loss of alleles, gain of new alleles, and reduction of rare alleles and increase of common alleles [3]. Significant changes in yield, 100 seed weight, maturity and leaf area were also reported [9]. Both studies concluded that changes in genetic and phenotypic makeup in *ex situ* subpopulations could be attributed to regeneration practices. Using ISSR markers similar results were found in Oca (*Oxalis tuberosa* Mol), a tuber crop species grown in the Andes [10].

The challenges of maintaining complementary samples of wild species using *ex situ* and *in situ* methods are also evident. RAPDs were used to examine three remaining natural populations and one *ex situ* (conserved in a botanical garden field bank) subpopulation of *Vatica guangxiensis* (X. L. Mo), a rare endemic of southwestern China. The *ex situ* subpopulation contained 88.31% of the total genetic variation and was thought to adequately represent extant natural variation [11]. In contrast, RAPDs were used to examine *Parashorea chinensis* H. Wang, a rare endemic timber tree of southwest China and adjacent areas of Laos and Vietnam. The *ex situ* subpopulation (conserved in a field bank) was found to contain only 77.1% of the total variation found in samples taken from seven *in situ* sites. Additional sampling was recommended [12]. RAPDs were also used to assess *ex situ* samples (cultivated as ornamentals) of *Berberidopsis corallina* Hook. f., a threatened vine endemic to southern Chile, and found to only represent the northern part of the natural range of the species [13]. All three studies emphasized the importance of adequate sampling to ensure *ex situ* accessions represent the genetic diversity of wild species. In contrast, *in situ* populations of *Agropyron cristatum* (L.) Gaertn., a widely distributed wind pollinated grass species, were compared with gene bank counterparts collected 28–30 years previously, using SSRs. Although there were individual loci differences for number of alleles and genetic diversity, when averaged across all loci there was no significant difference among recent recollections and *ex situ* stored accessions [14]. RAPDs were used to examine recollections of *Solanum jamesii* Torrey (diploid outcrosser) and *Solanum fendleri* A. Gray (tetraploid inbreeder), crop wild relative (CWR) species of potato (*Solanum tuberosum* L.). Subpopulations were originally sampled in 1958, and recollected in 1978 and 1992. Significant genetic differences were found between all *ex situ* and re-collected *in situ* subpopulations of *S. jamesii*, and 12 of 16 comparisons of *S. fendleri*. Relative differences were attributed to mating system, vulnerability of small populations (<100 plants) to genetic change, and difference in original sampling procedures [2].

These studies emphasize two factors that are important to obtain and maintain *ex situ* samples that reflect an accurate snapshot of *in situ* populations. One, comprehensive initial sampling is imperative at both population and taxon level; and two, changing selection regimes–from either the *ex situ* regeneration/field site, or *in situ* site itself, can contribute to the divergence of *in situ* and *ex situ* accessions. Although a large body of literature outlines collecting parameters to ensure adequate sampling of PGR [15–17], and regeneration strategies to minimize genetic change [18–20], the literature also reflects how challenging it is to obtain *ex situ* samples that represent *in situ* populations. Recollection, rather than regeneration, may be a useful approach. This has been proposed for wild species that are difficult to regenerate *ex situ*, because natural conditions are too costly to replicate, or the material itself is difficult to propagate [20]. An area that has had little investigation is the frequency that recollection should occur. In other words, how frequently should a snapshot be taken? Recollection frequency is usually driven by declining viability of *ex situ* samples. Should *ex situ* managers also consider how quickly the genetic structure of *in situ* PGR might be changing? Seven farmer varieties of rice, conserved *ex situ* and *in situ* (on farm) over a period of time when production practices changed, were compared using allozymes, agronomic, stress resistance and morphological traits. Although yield and genetic diversity remained the same, traits associated with adaptation to new production practices significantly changed between *ex situ* and on farm subpopulations in a period of seventeen years [21]. Considering that climate change is occurring at a faster pace than predicted [22] and that models predict environmental change will not only drive changes in distribution and species extinction [23], but impact genetic diversity [24]. It is timely investigating how quickly *in situ* populations change over time, and how this may impact *ex situ* resampling.

Trifolium thompsonii Morton ($2n = 2x = 16$) is a restricted endemic herbaceous species that grows on the east slopes of the Cascade Mountains, in central Washington State, USA. The species is ranked as globally imperiled [25] and wild clover species, especially native to the US, have been identified as important to conserve in the USDA National Plant Germplasm System collection [26]. The species is a dominant forb in early seral communities that are disturbed by fire or grazing [27]. Fire has been excluded for nearly a century in the Cascade Mountains and species rarity has been attributed to increased shading due to overstory trees and competition with ground-layer species [28–30]. In 1988 the Dinkleman Fire swept through much of the range of *T. thomsonii*, effectively returning the area consumed to an early seral community. We used this species as an insect-pollinated, strongly out-crossing model to study population genetic change, since it is diploid, occurs in diverse environments that are undergoing successional change, occurs in populations greater than 10,000 plants, and is readily accessed and easily sampled. The aims of our study were to examine four distinct populations of *T. thompsonii* using Amplified Fragment Length Polymorphism (AFLP) markers to i) understand general patterns of population genetic diversity and structure, including the relative influence of gene flow, selection and genetic drift, and ii) examine the extent that diversity and structure changed over time. Specifically, this information would help us understand how adequate our efforts have been to conserve *T. thomsonii* using *in situ* and *ex situ* methods and help us determine if *ex situ* resampling intervals should be driven by declining seed viability in storage, or the occurrence of significant change in the genetic structure of *in situ* populations.

Materials and Methods

Ethics Statement

Collecting permits for populations growing on US Forest Service land were obtained from the Entiat District Ranger of the Okanogan-Wenatchee National Forest. Verbal permission to collect on state land was obtained from the Natural Areas Ecologist of the Washington Department of Natural Resources. Verbal permission from the landowner (Weaver family) was obtained to collect from private land.

Species, Study Locations and Sampling

T. thompsonii is reported to occur at 9–14 sites within 187 km^2 on the east slopes of the Cascade Mountains, in southeastern

Chelan and adjacent Douglas County in central Washington State, USA [25]. In 1977, the USDA Forest Service (USFS) established the 81 ha Thompson Clover Research Natural Area (RNA) and in 1993, the Washington Department of Natural Resources (DNR) established the 777 ha Entiat Slope Natural Area Preserve (KEYSTONE). Four sites, including RNA and KEYSTONE were sampled (Fig. 1). Area of each study location was mapped using a GPS (Global Positioning System) in 1995. Study location latitude and longitude were reported in decimal degrees, and elevation, in meters above sea level (masl). Badger Mountain (BADGER) (47.534, −120.2057, 1068 masl) was the only site that lies east of the Columbia River and occurred on private land. RNA (47.566, −120.324, 687 masl) and TENAS (47.581, −120.299, 982 masl) were located due west and northwest (respectively) of BADGER, on the USFS Okanogan-Wenatchee National Forest. KEYSTONE (47.635, −120.278, 1176 masl) was the northern most site. Size of area, habitat classification, and estimated census size of the four populations can be found in Table 1.

In 1995 and again in 2004, seed was sampled between July 22 and August 10 from the four sites, and placed in *ex situ* storage (−18°C). In 2009, leaf tissue was sampled from *in situ* populations growing at the four sites. For the purposes of this study, each site was designated as a population, and samples collected from a specific site during a specific year were considered subpopulations. Seed and leaf tissue samples were collected by walking multiple haphazard linear transects of 150 m throughout the study site. Two inflorescence or three young and healthy leaves were sampled from a single plant if it occurred at 5 m intervals along thetransect. This distance was considered sufficient to minimize within family sampling based on observations of seed dispersal [28]. A total of 150 plants were sampled per site. Sampling protocols, which also included bulking seed from maternal lines, were in accordance to standard collecting protocols for efficient *ex situ* conservation of plant genetic resources. Seed was cleaned and stored in a −18°C freezer. For the analysis, seed was scarified and subjected to an 18 hour water rinse that we have found effective in overcoming seed dormancy. Seeds were germinated in petri dishes and fresh and healthy seedlings were transported to the molecular laboratory where seedling tissue was frozen at −80°C and lyophilized in a Vertis Lyo-Centre Freeze Dryer (SP Industries, Gardiner NY). A third of the 1995 seedlings were inadvertently frozen during transit and had to be discarded. The 2009 leaf samples were placed in envelops, kept at 4°C for 24 hours then placed in a −80°C freezer. Leaf tissue was lyophilized using the same procedure as the seedling tissue.

DNA Isolation and AFLP marker generation

Although less informative than codominant markers, AFLP markers have been used widely to investigate population genetic structure in insect-vectored out crossing species (i.e. [31–34]). As with any molecular marker system, care needs to be taken to

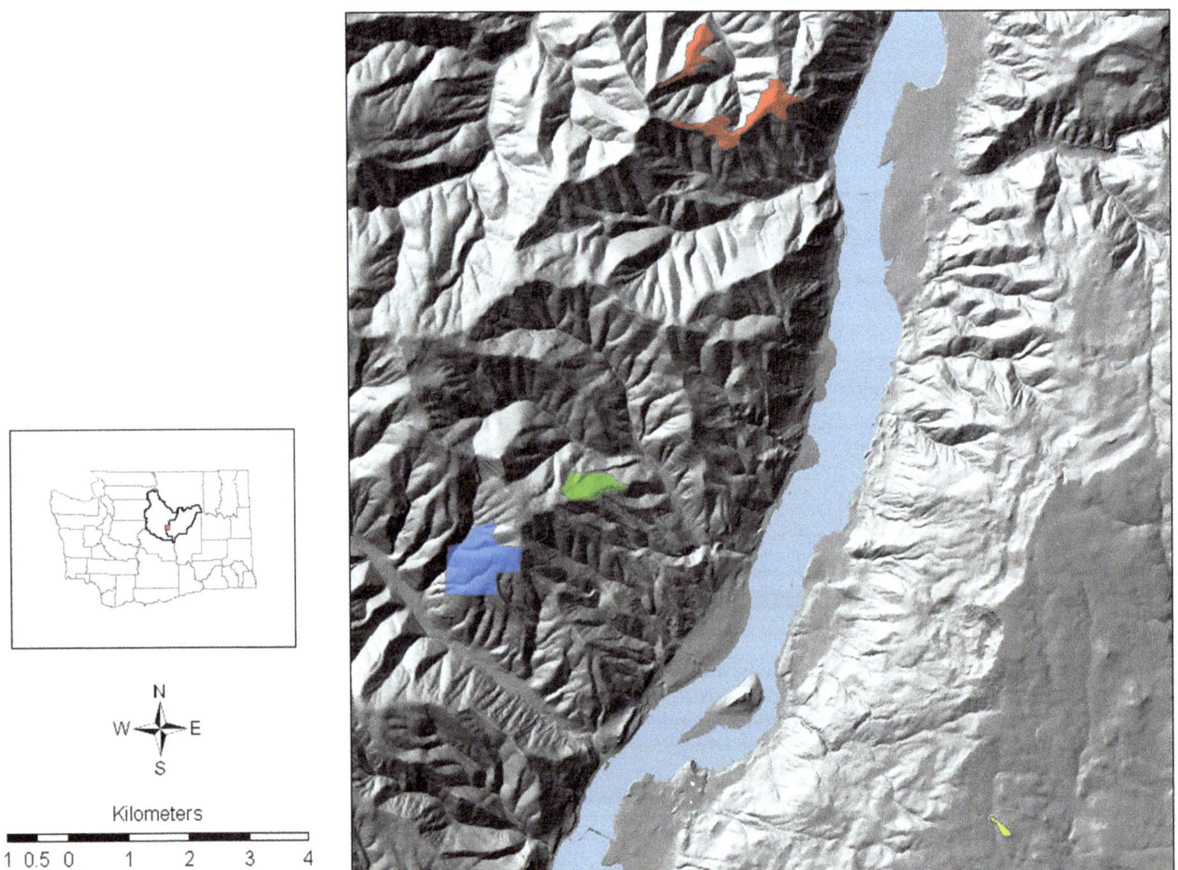

Figure 1. Location of studied populations of *Trifolium thompsonii* in Washington State, USA. Inset map shows Douglas and Chelan county, where Thompson clover is endemic. Four populations were studied: KEYSTONE (red), TENAS (green), RNA (blue) and BADGER (yellow, located east of the Columbia river, and south of RNA).

Table 1. Size and habitat of study locations, population census estimate based on mean plant density, and number of plants analyzed using AFLP markers.

Location	Area (km^2)	Habitat[a]	Population census		No. plants analyzed		
			<1998	2008	1995	2004	2009
BADGER	0.04	*Artemisia tridentata/Agropyron spicatum*	193,368[b]	166,666[d]	18	64	60
RNA	1.01	*Artemisia tridentata-vasayana/A. spicatum*	6,075828[c]	1,630,785[e]	19	63	56
TENAS	0.36	*Pseudotsuga menziesii/Calmagrostis rubescens*	7,174,060[c]	5,564,140[f]	0	64	60
KEYSTONE	0.82	*Artemisia tridentata-vasayana/A. spicatum*	1,481,738[b]	1,545,595[g]	35	64	48

[a]Habitat description based on classification of vegetation [27].
[b][27].
[c][30].
[d]Washington Natural Heritage Program Report *Trifolium thompsonii* EO 2342-SF 1856.
[e]Washington Natural Heritage Program Report *Trifolium thompsonii* EO 6503.
[f]Washington Natural Heritage Program Report *Trifolium thompsonii* EO 3235.
[g]Washington Natural Heritage Program Report *Trifolium thompsonii* EO 2342.

minimize genotyping errors so methods used to isolate DNA and generate AFLPs incorporated rigorous protocols, internal standards and replication to ensure genotypes determined by the marker analysis correspond to the real genotypes being examined [35].

Freeze-dried seedlings or leaf tissues were pulverized in a SPEX SamplePrep 2000 Geno/Grinder, and DNA extraction was automated using the Wizard Magnetic 96 DNA Plant System (Promega). The AFLP analysis was based on individual plants germinated from seed collected at the same locations in 1995 and 2004, and on leaf samples of individual plants collected from the same locations in 2009.

AFLP markers were generated using locally developed procedures based on technology by [36]. Double restriction digestion was done in a 25 ul reaction containing 250 ng of DNA, 1X NE Buffer 4, 1X Purified BSA and 5.0 U each of EcoRI and MseI restriction enzymes (New England BioLabs). Fifteen ul of the restriction digest reaction was run on a 1.5% agarose gel to verify the completion of digestion.

Adapter sequences (EcoRI-Fwd, 5′- ctc gta gac tgc gta cc; EcoRI-Rev, 5′- aat tgg tac gca gtc tac; MseI - Fwd, 5′- gac gat gag tcc tga g, and MseI-Rev, 5′- tac tca gga ctc at) were obtained through Eurofins MWG/Operon (Huntsville, Alabama). Each adapter pair was adjusted to 100 pM/ul (EcoRI) and 200 pM/ul (MseI), mixed in equal amounts, and allowed to anneal for 1 hour at 37°C and cooled to room temperature. The annealed pairs were then diluted to 5 pM/ul (EcoRI) and 50 pM/ul (MseI), aliquoted to 100 ul amounts and frozen for future use.

Ligation was done at 20°C for 2 hours in a 20 ul reaction containing 10 ul of the remaining restriction digest, 1X T4 Ligase Buffer (New England BioLabs), 5 pMoles EcoRI adapter, 50 pMoles MseI adapter, 0.5 mM ATP, and 80 cohesive end Units of T4-ligase (New England Biolabs). The completed ligation reaction was diluted 10:1 and used for the Pre-amplification. Pre-amplification and selective amplification were done on an ABI 9700 thermocycler using cycling programs prescribed by [36] but in 10 ul reactions. The pre-amplification product was diluted 10:1 and 2 ul used for selective amplification. Four separate primer pairs were used for selective amplification (Eacg/Mctg, Eaca/Mctc, EacaMcag, and EaggMctg, where the last 3 letters indicate the selective nucleotides following the EcoRI and MseI primer sequences) since they provided clear reproducible bands and were sufficiently polymorphic to show variation within and between populations. Marker fragments were separated on a LI-LOR 4300 DNA Analyzer (LI-COR Biosciences). Unambiguous bands were identified and scored as either present or absent. To estimate genotyping error, four replicates were run on eight individual plants each, from two contrasting sites. Genotyping error was estimated at less than 1%.

Data Analysis

Since we had unequal sample sizes (Table 1.), a subsampling approach was used to determine if uneven sample sizes would bias our analysis [37–38]. Sub sampling was performed in R (script available upon request) by modifying the Diversity function in AFLPdat [39] to sub sample a random sample of 18 individuals (the minimal sample size), from each subpopulation and calculate Nei's gene diversity and proportion of polymorphic markers. This was reiterated 100 times and means and standard deviation were calculated. Since observed and sub sampled data were consistently the same (Table S1.), and suggested that bias due to unequal sample size was low, the subsequent analysis was based on the observed data.

For each year, band patterns were calculated for overall population and subpopulations for the following: total number of bands (NB), number of common bands (CB) (frequency >0.05), number of rare bands (RB), (frequency <0.05), number of not widely shared bands (NWS), (common bands found in 25% or fewer populations) and number of private bands (PB). Allele frequencies were estimated using AFLPsurv 1.0 [40] based on a Bayesian method with non-uniform prior distribution [41]. Hardy-Weinberg equilibrium was assumed since Thompson clover is a highly out crossing wild species. Allele frequencies were used to estimate percent polymorphic loci at the 5% level and Nei's gene diversity and its standard error was estimated for each subpopulation. T tests were carried out to determine if genetic diversity was significantly different among the four populations for each sampled year (i.e BADGER04, RNA04, TENAS04, KEYSTONE04), and across the three sampled years for each population (i.e. BADGER95, BADGER04, BADGER09). A Bonferroni correction was made to ensure an overall critical p-value of 5%. To examine genetic differentiation, AFLPsurv 1.0 was used to estimate overall population Fst at the four locations. It was also used to test the null hypothesis that there was no genetic differentiation among the four populations for each sampled year, and across the three sampled years for each population. Five thousand permutations were used.

To further examine population differentiation and partitioning of genetic variance, a euclidean distance metric was estimated and used for an analysis of molecular variance (AMOVA) to calculate Phi statistics (which is analogous to Fst), using GENALEX 6.5 ([42–43]. Pairwise PhiPT was estimated among the four populations by combining the 2004 and 2009 data. Probability values were based on 999 permutations. The AMOVA was also used to examine the distribution of variation among location (PhiRT) (i.e. BADGER, RNA, KEYSTONE), and among subpopulations (i.e. 1995, 2004, 2009) within location (PhiPR). Since TENAS lacked data for 1995, it was not included in this analysis. Significance of the AMOVA was tested using 999 permutations.

To examine genetic distance among the spatially and temporally sampled populations, Nei's D_{Ain} [44] which does not involve an evolutionary model, was estimated using PowerMarker 3.25 [45]. A cluster analysis was conducted on all subpopulations using the neighbor joining algorithm to construct a tree from the distance matrix. The program was also used to calculate 1000 bootstrap distance matrices which were imported into the CONSENSE program of PHYLIP [46] to build a majority rule consensus tree.

Population structure was also examined using the software STRUCTURE v2.3.3 [47–50]. Ten replications with a burn-in of 20,000 iterations followed by 20,000 additional iterations were used at each K level until results indicated lowered and erratic values for P(X|K). The parameter set included the ADMIXTURE model with allele frequencies correlated, RECESSIVE ALLELES model, USEPOPINFO and STARTATPOPINFO turned on with 11 subpopulations labeled (each location/year), and the LOCPRIOR model. Average Q-plots over the ten replications were calculated using the ancillary software CLUMPP [51], and graphic displays of population structure were developed from the q-frequencies of the mean of ten runs using DISTRUCT software [51].

To examine gene flow on a current timescale we carried out two different assignment tests. The first used AFLPOP 2.0 [52] which uses a modified allocation method [53] to compute the log likelihood of an unknown individual's allelic phenotype in each subpopulation based upon the frequency of the dominant band (presence) at each locus. Each unknown individual is then allocated into a source subpopulation with the greatest log likelihood. The minimal log-likelihood difference for the allocation of an individual to a population was set at 0.5. This is a conservative threshold in that assignment of an individual to a subpopulation was not made unless the probability of the given assignment was 5 times more likely than the next most probable assignment. If this threshold was not met, the individual was not assigned to any subpopulation and were designated as 'criteria not met' (CNM). The CNM category does not imply that an individual does not belong to a subpopulation, but that there are two or more subpopulations with similar probabilities of assignment, and hence minimal log-likelihood difference is less than the designated threshold. We also used STRUCTURE v2.3.3 to identify potential migrants between populations taking into account the source population of the sample. We set GENSBACK = 2, which tests for evidence of ancestry for two generations.

To examine the presence of historic gene flow, we tested for isolation by distance among the 4 populations for each year separately using GENALEX 6.5 [43]. A geographic distance matrix was computed based on latitude and longitude coordinates and correlations with the genetic distance matrix were tested using a Mantel test with 9999 permutations [54].

Although genetic differentiation can result from limited gene flow, forces of selection can play an important role in determining the rate of population differentiation. Selection can be detected by identifying loci that have F_{st} values that fall outside the F_{st} range for neutral loci. We used the outlier method of BayeScan 2.1 [55], that uses a Bayesian method to infer the posterior probability of each locus being under selection by defining and comparing two alternative models; including and excluding the effects of selection [56–57]. We used 20 pilot runs of 2000 iterations followed by a burn-in of 50 000 iterations and 100,000 iterations, and used a thinning interval of 10. Loci were considered under selection if the posterior odds (PO) was greater than 10 [56]. Subpopulations were examined since the main assumption of the underlying model is that populations exchanging migrants contribute to a common pool of migrants [56]. This assumption would have been violated if we combined data across years. We also dropped out loci whose overall frequency was <0.05, as recommended by the manual. We also used DFDIST as a second outlier detection method, implemented in MCHEZA [58]. MCHEZA allows for the use of dominant markers and models neutral distribution through coalescent simulations under a symmetric island model parameterized by the observed data of neutral loci (i.e. non-neutral loci are dropped out). The simulated neutral distribution is then used to identify individual loci that fall outside the neutral range. For each program we conducted three independent runs and corrected for multiple testing by setting a false discovery rate (FDR) <0.1.

The occurrence of genetic drift can also influence genetic diversity, the rate unfavorable alleles are fixed, and the efficiency of selection [3]. Variance effective population size (Ne_V) is an important parameter used to assess genetic drift [59]. Since we had temporally spaced samples, we estimated contemporary Ne_V using a method developed for dominant markers by [60]. Fixed alleles, alleles identified as under selection, and those with a frequency ≤ 0.2 were eliminated prior to estimation of Ne_V [60]. Because Thompson clover requires several years of establishment prior to seed set, the samples were estimated to be 3 (1995–2004) and 2 (2004–2009) breeding generations apart. Since sample sizes were small and N_e reasonably large, $\hat{N}_e$ was estimated using $\hat{q}_z$ in all cases.

Results

Patterns of genetic diversity and differentiation

The four primer pairs revealed a total of a 129 readily discernible polymorphic bands. Table 2 summarizes the AFLP alleles from the *ex situ* samples collected in 1995 and 2004 and the *in situ* sample from the same populations in 2009. Across all populations, the number of common alleles was the same in all three sampling periods; however, more rare alleles were sampled in 2004 and 2009, compared to 1995. In the individual populations, the proportion of rare alleles was greater in TENAS and KEYSTONE, compared to BADGER and RNA, and this pattern was observed in all three sample years. The occurrence of private alleles was very low; however, there were the occurrence of common bands that were not widely shared among populations. KEYSTONE had the lowest proportion of polymorphic loci, followed by TENAS. BADGER and RNA had the highest percentage of polymorphic loci and were comparable. For all populations, percent polymorphic loci did not change over the fifteen years, with the exception of BADGER, where 2004 (79.8) was greater than 1995 (75.2) and 2009 (76.0).

Across all populations there was no significant change in gene diversity ($P<0.05$) from 1995 to 2009. Table 3 summarizes the gene diversity statistics for each subpopulation. In 1995, KEYSTONE had significantly less diversity than BADGER and RNA. TENAS and KEYSTONE also had significantly less diversity in 2004 and 2009 compared to BADGER and RNA. Interestingly, the BADGER 1995 subpopulation, based on eighteen individual plants, had significantly greater gene diversity than 2004 and 2009 subpopulations based on >50 individual plants. Diversity measurements were the same for RNA, TENAS and KEYSTONE regardless of the year sampled. The four populations showed significant but low genetic differentiation ($P<0.001$). Populations located close together had less differentiation than populations located further apart. The overall estimate for F_{st} was 0.05. Pairwise F_{st} between geographically close sites ranged from.03 in 1995 and 2004 to.04 in 2009. In contrast, F_{st} between BADGER and KEYSTONE, which were the furthest away, ranged from.07 in 1995, .05 in 2004 and.06 in 2009. Population pairwise PhiPT values showed a similar result (Table 4).

Spatial and temporal structure

Based on the AMOVA analysis, 90% of the variation occurred within populations. Seven percent of the variation occurred among populations and 3% occurred among subpopulations collected within each location (P-value <0.01 with 999 permutations) (Table 5). The neighbor-joining tree based on Nei's genetic distance is shown in Fig. 2. Similar to the AMOVA, the tree suggested that spatial structure was more significant than temporal structure. Subpopulations collected from the same location clustered together; however, populations sampled in 1995 were less similar to populations sampled in 2004 and 2009, which were more closely related. Genetic structure associated with geographic distance was also evident. Subpopulations collected at KEYSTONE, the northern-most site, were genetically more distant compared to the subpopulations collected at TENAS, RNA, and

Table 2. AFLP band patterns and % polymorphic loci for four populations and for the combined populations (POP), sampled for three time periods.

BAND	BADGER	RNA	TENAS	KEYSTONE	POP
1995					
NB	110	109	–	109	121
CB	110	109	–	102	109
RB	0	0		7	12
NWS	2	0	–	0	0
PB	1	0	–	0	1
%P	75.2	76.7	–	69.0	73.6
2004					
NB	111	118	112	114	126
CB	105	109	99	100	109
RB	6	9	13	14	17
NWS	2	0	3	1	0
PB	0	0	0	0	0
%P	79.8	77.5	72.9	69.0	74.8
2009					
NB	109	118	113	109	127
CB	107	112	102	97	109
RB	2	6	11	12	18
NWS	3	3	2	1	0
PB	0	0	0	0	1
%P	76.0	77.5	72.1	67.4	73.2

NB = total number of bands; CB = Common bands (number of different bands with a frequency >0.05); RB = Rare bands (number of different bands with a frequency < 0.05); NWS = Not widely shared bands, (common bands found in 25% or fewer populations); PB = number of private bands; %P = percent polymorphic loci at the 5% level, expressed as a percentage.

Table 3. Expected heterozygosity (H_e) under Hardy-Weinberg genotypic proportions for *ex situ* samples collected in 1995, recollected in 2004, and *in situ* populations sampled in 2009.

Location	1995	2004	2009
BADGER	0.292 aa	0.276 ab	0.275 ab
RNA	0.278 aa	0.271 aa	0.274 aa
TENAS	–	0.247 ba	0.250 ba
KEYSTONE	0.250 ba	0.250 ba	0.242 ba

Differences in first letter correspond to significant differences in H_e among the four locations within the same year. Differences in second letter correspond to significant differences in H_e within location across the three years sampled (Bonferroni correction for testing seven hypotheses based on overall critical p-value of 5% is ≤0.007).

BADGER. However, KEYSTONE subpopulations were more closely related to TENAS, which is closer, geographically, and least related to BADGER, which is the site furthest away.

The STRUCTURE analysis also supported strong spatial structure. The plot of P(X|K) indicated that K = 2 and K = 4 were the most likely groupings (Figure S1). Fig. 3 shows Q-plots for each K group which are useful for visualizing population genetic structure and the presence of admixture. At K = 2, the KEYSTONE population stood out as distinct from BADGER, RNA and TENAS, in all three years. As with the cluster analysis, geographic structure was apparent. Looking at the level of admixture, the proportion of alleles belonging to the KEYSTONE group decreased from TENAS to BADGER, reflecting increasing geographic distance. At K = 4, geographic differentiation between the four populations was apparent. Although not as distinct, variation across sampling year was apparent. For example in BADGER 1995, admixture with RNA and TENAS was apparent, but the proportion of alleles belonging to RNA was much less in BADGER 2004 and BADGER 2009. The proportion of alleles belonging to TENAS was somewhat less in BADGER 2004, and much less in BADGER 2009.

Evidence for gene flow

The assignment tests suggested that limited gene flow did occur between the four populations. With AFLPOP, (Table 6), individuals were assigned to the correct source population at a rate that ranged from 80 to 32%. Generally individuals from subpopulations collected at the same location were assigned with a similar level of success to the correct population. TENAS had the highest success, averaging 71% across 2004 and 2009. RNA had the lowest success rate, averaging 43% across the three sampling years. Seventy six percent of assignment events between locations were between locations that were adjacent. However, there was evidence for long distance gene flow from BADGER to RNA, RNA to KEYSTONE, BADGER to TENAS and TENAS to BADGER. Migration between populations did not appear symmetrical. There were twice as many genotypes from RNA assigned to BADGER, then from BADGER to RNA. There were twice as many genotypes from TENAS assigned to RNA, as there were from RNA to TENAS. Only one genotype from KEYSTONE was assigned to a location other than KEYSTONE. Mantel tests for isolation by distance found a significant and positive relationship between genetic distance and geographic distance ($P<0.001$, 9999 permutations) which suggested that gene flow between the four populations was limited, but did occur based on the evidence provided by the assignment tests.

Extent of selection and genetic drift

Outlier loci detection suggested that most of the markers we examined were neutral but there was some signature of selection. BayeScan 2.1 identified no significant outliers among the 4 K groups at either prior odds of 10:1 or 1:1 for all three years sampled. However, DFDIST showed a high probability (>95%) for divergent selection for 9, 7 and 9 loci in 1995, 2004 and 2009, respectively. In one instance, the same locus experienced selection in all three years, while a second locus experienced selection in 1995 and 2004, and a third, experienced selection in 2004 and 2009.

When we estimated effective population size, we found that NeV declined in RNA from 163 to 113. However, it increased in BADGER and KEYSTONE (Table 7). This coincided with the census population numbers (Table 1), that suggested a drop from 6 million plants in 1998 to slightly more than 1.5 million plants in 2008 at RNA while BADGER and KEYSTONE remained relatively stable.

Discussion

General patterns of genetic diversity and population structure

Although *T. thompsonii* is a narrowly distributed endemic species, our results showed that it did not have a narrow genetic

Table 4. Geographic distance (km) between each location (above diagonal) and population pairwise PhiPT values (below diagonal) based on samples collected in 2004 and 2009 (Values significant at $P<0.001$).

Population	BADGER	RNA	TENAS	KEYSTONE
BADGER	–	10.07	9.09	13.21
RNA	0.068	–	2.47	8.21
TENAS	0.081	0.046	–	6.31
KEYSTONE	0.099	0.087	0.085	–

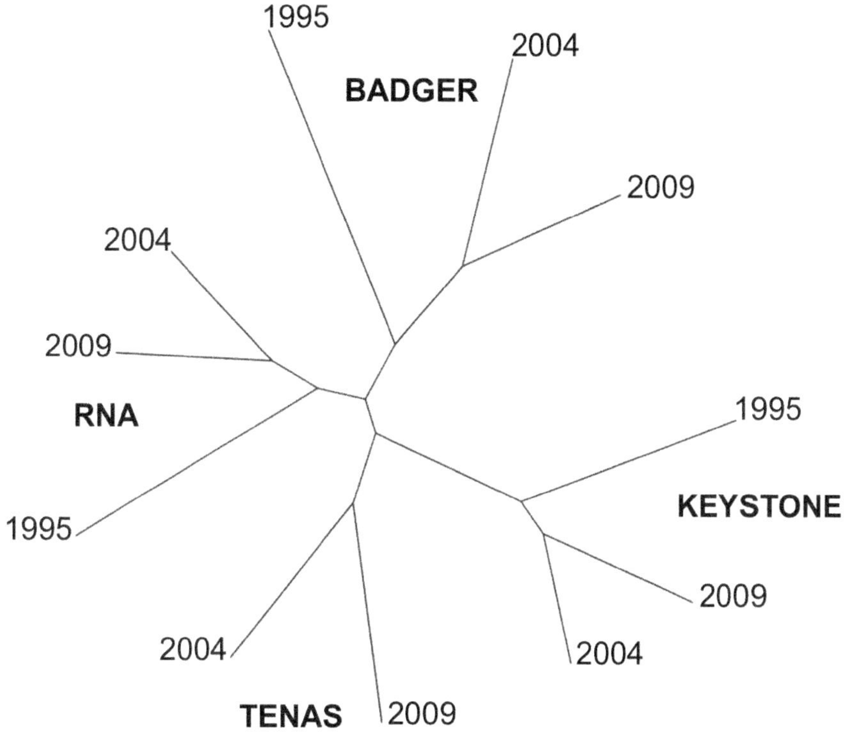

Figure 2. Neighbor-joining tree showing genetic relationship among Thompson clover. Relatedness of four populations of Thompson clover sampled in 1995, 2004, and 2009, based on Nei's genetic distance estimated using AFLPs.

base. The overall level of genetic diversity (0.26) was similar to comparable AFLP studies on widespread perennial insect-pollinated congeners, T. *montanum* L., (0.23), [61], and *T. alpinum* L., (0.24), [62], and higher than *T. repens*, (0.1) [63]. Average percent polymorphic loci was higher (73%) than reports for *T. montanum* (58%) [61] and *T. repens* (31%) [63]. Diversity estimates were also similar to comparable AFLP studies in non related herbaceous perennial insect pollinated species, such as *Silene chlorantha* (Willd.) (0.20) [34], *Abronia alpina* Brandegee (.28) [64], and *Echinacea laevigata* (Boynton and Beadle) Blake (0.26) [65].

Among the four populations, KEYSTONE and TENAS, the northern most populations, had less overall diversity and a lower percent of polymorphic loci compared to BADGER and RNA. They also had about twice as many rare bands. An explanation may be that these populations are relicts of the leading north edge of the species during the Pleistocene glaciations. These populations lie due south of the furthest extent of the Okanogan lobe of the Cordillera ice sheet, and although major valley glaciers extended eastward from the Cascade Range, the closest stopped at the upper Entiat river, approximately 10 km west of KEYSTONE [28,66]. Simulation studies have shown that neutral genetic variation can be lower along the leading edge of a species range due to founder effects and allele surfing [67,68].

BADGER was the only population that showed a significant decline in genetic diversity from 1995 to 2004. It was also the only population where a private band was detected. Considering that this population is located in open shrub steppe habitat, as opposed to the other three populations that are located in the forest transition zone, it may contain unique alleles that help it persist in a drier, hotter environment. Although resampling suggested that uneven sampling sizes did not bias our genetic estimates, we would expect if bias was present, sampling too few plants would have underestimated diversity. Although we sampled fewer plants in 1995, compared to 2004 and 2009, 20 individuals were determined to be adequate to assess genetic variation in *T. repens*, also an insect-pollinated outcrossing perennial species [69]. Our results were also consistent with field observations. Although the

Table 5. Nested analysis of molecular variance (AMOVA) based on 129 polymorphic loci based on populations sampled at three locations (BADGER, RNA, and KEYSTONE), for three different years (1995, 2004, 2009).

AMOVA	df	MS	Variation	% variation	Phi	P-value
Location	2	225.538	1.229	7	0.069	0.01
Year/location	6	40.166	0.529	3	0.032	0.01
Subpopulation	431	16.085	16.085	90	0.099	0.01
Total	439		17.844	100		

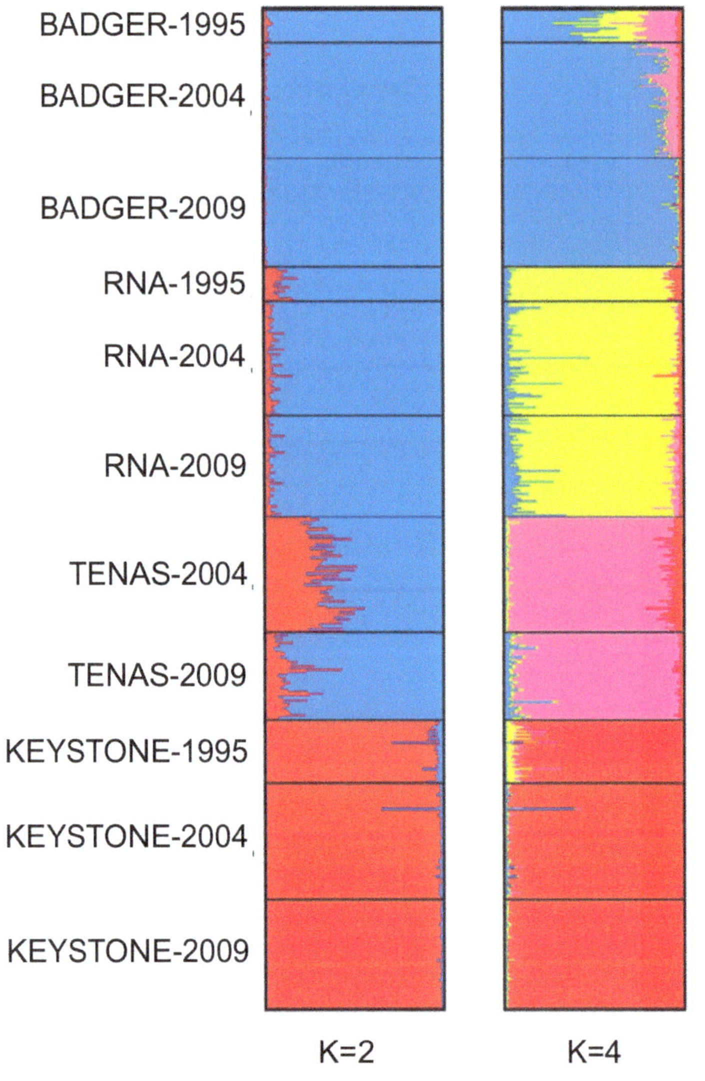

Figure 3. Q-plots based on AFLP analyses. Bayesian analysis using the software STRUCTURE suggested most likely groups of K = 2 and K = 4.

census population size of BADGER has remained relatively stable, we observed that the population was protected from grazing in 1995, but was not protected when we returned in 2004 and 2009. Because BADGER is in a unique habitat, compared to other populations, and does not have State or Federal protection, close monitoring is warranted to ensure this population does not continue to decline in the future.

Relative influence of gene flow, selection and genetic drift

Although genetic differentiation between the four populations was significant (albeit weak), as was our test of isolation by distance, there was sufficient evidence to suggest that gene flow did occur among the studied populations.

The assignment test suggested infrequent gene flow occurred, despite the distance between populations. The asymmetrical flow of genes was concordant with site specific characteristics, in terms of climate and topography. The occurrence of twice as many migration events from RNA to BADGER, compared to BADGER to RNA supported the hypothesis that prevailing northwesterly winds may disperse fruits of *T. thompsonii* or contribute to the long distance movement of insect pollinators, since RNA is located northwest of BADGER [28]. Gene flow also occurred more frequently from TENAS to RNA than from RNA to TENAS. Although separated by 2.5 km, TENAS and RNA occur along the same road. However, TENAS is at a higher elevation, so seed and pollen would more likely move downhill, from TENAS to RNA. Results from the assignment tests reflected the influence of distance as well. Gene flow appeared to have occurred between KEYSTONE and TENAS, separated by 6.31 km, but was restricted between KEYSTONE and RNA, (8.21 km), and did not occur between KEYSTONE and BADGER (13.21 km). The usefulness of topographic data to infer gene flow in wild *Trifolium* populations has been reported and has been concordant with data generated from morphologic [70], isozyme [71], RAPD [70] and AFLP [63] markers. This supports the use of maps and satellite images to infer gene flow due to the occurrence of landscape barriers or corridors that influence the connectivity between populations.

Table 6. Assignment of 551 individuals from seven subpopulations (columns) to source populations (rows) to detect gene flow between four locations over 15 years.

Allocated to	BAD95	BAD04	BAD09	RNA95	RNA04	RNA09	TENAS04	TENAS09	KEY95	KEY04	KEY09	Total
BAD 1995	11	0	0	0	0	0	0	0	0	0	0	11
BAD 2004	0	40	7	0	1	2	0	0	0	0	0	50
BAD 2009	0	8	33	0	2	1	0	1	0	0	0	45
RNA 1995	1	0	0	8	1	0	0	0	0	0	0	10
RNA 2004	0	1	0	3	31	9	2	0	0	0	0	46
RNA 2009	0	0	1	0	8	22	0	4	0	0	0	35
TEN 2004	1	1	0	0	0	0	51	1	1	0	0	55
TEN 2009	0	0	0	0	1	2	6	30	0	0	0	39
KEY 1995	0	0	0	0	0	0	1	1	19	4	0	25
KEY 2004	0	0	0	0	0	0	0	0	5	21	3	29
KEY 2009	0	0	0	0	1	0	0	0	0	14	35	50
CNM[a]	5	14	19	8	18	20	3	11	10	25	22	155
# IND	18	64	60	19	63	56	63	48	35	64	60	551

[a]Criteria not met (CNM) to meet minimal log-likelihood difference of 0.5 for the allocation of an individual to a population.

Although there was some evidence that selective forces were responsible for influencing Fst values for some of the markers we examined, it was only for a few. However one marker was identified as being under divergent selection in all three years examined, and may be associated with an adaptive trait. Although succession has been occurring in the area where RNA, TENAS and KEYSTONE are located since the 1988 fire, our marker set did not detect any overall temporal trends in changing selection pressure. Studies similar to [72] would be useful to examine this in further detail since it has direct relevance to recollecting intervals.

There was limited evidence for genetic drift although we did detect a decline in effective population size in RNA, which was concordant with consensus data. However, we did not detect a decrease in genetic diversity. Our study suggested that further monitoring is warranted in RNA to ensure population decline does not have a negative impact since the population is the largest and oldest *in situ* reserve of Thompson clover.

Altogether, we tracked the genetic structure of 4 different populations of Thompson clover over 15 years. With the exception of BADGER, the populations did not appear to have undergone substantial change during the time period of our study. Changes in genetic structure of the RNA, TENAS and KEYSTONE populations due to succession following the 1988 fire, were not evident. Although gene flow was evident, admixture tended to decline over time, especially in BADGER and KEYSTONE. This suggested that despite the occasional influx of migrants, individual populations tended to return to equilibrium, as unadapted genotypes were selected against. Overall, spatial differences among the 4 study locations accounted for more than twice the variation as temporal differences within study locations. These results suggested that recollecting intervals for *ex situ* samples of Thompson clover could be greater than 15–20 years.

Conclusion

In this paper, we report the spatial and temporal genetic variation and population substructure revealed by AFLP markers in T. *thompsonii* populations sampled from four diverse locations over a 15 year time period. Although there was some evidence for gene flow, selection and drift, there was no difference in overall diversity between *ex situ* samples collected in 1995 and 2004 and *in situ* samples collected in 2009. Only one population showed a decline in genetic diversity over the 15 years studied. Our results echo other reports that have found that perennial out crossing species generally showed less difference in genetic diversity between *in situ* and *ex situ* accessions compared to annual, self-pollinated species [73].

We can use the lessons we learned in our study to formulate some general guidelines to help ensure *ex situ* samples of insect-pollinated herbaceous perennial species provide clear snap shots of *in situ* populations. Our study suggested that a successful *ex situ* collecting strategy is to collect from a large number of plants (> 150 plants), equalize maternal contribution by sampling the same number of heads per plant, and sample along random transects with sufficient space between plants to minimize intrafamilial sampling. We also learned that long distance dispersal can occur, and appeared to play an important role in limiting the extent that our four populations diverged from one another. Long distance gene flow by insect pollinators not only maintained genetic connectivity, but can also help in maintaining and increasing effective population size [74,75]. Only one immigrant every other generation is needed to prevent differentiation due to genetic drift [76,77]. Long distance dispersal of pollen by insects can be highly variable since it is dependent upon plant species, pollination vector

Table 7. Estimated population sizes ($\hat{N}_e$) of the four locations over the three sampling years.

Years	BADGER	RNA	TENAS	KEYSTONE
Loci	66	77	83	77
1995–2004	30	163	-	105
2004–2009	44	113	57	156

and the environment [74], so the effective conservation of insect pollinated species requires conserving not only the plant species, but pollinator diversity and abundance [78]. Our results suggested that as a general guideline, for a given geographic area, the benefits of long distance gene flow may allow us to conserve the intraspecific diversity of insect pollinated species (especially populations that occur in similar habitats), with fewer populations, since populations are likely to be weakly divergent. In contrast, self pollinated species, especially annuals, may need more populations protected, since we would expect individual populations to be strongly divergent, due to isolation by distance.

The most important lesson we can apply to the effective *ex situ* and *in situ* conservation of PGR is the value of comparing the genetic structure of *ex situ* and *in situ* samples. Quantifying genetic change over time not only allows genetic resource managers to validate conservation protocols, develop appropriate recollection intervals, but provides them with an early detection mechanism for identifying problematic conditions that can be addressed to prevent the local extinction of vulnerable *in situ* populations.

Supporting Information

Figure S1 The plot of P(X|K) indicated that K = 2 and K = 4 were the most likely groupings.

Table S1 Results from resampling to determine unequal sample bias.

Acknowledgments

We gratefully acknowledge the assistance of Greg Sinton, Adam Smolenski, Andrew Bell, Alexandr Afonin, Francisco González, Jamie Calley, Estela Cervantes and Jesse Prieto in surveying and collecting. Thanks to Lisa Taylor for AFLP lab work and Gwen Pentecost for producing Fig. 1. We wish to thank Jinguo Hu for helpful comments and suggestions.

Author Contributions

Conceived and designed the experiments: SLG. Performed the experiments: SLG TJK. Analyzed the data: SLG TJK LXY MPQ. Contributed reagents/materials/analysis tools: MPQ LXY. Wrote the paper: SLG TJK.

References

1. Jensen HR, Dreiseitl A, Sadiki M, Schoen DJ (2012) The Red Queen and the seed bank: Pathogen resistance of *ex situ* and *in situ* conserved barley. Evol Appl 5: 353–367.
2. Del Rio AH, Bamberg JB, Huaman Z, Salas A, Vega SE (1997) Assessing changes in the genetic diversity of potato gene banks. 2. *In situ* vs *ex situ*. Theor Appl Genet 95: 199–204.
3. Negri V, Tiranti B (2010) Effectiveness of *in situ* and *ex situ* conservation of crop diversity. What a *Phaseolus vulgaris* L. landrace case study can tell us. Genetica 138: 985–998.
4. Maxted N, Kell S, Ford-Lloyd B, Dulloo E, Toledo Á (2012) Toward the systematic conservation of global crop wild relative diversity. Crop Sci 52: 774–785.
5. Maxted N, Ford-Lloyd BV, Kell SP (2008). Crop wild relatives: establishing the context. In: Maxted N, Ford-Lloyd BV, Kell, SP, Iriondo J, Dullo E, Turok J, editors. Crop wild relative conservation and use. Wallingford: CAB International. 3–30.
6. Bretting PK, Duvick DN (1997) Dynamic Conservation of Plant Genetic Resources. Advances in Agronomy 61: 1–51.
7. Soleri D, Smith SE (1995) Morphological and phenological comparisons of two hopi maize varieties conserved *in situ* and *ex situ*. Econ Bot 49: 56–77.
8. Rice EB, Smith ME, Mitchell SE, Kresovich S (2006) Conservation and change: A comparison of *in situ* and *ex situ* conservation of Jala maize germplasm. Crop Sci 46: 428–436.
9. Gómez OJ, Blair MW, Frankow-Lindberg BE, Gullberg U (2005) Comparative study of common bean (*Phaseolus vulgaris* L.) landraces conserved *ex situ* in genebanks and *in situ* by farmers. Genet Res Crop Evol 52: 371–380.
10. Malice M, Martin N, Pissard A, Rojas-Beltran JA, Gandarillas A, et al. (2007) A preliminary study of the genetic diversity of Bolivian oca (*Oxalis tuberosa* Mol.) varieties maintained *in situ* and *ex situ* through the utilization of ISSR molecular markers. Genet Res Crop Evol 54: 685–690.
11. Li Q, Xu Z, He T (2002) *Ex situ* genetic conservation of endangered *Vatica guangxiensis* (Dipterocarpaceae) in China. Biol Conserv 106: 151–156.
12. Li Q, He T, Xu Z (2005) Genetic evaluation of the efficacy of *in situ* and *ex situ* conservation of *Parashorea chinensis* (Dipterocarpaceae) in Southwestern China. Biochem Genet 43: 387–406.
13. Etisham-Ul-Haq M, Allnutt TR, Smith-Ramírez C, Gardner MF, Armesto JJ, et al. (2001) Patterns of genetic variation in in and *ex situ* populations of the threatened chilean vine *Berberidopsis corallina*, detected using RAPD markers. Ann Bot 87: 813–821.
14. Che YH, Yang YP, Yang XM, Li XQ, Li LH (2011) Genetic diversity between *ex situ* and *in situ* samples of *Agropyron cristatum* (L.) Gaertn. based on simple sequence repeat molecular markers. Crop Pasture Sci 62: 639–644.
15. Marshall DR, Brown AHD (1975) Optimal sampling strategies in genetic conservation In: Frankel OH, Hawkes JG, editors. Crop genetic resources for today and tomorrow, International Biological Programme 2. Cambridge:Cam-:Cambridge University Press. 53–80.
16. Marshall DR, Brown AHD (1995) A basic sampling strategy: theory and practice. In: Guarino L, Ramanathan Rao V, Reid R editors. Collecting Plant Genetic Diversity: Technical Guidelines UK:Cab International. 75–92.
17. Von Bothmer R, Seberg O (1995) Strategies for collecting wild species. In: Guarino L, Ramanathan Rao V, Reid R editors. Collecting Plant Genetic Diversity: Technical Guidelines UK: Cab International. 93–112.
18. Breese EL (1989) Regeneration and multiplication of germplasm resources in seed gene banks: the scientific background. Rome: IBPGR.
19. Sackville Hamilton NR, Chorlton KH (1997) Regeneration of accessions in seed collections: a decision guide. Handbook for genebanks No. 5. Rome:Interna-:International Plant Genetic Resources Institute. 75 p.
20. Brown AHD, Brubaker CL, and Grace J P (1997) Regeneration of Germplasm Samples: Wild Versus Cultivated Plant Species. Crop Sci 37: 7–13.
21. Tin HQ, Berg T, Bjørnstad Å (2001) Diversity and adaptation in rice varieties under static (*ex situ*) and dynamic (*in situ*) management. Euphytica 122: 491–502.
22. Osborn TJ, Briffa KR (2006) The spatial extent of 20th-century warmth in the context of the past 1200 years Science 311, 841–844.
23. Jarvis A, Lane A, Hijmans RJ (2008) The effect of climate change on crop wild relatives. Agric Ecosyst Environ 126: 12–23.
24. Pauls SU, Nowak C, Bálint M, Pfenninger M (2013) The impact of global climate change on genetic diversity within populations and species. Mol Ecol 22: 925–946.
25. NatureServe (2011) NatureServe Explorer: An online encyclopedia of life [web application]. Version 7.1. NatureServe, Arlington, Virginia. Available: http://www.natureserve.org/explorer. Accessed: 30 January 2012.
26. Morris JB, Greene SL (2001) Defining a multiple-use germplasm collection for the genus *Trifolium*. Crop Sci 41: 893–901.
27. Scherer G, Zamora B, Everett R, Roch B Jr. (1996) Habitat characteristics and morphological differences of *Trifolium thompsonii* populations. Northwest Sci 70: 242–251.

28. Canfield JE (1977) The ecological life history of Trifolium thompsonii Morton, with reference to its restricted distribution. [Masters Thesis]: University of Washington. 145p.
29. Harrod RJ, Halpern CB (2005) Effects of experimental shading and removal of competitors on vegetative and reproductive performance of *Trifolium thompsonii* Morton, a rare endemic herb of eastern Washington. Nat Areas J 25: 156–164.
30. Harrod RJ, Halpern CB (2009) Effects of experimental burning on individual performance and population structure of two rare plants of north central Washington. Restoration Ecol 17: 215–225.
31. Rivers MC, Brummitt NA, Lughadha EN, Meagher TR (2011) Genetic variation in *Delonix* s.l. (Leguminosae) in Madagascar revealed by AFLPs: Fragmentation, conservation status and taxonomy. Conserv Genet 12: 1333–1344.
32. Baskauf CJ, Burke JM (2009) Population genetics of *Astragalus bibullatus* (Fabaceae) using AFLPs. J Hered 100: 424–431.
33. Garrido JL, Fenu G, Mattana E, Bacchetta G (2012) Spatial genetic structure of Aquilegia taxa endemic to the island of Sardinia. Ann Bot 109: 953–964.
34. Lauterbach D, Burkart M, Gemeinholzer B (2012) Rapid genetic differentiation between *ex situ* and their *in situ* source populations: An example of the endangered *Silene otites* (Caryophyllaceae). Bot J Linn Soc, 168: 64–75.
35. Bonin A, Bellemain E, Eidesen PB, Pompanon F, Brochmann C, et al. (2004) How to track and assess genotyping errors in population genetics studies. Mol Ecol 13: 3261–3273.
36. Vos P, Hogers R, Bleeker M, Reijans M, van de Lee T, et al. (1995) AFLP: A new technique for DNA fingerprinting. Nucleic Acids Res 23: 4407–4414.
37. Bomblies K, Yan L, Laitinen RA, Kim S-T, Hollister JD, et al. (2010) Local-Scale Patterns of Genetic Variability, Outcrossing, and Spatial Structure in Natural Stands of Arabidopsis thaliana. PLoS Genet 6(3): e1000890. doi:10.1371/journal.pgen.1000890.
38. Gomaa NH, Montesinos-Navarro A, Alonso-Blanco C (2011) Temporal variation in genetic diversity and effective population size of Mediterrean and subalpine *Arabidopsis thaliana* populations. Mol Ecol 20: 3540–3554.
39. Ehrich D (2006) AFLPdat: a collection of R functions for convenient handling of AFLP data. Mol Ecol Notes 6: 603–604.
40. Vekemans X, Beauwens T, Lemaire M, Roldan-Ruiz I (2002) Data from amplified fragment length polymorphism (AFLP) markers show indication of size homoplasy and of a relationship between degree of homoplasy and fragment size. Mol Ecol 11: 139–151.
41. Zhivotovsky LA (1999) Estimating population structure in diploids with multilocus dominant DNA markers. Mol Ecol 8: 907–913.
42. Peakall R, Smouse PE (2006) GENALEX 6: genetic analysis in Excel. Population genetic software for teaching and research. Mol Ecol Notes 6: 288–295.
43. Peakall R, Smouse PE (2012) GENALEX 6.5: Genetic analysis in Excel. Population genetic software for teaching and research-an update. Bioinformatics 28: 2537–2539.
44. Nei M, Takezaki N (1983) Estimation of genetic distances and phylogenetic trees from DNA analysis. Proc. 5th World Cong. Genet. Appl. Livestock Prod. 21: 405–412.
45. Liu K and Muse SV (2005) PowerMarker: an integrated analysis environment for genetic marker analysis. Bioinformatics 21: 2128–2129.
46. Felsenstein J (1993) Phylogeny Inference Package (PHYLIP). Version 3.5. Seattle: University of Washington.
47. Pritchard JK, Stephens M, Donnelly P (2000) Inference of population structure using multilocus genotype data. Genetics 155: 945–959.
48. Falush D, Stephens M, Pritchard JK (2003) Inference of population structure using multilocus genotype data: Linked loci and correlated allele frequencies. Genetics. 164: 1567–1587.
49. Falush D, Stephens M, Pritchard JK (2007) Inference of population structure using multilocus genotype data: dominant markers and null alleles. Mol Ecol Notes 7: 574–578.
50. Evanno G, Regnaut R, Goudet J (2005) Detecting the number of clusters of individuals using the software STRUCTURE: A simulation study. Mol Ecol 14: 2611–2620.
51. Rosenberg NA (2004) DISTRUCT: a program for the graphical display of population structure. Mol Ecol Notes 4: 137–138.
52. Duchesne P, Bernatchez L (2002) AFLPOP: A computer program for simulated and real population allocation based on AFLP data. Mol Eco Notes 3: 380–383.
53. Paetkau D, Calvert W, Stirling I, Strobeck C (1995) Microsatellite analysis of population structure in Canadian polar bears. Mol Ecol 4: 347–354.
54. Mantel N (1967) The detection of disease clustering and a generalized regression approach. Cancer Res. 27: 209–220.
55. Foll M, Fischer MC, Heckel G, Excoffier L (2010) Estimating population structure from AFLP amplification intensity. Mol Ecol 19: 4638–4647.
56. Foll M, Gaggiotti O (2008) A Genome-Scan method to identify selected loci appropriate for both dominant and codominant markers: A Bayesian perspective. Genetics 180: 977–993.
57. Fischer MC, Foll M, Excoffier L, and Heckel G (2011) Enhanced AFLP genome scans detect local adaptation in high-altitude populations of a small rodent (*Microtus arvalis*). Mol. Ecol. 20: 1450–1462.
58. Antao T, Beaumont MA (2011) Mcheza: A workbench to detect selection using dominant markers. Bioinformatics 27: 1717–1718.
59. Leberg P (2005) Genetic approaches for estimating the effective size of populations. J Wildl Manage 69: 1385–1399.
60. Jorde PE, Palm S, Ryman N (1999) Estimating genetic drift and effective population size from temporal shifts in dominant gene marker frequencies Mol Ecol 8: 1171–1178.
61. Hahn T, Kettle CJ, Ghazoul J, Frei ER, Matter P, et al. (2012) Patterns of genetic variation across altitude in three plant species of semi-dry grasslands PLoS ONE, 7(8) art. no. e41608.
62. Meirmans PG, Goudet J, Gaggiotti OE (2011) Ecology and life history affect different aspects of the population structure of 27 high-alpine plants. Mol Ecol 20: 3144–3155.
63. Hargreaves S, Maxted N, Hirano R, Abberton M, Skøt L, et al. (2010) Islands as refugia of *Trifolium repens* genetic diversity. Conserv Genet: 1317–1326.
64. Jabis MD, Ayers TJ, Allan GJ (2011) Pollinator-mediated gene-flow forsters genetic variability in a narrow alpine endemic, *Abronia alpine* (Nyctaginaceae). Am J Bot 98: 1583–1594.
65. Peters MD, Xiang Q, Thomas DT, Stucky J, Whiteman NK (2009) Genetic analysis of the federally endangered *Echinacea laevigata* using amplified fragment length polymorphisms (AFLP)- inferences in population genetic structure and mating system. Conserv Genet 10: 1–14.
66. Waitt J, Thorson (1983) The Cordilleran Ice Sheet in Washington, Idaho, and Montana. In: Wright J. editor. Late-Quaternary Environments of the United States, Volume 1: The Late Pleistocene St Paul:University of Minnesota Press, 407 p.
67. Cobben MMP, Verboom J, Opdam PFM, Hoekstra RF, Jochem R, et al. (2011) Projected climate change causes loss and redistribution of genetic diversity in a model metapopulation of a medium-good disperser. Ecography 34: 920–932.
68. McInerny GJ, Turner JRG, Wong HY, Travis JMJ, Benton TG (2009) How range shifts induced by climate change affect neutral evolution. Proc R Soc Edinb Biol Sci 276: 1527–1534.
69. Khanlou KM, Vandepitte K, Asl LK, van Bockstaele E (2011) Towards an optimal sampling strategy for assessing genetic variation within and among white clover (*Trifolium repens* L.) cultivars using AFLP. Int J Genet Mol Biol 34: 252–258.
70. Greene SL, Gritsenko M, Vandemark G (2004) Relating morphologic and RAPD marker variation to collection site environment in wild populations of red clover (*Trifolium pratense* L.) Gen Res Crop Evol 51: 643–653.
71. Mosjidis JA, Greene SL, Klingler KA, Afonin A (2004) Isozyme diversity in wild red clover populations from the Caucasus. Crop Sci 44: 665–670.
72. Bothwell H, Bisbing S, Therkildsen NO, Crawford L, Alvarez N, et al. (2013) Identifying genetic signatures of selection in a non-model species, alpine gentian (*Gentiana nivalis* L.), using a landscape genetic approach. Conserv Genet, 14: 467–481.
73. Brütting C, Hensen I, Wesche K (2013) *Ex situ* cultivation affects genetic structure and diversity in arable plants. Plant Biol 15: 505–513.
74. Ellstrand N (1992) Gene flow by pollen: implications for plant conservation genetics. Oikos 63: 77–86.
75. Trakhtenbrot A, Nathan R, Perry G, Richardson DM (2005) The importance of long-distance dispersal in biodiversity conservation. Diver Distrib 11: 173–181.
76. Wright S (1931) Evolution in Mendelian populations. Genetics 16: 97–159.
77. Slatkin M (1985) Gene flow in natural populations. Annu Rev Ecol Syst 16: 393–430.
78. Steffan-Dewenter I, Westphal C (2008) The interplay of pollinator diversity, pollination services and landscape change. J Appl Ecol 45: 737–741.

19

Global Wild Annual *Lens* Collection: A Potential Resource for Lentil Genetic Base Broadening and Yield Enhancement

Mohar Singh[1]*, Ishwari Singh Bisht[1], Sandeep Kumar[1], Manoranjan Dutta[1], Kailash Chander Bansal[1], Moreshwar Karale[1], Ashutosh Sarker[2], Ahmad Amri[3], Shiv Kumar[4], Swapan Kumar Datta[5]

1 National Bureau of Plant Genetic Resources, Pusa, New Delhi, India, **2** International Centre for Agricultural Research in Dry Areas, South Asia and China Regional Programme, NASC Complex, DPS Marg, Pusa, New Delhi, India, **3** International Centre for Agricultural Research in Dry Areas, Tunisia, Tunis, **4** International Centre for Agricultural Research in Dry Areas, Rabat, Morocco, **5** Division of Crop Science, Indian Council of Agricultural Research, Krishi Bhawan, New Delhi, India

Abstract

Crop wild relatives (CWRs) are invaluable gene sources for various traits of interest, yet these potential resources are themselves increasingly threatened by the impact of climate change as well as other anthropogenic and socio-economic factors. The prime goal of our research was to cover all aspects of wild *Lens* genetic resource management like species characterization, agro-morphological evaluation, diversity assessment, and development of representative sets for its enhanced utilization in lentil base broadening and yield improvement initiatives. We characterized and evaluated extensively, the global wild annual *Lens* taxa, originating from twenty seven counties under two agro-climatic conditions of India consecutively for three cropping seasons. Results on various qualitative and quantitative characters including two foliar diseases showed wide variations for almost all yield attributing traits including multiple disease resistance in the wild species, *L. nigricans* and *L. ervoides* accessions. The core set developed from the entire *Lens* taxa had maximum representation from Turkey and Syria, indicating rich diversity in accessions originating from these regions. Diversity analysis also indicated wide geographical variations across genepool as was reflected in the core set. Potential use of core set, as an initial starting material, for genetic base broadening of cultivated lentil was also suggested.

Editor: Qin Zhang, China Agricultrual University, China

Funding: This research was funded by the DAC-ICAR-ICARDA collaborative research programme (Department of Agriculture and Cooperation Govt of India-Indian Council of Agricultural Research and International Centre for Agricultural Research in Dry Areas, Aleppo, Syria). The funders had no role in study design, data collection and analysis, decision to publish, or preparation of the manuscript.

Competing Interests: The authors have declared that no competing interests exist.

* Email: singhmohar_2003@yahoo.com

Introduction

Crop wild relatives (CWRs) are the reservoir of useful genes and alleles that can be used in breeding new and better adapted varieties resistant to biotic and abiotic stresses, and more importantly to adapt to adverse effect of climate change. [1,2]. Despite their potential value, many CWRs are, however, not adequately collected and conserved in gene banks across the world. Increasing threats to natural habitats and farming systems makes it imperative to collect, conserve and characterize CWRs in order to make them available for use in mitigating the impact of major biotic and abiotic stresses caused by climate change [1] and other factors. The genus *Lens* Miller is part of the family *Fabaceae* (*Leguminosae*), and is placed in either subfamily *Faboideae* tribe *Fabeae* [3] or in subfamily *Papilionaceae* tribe *Vicieae* [4]. Lentil is an annual self-pollinating true diploid ($2n = 2x = 14$) species with an estimated genome size of 4063 Mbp/C [5]. The cultivated lentil (*Lens culinaris* Medikus ssp. *culinaris*) encompasses two groups on the basis of agro-morphological traits, the small-seeded (*microsperma*), large-seeded (*macrosperma*) and *L. culinaris* ssp. *orientalis* (Boiss.) Ponert, is considered its immediate wild progenitor. The other sub-species and species are *L. culinaris* ssp. *tomentosus* (Ladiz.), *L. culinaris* ssp. *odemensis* (Ladz.), *L. ervoides* (Brign.), *L. nigricans* (Bieb.) and *L. lamottei* Czefr. [6]. Globally lentil productivity has increased from an average yield of 560 kg ha^{-1} in 1961–63 to 950 kg ha^{-1} during 2010–2011. Despite this increase, the current yields are much lower as compared to many other grain legume species, because of the limited yield potential of modern lentil cultivars. Major yield limiting factors include, poor seedling vigour, slow leaf area development, high rate of flower drop, low pod setting, poor dry matter, low harvest index, lack of lodging resistance, and exposure to several biotic and abiotic stresses [7]. Further, lentil breeding programme in the past has been primarily depended on landraces of cultivated lentil. Recently developed lentil varieties, however, have superiority over traditional cultivars in terms of their yield potential and disease resistance. For the development of these modern varieties, a small number of improved local landraces have been utilized mainly through pure-line selection following hybridization between lines adapted to specific environmental conditions. Substantial improvement in the lentil productivity could not be achieved due to the loss of invaluable alleles for higher productivity including resistance/tolerance to major biotic and abiotic stresses and low extent of genetic variation in lentil cultivars [8–11]. Wild relatives of lentil are good sources of disease resistance as well as many useful agronomic characters like, high number of pods plant^{-1}, high number of pods cluster^{-1}, high number of branches plant^{-1}, and short internodes [12,13]. Furthermore, substantial genetic diversity is currently not available in *ex-situ* conserved collections. The treasure of lentil genetic resource is held up in populations poorly or incompletely sampled or completely unsampled wild *Lens* taxa. The gap not only limits its use, it also renders precious genetic diversity inaccessible and vulnerable to extinction [14]. Maxted et al. (2012a) [15] suggested a comprehensive strategy for gap analysis on CWRs including lentil for germplasm conservation and how existing georeferenced passport data associated with the collections of *Lens* species can be utilized in identifying gaps in the current *ex-situ* holdings and develop reliable *in-situ* conservation strategy.

The efficient utilization of desired diversity of crop wild relatives will be the solution to achieve substantial gains in productivity of cultivated gene pool. Maintaining and evaluating a large set of accessions of wild *Lens* species is often difficult due to various reasons like inadequate seed germination, high pod shattering and photoperiod sensitivity and certain wild *Lens* species do not flower under normal winter cropping season. These photoperiod sensitive accessions would require artificially lengthened days in the greenhouse or growing them under long days in summer season.

Establishing a core collection of the existing holding is a favoured approach for efficient utilization of novel variations in wild gene pool with a manageable number. The core collection is a limited set of accessions representing, with minimum repetitiveness, the genetic diversity of a crop and its wild relatives [16–19]. Genebank curators have the responsibility for conservation, regeneration, safety duplication, documentation, characterization and evaluation as well as facilitating utilization of the genetic resources to the breeder for crop improvement. The core collection, being smaller in size in comparison to the whole collection, can be used very effectively as a starting point for research involving screening of the germplasm collection for sources of desirable traits. Since, the inception of the idea of core collections, many approaches for selecting core collections have been proposed and used e.g. M-Strat [20], Genetic Distance Sampling [21], PowerCore [22] and Core Hunter [23]. Similarly, core has been developed using several kind of data ranging from genealogical data in the Czech spring wheat [24], agronomic data in groundnut [25,26] and molecular data or integration of data, in bread wheat [27] and in rice [28,29]. The core collection of wild *Lens* gene pool could serve as a working collection for the users, which then needs to be extensively evaluated and used in lentil improvement. Focused Identification of Germplasm Strategy (FIGS approach), based on co-evolution between the plant and the environment conditions, uses trait and environmental data (climate data and phenological data) to develop *a priori* information based on the quantification of the trait-environment relationship. This relationship when detected can be used to develop subsets of accessions with a high probability of containing the sought after traits. This approach was used efficiently in wheat, barley and faba bean to select novel sources of resistance and tolerance to diseases, insect, drought and salinity [30,31].

In view of the above, the objectives of the present study were (1) to systematically characterize and evaluate global wild annual *Lens* collection for different morphological and agronomic traits including major foliar biotic stresses, (2) to develop core set of wild *Lens* taxa for enhanced utilization in lentil genetic base broadening and yield enhancement, (3) to assess the diversity available in global wild annual *Lens* collection, and compare it with that of core set and (4) to analyse the geographical distribution of wild *Lens* taxa representing maximum diversity, and suggest potential areas for future collecting and conserving diversity *in-situ* on-farm.

Materials and Methods

Plant material

A total of 405 diverse global wild annual *Lens* accessions comprising 4 of cultivated *L. culinaris* ssp. *culinaris*, 171 of *L. culinaris* ssp. *orientalis*, 42 of *L. culinaris* ssp. *odemensis*, 20 of *L. culinaris* ssp. *tomentosus*, 35 of *L. nigricans*, 124 of *L. ervoides* and 9 of *L. lamottei* (Table 1) were studied encompassing various aspects of genetic resource management. These include species characterization and agro-morphological evaluation, development of core sets for its enhanced utilization in lentil genetic base broadening, diversity assessment, and geographical distribution of *Lens* taxa representing maximum diversity. The wild *Lens* accessions were obtained from the Genebank of the International Centre for Agricultural Research in the Dry Areas (ICARDA) Aleppo, Syria. The original identity of accessions in the present

Table 1. List of wild annual *Lens* accessions introduced from ICARDA, Aleppo, Syria.

S. No.	Species	Accessions	Origin
1	*Lens culinaris* ssp. *culinaris*	04	India, Israel
2	*Lens culinaris* ssp. *orientalis*	171	Turkey, Iran, Cyprus, Israel, Syria, Jordan, Uzbekistan, Tajikistan, Turkmenistan, Lebanon, Armenia, Azerbaijan, Russia, Czech Republic
3	*Lens culinaris* ssp. *odemensis*	42	Syria, Palestine, Turkey, Israel
4	*Lens culinaris* ssp. *tomentosus*	20	Turkey, Syria
5	*Lens nigricans*	35	Syria, Italy, France, Spain, Croatia, Bahrain, Turkey, Ukraine, Montenegro, Croatia
6	*Lens ervoides*	124	Ukraine, Turkey, Croatia, Italy, Former Soviet Union, Bahrain, Israel, Palestinian Territories, Syrian, Jordan, Tajikistan, Lebanon, Former Yugoslavia, Montenegro, Armenia, Azerbaijan
7	*Lens lamottei*	09	Spain, Turkey, France

Table 2. Frequency distribution of qualitative plant characteristics in different annual *Lens* species.

Character	Class (descriptor state)	Frequency						
		L. culinaris ssp. *culinaris*	*L. culinaris* ssp. *orientalis*	*L. culinaris* ssp. *odemensis*	*L. culinaris* ssp. *tomentosus*	*L. nigricans*	*L. ervoides*	*L. lamottei*
Seedling stem pigmentation	0: Absent	1.00	0.09	0.03	0.11	0.54	0.15	0.43
	1: Present	0.00	0.91	0.97	0.89	0.46	0.85	0.57
Leaf pubescence	0: Absent	0.00	0.03	0.00	0.00	0.14	0.30	0.29
	3: Slight	0.75	0.50	0.35	0.21	0.64	0.66	0.71
	7: Dense	0.25	0.47	0.65	0.79	0.22	0.04	0.00
Leaflet size	3: Small	0.00	0.53	0.30	0.16	0.82	0.37	0.29
	5: Medium	0.50	0.46	0.68	0.79	0.14	0.49	0.71
	7: Large	0.50	0.01	0.02	0.05	0.04	0.14	0.00
Tendril length	1: Rudimentary	0.17	0.29	0.14	0.11	0.25	0.33	0.71
	2: Prominent	0.83	0.71	0.86	0.89	0.75	0.67	0.29
Pod shedding	0: None	1.00	0.00	0.00	0.00	0.00	0.00	0.00
	3: Low	0.00	0.11	0.20	0.00	0.23	0.11	0.50
	5: Medium	0.00	0.28	0.60	0.36	0.54	0.26	0.33
	7: High	0.00	0.61	0.20	0.64	0.23	0.63	0.17
Pod dehiscence	0: None	1.00	0.00	0.00	0.00	0.00	0.00	0.00
	3: Low	0.00	0.00	0.20	0.00	0.18	0.11	0.33
	5: Medium	0.00	0.90	0.80	0.64	0.64	0.56	0.50
	7: High	0.00	0.10	0.00	0.36	0.18	0.33	0.17
Flower colour	1: White	0.17	0.09	0.00	0.00	0.22	0.00	0.29
	2: Yellow	0.00	0.00	0.00	0.00	0.00	0.00	0.00
	3: Red	0.00	0.00	0.00	0.00	0.00	0.00	0.00
	4: Purple	0.83	0.91	1.00	1.00	0.78	1.00	0.71
Ground colour of testa	1: Green	0.00	0.00	0.00	0.00	0.00	0.00	0.00
	2: Grey	0.00	0.07	0.35	0.06	0.37	0.26	0.57
	3: Brown	1.00	0.93	0.65	0.94	0.63	0.74	0.43
	4: Black	0.00	0.00	0.00	0.00	0.00	0.00	0.00
	5: Pink	0.00	0.00	0.00	0.00	0.00	0.00	0.00
Pattern of testa	0: Absent	0.50	0.01	0.00	0.06	0.00	0.04	0.00
	1: Dotted	0.50	0.17	0.05	0.00	0.11	0.08	0.00
	2: Spotted	0.00	0.12	0.11	0.06	0.15	0.08	0.14
	3: Marbled	0.00	0.58	0.16	0.76	0.59	0.74	0.57
	4: Complex	0.00	0.12	0.68	0.12	0.15	0.06	0.29
Cotyledon colour	1: Yellow	0.25	0.04	0.78	0.00	0.19	0.10	0.71
	2: Orange	0.75	0.96	0.22	1.00	0.81	0.90	0.29

Table 2. Cont.

Character	Class (descriptor state)	Frequency						
		L. culinaris ssp. *culinaris*	*L. culinaris* ssp. *orientalis*	*L. culinaris* ssp. *odemensis*	*L. culinaris* ssp. *tomentosus*	*L. nigricans*	*L. ervoides*	*L. lamottei*
	3: Olive-green	0.00	0.00	0.00	0.00	0.00	0.00	0.00

Table 3. Range of variations for different quantitative agro-morphological traits in annual *Lens* species.

Character	*Lens culinaris* ssp. *culinaris*			*Lens culinaris* ssp. *orientalis*			*Lens culinaris* ssp. *odemensis*		
	Range	Mean ± SE	CV (%)	Range	Mean ± SE	CV (%)	Range	Mean ± SE	CV (%)
Days to 50% flowering	34.50–43.00	39.75±1.89	9.53	27.00–135.00	93.76±2.35	32.81	88.00–143.00	126.42±1.81	9.30
Days to maturity	72.00–86.00	79.50±2.93	7.38	76.00–168.00	130.15±1.97	19.80	124.00–174.00	155.02±1.50	6.29
Plant height (cm)	17.00–24.00	20.50±1.55	15.12	12.33–95.50	26.58±0.68	33.48	20.33–59.94	33.38±1.59	30.94
No. of branches plant^{-1}	6.50–9.50	8.12±0.80	19.70	4.00–35.00	13.47±0.68	66.07	3.50–22.00	14.69±0.32	29.06
No. of pods plant^{-1}	18.50–208.00	92.50±44.50	96.22	3.00–250.00	59.58±3.68	80.89	5.50–185.50	74.75±6.95	60.29
No. of seeds pod^{-1}	1.51–1.83	1.63±0.07	8.58	0.74–2.33	1.41±0.02	23.40	1.00–1.96	1.78±0.03	12.35
No. of seeds plant^{-1}	28.50–316.50	147.37±67.62	91.76	4.00–409.00	82.53±5.1	80.84	6.00–357.50	136.26±13.83	65.77
100-seed weight (g)	0.75–2.50	1.62±0.38	47.53	0.15–4.00	0.91±0.03	56.04	0.40–1.10	0.73±0.02	24.65
Seed yield plant^{-1} (g)	0.25–4.25	1.85±0.89	96.75	0.01–3.26	0.80±0.04	80.00	0.06–3.30	1.05±0.11	68.57
Biological yield plant^{-1} (g)	6.35–13.90	8.92±1.70	38.22	0.50–18.50	3.73±0.15	54.95	0.25–10.55	5.45±0.29	35.59

Table 4. Range of variations for different quantitative agro-morphological traits in annual *Lens* species.

Character	*Lens culinaris* ssp. *tomentosus*			*Lens nigricans*			*Lens ervoides*			*Lens lamottei*		
	Range	Mean ± SE	CV (%)	Range	Mean ± SE	CV (%)	Range	Mean ± SE	CV (%)	Range	Mean ± SE	CV (%)
Days to 50% flowering	120.00–132.00	129.65±0.66	2.28	92.00–139.00	123.88±2.60	12.40	60.00–141.00	118.88±1.99	18.67	92.00–130.00	119.88±5.29	13.24
Days to maturity	20.00–173.00	162.75±0.78	2.16	130.00–176.00	160.42±2.41	8.90	93.00–173.00	153.63±1.97	14.38	130.00–165.00	154.11±4.65	9.05
Plant height (cm)	20.75–50.00	33.27±1.85	24.94	19.66–67.00	40.23±1.95	28.66	2.00–59.00	37.51±0.94	28.15	27.25–54.00	40.90±2.99	21.95
No. of branches plant^{-1}	10.50–19.50	15.26±0.60	17.62	2.00–22.75	13.57±0.85	37.14	4.25–52.00	16.99±1.67	109.53	6.00–19.80	14.20±1.54	32.53
No. of pods plant^{-1}	4.50–218.00	74.05±11.50	68.84	5.50–153.00	58.30±7.58	76.92	3.00–1002.00	111.78±11.74	116.94	8.00–185.50	63.97±18.80	88.19
No. of seeds pod^{-1}	1.00–1.97	1.70±0.06	17.05	1.00–2.00	1.64±0.04	14.63	1.00–2.22	1.72±0.02	14.53	1.68–2.10	1.85±0.04	7.02
No. of seeds plant^{-1}	6.00–420.00	133.00±22.30	74.97	7.50–275.00	100.45±14.01	82.46	3.00–1425.00	196.31±18.72	106.16	13.75–364.00	121.10±36.89	91.38
100-seed weight (g)	0.10–1.10	0.69±0.06	44.92	0.25–1.20	0.57±0.03	38.59	0.10–2.00	0.49±0.02	51.02	0.60–1.85	0.92±0.15	48.91
Seed yield plant^{-1} (g)	0.01–3.30	1.01±0.17	79.20	0.02–2.30	0.62±0.09	93.54	0.04–5.70	0.97±0.09	104.12	0.10–2.75	0.99±0.29	89.89
Biological yield plant^{-1} (g)	1.90–10.00	5.79±0.40	31.43	1.70–15.30	5.37±0.47	52.32	1.20–29.50	5.87±0.35	66.78	3.65–9.90	5.55±0.71	38.55

study was the same as established by the ICARDA Genetic Resources Unit.

Species characterization and agronomic evaluation

All 405 global annual *Lens* accessions including four commercial lentil cultivars were grown at two agro-ecological regions of India, the National Bureau of Plant Genetic Resources (NBPGR), New Delhi, India during the winter (*rabi*) season of 2011–2012, 2012–2013 and 2013–2014 for characterization and agro-morphological evaluation and the summer Himalayan nursery (*summer* 2012 and 2013) at Himachal Pradesh Agricultural University (HPAU), Research Station (RS), Sangla. The permission to conduct experiments at Sangla was granted by worthy Vice-Chancellor of the University. The location of NBPGR lies in the northern plain between 28^0 35′ N latitude, 70^0 18′ E longitude, 226 m asl altitude, whereas the HPAU, RS, Sangla lies between latitudes 31^0 55′ and 32^0 20′ N, and longitudes 77^0 00′ and 79^0 50′ E, in the northwestern Himalayas. Each accession was sown by hand in three replications of 3 m long rows spaced 30 cm apart. The soil type was sandy loam at both the locations. All these annual *Lens* species were characterized for qualitative morphological traits viz; seedling stem pigmentation, leaf pubescence, leaflet size, tendril length, pod shedding, pod dehiscence, flower colour, ground colour of testa, pattern of testa and cotyledon colour using standard lentil descriptors [32]. Observations were also recorded on ten plants accession^{-1} in each replication for ten quantitative morphological traits viz; days to flowering, days to maturity, plant height (cm), number of branches plant^{-1}, number of pods plant^{-1}, number of seeds pod^{-1}, number of seeds plant^{-1}, 100-seed weight (g), seed yield plant^{-1} (g) and biological yield plant^{-1}(g). The data were subjected to statistical analysis using MS Ofiice Excel and SAS software.

Screening against foliar biotic stresses

Rust (Uromyces fabae (Grev.) Fuckel). Reaction to *Uromyces fabae* was conducted at the Himachal Pradesh Agricultural University Research and Extension Centre (HPAUREC), Dhaulakuan (30^0 4′ N latitude, 77^0 5′ longitudes, 468 m asl) located in the foot hills of northwestern Himalayas. All the accessions were grown in 25 cm diameter plastic pots in soil and farmyard manure mixture (10:1). Each accession had two replications. Pots with plants at the vegetative stage were transferred in the field and were artificially inoculated by frequently spraying with ascosporic suspension (1×10^6 spores/ml) using the isolate of *Uromyces fabae*. The data were recorded on terminal disease reaction at vegetative and reproductive plant stages using 1–9 rating scale as suggested by [33]. Based on disease score, the wild accessions were categorized for their reaction to rust infection as: 1 = asymptomic (Free); 3 = resistant (R); 5 = moderately resistant (MR); 7 = susceptible (S); and 9 = highly susceptible (HS).

Powdery mildew (*Erysiphe polygoni DC.*). Powdery mildew is an economically important foliar disease of lentil especially in those areas where warm and humid weather persists. For the screening of wild *Lens* species, all accessions were grown in 25 cm diameter plastic pots in soil and farmyard manure mixture (10:1), and planted pots were kept and maintained in the glasshouse. The disease appeared at flower initiation stage, which was further allowed to develop. The pots with heavily infected plants were shifted near to test entries grown in the glass house. The heavily infected plants were shaken well over the test entries in the evening time to disperse conidia, which helps in spreading the disease on the test entries as suggested by [34]. Data on disease incidence were recorded using 1–9 rating scale at vegetative and reproduc-

Table 5. Species-wise representation of entire *Lens* collection and representation of accessions in core sets derived from different data sets.

Lens species	Accessions in entire collection	Accessions in core sets		
		I	II	III
L. culinaris ssp. *culinaris*	4	4	4	3
L. culinaris ssp. *orientalis*	171	21	19	7
L. culinaris ssp. *odemensis*	42	15	12	4
L. culinaris ssp. tomentosus	20	12	9	6
L. nigricans	35	18	17	7
L. ervoides	124	20	19	5
L. lamottei	9	6	6	4
Total	405	96 (23.7%)	86 (21%)	36 (8.9%)

Core sets based on I = combined data on both quantitative and qualitative variables; II = data only on quantitative variables, and III = data only on qualitative variables, respectively.

tive stages as: 1 = asymptomic (Free); 3 = resistant (R); 5 = moderately resistant (MR); 7 = susceptible (S); and 9 = highly susceptible (HS).

Development of core set

The entire set of *Lens* species was used for developing a core collection using "PowerCore" (http://genebank.rda.go.kr/powercore/) software developed at the Rural Development

Table 6. Comparison of SDI of entire collection to core sets derived from different data sets.

S. No.	Traits	Entire collection	Core sets (Using PowerCore)*			Core set derived through PCSS
			I	II	III	
Quantitative characters						
1	Days to flowering	0.733	0.819	0.843	0.750	0.691
2	Days to maturity	0.795	0.851	0.881	0.685	0.729
3	Plant height	0.964	0.777	0.725	0.874	0.664
4	No. of branches plant^{-1}	0.035	0.116	0.127	0.987	0.116
5	No. of pods plant^{-1}	0.200	0.406	0.440	0.760	0.511
6	No. of seeds plant^{-1}	0.297	0.533	0.635	0.632	0.716
7	No. of seeds pod^{-1}	0.806	0.817	0.830	0.958	0.837
8	100-seed weight	0.725	0.883	0.914	0.766	0.710
9	Seed yield plant^{-1}	0.735	0.745	0.698	0.816	0.698
10	Biological yield plant^{-1}	0.637	0.972	0.641	0.787	0.821
11	Resistance to rust	0.942	0.979	0.973	0.849	0.908
12	Resistance to powdery mildew	0.886	0.953	0.948	0.880	0.843
Qualitative characters						
13	Seedling stem pigmentation	0.475	0.613	0.613	0.687	0.525
14	Leaf pubescence	0.963	0.917	0.952	0.966	0.982
15	Leaflet size	0.871	0.926	0.858	0.935	0.890
16	Tendril length	0.614	0.584	0.581	0.654	0.562
17	Flower colour	0.235	0.391	0.359	0.572	0.311
18	Ground colour of testa	0.543	0.679	0.629	0.787	0.562
19	Pattern of testa	0.790	0.915	0.880	0.970	0.805
20	Cotyledon colour	0.500	0.621	0.622	0.637	0.562
21	Pod shedding	0.956	0.833	0.819	0.921	0.809
22	Pod dehiscence	0.837	0.835	0.833	0.870	0.970

Core sets based on I = combined data on both quantitative and qualitative variables; II = data only on quantitative variables, and III = data only on qualitative variables, respectively.

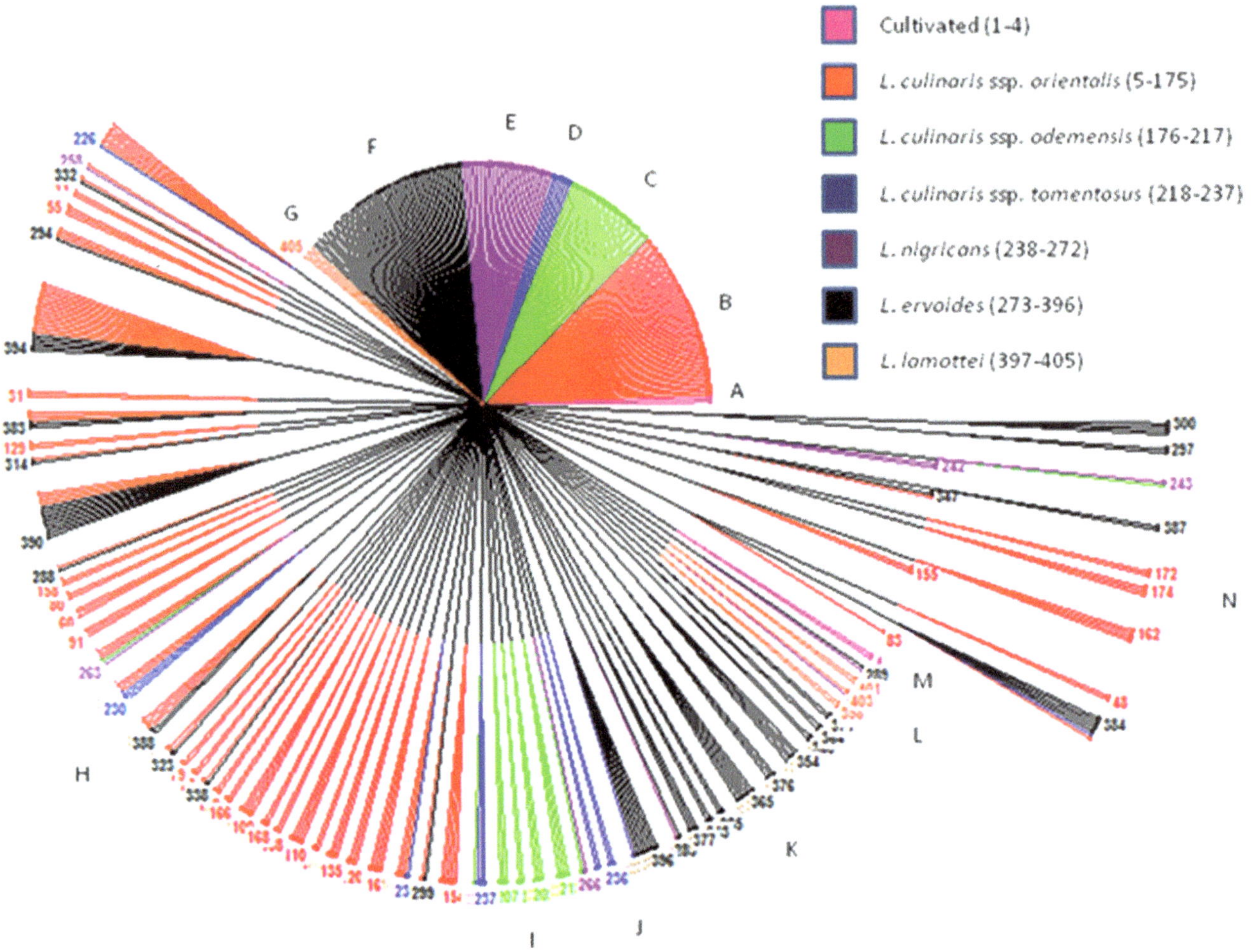

Figure 1. Hierarchical clustering of *Lens* accessions based on qualitative phenotypic data.

Administration (RDA), South Korea [22]. Separate sets were developed using both quantitative and qualitative data in combination as well as in isolation. "PowerCore" creates subsets representing all alleles or observation classes, with the least allelic redundancy, and ensures a highly reproducible list of entries. The Shannon Diversity Index [35] was used to compare the representativeness of the entire holding and the accessions selected as core entries.

Diversity analysis

Diversity analysis was conducted using both qualitative and quantitative phenotypic data of complete set of 405 accessions as well as of the core set developed from it. Genetic relationships among accessions were determined by means of Dice's coefficient [36] in case of qualitative data. Phenotypic relationships based on quantitative data were determined from the Euclidean distances. Trees were constructed using hierarchical clustering method by means of Jaccard's coefficient with UPGMA mode with the DARwin 5.0 software [37,38].

Results

All 405 accessions of global annual *Lens* taxa including four commercial cultivars were characterized and evaluated for morphological and agronomic traits as well as for two major fungal diseases under two agro-climatic regions in the north-western Indian conditions.

Species characterization using qualitative traits

The frequency distribution of seven annual *Lens* species showed a wide range of intraspecific variation for majority of the important morphological plant characteristics (Table 2). Seedling stem pigmentation exhibited variations in all the *Lens* species except, *L. culinaris* ssp. *culinaris*, where it was absent. Likewise, leaf pubescence, leaflet size and tendril length also revealed remarkable variation in all seven annual *Lens* taxa. There was substantial variation in pod shedding and dehiscence for majority of the *Lens* species except, *L. culinaris* ssp. *culinaris*. Flower colour was white and purple in majority of the *Lens* species like, *L. culinaris* ssp. *culinaris*, *L. culinaris* ssp. *orientalis*, *L. nigricans* and *L. lamottei* and in *L. culinaris* ssp. *odemensis*, *L. culinaris* ssp. *tomentosus*, and in *L. ervoides*, it was purple only. The ground colour of testa was mostly grey and brown in all the *Lens* species except, *L. culinaris* ssp. *culinaris*, where it was all brown. Substantial variations were also found in the pattern of testa for all the annual *Lens* species. All the annual *Lens* species had yellow and orange cotyledon colour except, *L. culinaris* ssp. *tomentosus*, where it was all orange (Table 2).

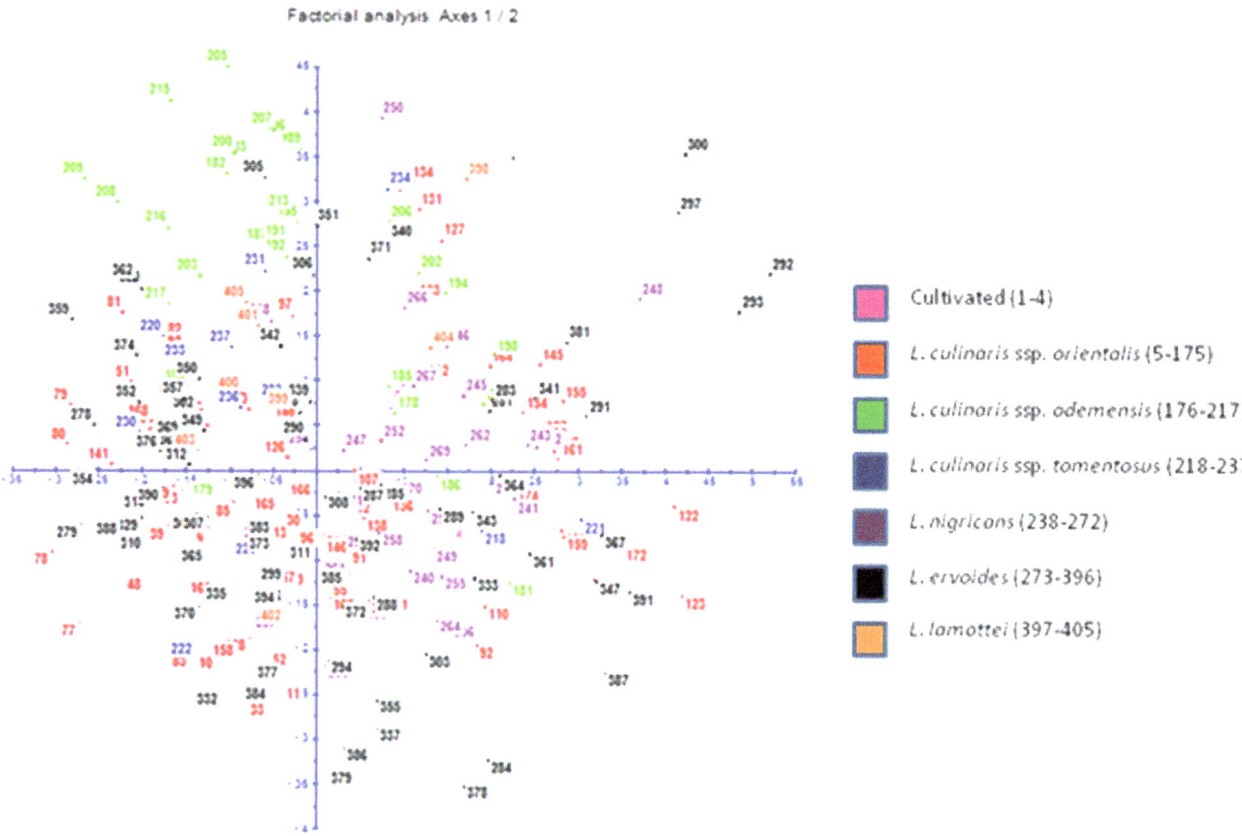

Figure 2. Factorial analysis of *Lens* accessions based on qualitative phenotypic data.

Agronomic evaluation using quantitative traits

All the annual *Lens* accessions were evaluated for various agro-morphological quantitative traits. The results revealed significant variations ($p = 0.05$) as was evident from the analysis of variance as also the range, mean and coefficient of variations for all traits studied (Table 3 and 4). In general, the coefficient of variation ranged from low (days to flowering and maturity) to high (number of branches^{-1}, number of pod plant^{-1} and number of seeds pod^{-1}) for important agro-morphological traits in all the cropping seasons. However, the mean performance for days to flowering and maturity was lowest in *L. culinaris* ssp. *culinaris*, followed by *L. culinaris* ssp. *orientalis*. Likewise, maximum plant height was recorded in *L. lamottei and L. nigricans* and lowest in *L. culinaris* ssp.*culinaris*. The mean number of branches plant^{-1} was highest in *L. ervoides* followed by *L. culinaris* ssp. *tomentosus, L. culinaris* ssp. *odemensis*, *L. lamottei* and *L. culinaris* ssp. *orientalis*, and lowest in *L. culinaris* ssp. *culinaris*. There was also substantial variation in the range for number of pods plant^{-1} with maximum number of pods plant^{-1} recorded in *L. ervoides* followed by *L. culinaris* ssp. *culinaris*, *L. culinaris* ssp. *odemensis*, *L. culinaris* ssp. *tomentosus* and *L. culinaris* ssp. *orientalis*. The number of seeds pod^{-1} was >1 in all the annual *Lens* species. As expected, cultivated species showed superior grain yield related agronomic performance as compared to the wild *Lens* species. But, several useful yield attributing traits like high number of pods plant^{-1}, high number of seeds plant^{-1}, high number of branches plant^{-1}, and greater plant height as well as resistance to major foliar diseases was observed in various wild annual *Lens* species.

Screening against foliar biotic stresses

Rust and powdery mildew. Many of the wild *Lens* accessions were resistant and moderately resistant to rust and powdery mildew during all the cropping seasons. Among accessions ILWL247 of *L. culinaris* ssp. *orientalis*; ILWL165, ILWL167 and ILWL238 of *L. culinaris* ssp. *odemensis*; ILWL37, ILWL22 and ILWL38 of *L. nigricans*; and ILWL40, ILWL41, ILWL54 and ILWL58 of *L. ervoides* recorded a score of 2 and identified as highly resistant to rust. While, ILWL350, ILWL369, and ILWL381 of *L. culinaris* ssp. *orientalis*; ILWL39 and ILWL81 of *L. culinaris* ssp. *odemensis*; ILWL191 of *L. nigricans*, and ILWL418, ILWL398, ILWL292, ILWL91 and ILWL294 of *L. ervoides* were found resistant to powdery mildew. Some of the accessions viz., ILWL37, ILWL38, ILWL22 and ILWL191 of *L. nigricans*, and ILWL269 of *L. ervoides* were rated as resistant to both rust and powdery mildew.

Development of core set

"PowerCore" statistical programme selected a set of 96 accessions (24% of the entire collection), when data on both quantitative and qualitative variables were used in combination (Table 5). A total of 86 accessions (21%) were selected, when data only on quantitative variables were used, whereas 36 accessions (about 9%) could be selected, when data only on qualitative characters were used for analysis. Compared with the whole collection, the Shannon Diversity Indices (SDIs) were always higher in the core sets derived from different data sets indicating better representation of the existing diversity in each core set. Better representation of diversity was observed using "Power-

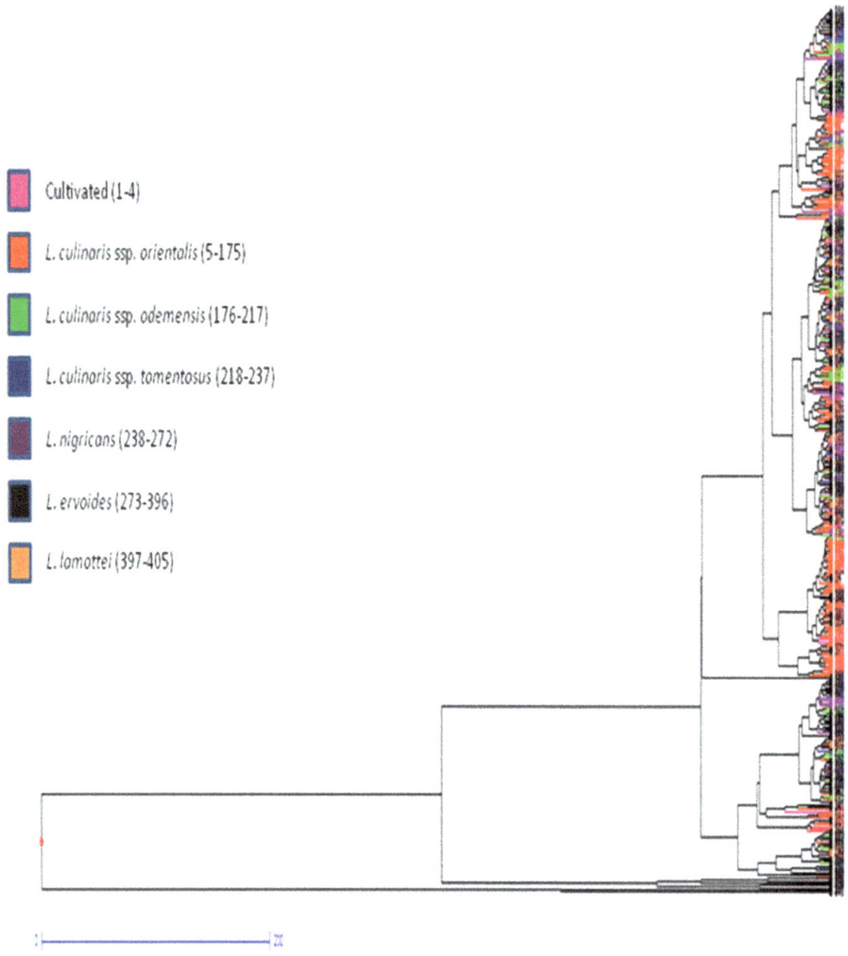

Figure 3. Hierarchical clustering of *Lens* accessions based on quantitative phenotypic data.

Core" as compared to Principal Components Score Strategy (PCSS, Table 6).

Diversity analysis of global *Lens* collection

A Dice similarity matrix was generated based on the morphological qualitative data, which was further used to construct phylogenetic tree for deciphering the diversity of wild *Lens* species. Based on the Dice dissimilarity matrix, an average dissimilarity coefficient value of 0.5 was recorded. The dissimilarity coefficient ranged from 0.0–1.0 indicating that accessions showed 100% similarity to 100% dissimilarity (Table S1). Maximum dissimilarity was assessed between ILWL429 (*L. lamottei*, Spain) and ILL10829 (*L. culinaris* ssp. *culinaris*, Syria); ILWL126 (*L. ervoides*, Syria) and ILL10829 (*L. culinaris* ssp. *culinaris,* Syria), whereas ILWL65 (*L. ervoides*, Turkey) showed 100% dissimilarity to the accessions of *L. culinaris* ssp. *orientalis*, namely ILWL378 (Turkmenistan), ILWL241 (Syria), ILWL231 (Syria) and ILWL230 (Syria); ILWL55 (*L. ervoides*, Israel) to *L. culinaris* ssp. *orientalis* accessions namely ILWL378 (Turkmenistan), ILWL241 (Syria), ILWL231 (Syria) and ILWL230 (Syria); ILWL55 (*L. ervoides*, Israel) to ILWL378 (*L. culinaris* ssp. *orientalis*, Turkmenistan); EC718264 (*L. nigricans*, Spain) to ILWL238 (*L. culinaris* ssp. *odemensis*, Syria); ILWL23 (*L. nigricans*, Italy) to ILWL241 (*L. culinaris* ssp. *orientalis*, Syria); ILWL97 (*L. culinaris* ssp. *tomentosus,* Turkey) to ILWL125 (*L. culinaris* ssp. *orientalis,* Syria); ILL10829 (*L. culinaris* ssp. *culinaris,* Syria) to *L. culinaris* ssp. *odemensis* from Syria including ILWL357, ILWL238 and ILWL167; ILWL371 (*L. culinaris* ssp. *orientalis*, Syria) to *L. culinaris* ssp. *orientalis* ILWL62 (11) and ILWL8 (9); ILWL227 (*L. culinaris* ssp. *orientalis*, Syria) to accessions of the same species ILWL62 (11) and ILWL8 (9). After exclusion of accessions, which showed 100% similarity to each other with dissimilarity matrix value of 0.0 (Table S1), the minimum dissimilarity was exhibited by ILWL247 (*L. culinaris* ssp. *orientalis,* Syria) to *L. ervoides* namely EC718432 (HRV), EC718429 (HRV), EC718427 (Ukraine), EC718422 (Turkey), ILWL67 (Turkey), and ILWL121 (*L. culinaris* ssp. *tomentosus*, Syria); ILWL443 (*L. culinaris* ssp. *orientalis,* Turkey) to other accessions of *L. culinaris* ssp. *orientalis,* ILWL469 (Syria), ILWL416 (Syria), ILWL402 (Lebanon), ILWL385 (Uzbekistan) and ILWL384 (Tajikistan); ILWL247 (*L. nigricans*, Syria) to *L. culinaris* ssp. *orientalis* accessions ILWL86 (Turkey), ILWL95 (Turkey) and ILWL117 (Syria), with dissimilarity coefficient of 0.053. The hierarchical clustering of accessions resulted into various clusters A–N, where clusters A–G, J and M were occupied by accessions of respective individual species, whereas other clusters included accessions from two or more species as shown in Fig. 1. The grouping pattern of different accessions in hierarchical clustering is listed along with their species and respective groups (Table S2). Clustering pattern of the various accessions obtained through factorial analysis, where

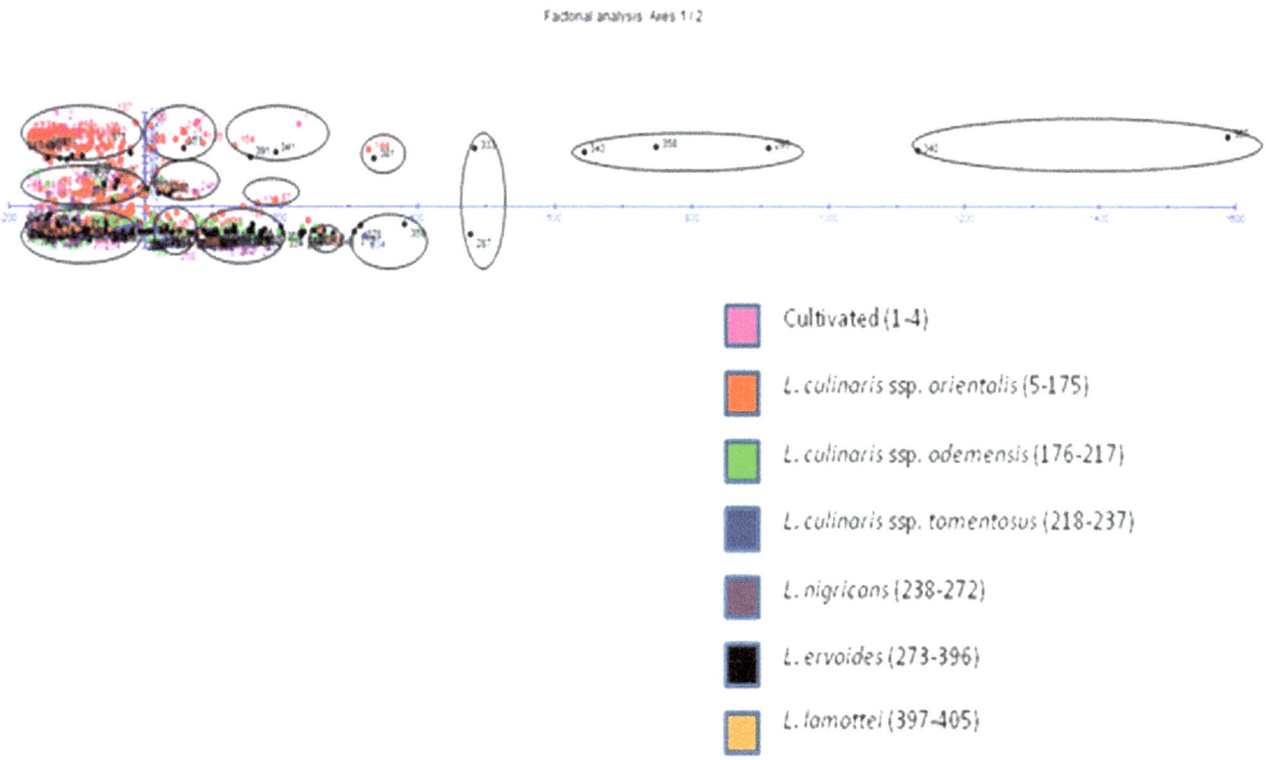

Figure 4. Factorial analysis of *Lens* accessions based on quantitative phenotypic data.

first and second principal coordinates explained only 10.17% and 8.8% variations, respectively (Fig. 2). In general, the clustering pattern of accessions, in both the hierarchical and factorial analysis was not according to their geographical origin.

In case of quantitative data, the Euclidean dissimilarity matrices varied from 2.67 between ILWL159 (Syria) and ILWL67 (Turkey) to 1740.53 between ILWL418 (Syria) and ILWL276 (Turkey) of *L. ervoides* (Table S3). In hierarchical clustering, two major clusters were formed in which major cluster I was occupied by only two accessions namely ILWL292 and ILWL418 of *L. ervoides* from Turkey and Syria, respectively, whereas cluster II included others. Major cluster II further divided into two clusters A and B, where A was occupied by accessions of *L. ervoides*, ILWL401 (Lebanon), ILWL321, ILWL58 and ILWL269 (Turkey), and ILWL51 (Montenegro), whereas cluster B was occupied by accessions from all seven species included in the study (Fig. 3). Cluster B was further divided into sub-clusters (B1 and B2), groups (a, b, c and d) outgroups (a1 and a2, b1 and b2, d1 and d2), clutches (b2a and b2b, d1a and d1b, d2a and d2b) and finally to sub-clutches (i-xxiii) as shown in Fig. 3 (Table S4). Cluster II was reported to be more heterogeneous with different sub-clusters and few individual accessions forming separate groups. Factorial analysis verified the results obtained *via* hierarchical clustering as it resulted into almost similar clustering pattern as shown in Fig. 4. In factorial analysis, first two principal coordinates explained 98.26 per cent cumulative variations with 92.94 and 5.32 per cent dissimilarity, respectively. The clustering pattern in both, the hierarchical and factorial analysis, grouping of accessions was not according to their geographical origin.

Diversity analysis of *Lens* core collection

Based on Dice dissimilarity matrix, the maximum dissimilarity matrix values ranged from 0.1–1.0 with an average of 0.55. No two genotypes showed 100% similarity to each other (Table S5). Minimum diversity was observed between accessions ILWL13 (*L. nigricans*, Italy) and ILWL20 (*L. culinaris* ssp. *odemensis*, Palestine); ILWL437 (*L. lamottei*, Turkey) and EC718266 (*L. nigricans*, Italy), with dissimilarity matrix value of 0.1. Maximum dissimilarity i.e. 100% was found between ILWL65 (*L. ervoides*, Turkey) and ILWL378 (*L. culinaris* ssp. *orientalis*, Turkmenistan); and ILWL227 (*L. culinaris* ssp. *orientalis*, Syria) and ILWL62 (*L. culinaris* ssp. *orientalis*, Turkey). In hierarchical clustering of the accessions, ILWL20 (*L. culinaris* ssp. *odemensis*, Palestine), ILWL13 (*L. nigricans*, Italy), ILWL15 (*L. lamottei*, Turkey) and ILWL18 (*L. nigricans*, France) each constitute a separate group alone as shown in Fig. 5. EC718266 of *L. nigricans* (Italy) clustered with ILWL437 of *L. lamottei* (Turkey), whereas ILWL97 of *L. culinaris* ssp. *tomentosus* (Turkey) clustered with EC718275 of *L. nigricans* (Turkey). ILWL203 (*L. culinaris* ssp. *odemensis*, Turkey), ILWL282, ILWL305 and ILWL307 (*L. culinaris* ssp. *tomentosus*, Turkey), and EC718267 (*L. nigricans*, Montenegro) occupied the same cluster. Further, ILWL343 and ILWL227 from *L. culinaris* ssp. *orientalis* (Syria) formed group with L830 of *L. culinaris* ssp. *culinaris* (India) and ILWL91 of *L. ervoides* (Turkey), respectively. Factorial analysis was also carried out as shown in Fig. 6, where PC1 and PC2 explained only 12.99 and 12.53% variations respectively. In general, first 12 principal coordinates had Eigen values >1.0. The clustering pattern in both, hierarchical and factorial analysis, did not reflect their geographical origin.

In case of quantitative data, the Euclidean dissimilarity matrices ranged from 7.15 between accessions ILWL282 (*L. culinaris* ssp.

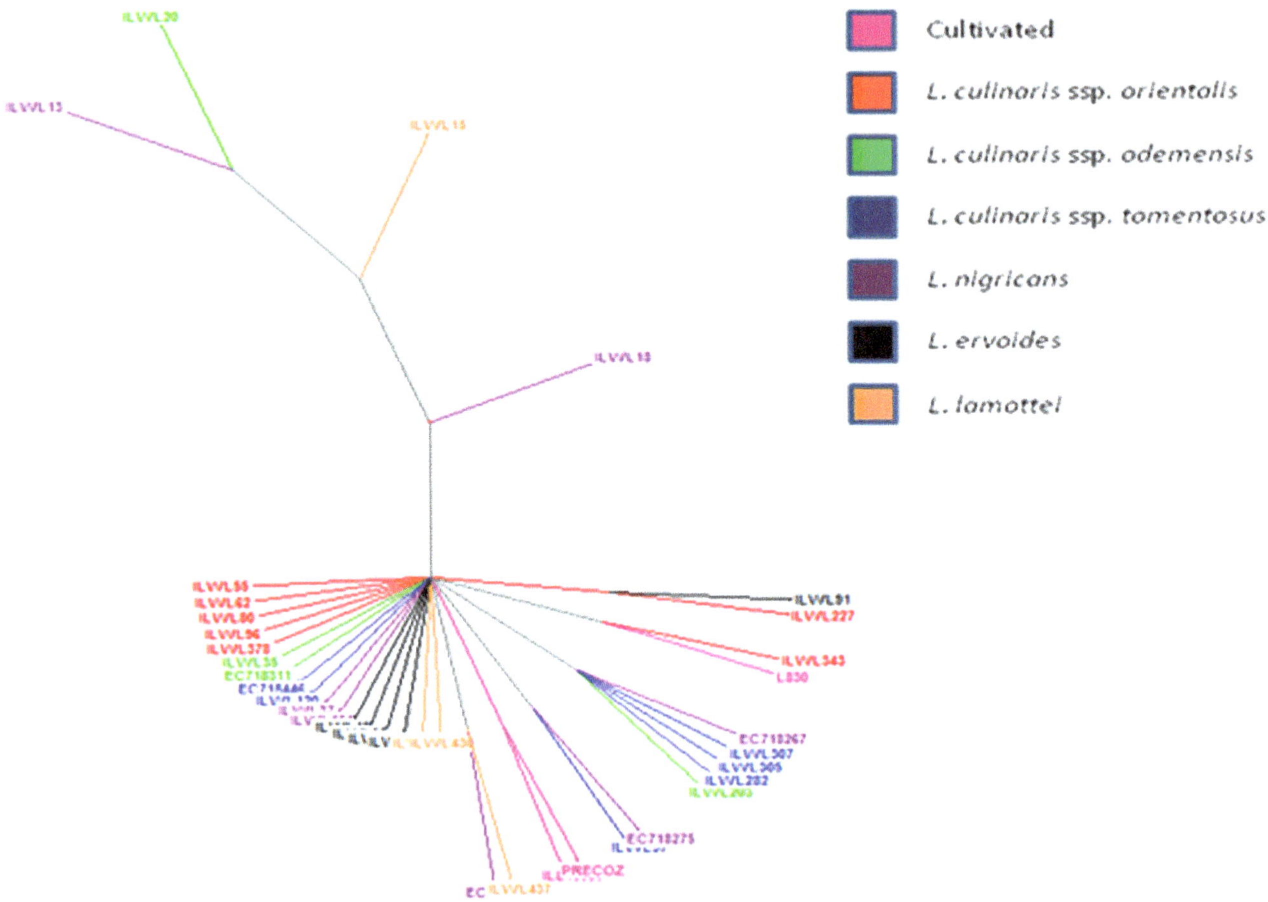

Figure 5. Hierarchical clustering of *Lens* core set accessions based on qualitative phenotypic data.

tomentosus, Turkey) and ILWL227 (*L. culinaris* ssp. *orientalis*, Syria) to 635.85 between accessions ILWL276 and ILWL269 of *L. ervoides* from Turkey (Table S6). The hierarchical clustering of accessions, where all the outgroups were occupied by accessions from different species excepting one containing Precoz (*L. culinaris* ssp. *culinaris*, Brazil) and ILL10829 (*L. culinaris* ssp. *culinaris,* Syria) as shown in Fig. 7. Factorial analysis resulted into first two principal coordinates having Eigen values >1 where first and second principal coordinates explained 89.6 and 8.85% variability respectively (Fig. 8). The clustering pattern in both, hierarchical and factorial analysis was not according to their geographical origin.

Discussion

Crop wild relatives (CWRs) can contain far greater diversity than their domestic counterparts and may hold increased potential to adapt against crop pests and diseases, adverse weather conditions and longer term changes in climate. One way to improve the resilience of our crop plants is to harness the genetic diversity found in their wild relatives. By introducing CWRs into breeding programs, the useful traits they contain, such as high yield and disease resistance, can be passed into cultivated crops. The introduction, characterization, evaluation, maintenance and utilization of CWRs for important morphological and agronomic traits, including adaptation to biotic and abiotic stresses, are often prerequisites for conducting a successful crop breeding research in annual plant species with a view to introgress novel trait of interest [39–41]. Effective evaluation and screening of diverse germplasm, especially unadapted gene pool [42–43] may prove useful to retrieve invaluable phenotypes to combat emerging agricultural challenges [44]. A strong pre-breeding program is needed to introgress genes from distant species and supply breeding pools with adequate parental germplasm.

Species characterization and agronomic evaluation

The results of the present study showed that global wild annual *Lens* collection of diverse origin had significant variation with respect to morphological and agronomic characters. Most of the wild annual *Lens* species had wide variations to seedling stem pigmentation and low to dense leaf pubescence. Dense leaf pubescence character can contribute towards better tolerance to insects and is a subject of further detailed investigation. More importantly to note that variation was observed within a species for leaflet size, tendril length, pod shedding, pod dehiscence, flower colour, ground colour of testa, pattern of testa and cotyledon colour (Table 2). These morphological plant characteristics can help to distinguish distinct phenotypic classes and their mode of inheritance can be studied using principles of classical genetics. Some of these traits are considered as descriptors for the taxonomic classification of taxon into their respective species and sub-species. It also allows phenotypic identification of specific alleles for specific loci [45–46]. Further, there were substantial variations exhibited in *Lens* species for important agro-morphological traits as reflected by their range, mean and coefficient of

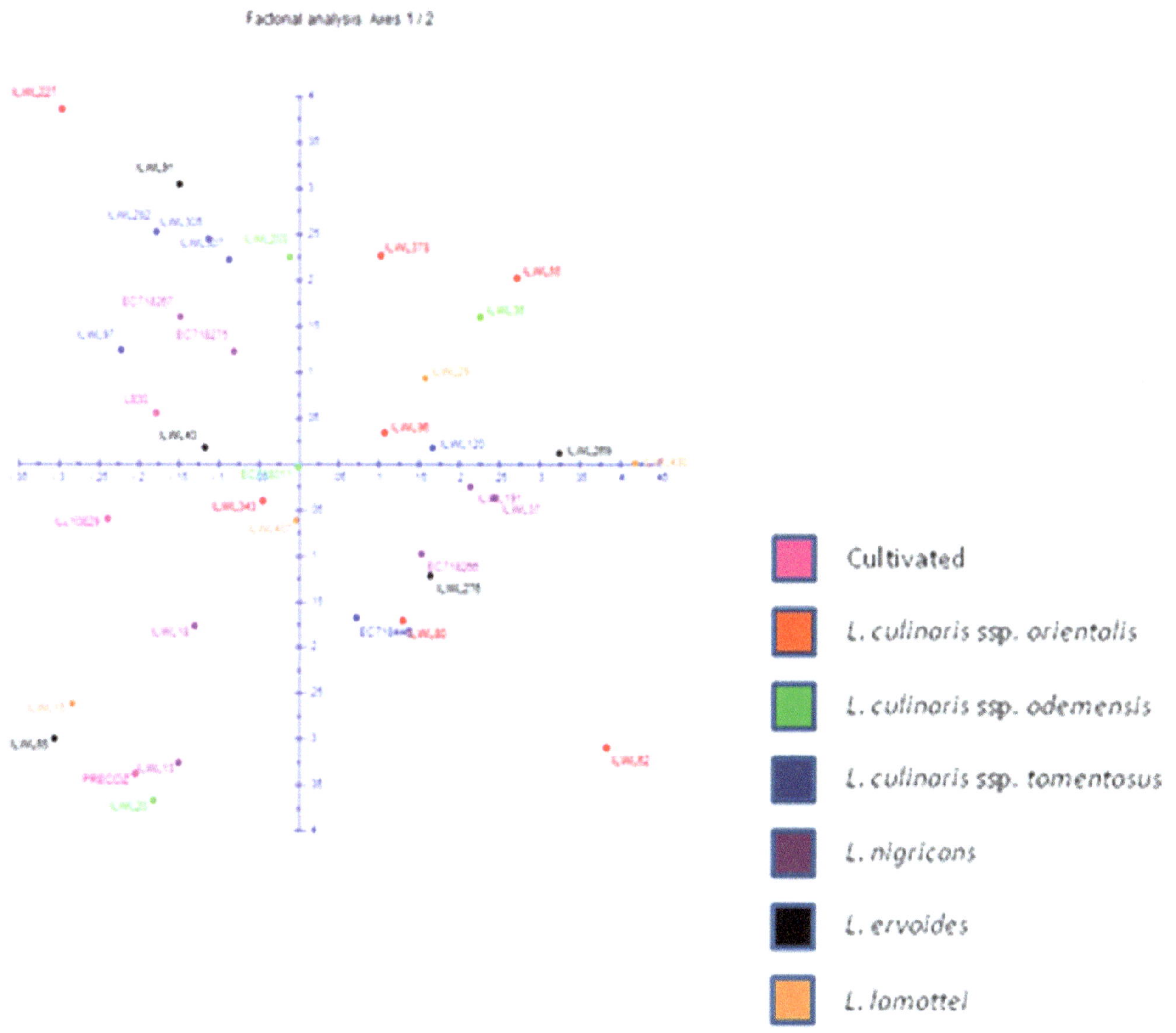

Figure 6. Factorial analysis of *Lens* core set accessions based on qualitative phenotypic data.

variation (Table 3 and 4), suggesting that it could be due to the diverse origin of these species and to their taxonomic identification. However, seed yield plant^{-1} and biological yield plant^{-1} - were higher in magnitude in the cultivated varieties, including Precoz, L830, ILL10829 and ILL8006 in both the agro-climatic regions. By contrast, in wild annual *Lens* species, the range and mean performance found promising for useful agro-morphological traits, including high number of pods plant^{-1} in accessions of *L. ervoides* viz., ILWL418 (1002 pods plant^{-1}), collected from Syria and ILWL321 (403 pods plant^{-1}), ILWL401(478 pods plant^{-1}), ILWL292 (697 pods plant^{-1}), ILWL369 (355 pods plant^{-1}) and ILWL308 (218 pods plant^{-1}) collected from Turkey. Other important accessions for high number of pods plant^{-1} and resistance to rust and powdery mildew in *L. nigricans* included, ILWL191 (141 pods plant^{-1}), ILWL40 (113 pods plant^{-1}), ILWL41 (147 pods plant^{-1}) and ILWL42 (143 pods plant^{-1}), also collected from Turkey. It is important to note that, wild accessions collected from Turkey exhibited substantial variations for useful agro-morphological traits including multiple resistance against major foliar diseases. These promising accessions of *L. nigricans* and *L. ervoides* from the secondary genepool can be exploited for broadening the genetic base of cultivated lentil. Further, rust and powdery mildew resistance gene sources identified can be of great significance for exploiting in breeding resistant lentil varieties, especially for the epidemiologically important regions in the lentil growing areas of the world. The list of some important accessions selected from different *Lens* species carrying useful traits of interest is presented in Table 7. The present study further strongly suggests revisiting of Turkey and other areas for more collections through explorations. In order to make quality *ex-situ* collection, it is necessary to know the geographical distribution of the target *Lens* taxa and where the gaps in the seed collection occur. Knowledge of current ICARDA genebank holding and understanding the gaps in collection would help prioritize the species and locations of wild *Lens*, that are in most need of collecting and conserving them in gene banks.

Assessment of among and within population diversity will also help to establish the potential of *in-situ* conservation of wild *Lens* species. Many areas of greatest interest for *in- situ* conservation (e.g. Turkey and other Mediterranean countries) are suffering

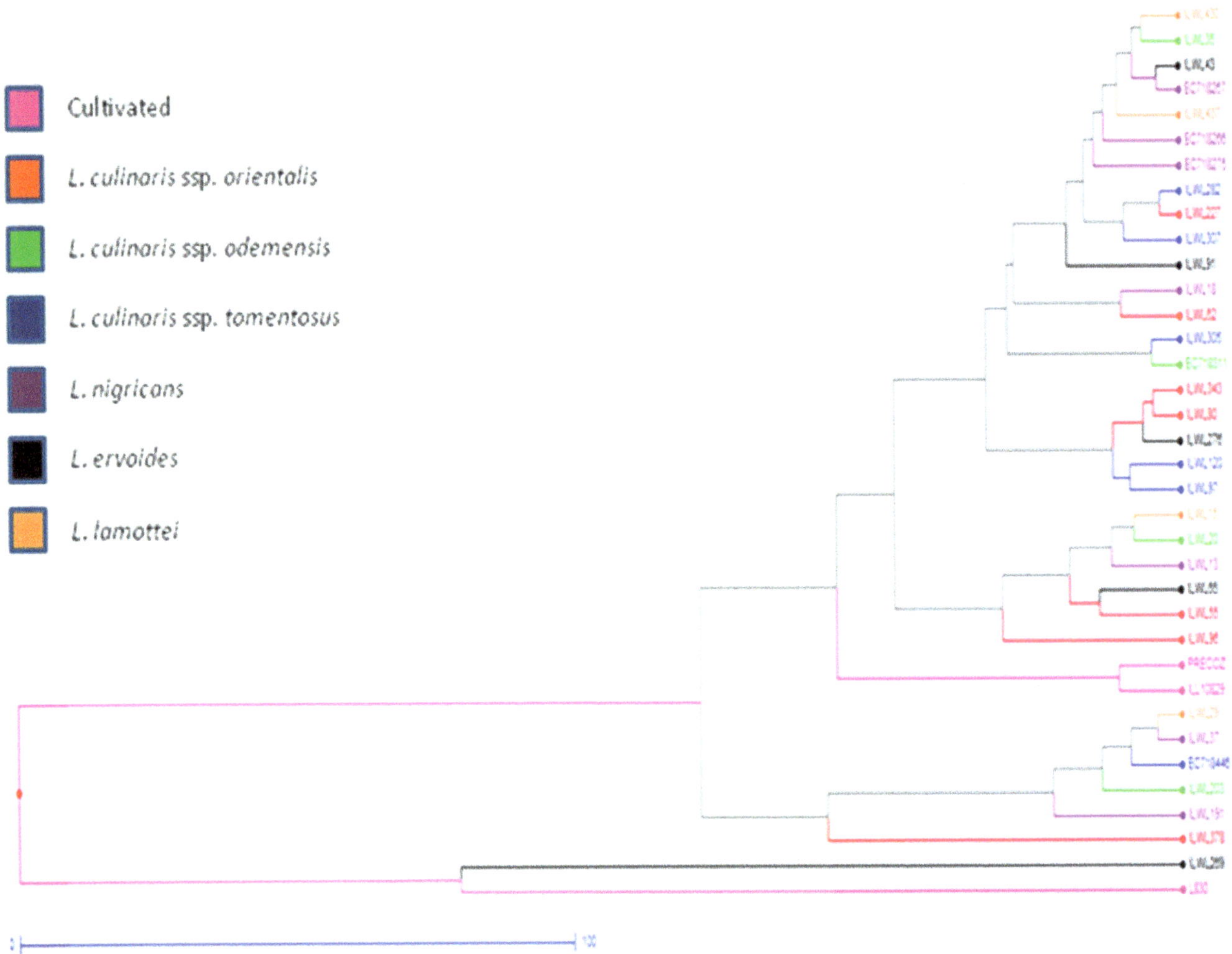

Figure 7. Hierarchical clustering of *Lens* core set accessions based on quantitative phenotypic data.

from rapid loss of invaluable genetic resources due to habitat destruction [47]. Important areas to target for *in-situ* conservation includes west Turkey for *L. nigricans*, Southwest Turkey, North-west Syria, south Syria and Jordan for *L. culinaris* spp. *orientalis*, south Syria for *L. culinaris* ssp. *odemensis* and the coastal border region between Turkey and Syria stretching along the Syrian coast for *L. ervoides* [48]. The accumulated information on this global collection of wild *Lens* will be used to conduct a gap analysis using appropriate approaches to guide future collecting missions to fill the gaps, but also to conduct traits-targeted collecting missions to sample new accessions with potential resistance/tolerance to major biotic and abiotic stresses. The same information can be used to define natural habitats that can allow effective *in-situ* conservation of the species richness as well as populations with potential sought traits. Maxted et al. (2012b) [49] proposed the Fertile Crescent, mainly the Aegean and South-western region of Turkey, North Syria and Lebanon as appropriate sites for *in-situ* conservation of most of temperate legume species.

Lens core collection

"PowerCore" is a rapid approach for developing core collection, which effectively simplifies the generation process of core set with reduced number of core entries, but maintaining high percent of diversity as compared to other methods used [22,50]. For larger genebank holdings, the core sets identified using "PowerCore" normally are small in size with greater diversity captured as compared to traditional clustering procedures. However, in the present study, as the number of accessions representing different *Lens* taxa in the entire collection were low, relatively more number of accessions were selected when quantitative variables were used either in combination with qualitative characters or alone. Using qualitative variables alone, a smaller core size of about 9% was achieved.

Majority of the wild *Lens* taxa in the entire collection belongs to Syria (37%) and Turkey (28%). However, maximum representation of accessions in core collections, derived from all data sets was from Turkey, followed by Syria, indicating rich diversity for many of the wild relatives of cultivated lentil. It is only in Aegean and the South-western region that the distribution of all four wild taxa of the genus *Lens* overlaps [51]. Unfortunately, Turkey, like other Mediterranean countries is suffering the rapid loss of many of its valuable genetic resources. The Aegean and South-western region of Turkey is, therefore, particularly important for exploration and collection of wild taxon for conservation *ex-situ* and also for analysis of within population variations in order to assess the threat of genetic erosion to each species and potential for *in-situ* conservation. Ferguson et al. (1996) [51] reported that *L. culinaris* ssp. *odemensis* and *L. ervoides* are the taxa most threatened by

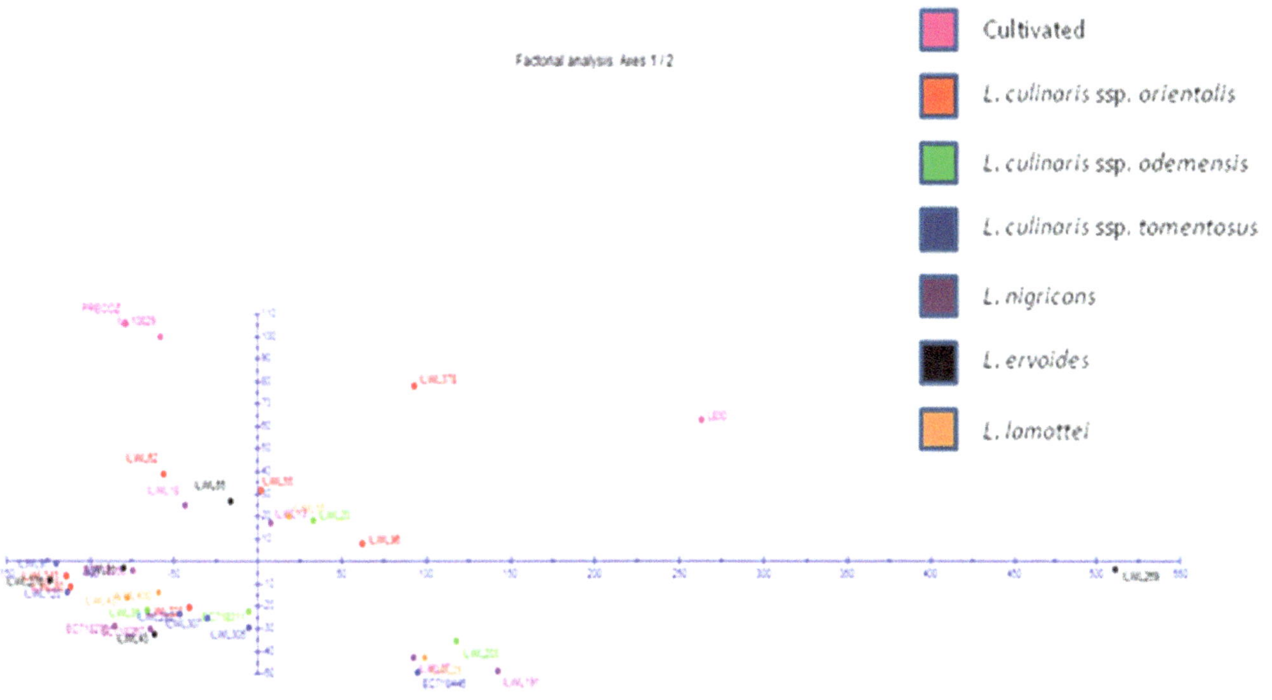

Figure 8. Factorial analysis of *Lens* core set accessions based on quantitative phenotypic data.

genetic erosion. Selection of core entries using different data sets gives a choice to the users as representation of common accessions across different core sets varies with the data sets used. A total of 66 and 25 accessions were common between the core sets developed using combined data on qualitative and quantitative variables and that obtained from quantitative and qualitative variables alone, respectively. Only 15 accessions were common between the core set obtained from quantitative and qualitative variables separately. The user, therefore, has a choice to use his experience and decide which core sets better serves the specific purpose.

Of the 46 accessions identified as promising for various desired traits from the entire *Lens* accessions (Table 7), a total of 21, 15 and 10 accessions, respectively, were represented in core sets derived from different data sets i.e. quantitative and qualitative traits combined, and quantitative and qualitative traits individually. Representation of proportionately greater number of desired accessions in the core set derived from qualitative morphological data could be the existence of good polymorphism with different descriptor states for qualitative traits with major gene effect. The core sets derived from different data sets gives us an opportunity to precisely estimate diversity at genetic/molecular level for use by the germplasm curators and the users. As majority of accessions in the core set derived from qualitative data originate from Turkey, this indirectly indicates greater polymorphism occurring in accessions originating from parts of Turkey (Table 8). Therefore, it is imperative that the wild gene pool of Turkey needs to be further explored, collected, conserved, characterized, and used in crop improvement. As has been emphasized, for broadening the genetic base of crop productions, where domestication is the bottleneck, protecting the wild gene pool is an utmost necessity in order to reconstructing the early evolutionary stages [52] to facilitate evolution. When selecting the species to conserve *in-situ*, the highest emphasis should be given to most genetically distinct groups of taxa (the widespread species with a wide range of adaptation). *L. culinaris* ssp. *orientalis* are distributed in Turkey throughout the northern belt of Fertile Crescent and share different types of habitat in different ecosystems, and they have a wide range of adaptability [53]. A few promising common accessions found across all core sets include ILWL29 (*L. lamottei*) from Spain; ILWL37 (*L. nigricans*) from Turkey; ILWL97 (*L. culinaris* ssp. *tomentosus*) from Turkey; ILWL269 (*L. ervoides*) from Turkey, and ILWL378 (*L. culinaris* ssp. *orientalis*) from Syria (Table 9). Besides possessing disease resistance, these accessions were also promising for yield related traits as high number of pods plant^{-1} and have significant breeding potential for yield enhancement.

Diversity assessment

The present study depicted a wide genetic variation in the wild annual *Lens* collection for agro-morphological traits as revealed by the dendrogram generated from the dissimilarity matrix and factorial analysis using DARwin 5 approach. In case of entire collection, qualitative data analysis revealed that maximum variability among accessions was interspecific and not limited to a particular country and this could be partly explained by the taxonomic differences. But in case of quantitative data, minimum as well as maximum diversity was reported among *L. ervoides* accessions collected from Turkey and Syria. Among core set accessions, qualitative data analysis revealed maximum interspecific as well as intraspecific variability between *L. ervoides* and *L. culinaris* ssp. *orientalis*, and among *L. culinaris* ssp. *orientalis*, accessions, respectively. Further, quantitative data revealed maximum diversity among *L. ervoides* accessions from Turkey. The diversity could mainly be attributed to diverse agro-ecological conditions as the current study included global wild annual *Lens* collection. The Euclidian distances calculated based on the quantitative data manifested large genetic distances among the accessions. Vieira et al. (2007) [54] analysed 19 wheat accessions for 17 phenotypic characters and reported genetic distances upto

Table 7. Identification of promising accessions for few agronomic and major biotic traits for their introgression in lentil enhancement selected from the entire *Lens* collection.

S. No.	Accessions	Species	Gene pool	Trait of Interest	Origin
1.	ILWL 247, 255, 476,	*L. culinaris* ssp. *orientalis*	Primary	High number of pods plant^{-1}, resistant to rust	Syria, Turkey
2.	ILWL 346, 350, 369, 378, 490	*L. culinaris* ssp. *orientalis*	Primary	High number of pods plant^{-1}, resistant to powdery mildew	Syria, Armenia
3.	ILWL 372	*L. culinaris* ssp. *orientalis*	Primary	High number of pods plant^{-1}, resistant to rust and powdery mildew	Uzbekistan
4.	ILWL 80, 381	*L. culinaris* ssp. *orientalis*	Primary	Resistant to rust and powdery mildew	Tajikistan
5.	ILWL 456	*L. culinaris* ssp. *orientalis*	Primary	High 100-seed weight and resistant to rust	Turkey
6.	ILWL 39	*L. culinaris* ssp. *odemensis*	Primary	Resistant to rust and powdery mildew	Turkey
7.	ILWL 165, 203, 237, 238	*L. culinaris* ssp. *odemensis*	Primary	High number of pods plant^{-1}, resistant to rust	Syria
8.	ILWL 167	*L. culinaris* ssp. *odemensis*	Primary	Resistant to rust and powdery mildew	Syria
9.	ILWL 97	*L. culinaris* ssp. *tomentosus*	Primary	Resistant to powdery mildew	Turkey
10.	ILWL 121, EC-718446	*L. culinaris* ssp. *tomentosus*	Primary	High number of pods plant^{-1}	Syria
11.	ILWL 23, 24, 25	*L. nigricans*	Secondary/ Tertiary	Resistant to rust	Italy, HRV
12.	ILWL 22	*L. nigricans*	Secondary/ Tertiary	Resistant to rust and powdery mildew	Italy
13.	ILWL 37, 191	*L. nigricans*	Secondary/ Tertiary	High number of pods plant^{-1}, resistant to rust and powdery mildew	Turkey
14.	ILWL 40, 41, 42, 159, 269, 292, 294, 321, 398, 418, EC-718428, 718439	*L. ervoides*	Secondary/ Tertiary	High number of pods plant^{-1}, resistant to rust and powdery mildew	Ukraine, Turkey, Italy, Syria,
15.	ILWL 50, 54, 56, 58	*L. ervoides*	Secondary/ Tertiary	High number of pods plant^{-1}, resistant to rust	HRV, Israel, PSE, Turkey
16.	ILWL 91	*L. ervoides*	Secondary/ Tertiary	Resistant to powdery mildew	Turkey
17.	ILWL 29, EC-718692	*L. lamottei*	Secondary/ Tertiary	High number of pods plant^{-1}	Spain, France

196.61. A large genetic distance between heterotic germplasm can be useful for developing lines with good combining ability in heterosis breeding [55–56]. The knowledge of genetic diversity among accessions provides clues about the heterotic potential that can be exploited while making crosses in order to create polymorphic populations. The clustering of accessions on the basis of phenotypic multivariate data, as compared to the molecular classification, was observed to be a much better procedure to explain the genotypic effects, since, the percentage of sum of squares between phenotypic sub-populations was higher in magnitude than that of between genetic sub-populations for all the traits analysed [57].

When genetic diversity was compared based on qualitative traits between entire set and core set, the core set included accessions which explained reasonable variability. In case of wild taxa namely *L. culinaris* ssp. *culinaris*, *L. culinaris* ssp. *orientalis*, *L. culinaris* ssp. *odemensis* and *L. lamottei*, core set included representative accessions from each group categorised based on entire set qualitative data excluding two to three accessions from *L. nigricans* and *L. culinaris* ssp. *tomentosus*, each. Further, accessions of various species namely *L. culinaris* ssp. *odemensis* from Palestine, *L. nigricans* from Italy and France, and *L. lamottei* from Turkey geographical origins constituted separate groups alone. Thus overall, it can be suggested that based on qualitative data, core set accessions revealing substantial diversity as available in the entire global collection. In case of quantitative data, core set included accessions from all the species and from majority of outgroups formed in genetic diversity analysis of entire set. Wang et al. (2006) [58] developed a core set capturing the genetic diversity of large soybean collection using 2% of total accessions to represent about 70% of the diversity from a whole sample set.

The tree formed from Dice Co-efficient and Euclidean distances of the standardized phenotypic means showed that genetic relatedness rarely matched analogous geographical origins in general. Plants that are phenotypically similar are not necessarily genetically so, as different gene pools could result in similar phenotypes. The reason might be continuous gene flow among wild species occurring in nature. Among wild taxa, one important aspect for species prioritization is the degree of relatedness, as it determines the actual potential for introducing useful traits from

Table 8. Agro-ecological representation of species-wise accessions in entire *Lens* collection and core sets derived from different data sets.

S. No.	Origin	Accessions in entire collection	Accessions in core sets		
			I	II	III
1	Alpes-Cote d'Azur	1	1	1	–
2	Armenia	2	1	1	–
3	Azerbaijan	1	–	–	–
4	Bahrain	4	–	1	–
6	Cyprus	3	–	–	–
7	Former Soviet Union	2	–	–	–
8	Former Yugoslavia	2	–	–	–
9	France	3	3	2	1
10	HRV	14	2	2	1
11	Iran	2	–	–	–
12	Israel	10	2	1	1
13	Italy	8	2	3	2
14	Jordan	11	1	1	–
15	Lebanon	9	2	2	–
16	Montenegro	3	–	1	1
17	Palestine	1	1	–	1
18	PSE	3	–	–	–
19	Russia	1	–	–	–
20	Spain	12	6	5	2
21	Syria	152	20	24	3
22	Tajikistan	4	2	1	–
23	Turkmenistan	2	1	1	1
24	Turkey	113	28	22	14
25	UBA	1	–	–	–
26	Ukraine	3	1	1	–
27	Uzbekistan	8	2	–	–
28	Unknown	30	21	17	9
	Total	**405**	**96**	**86**	**36**

Table 9. Identification of some promising accessions for few agronomic and major biotic traits for their introgression in lentil enhancement selected from the qualitative core set.

S. No.	Accessions	Species	Gene pool	Trait of Interest	Origin
1.	ILWL 80	*L. culinaris* ssp. *orientalis*	Primary	Resistant to rust and powdery mildew	Tajikistan
2.	ILWL 378	*L. culinaris* ssp. *orientalis*	Primary	High number of pods plant^{-1}, resistant to powdery mildew	Turkmenistan
3.	ILWL 203	*L. culinaris* ssp. *odemensis*	Primary	High number of pods plant^{-1}, resistant to rust	Syria
4.	EC-718446	*L. culinaris* ssp. *tomentosus*	Primary	High number of pods plant^{-1}	Syria
5.	ILWL 97	*L. culinaris* ssp. *tomentosus*	Primary	Resistant to powdery mildew	Turkey
6.	ILWL 37	*L. nigricans*	Secondary/Tertiary	High number of pods plant^{-1}, resistant to rust and powdery mildew	Turkey
7.	ILWL 191	*L. nigricans*	Secondary/Tertiary	High number of pods plant^{-1}, resistant to rust and powdery mildew	Turkey
8.	ILWL 91	*L. ervoides*	Secondary/Tertiary	Resistant to powdery mildew	Turkey
9.	ILWL 269	*L. ervoides*	Secondary/Tertiary	High number of pods plant^{-1}, resistant to rust and powdery mildew	Turkey
10.	ILWL 29	*L. lamottei*	Secondary/Tertiary	High number of pods plant^{-1}	Spain

the wild into cultigen. Relatedness information derived from a combination of biosystematics based upon traditional morphological and genotypic data, which are being used increasingly as well as information derived from plant breeders attempting crosses between CWRs and crops [59] are extremely useful for planning future study for the introgression of crop wild relatives.

Conclusions

The present study has helped in identifying the useful gene sources viz, earliness, high number of branches plant^{-1}, high number of pods plant^{-1} and, multiple disease resistance etc. in different *Lens* taxa across gene pools. Diversity analysis suggested ample scope for future germplasm collecting from the hotspots in Turkey and Syria. The promising trait-specific accessions reported particularly in *L. nigricans* and *L. ervoides* need to be considered while planning future lentil breeding programme for introgressing gene of interests from wild *Lens*. Further, certain accessions belonging to the secondary gene pool of *Lens* and also extracted in core set developed, can be used as a starting material aimed at large scale base broadening of cultivated lentil.

Supporting Information

Table S1 Dissimilarity matrices of *Lens* accessions based on qualitative morphological data.

Table S2 Clustering of entire *Lens* collection based on qualitative morphological data.

Table S3 Dissimilarity matrices of *Lens* accessions based on quantitative agro-morphological data.

Table S4 Clustering of entire *Lens* collection based on quantitative agro-morphological data.

Table S5 Dissimilarity matrices of *Lens* accessions constituting the core set based on qualitative morphological data.

Table S6 Dissimilarity matrices of *Lens* accessions constituting the core set based on quantitative agro-morphological data.

Acknowledgments

The authors acknowledge the Genetic Resources Unit at the International Centre for Agricultural Research in Dry Areas (ICARDA), Aleppo, Syria, for providing global wild *Lens* collection. The authors also thank the Vice Chancellor, HPAU, Palampur, Himachal Pradesh, India for allowing use of the hot spot facility at Sangla and Dhaulakuan centre.

Author Contributions

Conceived and designed the experiments: MS. Performed the experiments: MS ISB. Analyzed the data: Sandeep Kumar MD KCB MK. Contributed reagents/materials/analysis tools: AS AA Shiv Kumar. Contributed to the writing of the manuscript: MS ISB Sandeep Kumar. Provided necessary support for inter-institutional collaboration: SKD.

References

1. Lane A, Jarvis A (2007) Changes in climate will modify the geography of crop suitability: Agricultural biodiversity can help with adaptation. Paper presented at ICRISAT/CGIAR 35th Anniversary Symposium. "Climate Proofing Innovation for Poverty Reduction and Food Security" 22–24 November 2007, ICRISAT, Patancheru, India.
2. Maxted N, Ford BV, Kell SP, Iriondo JM, Dulloo ME, et al. (2008) Crop wild relative conservation and use. Wallingford, UK, CABI publishing.
3. Soltis DE, Smith SA, Cellinese N, Wurdack KJ, Tank DC, et al. (2011) Angiosperm phylogeny: 17 genus, 640 taxa. Amer J Bot 98 (4): 704–730.
4. Sonnante G, Hammer K, Pignone D (2009) From the cradle of agriculture a handful of lentils. History of domestication. Rendiconti. Lincei 20 (1): 21–37.
5. Arumuganathan K, Earle ED (1991) Nuclear DNA content of some important plant species. Plant Mol Biol Rep 9(3): 208–218.
6. Ferguson ME, Maxted N, Slageren MV, Robertson LD (2000) A re-assessment of the taxonomy of *Lens* Mill. (Leguminosae, Papilionoideae, Vicieae). Bot J Linn Soc 133(1): 41–59.
7. Erskine W, Muehlbauer FJ, Sarker A, Sharma B (2009) The Lentil: Botany, Production and Uses. Wallingford, UK, CABI publishing.
8. Muench DG, Slinkard AE, Scoles GJ (1991) Determination of genetic variation and taxonomy in lentil species by chloroplast DNA polymorphism. Euphytica 56(3): 213–218.
9. Alveraz MT, Garcia P, Perez de la Vega M (1997) RAPD polymorphism in Spanish lentil landraces and cultivars. J Genet Breed 51(2): 91–96.
10. Ford R, Pang ECK, Taylor PWJ (1997) Diversity analysis and species identification in *Lens* using PCR generated markers. Euphytica 96(2): 247–255.
11. Duran Y, Fratini R, Garcia P, Perez de la Vega M (2004) An inter sub-specific genetic map of *Lens*. Theor Appl Genet 108: 1265–1273.
12. Gupta D, Sharma SK (2006) Evaluation of wild *Lens* taxa for agro-morphological traits, fungal diseases and moisture stress in North-western Indian Hills. Genet Resour Crop Evol 53: 1233–1241.
13. Singh M, Bisht IS, Dutta M, Kumar K, Gautam NK, et al. (2011) Widening genetic base through pre-breeding in lentil and chickpea for faster utilization of genetic resources at the National Seminar on contemporary approaches to crop improvement, May 22–25 at Bangalore, India pp 365–366.
14. Scott JM, Davis F, Csuti B, Noss R, Butterfield B, et al. (1993) Gap analysis: a geographic approach to protect of biological diversity. Wildlife Monographs 123: 3–41.
15. Maxted N, Hargreaves S, Kell SP, Amri A, Street K, et al. (2012a) Temperate forage and pulse legume genetic gap analysis. Bocconea 24: 5–36.
16. Frankel OH (1984) Genetic perspectives of germplasm conservation. In: Arber WK, Llimensee K, Peacock WJ, Starlinger P, ed. Genetic Manipulation: Impact on Man and Society. Cambridge: Cambridge University Press. pp 161–170.
17. Frankel OH, Speer AHD (1984) Current plant genetic resources: a critical appraisal. In: Chopra VL, Joshi BC, Sharma RP, Sharma HC, eds. Genetics: New Frontiers. Vol. IV. Oxford & IBH Publ. Co. New Delhi, India.
18. Brown AHD (1989a) Core collections: A practical approach to genetic resources management. Genome 3: 818–824.
19. Brown AHD (1989b) The case for core collections. In: Brown AHD, ed. The use of plant genetic resources. Cambridge University Press, Cambridge, England. pp 136–155.
20. Gouesnard B, Bataillon TM, Decoux G, Rozale C, Schoen DJ, et al. (2001) M-Strat: an algorithm for building germplasm core collection by maximizing allelic or phenotypic richness. J Hered 92: 93–94.
21. Jansen J, van Hintum TJL (2007) Genetic distance sampling: a novel sampling method for obtaining core collections using genetic distances with an application to cultivated lettuce. Theor Appl Genet 114: 421–428.
22. Kim KW, Chung HK, Cho GT, Ma KH, Chandrabalan D, et al. (2007) Power Core: a program applying the advanced M strategy with a heuristic search for establishing core sets. Bioinformatics 23: 2155–2162.
23. Thachuk C, Crossa J, Franco J, Dreisigacker S, Warburton M, et al. (2009) Core Hunter: An algorithm for sampling genetic resources based on multiple genetic measures. BMC Bioinformatics pp 10: 243.
24. Zdenik S, Iva F, Ladislav D, Sergey M, Dobrotvorskaya T (2006) Genealogical analysis in the Czech Spring Wheat collection and its use for the creation of core collection. Czech J Genet Breed 42: 117–125.
25. Upadhyaya HD (2003) Phenotypic diversity in groundnut core collection assessed by morphological and agronomical evaluations. Genet Resour Crop Evol 50: 539–550.
26. Upadhyaya HD, Ortiz R, Bramel PJ (2003) Development of a groundnut core collection using taxonomical, geographical and morphological descriptors. Genet Resour Crop Evol 50: 139–148.
27. Francois B, Valérie R, Pjotr S, Florence EV, Pierre S, et al. (2007) A worldwide bread wheat core collection arrayed in a 384-well plate. Theor Appl Genet 114: 1265–1275.
28. Borba TCO, Brondani RPV, Rangel PHN, Brondani C (2009) Microsatellite marker– mediated analysis of the EMBRAPA rice core collection genetic diversity. Genetica 137: 293–304.
29. Yan W, Rutger JN, Bryant RJ, Bockelman HE, Fjellstrom RG, et al. (2007) Development and evaluation of a core subset of the USDA rice germplasm collection. Crop Sci 47: 869–878.

30. Bari A, Street K, Mackay M, Endresen DTF, De Pauw E, et al. (2012) Focused Identification of germplasm strategy (FIGS) detects wheat stem rust resistance linked to environmental variables. Genet Resour Crop Evol 59: 1465–1481.
31. Khazaei H, Street K, Bari A, Mackay M, Stoddard FL (2013) The FIGS (Focused Identification of Germplasm Strategy) approach identifies traits related to drought adaptation in *Vicia faba* genetic resources. PLoSONE 8(5): e63107. doi:10.1371/journal.pone.0063107.
32. IBPGR/ICARDA (1993) Lentil descriptors: IBPGR Secretariat, Rome, Italy pp 1–15.
33. Mayee CD, Datar VV (1986) Phytopathometry. Technical Bulletin-1, Marathwada Agricultural University, Parbhani, p 218.
34. Tiwari KR, Penner GA, Warkentin TD, Rashid KY (1997) Pathogenic variation in *Erisiphe polygoni*, the casual organism of powdery mildew of lentil. Can J Plant Pathol 19: 267–271.
35. Shannon CE, Weaver W (1949) The Mathematical Theory of Communication. University of Illinois Press, Urbana.
36. Dice LR (1945) Measurements of the amount of ecologic association between species. Ecology 26: 297–302.
37. Perrier X, Flori A, Bonnot F (2003) Data analysis methods In: P. Hamon, M. Seguin, X. Perrier, J.C. Glaszmann, eds. Genetic diversity of cultivated tropical plants. Enfield, Science Publishers. Montpellier pp 43–76.
38. Perrier X, Jacquemoud-Collet JP (2006) DARwin software. Available: http://darwin.cirad.fr/darwin.
39. Duvick DN (1984) Genetic diversity in major farm crops on the farm and in reverse. Eco Bot 38: 151–178.
40. Lazaro A, Ruiz M, Rosa L, Martin I (2001) Relationship between agro-morphological characters and climate parameters in Spanish landraces of lentil. Genet Resour Crop Evol 48: 239–249.
41. Naghavi MR, Jahansouz MR (2005) Variation in the agronomic and morphological traits of lentil accessions. J Integ Plant Biol 47: 375–379.
42. Tanksley SD, McCouch SR (1997) Seed banks and Molecular maps: unlocking genetic potentials from the wild. Science 277: 418–423.
43. Gur A, Zamir D (2004) Unused natural variation can lift yield barriers in Plant Breeding. PLoS Biol 2: e245.
44. Kumar GR, Sakthivel K, Sundaram RM, Neeraja CN, Balachandran SM, et al. (2010) Allele mining in crops: prospects and potential. Biotech Adv 28: 451–461.
45. Muehlbauer FJ, Singh KB (1987) Genetics of chickpea. In: Saxena MC, Singh KB, ed. The chickpea. CAB International, Cambridge pp 99–126.
46. Pundir RPS, Rao NK, van der Maesen LJG (1985) Distribution of qualitative traits in the world germplasm of chickpea. Euphytica 34(3): 697–703.
47. Solh M, Erskine W (1981) In: Webb C, Hawtin GC, eds. Commonwealth Agricultural Bureaux, Slough, UK. Plant Genet Resour pp 54–67.
48. Ferguson ME, Newbury HJ, Maxted N, Ford-Lloyd BV, Robertson LD (1998) Population genetic structure in *Lens* taxa revealed by isozyme and RAPD analysis. Genet Resour Crop Evol 45: 549–559.
49. Maxted N, Kell S, Ford-Lloyd B, Dulloo E, Toledo A (2012b) Towards the systematic conservation of global crop wild relative diversity. Crop Sci 52(2): 774–785.
50. Gowda J, Krishanappa M, Pathak N, Mathur PN, Seetharam A (2012) Use of Heuristic Approach for the Development of a Core Set from Large Germplasm Collection of Foxtail Millet. Indian J Plant Genet Resour 26: 13–18.
51. Ferguson MN, Açikgöz A, Ýsmaýl AS, Cýnsoy B (1996) An eco-geographic survey of wild *Lens* species in Aegean and South West Turkey. Ana J Agric Sci 6: 159–166.
52. Cooper DC, Spillane C, Hodgkin T (2000) Broadening the genetic bases of crop production, IPGRI, FAO: Rome, CABI: Wallingford 480.
53. Tan A, Tan AS (2002) *In-situ* conservation of wild species related to crop plants: the case of Turkey. Managing Plant Genetic Diversity. In: Engels JMM, Ramanatha Rao V, Brown AHD, Jackson MT, eds. CABI Publishing, UK pp 195–204.
54. Vieira EA, de Carvalho FIF, Bertan I, Kopp M, Zimmer PD, et al. (2007) Association between genetic distances in wheat (*Triticum aestivum* L.) as estimated by AFLP and morphological markers. Genet Mol Biol 30(2): 392–399.
55. Melchinger AE (1999) Genetic diversity and heterosis. In: JG Coors, and S Pandey, eds. The genetics and exploitation of heterosis in crops, ASA, CSSA, and SSSA, Madison, WI. pp 99–118.
56. Moll RH, Longquist JH, Fortuna JV, Johnson EC (1965) The relation of heterosis and genetic divergence in maize. Genetics 52: 139–144.
57. Royo C, Maccaferri M, Alvaro F, Moragues M, Sanguineti MC, et al. (2010) Understanding the relationships between genetic and phenotypic structures of a collection of elite durum wheat accessions. Field Crops Res 119: 91–105.
58. Wang L, Guan Y, Guan R, Li Y, Ma Y, et al. (2006) Establishment of Chinese soybean (*Glycine max*) core collections with agronomic traits and SSR markers. Euphytica 151: 215–223.
59. Singh M, Rana MK, Kumar K, Bisht IS, Dutta M, et al. (2013) Broadening the genetic base of lentil cultivars through inter-sub-specific and interspecific crosses of *Lens* taxa. Plant Breed 132 (6): 667–675.

Conservation Genetics of the Philippine Tarsier: Cryptic Genetic Variation Restructures Conservation Priorities for an Island Archipelago Primate

Rafe M. Brown[1]*, Jennifer A. Weghorst[2], Karen V. Olson[1], Mariano R. M. Duya[3], Anthony J. Barley[1], Melizar V. Duya[4], Myron Shekelle[5], Irene Neri-Arboleda[6], Jacob A. Esselstyn[7], Nathaniel J. Dominy[8], Perry S. Ong[9], Gillian L. Moritz[10], Adrian Luczon[9], Mae Lowe L. Diesmos[11], Arvin C. Diesmos[12], Cameron D. Siler[13]

1 Biodiversity Institute and Department of Ecology and Evolutionary Biology, University of Kansas, Lawrence, Kansas, United States of America, **2** Biodiversity Institute, University of Kansas, Lawrence, Kansas, United States of America, **3** Institute of Biology, University of the Philippines Diliman, Quezon City, Philippines, **4** Diliman Science Research Foundation Inc., University of the Philippines Diliman, Quezon City, Philippines, **5** Research Institute of EcoScience, Ewha Womans University, Seoul, Republic of Korea, **6** Thomas Glen Erin Animal Hospital, Mississauga, Ontario, Canada, **7** Museum of Natural Science & Department of Biological Sciences, Louisiana State University, Baton Rouge, Louisiana, United States of America, **8** Department of Anthropology, Dartmouth College, Hanover, New Hampshire, United States of America, **9** Institute of Biology, University of the Philippines, Diliman, Quezon City, Philippines, **10** Department of Biological Sciences, Dartmouth College, Hanover, New Hampshire, United States of America, **11** University of Santo Tomas, Espana, Manila, Philippines, **12** Herpetology Section, Zoology Division, Philippine National Museum, Ermita, Manila, Philippines, **13** Sam Noble Oklahoma Museum of Natural History and Department of Biology, University of Oklahoma, Norman, Oklahoma, United States of America

Abstract

Establishment of conservation priorities for primates is a particular concern in the island archipelagos of Southeast Asia, where rates of habitat destruction are among the highest in the world. Conservation programs require knowledge of taxonomic diversity to ensure success. The Philippine tarsier is a flagship species that promotes environmental awareness and a thriving ecotourism economy in the Philippines. However, assessment of its conservation status has been impeded by taxonomic uncertainty, a paucity of field studies, and a lack of vouchered specimens and genetic samples available for study in biodiversity repositories. Consequently, conservation priorities are unclear. In this study we use mitochondrial and nuclear DNA to empirically infer geographic partitioning of genetic variation and to identify evolutionarily distinct lineages for conservation action. The distribution of Philippine tarsier genetic diversity is neither congruent with expectations based on biogeographical patterns documented in other Philippine vertebrates, nor does it agree with the most recent Philippine tarsier taxonomic arrangement. We identify three principal evolutionary lineages that do not correspond to the currently recognized subspecies, highlight the discovery of a novel cryptic and range-restricted subcenter of genetic variation in an unanticipated part of the archipelago, and identify additional geographically structured genetic variation that should be the focus of future studies and conservation action. Conservation of this flagship species necessitates establishment of protected areas and targeted conservation programs within the range of each genetically distinct variant of the Philippine tarsier.

Editor: Adam Stow, Macquarie University, Australia

Funding: Support for fieldwork was provided by the University of Kansas Biodiversity Institute and Department of Ecology and Evolutionary Biology (to RMB, CDS, and JAE), the National Geographic Society (NGS 8446-08 to RMB, JAW, MS and INA), funds from the Primate Action Fund (to MS), Ewha Womans University (Ewha Global Top5 Grant 2013 to MS), the David and Lucile Packard Foundation (Fellowship in Science and Engineering no. 2007-31754, to NJD), and U.S. National Science Foundation (DEB 0743491 to RMB). The funders had no role in study design, data collection and analysis, decision to publish, or preparation of the manuscript.

Competing Interests: The authors have declared that no competing interests exist.

* Email: rafe@ku.edu

Introduction

Biodiversity-rich tropical forests are being degraded worldwide, and the pace of forest destruction is exceptionally rapid in insular Southeast Asia [1]. With only 4–8% of its original forest remaining [2], the Philippines has been designated as both a global conservation biodiversity hotspot [3] and a Megadiverse nation [4]—a distinction shared only with Madagascar. Within this archipelago, the Philippine tarsier, a small endemic nocturnal primate, has been enlisted as a flagship species for an emerging societal conservation movement and an expanding ecotourism industry [1,5].

Traditionally, taxonomy and conservation have been inextricably linked and most conservation strategies have targeted formally named taxonomic units: species or subspecies [6]. Although most conservation efforts have targeted these taxonomic entities, conserving finer-grained genetic diversity across a species' range [7,8] is essential to preserving metapopulation dynamics, preventing inbreeding depression, avoiding population collapses, and ultimately ensuring against extinction [9–11]. Unique evolutionary lineages or genetically defined "Evolutionarily Significant Units" [12] are appropriate targets of conservation programs aimed at preserving genetic diversity among and within species; targeting empirically defined distinct evolutionary lineages has the added benefit of potentially removing the subjectivity sometimes associated with traditional taxonomy [13].

Despite being the focus of a disproportionate number of intensely focused studies, multiple lines of evidence suggest that numerous primate taxa await discovery and formal taxonomic description. For example, between the 1975 and 1999 nocturnal primate species diversity grew worldwide by 2.85-fold increase [14–18], and has since climbed by an additional 1.69-fold increase [19]. Unique among nocturnal primates, tarsiers are found only in insular Southeast Asia (See Appendix S1). Ten species are recognized currently, with several new taxa recently proposed through bioacoustic analysis and molecular data [15]. The Philippine tarsier (Fig. 1) has been the focus of recent attempts to understand morphological variation [16], clarify taxonomy [17], and establish the conservation status of populations [18,19]; these efforts have met with limited success and left unanswered the questions of appropriate targets for conservation action.

In the Philippines, many morphologically indistinguishable yet genetically differentiated "cryptic" species [20–22] have recently been discovered, suggesting that the codistributed Philippine tarsier might also harbor hidden evolutionary lineages (unrecognized species, genetic variants, putative taxa, etc.), which may be unprotected and not incorporated into current conservation planning.

We assessed the conservation genetics of Philippine tarsier to determine (1) how many distinct evolutionary lineages can be identified, (2) whether genetic structure conforms to biogeographical predictions and/or (3) expectations derived from current taxonomy (three subspecies). Due to limited natural history data or consensus from other sources of information (morphology, bioacoustics, ecology), we argue that genetic data should be used to distinguish evolutionarily distinct tarsier population groups as objective, empirically defined conservation priorities.

Materials and Methods

Mitochondrial data collection

Genetic material from *T. syrichta* was nondestructively sampled (ear and tail-tip biopsies) from throughout as much of the species' range as feasible (targeting ranges of all described subspecies; Appendix S1) on the large islands of Mindanao, Samar, Leyte, Bohol, and Dinagat, Philippines (Figs. 1, 2; Appendix S2). Deceased animals were salvaged from illegal roadside animal dealers, local Provincial or City Environmental Natural Resources (PENRO, CENRO) officers or university administrators (confiscations of poached animals by students, animal traders, and bush meat hunters), and from indigenous hunters in forested areas. Salvaged animals were prepared as museum specimens and deposited at the Leyte State University (ViSCA) Natural History Museum, the National Museum of the Philippines, or the University of Kansas Biodiversity Institute (Appendix S2). We sampled 77 individuals of *T. syrichta* from 17 localities for genetic material from most of the large islands within the Mindanao Pleistocene Aggregate Island Complex (PAIC) [2], including representatives of all recognized subspecies. We sequenced the mitochondrial 12S ribosomal RNA (12S), Cytochrome B (CytB), and NADH Dehydrogenase Subunit 2 (ND2) gene fragments and nine nuclear microsatellites.

Genomic DNA was extracted from tissues following Fujita's guanidine thiocyanate method [23]. Molecular data were derived from all available genetic samples and included efforts to determine the sequence of the 12S mitochondrial gene fragment, the ND2, and the CytB gene region. Sequence identity was confirmed via examination of amino acid translations for internal stop codons for Cytb and ND2 sequence alignments, GenBank BLAST searches, and alignment comparisons with annotated, published comparative sequence data available on GenBank. All sequences were deposited in GenBank under accession numbers KM217270–99 (12S), KM217300–23 (CytB), and KM217324–49 ND2 (Appendix S2). Primers used for amplifying mitochondrial gene sequences include: 12S (Tars12s_AJB.F1, 5′–TCA-CAACGTCTTGCTCAACC–3′ and Tars12s_AJB.R2, 5′–TTGAGGAGGGTGACGGGCGG–3′); ND2 (TarND2_AJB.f2, 5′–ACTTTCTAATTCAAGCGACAGCCTCC–3′ and TarND2_AJB.r2, 5′–TGGGGGATATGGGTAAAAGTAGGGTGG–3′); CytB (TarsCytB.F1, 5′–CACATCTGCCGAGACGTAAA–3′ and TarsCytB.R1, 5′–TGGGGTGGAGTGTTTAGAGG–3′). We used the following thermal profiles for the mitochondrial fragment amplification: 1 min at 95°C followed by 38 cycles of 94°C for 30 sec, 30 sec at the primer specific annealing temperature (60°C for 12 s; 61°C for ND2; 62°C for CytB), and 72°C for 1 min, and a final extension phase at 72°C for 10 min.

Amplified products were visualized on 1.5% agarose gels and PCR products were purified with 1 µL of a 20% dilution of ExoSAP-IT (US78201, Amersham Biosciences, Piscataway, NJ) on the following thermal profile: 30 min at 37°, followed by 15 min at 80°. Cycle sequencing reactions were run using ABI Prism BigDye Terminator chemistry (Ver. 3.1; Applied Biosystems, Foster City, CA), and purified with Sephadex Medium (NC9406038, Amersham Biosciences, Piscataway, NJ) in Centri-Sep 96 spin plates (CS-961, Princeton Separations, Princeton, NJ). Purified products were analyzed using an ABI Prism 3130xl Genetic Analyzer (Applied Biosystems). Sequence contigs were assembled and edited using Sequencher 4.8 (Gene Codes Corp., Ann Arbor, MI).

Sequencing proved highly problematic for our many degraded tissue samples and approximately half our samples would not amplify for any targeted gene regions. Of the half that could be amplified for >2 target regions, initial sequence alignments were produced in Muscle [24] and manual adjustments were made in MacClade 4.08 [25]. We estimated the phylogeny for each mitochondrial gene fragment independently using likelihood and concatenated mitochondrial sequences following no observation of statistically significant incongruence between datasets. Exploratory analyses of the combined dataset of 36 individuals inferred relationships that did not strongly conflict with topologies inferred from individual gene fragments.

Microsatellite data collection

We sampled a total of 66 individuals from 17 localities of *Tarsius syrichta* across the southern Philippines. Each locality was represented by 1–27 individuals (mean = 3.88). DNA was extracted using QIAGEN DNeasy extraction kits and we used nine microsatellite makers previously developed for *Tarsius syrichta*: T5, T6, T22, T34, T35, T43, T69 [26] and *Tarsius* spp.: T42, T50 [27]. PCR reaction volumes and conditions for the PCR amplifications followed the protocol of Schuelke [28]. Conditions of the PCR amplification were as follows: 94°C for 30 s, 56°C for 45 s, 72°C for 45 s, followed by 8 cycles at 94°C for 30 s, 53°C for 45 s, 72°C for 45 s, and a final extension at 72°C for 10 min. Primers T42, T43 and T45 were fluorescently labeled with 6-FAM by Integrated DNA Technologies (Coralville, Iowa) for PCR. Primer annealing temperatures for these three loci were as follows: T42 at 53°C [27], T43 at 54°C and T45 at 57°C [26].

Philippines
Current Dry Land
120 m isobath
Luzon
Luzon PAIC
15°
10°
Samar
Leyte
Dinagat
Siargao
Bohol
Zamboanga Peninsula
Mindanao
Borneo
Mindanao PAIC
121°
125°
200 km
Bohol
29—Leyte
13—Bohol
8—Bohol
27—Bohol
26—Bohol
32—Leyte
24—Bohol
25—Bohol
4—Bohol
1—Bohol
30—Leyte
31—Leyte
39—Samar
37—Samar
NC012774
AB371090
28—Leyte
Samar-Leyte
Dinagat-Caraga
Surigao del Norte—42
Dinagat—40
Surigao del Norte—53
Surigao del Norte—55
Surigao del Norte—54
Surigao del Norte—51
0.0070
Agusan del Sur—59
Misamis Oriental—61
Agusan del Sur—58
Sarangani—64
Sarangani—63
Agusan del Sur—57
Misamis Oriental—60
Zamboanga—65
Zamboanga—66
Mindanao
Zamboanga

Figure 1. Phylogeographic relationships of *Tarsius syrichta* (see Appendix S1 for taxonomic summary) estimated from a combined, partitioned, RAxML ML analysis of mitochondrial (12S, CytB, ND2) gene fragments. Black circles at nodes correspond to ML bootstraps ≥ 70% and Bayesian PP ≥95%.

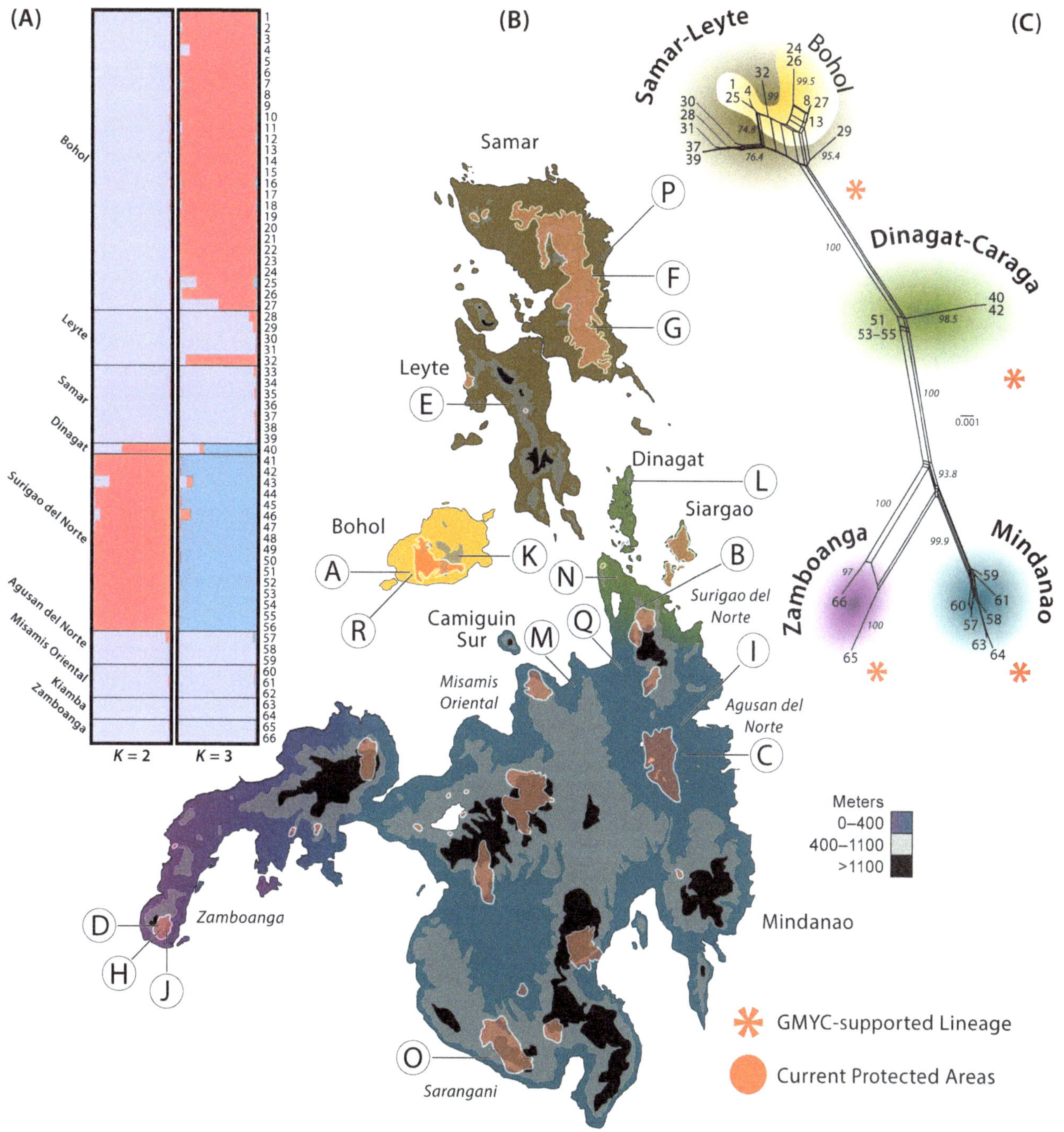

Figure 2. DISTRUCT visualization of STRUCTURE analyses (A) assigning individuals to major population groupings (genetically distinct evolutionary lineages) for Philippine tarsier demes (K=2 and 3 populations). Mindanao faunal region (B; see Fig. 1, inset) with sampling (17 sites, 66 individuals) labeled with letters corresponding to full localities listed in Appendix S2, protected areas shaded red. SplitsTree gene network (C; numbers at internodes = ML bootstrap replicates), and results of GMYC analyses (red asterisks denote lineages delineated by the Yule-coalescent), with numbers at tips corresponding to individual samples in Structure plots (A) and cluster shading corresponding to islands on map (B).

Subsequently, 2.1 µl of the PCR product was added to 9 µl of formimide and .15 µl was electrophoresed with the LIZ-500 size standard and analyzed on a 3730 Genetic Analyzer (Applied Biosystems). Fragments were sized with GeneMapper version 4.1 (Applied Biosystems). We replicated data collection for all samples a minimum of two times per locus in order to monitor for allelic dropout and false alleles. In cases where we detected allelic dropout or inconsistent genotypes (n = 2), the sample was replicated once more in order to minimize genotyping errors. In a few instances of ambiguous genotypes, despite these efforts, samples were simply excluded from subsequent analyses (n = 8). Microsatellite data were archived at Dryad (doi:10.5061/dryad.r7468).

Phylogenetic analyses and lineage delimitation: mitochondrial data

Partitioned Bayesian analyses of the combined dataset were conducted in MrBayes v3.1.2 [29]. As the mtDNA genes sampled in this study included incomplete sections for some targeted genes, datasets were partitioned by gene only. The Akaike Information Criterion (AIC) implemented in jModeltest [30] was used to select the model of nucleotide substitution for each subset (Table 1). We ran four independent Metropolis-coupled MCMC analyses, each with four chains and the default heating scheme (temp = 0.2). All analyses were run for 20 million generations, sampling every 1000 generations. To assess stationarity, all sampled log-likelihoods and parameter values from the cold Markov chain were plotted against generation time and compared among independent runs using Tracer v1.4 [31]. These runs demonstrated patterns consistent with stationarity after 4 million generations, hence the first 20% of samples were discarded as burn-in.

Partitioned maximum likelihood (ML) analyses were conducted in RAxMLHPC v7.0 [32] for the combined dataset, using the same partitioning strategy as implemented in the Bayesian analyses. The more complex model (GTR + I + Γ) was used for all subsets, and 100 replicate ML inferences were performed for each run. Each inference was initiated with a random starting tree, employed the rapid hill-climbing algorithm and support was assessed with 1000 bootstraps [33].

To generate an ultrametric phylogeny for our Generalized Mixed Yule-Coalescent (GMYC) analysis, we ran a strict clock divergence time analysis of the mtDNA data, partitioning the data by gene, and fixing the mean substitution rate to 1.0 in BEAST v1.7.5 [34]. We first removed all individuals with identical sequences, however, since zero-length branches can cause the GMYC analysis to over-partition the dataset. BEAST analyses were run for 100 million generations, sampling every 10,000. Convergence was assessed with Tracer as described above. We then used the ultrametric phylogeny with the "single threshold" GMYC model [35] to determine the number of evolutionary lineages in the mtDNA dataset.

Population structure and gene flow: nuclear data

As currently recognized, *Tarsius syrichta* spans the Mindanao PAIC in the southern Philippines. As an initial starting point, our *a priori* hypothesis of population structure, given our understanding that all these islands are part of the same faunal region [2], were that genetic diversity might be partitioned among individual islands. In order to test these expectations, we used the Bayesian clustering method of the program Structure v2.3.3 [27,28,36–39] with allelic data from nine nuclear loci, to estimate the number of populations among *T. syrichta* samples, as well as infer the probabilities of individuals belonging to each of the estimated populations. We considered individuals with q values between 0.10 and 0.90 to be admixed [39–41]. Due to the absence of *a priori* knowledge of inter-locus relationships, analyses were run under the correlated allele frequency model and the admixture ancestry model [36,40]. All other parameters were left with default settings. Taking a conservative approach to evaluating population boundaries for this study, we considered the number of possible populations, represented by K, to range from one (a single geographically non-structured species) to seven (each island sampled in this study, and allowing for additional population genetic complexity that may be present on the large island of Mindanao [2]). We conducted 10 independent runs in Structure for each of the seven values of K (1–7). We ran each analysis for 10 million iterations, with a burn-in of 500,000. The rate of change of K (ΔK) was evaluated in Structure Harvester (Web v0.6.93) [42] following the method of Evanno et al. [43] to determine the number of preferred genetic clusters. The results of structure analyses were visualized using the program *Distruct* v1.1 (Fig. 2) [44].

To assess whether there is evidence of recent migration among populations, we estimated the level of historical gene flow within a Bayesian framework in the program BAYESASS v1.3 [45] for the four evolutionary lineages supported in phylogenetic analyses (Fig. 1). Four independent runs were conducted, each with 20,000,000 iterations, a sampling frequency of 2,000, and a burn-in of 2,000. Trace files were assessed using the program Tracer v1.5 [31] for evidence of convergence prior to result summary.

To visualize population genetic structure, we generated phylogenetic networks for the mitochondrial dataset by employing the Neighbor-Net algorithm [46] in the program SplitsTree v4.10 [47] (Fig. 2). To assess the support for inferred splits in the network, a bootstrap analysis was conducted with 1000 pseudoreplicates.

Results

Genetic variation and gene networks

Of the 77 samples collected, 66 could be characterized for microsatellite variation (Tables 2, 3) but, due to sample degradation, only 36 could be sequenced for mitochondrial DNA. As

Table 1. Models of evolution selected by AIC and applied for partitioned, phylogeographic analyses[1].

Partition	AIC Model	Number of Characters
12S	GTR + Γ	618
Cytb	HKY + Γ	557
ND2	HKY + Γ	538

[1]The model GTR + Γ was used for partitioned RAxMLHPC analyses.

Table 2. Microsatellite variation in 66 individuals of *Tarsius syrichta* from 17 localities (Fig. 2).

Population	N	H_O	H_E	L_P	F_{IS}	P value
Bohol	27	0.58674±0.27641	0.74946±0.16146	9	−0.12254	0.859238
Samar-Leyte	12	0.53519±0.24444	0.80535±0.07257	6	0.10995	0.214076
Dinagat-Caraga	17	0.53184±0.19296	0.63060±0.12853	4	0.0764	0.304008
E. Mindanao	8	0.63294±0.17635	0.83488±0.12227	7	0.48611	0.008798
W. Mindanao	2	0.56250±0.32043	0.75000±0.17817	8	0.42857	0.339198

Abbreviations include: N, number of samples; H_O, observed heterozygosity; H_E, expected heterozygosity; L_P, number of polymorphic loci; F_{IS} inbreeding coefficient.

Table 3. Locus-specific microsatellite variation in 66 individuals of *Tarsius syrichta* from 17 localities (Fig. 2).

		Bohol	Samar-Leyte	Dinagat-Caraga	E. Mindanao	W. Mindanao
Locus T22	H_O	0.62963	0.75	0.64286	0.5	0.5
	H_E	0.76799	0.80435	0.5291	0.83333	0.83333
	A_N	6	9	6	6	3
	HWE	**0.0064**	0.335	1	0.0562	0.3301
Locus T35	H_O	0.69231	0.75	0.6875	0.5	n/a
	H_E	0.7911	0.7971	0.68347	0.53333	n/a
	A_N	8	8	5	5	n/a
	HWE	**0.0211**	0.7219	0.4304	0.279	n/a
Locus T5	H_O	0.84615	0.58333	0.41176	0.375	0.5
	H_E	0.89819	0.84783	0.58824	0.84167	0.5
	A_N	10	9	5	5	2
	HWE	0.2585	**0.0039**	0.1024	**0.0046**	n/a
Locus T6	H_O	0.46154	0.08333	0.35294	0.5	0
	H_E	0.79864	0.82246	0.67023	0.93333	0.66667
	A_N	11	7	7	9	2
	HWE	**0.0**	**0.0**	**0.0134**	**0.0**	0.3346
Locus T34	H_O	0.57692	0.4	0.23529	0.875	0.5
	H_E	0.73454	0.73684	0.36898	0.925	0.83333
	A_N	6	5	5	10	3
	HWE	**0.0161**	**0.0204**	**0.0245**	0.1925	0.3305
Locus T42	H_O	0.85185	0.66667	0.53846	0.85714	0.5
	H_E	0.8833	0.86928	0.77231	0.89011	0.5
	A_N	11	8	5	8	2
	HWE	**0.0**	**0.0064**	0.4807	0.4357	1
Locus T69	H_O	0.18519	0.33333	0.45455	0.75	1
	H_E	0.53319	0.66013	0.58009	0.79167	0.83333
	A_N	5	4	6	5	3
	HWE	**0.0009**	0.0862	**0.048**	0.2565	n/a
Locus T43	H_O	0.88889	0.83333	0.875	0.625	0.5
	H_E	0.89727	0.9058	0.7379	0.90833	0.83333
	A_N	13	11	6	9	3
	HWE	0.562	0.0843	0.1855	**0.0**	0.3384
Locus T45	H_O	0.14815	0.41667	0.58824	0.71429	1
	H_E	0.44095	0.80435	0.7451	0.85714	1
	A_N	6	7	8	6	4
	HWE	**0.0**	**0.0037**	0.0542	0.1938	1

Abbreviations include: H_O, observed heterozygosity; H_E, expected heterozygosity; A_N, number of alleles; *HWE*, deviation from Hardy–Weinberg equilibrium (p-value).

anticipated, the results of network analyses corroborate the major results observed from phylogenetic analyses (Figs. 1, 2); these same mitochondrial clades were identified as distinct in the GMYC lineage delineation procedure.

Phylogenetic and lineage delimitation analyses

GMYC analysis of the mitochondrial phylogeny identified three putative evolutionary lineages; each of these three lineages was supported in Bayesian and ML phylogeographic analyses and phylogenetic networks of mitochondrial sequence data, and were provisionally adopted as the following distinct evolutionary groups (Fig. 1): (1) Bohol, Samar, and Leyte island, (2) Dinagat and Northeast Mindanao (Caraga Region) island, and (3) Mindanao island, composed of (*a*) eastern and (*b*) western subclades; substantial genetic divergence was found among groups (2.1–4.7% in mtDNA; Table 4), with markedly less genetic variation within groups (0.0–1.1%) and minimal gene flow between most groups (Table 5).

Assignment of individuals to population clusters

The *ad hoc* ΔK test of the results of Structure analyses of microsatellite loci preferred a K of two (LnP[K] = −2341.58; Dinagat-Caraga vs. all remaining samples), suggesting an initial estimate of only two distinct populations as defined by our nuclear data alone. All individuals were assigned to one of these two genetic clusters with q values above 0.90. With the exception of populations from Dinagat Island and the northeastern peninsula of Mindanao Island (Surigao del Norte Province; Fig. 2), all sampled populations were inferred to be part of one cohesive genetic group. Evaluating a three-population model in structure reveals a similar pattern of unambiguous individual assignments to populations; however, the model of $K = 3$ which was close to maximally preferred included additional support for the distinctiveness of populations on Bohol (Fig. 2). All Structure analyses provided support for the cryptic Dinagat-Caraga population (Fig. 2).

Discussion

New Philippine tarsier evolutionary lineages: novel conservation priorities

The unambiguous support across all analyses for a genetically distinct Dinagat-Caraga tarsier lineage, mirrored in all analyses of all loci, is novel, and identifies a range-restricted "cryptic" evolutionary entity with unique conservation concerns. Although moderately forested, Dinagat and northeast Mindanao are impoverished economically, lack low-elevation protected areas (Fig. 2), and have become the focus of particularly intensive mining operations—all of which threaten the remaining suitable habitat of this newly documented evolutionary lineage. The Dinagat-Caraga tarsier should therefore be regarded as tantamount to the conservation importance of celebrated Philippine flagship species (e.g., Philippine eagle, tamaraw, golden-spotted monitor lizard, etc.: [2]). Nearby Siargao Island is protected and may harbor the same genetic variant identified here from Dinagat; intensive studies of these populations are urgently needed.

We anticipate future studies will further refine the known distribution of the Dinagat-Caraga tarsier and extend its range to nearby Siargao Island. Additionally it is not clear if the Surigao del Norte (NE Mainland Mindanao, Caraga Region) samples identified here occur naturally on Mindanao or are the result of Dinagat and/or Siargao island animals confiscated from smugglers by regional wildlife enforcement and subsequently released on Mindanao (PSO, *personal communication* with Surigao del Norte Provincial Environmental Natural Resources Office [PENRO] staff).

Geographical distribution of genetic variation

Our empirical findings are more complex than is reflected in current taxonomy [18,48], stand in contrast to past biogeographic paradigms [2,49], uncover a novel conservation target (Dinagat-Caraga lineage; Figs. 1, 2), and identify a conservation research priority of urgent concern (potential differentiation between western and eastern Mindanao). The moderate level of sequence divergence (2.4%; Table 4) detected between eastern and western Mindanao (both supported as distinct lineages in the GMYC analysis, but with marginally significant migration rates detected in analysis of microsatellite data; Table 4) may indicate that separate tarsier conservation programs for these populations are warranted if future studies confirm their distinctiveness. An east–west sampling transect across Mindanao will be necessary to investigate the genetic relationships of these populations which, given the gap in our sampling, may simply be the extremes of natural geographically based genetic structure.

Our results from analyses of mitochondrial data do not differentiate the populations on Bohol from those on Samar-Leyte (currently recognized as separate subspecies: [48]). Considering the suboptimal $K = 3$ results, Structure analyses of microsatellites identified Bohol potentially as distinct. However, given that these analyses potentially are sensitive to unequal sampling, and that we possessed many more samples from Bohol than from Samar-Leyte, we suspect this result may be an artifact; future studies with better sampling from Samar and Leyte islands, and additional microsatellite loci will be necessary to investigate further genetic relations within the Bohol-Samar-Leyte clade. At present, in our efforts to adhere to objective, empirically defined evolutionary lineages, we stop short of defining Bohol as a distinct conservation target, and

Table 4. Uncorrected mitochondrial sequence divergence (%) among Philippine tarsier (*Tarsius syrichta*) evolutionary lineages shown below diagonal.

	Bohol	Samar-Leyte	Dinagat-Caraga	E. Mindanao	W. Mindanao
Bohol	0.0–0.9	0.018	0.015	0.018	0.026
Samar-Leyte	0.4–1.4	0.0–1.1	0.019	**0.065**	**0.056**
Dinagat-Caraga	2.1–3.5	2.1–3.4	0.0–2.3	0.021	0.029
E. Mindanao	3.4–4.6	3.5–4.6	2.7–4.6	0.1–0.8	**0.051**
W. Mindanao	3.8–4.6	3.7–4.6	3.0–4.7	2.0–2.4	0.0–1.0

Percentages on the diagonal represent intraspecific (or within clade) genetic diversity. Mean inferred migration rates (inferred in BAYESASS) between regionally partitioned diversity shown above diagonal. Migration rates above the 0.05 threshold bolded for emphasis.

Table 5. Identities by geographical region and numbers (in parentheses) of Philippine tarsier (*Tarsius syrichta*) evolutionary lineages, putative taxa, or conservation targets inferred (new data and analyses) or predicted (biogeography, taxonomy) from each of the five available sources of information.

Divergent mtDNA lineages & GMYC results (4)	Distinct nuDNA groups (3)	Islands (5)	Taxonomy (3)
Bohol-Samar-Leyte	Dinagat-Caraga	Bohol	Bohol
Dinagat-Caraga	Bohol- Samar-Leyte-east Mindanao-west Mindanao	Samar	Samar-Leyte
Mindanao (a: east Mindanao, b: west Mindanao)		Leyte	Caraga-Mindanao-Zamboanga
		Dinagat	
		Mindanao	

point to the nested nature of the Bohol samples within the Bohol-Samar-Leyte tarsier group (Figs. 1, 2).

Combining mitochondrial and nuclear gene signals to establish targets for conservation action

Under current implementation of international conservation status assessment (IUCN, 2014), prioritization of populations for conservation action follows the recognition of named taxonomic units, i.e., species and subspecies. In the face of differing taxonomic arrangements, biogeographic expectations, and differing geographic patterns of nuclear versus mitochondrial genetic variation, how should the Philippine tarsier be prioritized for implementation of applied conservation measures? Currently, one tarsier sanctuary has been established on Bohol Island and another is under consideration for construction on Leyte. Would these two, and only these two efforts adequately conserve genetic components of Philippine tarsier diversity? We argue that they would not. Would establishment of conservation programs based on current tarsier taxonomy (one effort on Bohol, another on Samar-Leyte, and a third on Mindanao) properly conserve the genetic variation we have documented? Similarly, we argue that such an approach would fail to conserve the genetic variation elucidated here.

Brandon-Jones et al. [17] characterized the three subspecies (one from Bohol, another from Leyte-Samar, and a third from Mindanao) as "dubious" taxa. They wrote: "Groves (2001) recognized no subspecies for *Tarsius syrichta*. Hill (1955) recognized *T. s. syrichta/carbonarius/fraterculus* as a poorly defined subspecies, perhaps synonymous, with *T. s. syrichta*. Museum specimen variation seemed insignificant to Niemitz (1984), but an inadequate basis for judgment, according to Musser, and Dagosto (1987)." Our study sheds some light on the uncertainty of Brandon-Jones et al. [17] and suggests that a detailed study employing other lines of evidence is still needed to revise Philippine tarsier taxonomy. One possible outcome of further study could be that our evolutionary groups are not sufficiently distinct based on new lines of evidence (morphology, bioacoustics, whole genome data) to warrant taxonomic separation, and that Philippine tarsiers should therefore be classified as a single taxon, *T. syrichta* [16,50,51]. Alternatively, further investigation might find two or more of our conservation priority groups warrant taxonomic separation, in which case, the epithets applied by Hill [52] will be available (as species or subspecies). Three names are available for a subset of the evolutionary lineages we identify: (1) Samar-Leyte, (2) Bohol, and (3) eastern Mindanao (but not including western Mindanao [Zamboanga], or the novel Dinagat-Caraga lineage). For the immediate future, we argue that applied conservation efforts based upon the combined intersection of our variable genetic results are superior to those based upon existing taxonomy.

A number of genetic and social system phenomena could conceivably account for the differences in lineage-specific support we have inferred between mitochondrial and nuclear loci. Possibilities include persistence of genetic polymorphism, lineage sorting, nuclear gene flow between distinct mitochondrial lineages, sex-biased dispersal, mitochondrial gene sweeps, or an undocumented Philippine tarsier mating system. Given the present sampling, we are unable to distinguish between these possibilities, which may provide intriguing questions for future research. However, from a practical and applied conservation perspective, we argue that a single, clearly optimal, and objective solution results from the combination of patterns observed here in mitochondrial and nuclear gene loci.

Given our variable inferences, we advocate a combined, cumulative approach towards identification and recognition of tarsier evolutionary lineages for conservation planning. Thus, we argue that in order to maximally preserve genetic variation, it is the *combined* results of our variable sources of information that likely will have the best chance of maximally preserving genetic variation in the Philippine tarsier. As such, we advocate a multi-tiered conservation program involving conservation programs and protected areas in the ranges of each distinctive population lineages identified here, minimally including (1) Bohol-Samar-Leyte, (2) Dinagat-Caraga, and (3) Mindanao. Until a comprehensive taxonomic study using multiple lines of evidence (molecular, morphological and/or bioacoustic data) can be undertaken, we find that the existing evidence provides an inadequate basis for distinguishing between taxonomic alternatives. Instead, we emphasize that the archipelago's tarsier populations are partitioned into at least the three genetic variants, which we have empirically defined here for practical conservation purposes. We caution primatologists from taking taxonomic action until multiple lines of evidence converge on a meaningful solution [20,22], ideally with information from throughout the genome [53] and more robust sampling along multiple transects from throughout the range of this species (i.e., eastern versus western Mindanao). Nevertheless, our results necessitate a refined conservation strategy in order to effectively preserve the geographical distribution of genetic variation in this flagship species. Such an approach will greatly enhance the prospects for continued survival of this endemic primate and, combined with many other recent discoveries in the country, will contribute to the recognition of the archipelago as a globally significant biodiversity conservation priority [2,5,49].

Acknowledgments

We thank the Philippine Department of Environment and Natural Resources for facilitating research, sample collection, and export permits (in particular T. M. Lim, C. Custodio, J. deLeon, and A. Tagtag) necessary for this and related research. Sampling protocols were approved by the University of Kansas Animal Care and Use Committee (IACUC 158-01 to RMB) and protocols to capture, sedate, and harvest ear biopsies from wild tarsiers were approved by the Dartmouth Animal Care and Use Committee (IACUC 10-11-12 and 11-06-06AT to NJD). Thanks are due to J. Quilang for assistance with data and comments on the manuscript. We thank N. Antoque, J. Cantil, and V. Yngente for assistance in the field and anonymous reviewers for comments on previous versions of the manuscript.

Author Contributions

Conceived and designed the experiments: RMB JAW CDS JAE MS MLD ACD. Performed the experiments: RMB MRMD LVD INA JAE NJD PSO AL MLD ACD CDS. Analyzed the data: KVO AJB CDS. Contributed reagents/materials/analysis tools: RMB KVO AJB CDS JAE LVD GLM. Wrote the paper: RMB JAW CDS JAE MS GLM. Obtained permission and executed field surveys: RMB MRMD LVD MS INA JAE NJD PSO GLM AL MLD ACD CDS.

References

1. Sodhi NS, Koh LP, Brook BW, Ng PK (2004) Southeast Asian biodiversity: an impending disaster. Trends Ecol Evol 19: 654–660.
2. Brown RM, Diesmos AC (2009) Philippines, Biology. In: Gillespie R, Clague D, editors. Encyclopedia of islands. Berkeley: University of California Press. pp. 723–732.
3. Myers N, Mittermeier RA, Mittermeier CG, da Fonseca GAB, Kent J (2000) Biodiversity hotspots for conservation priorities. Nature 403: 853–858.
4. Mittermeier RA, Gil PR, Mittermeier CG (1997) Megadiversity: Earth's biologically wealthiest nations. Washington, DC: Conservation International, Cemex.
5. Posa MR, Diesmos AC, Sodhi NS, Brooks TM (2008) Hope for threatened tropical biodiversity: lessons from the Philippines. Bioscience 58: 231–240.
6. Mace G (2004) The role of taxonomy in species conservation. Proc Royal Soc London B 359: 711–719.
7. Evans BJ, Supriatna J, Andayani N, Setiad MI, Cannatella DC, et al. (2003) Monkeys and toads define areas of genetic endemism on the island of Sulawesi. Evolution 57: 1436–1443.
8. Melnick DJ, Morales JC, Honeycutt RL (1999) Conservation genetics: applying molecular methods to maximize the conservation of taxonomic and genetic diversity. In: Raven P, Williams T, editors. Nature and human society: The quest for a sustainable world. Washington, DC: National Academies Press. pp. 264–275.
9. Brook BW, Tonkyn DW, O'Grady JJ, Frankham R (2002) Contribution of inbreeding to extinction risk in threatened species. Conserv Ecol 6: 16. Available: http://www.consecol.org/vol6/iss1/art16. Accessed 11 December 2013.
10. Hanski I (1998) Metapopulation dynamics. Nature 396: 41–49.
11. Saccheri I, Hanski I (2006) Natural selection and population dynamics. Trends Ecol Evol 21: 341–347.
12. Ryder OA (1986) Species conservation and systematics—the dilemma of subspecies. Trends Ecol Evol 1: 9–10.
13. Moritz C (1994) Defining 'evolutionary significant units' for conservation. Trends Ecol Evol 9: 373–375.
14. Bearder SK (1999) Physical and social diversity among nocturnal primates: a new view based on long-term research. Primates 40: 267–282.
15. Groves C, Shekelle M (2010) The genera and species of Tarsiidae. Int J Primatol 31: 1071–1082.
16. Groves C (2001) Primate taxonomy. Washington, DC: Smithsonian Institution Press.
17. Brandon-Jones D, Eudey AA, Geissmann T, Groves CP, Melnick DJ, et al. (2004) Asian primate classification. Int J Primatol 25: 97–164.
18. Dagosto M, Gebo DL, Dolino CN (2003) Natural history of the Philippine tarsier. In: Wright P, Simons E, Gursky S, editors. Tarsiers: Past, present, and future. New Brunswick: Rutgers University Press. pp. 237–259.
19. Shekelle M, Arboleda I (2008) *Tarsius syrichta*. In: IUCN 2014. IUCN red list of threatened species. Version 2014.1. Available: www.iucnredlist.org. Accessed 6 June 2014.
20. Barley A, White J, Diesmos AC, Brown RM (2013) The challenge of species delimitation at the extremes: diversification without morphological change in Philippine sun skinks. Evolution 67: 3556–3572. doi:10.1111/evo.12219.
21. Esselstyn JA, Evans BJ, Sedlock JL, Khan FAA, Heaney LR (2012) Single-locus species delimitation: a test of the mixed Yule–coalescent model, with an empirical application to Philippine round-leaf bats. Proc Royal Soc London B 279: 3678–3686. doi:10.1098/rspb.2012.0705.
22. Welton LJ, Siler CD, Oaks JR, Diesmos AC, Brown RM (2013) Multilocus phylogeny and Bayesian estimates of species boundaries reveal hidden evolutionary relationships and cryptic diversity in Southeast Asian monitor lizards. Molec Ecol 22: 3495–3510.
23. Esselstyn JA, Garcia HJD, Saulog MG, Heaney LR (2008) A new species of *Desmalopex* (Pteropodidae) from the Philippines, with a phylogenetic analysis of the Pteropodini. J Mammal 89: 815–825.
24. Edgar RC (2004) MUSCLE: multiple sequence alignment with high accuracy and high throughput. Nuc Acid Res 32: 1792–1797.
25. Maddison DR, Maddison WP (2005) MacClade: analysis of phylogeny and character evolution, Version 4.08. Sunderland: Sinauer Associates.
26. Merker S, Driller C, Neri-Arboleda I, Zischler H (2008) Polymorphic microsatellite markers for Philippine tarsiers (*Tarsius syrichta*). Conserv Genet 9: 1349–1351.
27. Merker S, Driller C, Perwitasari-Farajallah D, Zahner R, Zischler H (2007) Isolation and characterization of 12 microsatellite loci for population studies of Sulawesi tarsiers (*Tarsius* spp.). Mol Ecol Notes 7: 1216–1218.
28. Schuelke M (2000) An economic method for the fluorescent labeling of PCR fragments. Nat Biotechnol 18: 233–234.
29. Ronquist F, Huelsenbeck JP (2003) MRBAYES 3: Bayesian phylogenetic inference under mixed models. Bioinformatics 19: 1572–1574.
30. Posada D (2008) jModeltest: Phylogenetic model averaging. Molec Biol Evol 25: 1253–1256.
31. Rambaut A, Drummond AJ (2007) Tracer v1.4. Available: http://beast.bio.ed.ac.uk/Tracer.
32. Stamatakis A (2006) RAxML-VI-HPC: maximum likelihood-based phylogenetic analyses with thousands of taxa and mixed models. Bioinformatics 22: 2688–2690.
33. Stamatakis A, Blagojevic F, Nikolopoulos D, Antonopoulos C (2007) Exploring new search algorithms and hardware for phylogenetics: RAxML meets the IBM cell. J VLSI Signal Proc 48: 271–286.
34. Drummond AJ, Rambaut A (2007) BEAST: Bayesian evolutionary analysis by sampling trees. BMC Evol Biol 7: 214. Available: http://www.biomedcentral.com/1471-2148/7/214. Accessed 2013 December 13.
35. Fujisawa T, Barraclough TG (2013) Delimiting species using single-locus data and the generalized mixed Yule coalescent approach: a revised method and evaluation on simulated data sets. Syst Biol 62: 707–724.
36. Falush D, Stephens M, Pritchard JK (2003) Inference of population structure using multilocus genotype data: linked loci and correlated allele frequencies. Genetics 164: 1567–1587.
37. Falush D, Stephens M, Pritchard JK (2007) Inference of population structure using multilocus genotype data: dominant markers and null alleles. Molec Ecol Notes 7: 574–578. doi: 10.1111/j.1471-8286.2007.01758.x.
38. Hubisz MJ, Falush D, Stephens M, Pritchard JK (2009) Inferring weak population structure with the assistance of sample group information. Mol Ecol Resources 9: 1322–1332.
39. Pritchard JK, Stephens M, Donnelly P (2000) Inference of population structure using multilocus genotype data. Genetics 155: 945–959.
40. Spinks PQ, Thomson RC, Barley AJ, Newman CE, Shaffer HB (2010) Testing avian, squamate, and mammalian nuclear markers for cross amplification in turtles. Conserv Genet Resour 2: 127–129. doi: 10.1007/s12686-010-9184-7.
41. Vähä J-P, Primmer CR (2006) Efficiency of model-based Bayesian methods for detecting hybrid individuals under different hybridization scenarios and with different numbers of loci. Molec Ecol 15: 63–72. doi: 10.1111/j.1365-294X.2005.02773.x.
42. Earl DA, vonHoldt BM (2012) STRUCTURE HARVESTER: a website and program for visualizing STRUCTURE output and implementing the Evanno method. Conserv Genet Resour 4: 359–361. doi: 10.1007/s12686-011-9548-7.
43. Evanno G, Regnaut S, Goudet J (2005) Detecting the number of clusters of individuals using the software STRUCTURE: a simulation study. Molec Ecol 14: 2611–2620.
44. Rosenberg NA (2004) Distruct: a program for the graphical display of population structure. Molec Ecol Notes 4: 137–138.

45. Rannala B, Mountain JL (1997) Detecting immigration by using multilocus genotypes. P Natl Acad Sci USA 94: 9197–9201.
46. Bryant D, Moulton V (2004) Neighbor-Net: an agglomerative method for the construction of phylogenetic networks. Mol Biol Evol 21: 255–265.
47. Huson DH, Bryant D (2006) Application of phylogenetic networks in evolutionary studies. Mol Biol Evol 23: 245–267.
48. Shekelle M, Gursky-Doyen S, Richardson MC (2013) Family Tarsiidae (Tarsiers). In: Mittermeier RA, Rylands AB, Wilson DE, editors. Handbook of the mammals of the world - Volume 3. Washington DC: Lynx Edicions. pp. 63–64.
49. Brown RM, Siler CD, Oliveros CH, Esselstyn JA, Diesmos AC, et al. (2013) Evolutionary processes of diversification in a model island archipelago. Ann Rev Ecol Evol Syst 44: 411–435. doi: 10.1146/annurev-ecolsys-110411-160323.
50. Groves CP (1998) Systematics of tarsiers and lorises. Primates 39: 13–27.
51. Niemitz C (1984) Taxonomy and distribution of the genus *Tarsius* Storr, 1780. In: Niemitz C, editor. The biology of tarsiers. New York: Gustav Fischer Verlag. pp. 1–16.
52. Hill WCO (1955) Primates: comparative anatomy and taxonomy. II. Haplorhini: Tarsioidea. Edinburgh: Edinburgh University Press.
53. Funk CW, McKay JK, Hohenlohe PA, Allendorf FW (2012) Harnessing genomics for delineating conservation units. Trends Ecol Evol 27: 489–496.

21

Effects of Traditional Flood Irrigation on Invertebrates in Lowland Meadows

Jens Schirmel*, Martin Alt, Isabell Rudolph, Martin H. Entling

Institute of Environmental Science, University of Koblenz-Landau, Landau, Germany

Abstract

Lowland meadow irrigation used to be widespread in Central Europe, but has largely been abandoned during the 20th century. As a result of agri-environment schemes and nature conservation efforts, meadow irrigation is now being re-established in some European regions. In the absence of natural flood events, irrigation is expected to favour fauna typical of lowland wet meadows. We analysed the effects of traditional flood irrigation on diversity, densities and species composition of three invertebrate indicator taxa in lowland meadows in Germany. Unexpectedly, alpha diversity (species richness and Simpson diversity) and beta diversity (multivariate homogeneity of group dispersions) of orthopterans, carabids, and spiders were not significantly different between irrigated and non-irrigated meadows. However, spider densities were significantly higher in irrigated meadows. Furthermore, irrigation and elevated humidity affected species composition and shifted assemblages towards moisture-dependent species. The number of species of conservation concern, however, did not differ between irrigated and non-irrigated meadows. More variable and intensive (higher duration and/or frequency) flooding regimes might provide stronger conservation benefits, additional species and enhance habitat heterogeneity on a landscape scale.

Editor: Judi Hewitt, University of Waikato (National Institute of Water and Atmospheric Research), New Zealand

Funding: The study was funded by the research initiative 'AufLand' of the federal state of Rhineland-Palatinate' (Germany) and the research and development fund of the University of Koblenz-Landau. The funders had no role in study design, data collection and analysis, decision to publish, or preparation of the manuscript.

Competing Interests: The authors have declared that no competing interests exist.

* Email: schirmel@uni-landau.de

Introduction

Semi-natural grasslands are key habitats for biodiversity conservation and an integral part of the Central European cultural landscape [1–3]. They are among the most species-rich habitats and serve as refuges for several rare and endangered species [2–4]. Regular disturbance due to traditional management permits the coexistence of numerous species in semi-natural grasslands [2]. During the last decades however, semi-natural grassland have dramatically declined in Central Europe and further declines to less than 50% of the current area are predicted [5,6]. Major causes are agricultural intensification and the abandonment of traditional management. The latter is mainly due to the reduced cost-effectiveness of traditional land use practices [5,7]. Agricultural intensification practices for semi-natural meadows include higher fertilizer and herbicide applications, earlier and more cuts per year, and the use of modern mowing techniques [8]. This results in eutrophic, structurally poor, and homogeneous meadows with negative impacts on diversity, species composition, and ecosystem processes [9].

Until the early 20th century, meadow irrigation was widespread in Central Europe to increase hay yield [10]. For example, around 1900 in some regions of Germany, irrigated meadows made up about 60% of the total grassland [11]. The main effects of irrigation were nutrient input, topsoil humidification, and extension of the vegetation period. Nowadays irrigation practices are mostly abandoned and traditionally irrigated meadows with their associated species are restricted to few remnant areas [12]. Thanks to agri-environment schemes (e.g. in form of compensation payments), nature conservation efforts, and due to mitigation and compensation measures, the traditional irrigation practices could be maintained or re-established in some European regions [13]. However, the value of agri-environment schemes is under debate and further analyses of management strategies are necessary [14,15]. Therefore, it is of growing interest to determine, how traditional irrigation practices affect biological diversity. In this context, Riedener et al. [16] recently showed that changes in irrigation techniques have influenced some aspects of plant and gastropod diversity in Swiss mountain hay meadows. However, knowledge of the influence of traditional meadow irrigation on invertebrate diversity and composition is still poorly understood, and this is especially true for flood irrigation in lowland regions.

The objective of this study was to analyse whether traditional flood irrigation in lowland meadows has an effect on invertebrate diversity and species composition. Irrigation is assumed to create small-scale differences in moisture and sediment conditions which may increase microhabitat and vegetation heterogeneity [16]. In accordance to the habitat-heterogeneity-hypothesis irrigation might therefore have positive effects on local species richness [17–20]. Moreover, flood irrigation in our study area is conducted in a similar way among irrigated sites, but differs in timing and intensity. This may lead to non-uniform moisture conditions among irrigated meadows with heterogeneous species compositions and higher beta diversity. To investigate these predictions we conducted a field survey in the 'Queichtal' in Rhineland-Palatine, Germany. We compared traditionally flood-irrigated meadows with meadows, where there has been no irrigation for at least thirty years. We focused on orthopterans, carabids and spiders, which are found at different trophic levels within grassland food-webs and occur in different vegetation layers. Orthopterans (Orthoptera) are mostly grass-dwelling herbivores, where they are often both the main invertebrate consumers and the main food source [21]. Most carabids (Coleoptera: Carabidae) are ground-dwelling predators, but some are scavengers and herbivores [22]. Spiders (Araneae) inhabit both the ground and field layer, often in high abundances, and are predatory [23]. All three arthropod groups have been used as indicators of ecosystem conditions and habitat quality (orthopterans: [24,25]; carabids: [26,27]; spiders: [28,29]).

We addressed the following hypotheses: (i) Flood irrigation increases the local diversity of orthopterans, carabids, and spiders compared to non-irrigated lowland meadows. (ii) Flood irrigation leads to higher beta diversity relative to non-irrigated meadows. (iii) Flood irrigation shifts species assemblages towards more moisture-dependent species and those of higher conservation concern than species in non-irrigated meadows. Based on our findings we discuss if traditional flood irrigation can be useful for conserving biodiversity of semi-natural grassland species.

Materials and Methods

Ethics statement

Invertebrates were collected with the permission 42/553-254 from the Struktur- und Genehmigungsdirektion Süd (federal state authority of Rhineland-Palatine, Germany). Additionally, we obtained permissions from all private farmers and landowners to conduct the field work on their meadows.

Study sites

The study was conducted in 2012 in the 'Queichtal' in the Upper Rhine valley in Rhineland-Palatine, Germany (Fig. 1). With a length of ~51 km, the river Queich is an important drainage system of the adjacent low mountain range 'Pfälzerwald' into the Rhine. Soils of the alluvial sediments are sandy to loamy [30]. Annual mean temperature in this region is 10.5°C (station Neustadt) and annual mean precipitation is 667 mm (station Landau; German Weather Service). The studied section of the Queichtal covers about 700 ha, is part of the NATURA 2000 network, and is thus protected by the EU habitats Directive [31].

Due to the predominance of moist soils with low cation availability, land use in the Queichtal is dominated by forest and grassland with different management and irrigation regimes. The formerly widespread traditional flood irrigation of lowland meadows was almost totally abandoned after the Second World War and is nowadays only applied in a few remnant areas. For flood irrigation the water of the river Quеich (or the tributaries 'Spiegelbach' and 'Fuchsbach') is dammed by weirs (Fig. 1) and delivered to the meadows by open ditches where it slowly flows over the ground ('lowland irrigation type'; [10]). Meadows are irrigated on average four times per year between April and August and each irrigation event lasts for 1–3 days.

A total of 32 meadows were selected stratified to meadow irrigation practice (yes or no) and fertilization (yes or no) (Table S1). Half of the meadows were traditionally irrigated and days of irrigation ranged from 4 to 12 days per year. On the other 16 meadows irrigation ceased more than 30 years ago. Half of the irrigated and not irrigated meadows were fertilized (with a maximum of ~60 kg N•ha-1•yr-1). Meadows were normally mown twice per year and extensive winter grazing by sheep occured on all meadows.

On each meadow we selected a 50×50 m plot with a minimum distance of 100 m from the nearest plot and 10 m to the next ditch to minimise edge effects. Irrigated and non-irrigated meadows did not differ significantly in mean distance to the nearest forest (t-test: $t_{30} = 0.563$, $P = 0.578$) and to the nearest permanent water (t-test: $t_{30} = 0.529$, $P = 0.601$). Permanent water was defined as any standing and flowing water body which permanently contained water.

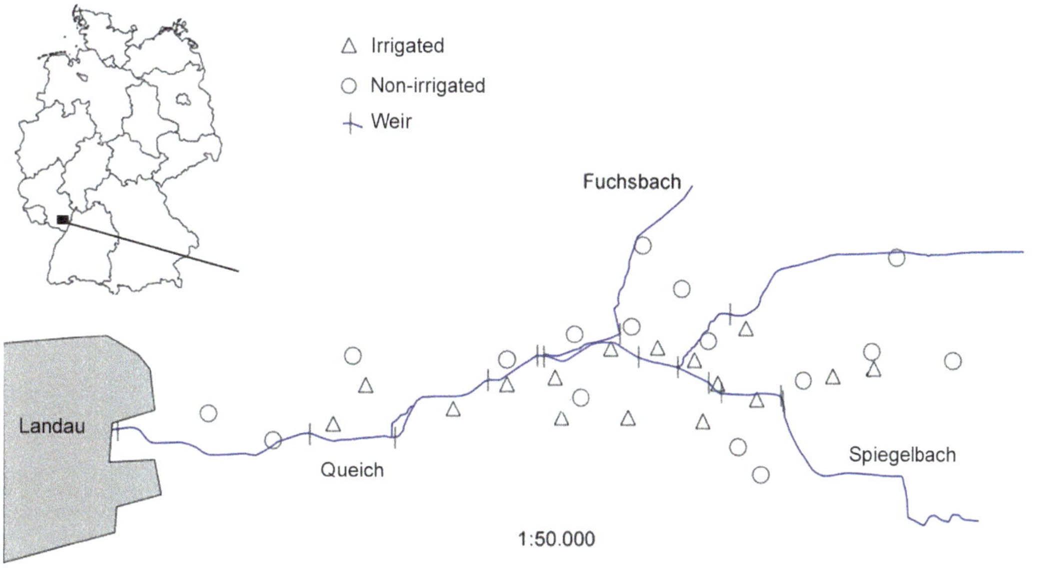

Figure 1. Position of the study area 'Queichtal' in Germany (small figure) and of the 32 study sites.

Management and environmental parameters

Land use data on irrigation practice (yes or no) and fertilization (yes or no) were collected through on-site observations and interviews with landowners and local farmers. Plant species were recorded in three randomly selected 3×3 m subplots per plot in May/June before the first cut (unpublished data). For subsequent analyses, data of the three subplots were averaged and we calculated the unweighted mean Ellenberger indicator values for moisture (in the following 'humidity' to avoid confusion with animal species moisture indicator values) and nitrogen for a description of local habitat conditions. As species can be influenced by patch isolation [32], we calculated the distance (m) to permanent water and to forest for each sampling location in Google Earth [33].

Invertebrate sampling

Orthopterans were sampled once per plot during their main activity period in August with a box-quadrat. The box-quadrat is a very effective method for sampling orthopteran densities [34]. The box-quadrat we used had an area of 2 m^2 (1.41×1.41 m) with white gauze-covered sides of 0.8 m height and was randomly dropped at 20 different locations per plot (total sampled area = 40 m^2 per plot). Collected individuals were determined to species level directly in the field using Bellmann [35] and then released.

Carabids and spiders were sampled using pitfall traps (6.5 cm in diameter, 7 cm deep) filled to one third with a 50% propylene-glycol solution. Per plot, four pitfall traps (N = 160 traps) were randomly installed with a minimum distance of 5 m to each other. Traps were exposed for two sampling periods from 03 to 24 April and again from 12 to 28 June. Carabids and spiders were determined to species level using the identification keys of Müller-Motzfeld [36] (carabids) and Roberts [37] (spiders). The four traps per plot were treated as a unit and data from both sampling periods were combined to obtain one dataset for further analyses. Due to loss and damage of some pitfall traps, we finally included 28 plots (N = 14 per meadow type each with N = 7 fertilized) in the data analyses of carabids and spiders.

Data analysis

Species were classified as species of conservation concern when they were listed in regional red lists (all species belonging to the categories '1', '2', '3', '4', and 'V'; orthopterans: [38]; carabids: [39]; spiders: [40]). For species moisture dependence we used published moisture indicator values. For orthopterans, transformed moisture values were obtained from Maas et al. [41] (Table S2). The values range from '1' (strongly xerophilic) to '5' (strongly hygrophilic). For carabids, moisture values range from '0' (most xerophilic) to '9' (most hygrophilic) according to Irmler and Gürlich [42]. For spiders, we used the moisture values of Entling et al. [43]. For a better comparison to the other taxa we transformed values by $1-x$, i.e. species with the lowest value '0' are most xerophilic and species with the highest value '1' are most hygrophilic. For each species we calculated the spearman rank correlation coefficient between species density and irrigation to express their 'species irrigation affinity' for our study area. Relationships between species irrigation affinities and species moisture indicator values (based literature data) were analysed using linear models.

The effect of irrigation on species richness and densities of orthopterans, carabids, and spiders were analysed using Poisson GLM's for count data. Similarly, the irrigation effect on the combined number of species of conservation concern of all taxa (N = 28 sites) was analysed. In cases of overdispersion, we corrected the standard errors using a quasi-GLM model [44]. Differences in Simpson diversity ($1-D$) between irrigated and non-irrigated meadows were tested with non-parametric Wilcoxon rank sum test, because assumptions for a t-test were violated.

Community differentiation (beta diversity) among irrigated and non-irrigated meadows was analysed using the homogeneity of multivariate dispersions based on the Sørensen similarity of species presence-absence data (using the command 'betadisper' in the R package 'vegan') [45]. For each taxon, an ANOVA was used to test for differences between the multivariate dispersions of both meadow types.

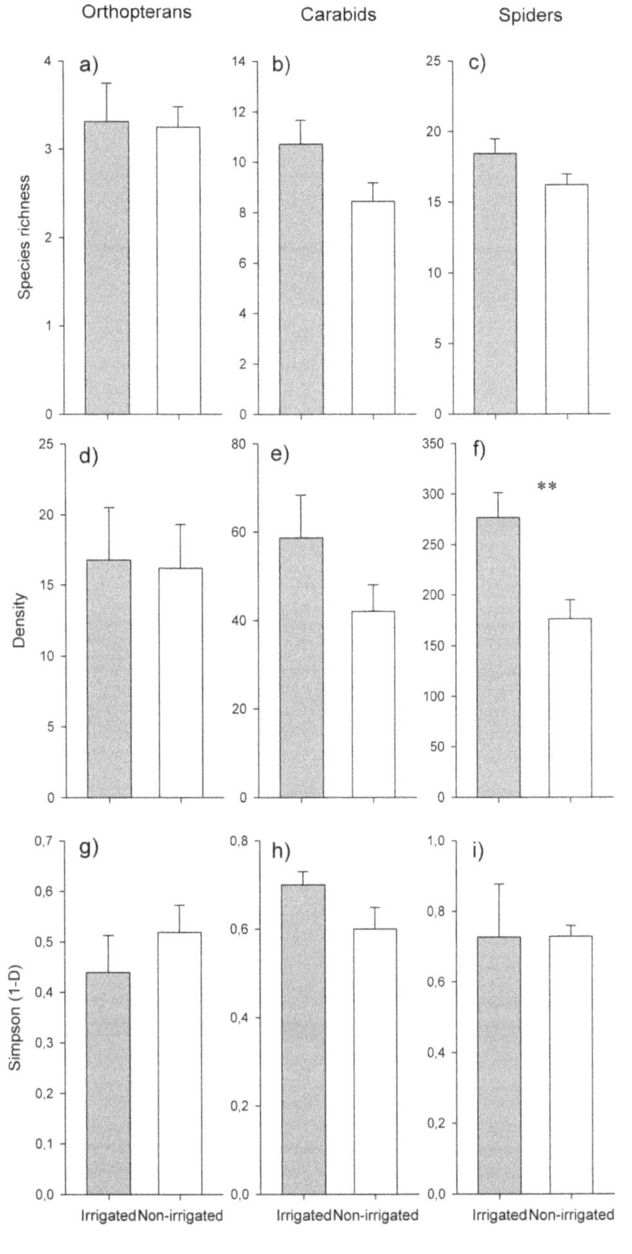

Figure 2. Comparison of species richness (a–c), densities (d–f), and Simpson diversity (g–i) of orthopterans, carabids, and spiders between irrigated and non-irrigated meadows (mean and SE). Differences of species richness and densities were tested with Poisson GLM's and of Simpson diversity ($1-D$) with non-parametric wilcoxon rang sum tests.

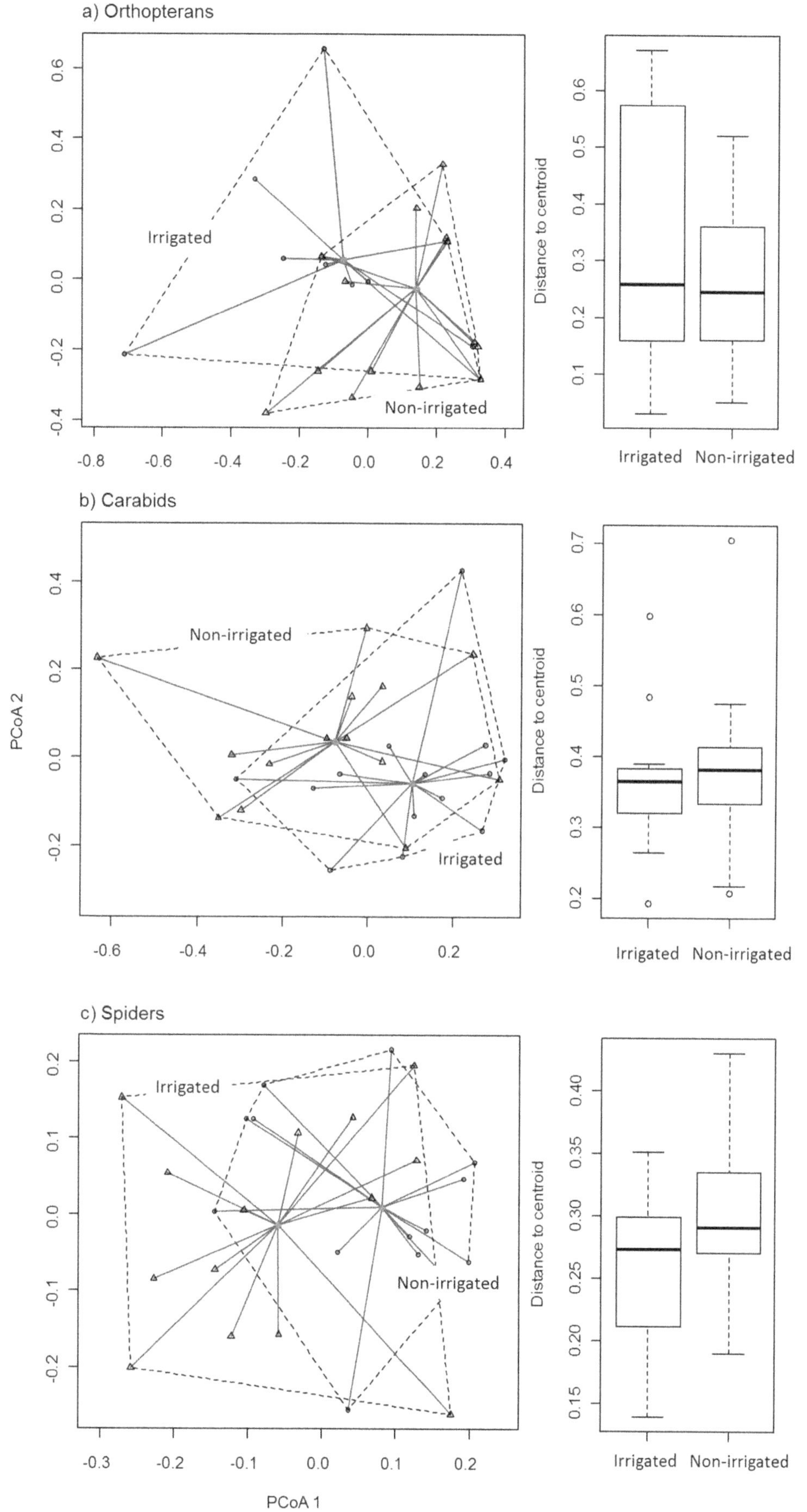
a) Orthopterans
Irrigated
Non-irrigated
Distance to centroid
Irrigated Non-irrigated
b) Carabids
Non-irrigated
Irrigated
PCoA 2
Distance to centroid
Irrigated Non-irrigated
c) Spiders
Irrigated
Non-irrigated
Distance to centroid
Irrigated Non-irrigated
PCoA 1

Figure 3. Beta diversity (multivariate homogeneity of dispersions) of a) orthopterans, b) carabid and c) spider assemblages of irrigated and non-irrigated meadows. Distances (Sørensen similarity) are reduced to principal coordinates and distances to group centroids (irrigated or non-irrigated) are shown. Differences of mean distances between meadow types were tested by ANOVA.

Effects of the management and environmental variables on species composition of orthopterans, carabids, and spiders were analysed with a permutational multivariate ANOVA (command 'adonis' in R package 'vegan'; [46]). Predictor variables were the two factors irrigation (yes or no) and fertilization (yes or no), the two local habitat parameters humidity and nitrogen (mean Ellenberger indicator values), and the two landscape parameters distance to permanent water and distance to forest. As a distance measure the Bray-Curtis distance was used. Significance of environmental variables was tested with permutation tests (999 permutations) with pseudo-F ratios. Variation of species compositions were visualised using nonmetric multidimensional scaling (NMDS) with the command 'metaMDS' in R package 'vegan'. Again, the Bray-Curtis distance was used as a distance measure. All statistical analyses were done in R 2.12.2 [47].

Results

General results

A total of 7 orthopteran species (528 individuals), 47 carabid species (1,410 individuals), and 56 spider species (6,347 individuals) were found (Tables S3–S6). All 7 orthopteran species were detected in both meadow types (Fig. 2a). A total of 40 carabid species were found in irrigated meadows compared to 32 species in non-irrigated meadows (Fig. 2b). In total 49 spider species could be detected in irrigated and 46 species in non-irrigated meadows (Fig. 2c).

Local diversity

There was no significant effect of irrigation on species richness of orthopterans ($z = 0.098$, $P = 0.922$, Fig. 2a), carabids ($t = 1.950$, $P = 0.051$; Fig. 2b), and spiders ($z = 1.407$, $P = 0.160$, Fig. 2c). While densities of orthopterans ($t = 0.130$, $P = 0.898$, Fig. 2d) and carabids ($t = 1.484$, $P = 0.150$, Fig. 2e) did not significantly differ between both meadow types densities of spiders were significantly higher in irrigated meadows ($t = 3.266$, $P = 0.003$, Fig. 2f). Similar to species richness, Simpson diversity did not differ for orthopterans ($W_{30} = 147.5$, $P = 0.4731$, Fig. 2g), carabids ($W_{26} = 62$, $P = 0.104$, Fig. 2h), and spiders ($W_{26} = 87$, $P = 0.629$, Fig. 2i).

Community differentiation (beta diversity)

Beta diversity (multivariate dispersion) of all investigated taxa was not influenced by irrigation. Mean distances to centroids did not differ significantly between irrigated and non-irrigated meadows for orthopterans ($F = 1.237$, $P = 0.275$, Fig. 3a), carabids ($F = 0.287$, $P = 0.596$, Fig. 3b), and spiders ($F = 4.023$, $P = 0.055$, Fig. 3c).

Species composition

Irrigation (yes or no) and humidity were the only variables having a significant effect on species composition, while fertilization, nitrogen availability, distance to permanent water and distance to forest had no effect (Table 1, Fig. 4). Orthopteran species composition was significantly affected by irrigation ($F = 2.51$, $R^2 = 0.073$, $P = 0.019$) and humidity ($F = 2.93$, $R^2 = 0.085$, $P = 0.011$). Carabid species composition was affected only by humidity ($F = 2.49$, $R^2 = 0.088$, $P = 0.024$) while spider species composition was influenced by irrigation ($F = 2.31$, $R^2 = 0.080$, $P = 0.041$).

As hypothesised, irrigation favoured the occurrence of moisture-dependent species. For carabids ($r = 0.48$, $P = 0.002$, Fig. 5b) and spiders ($r = 0.44$, $P<0.001$, Fig. 5c) there was a significant positive relationship between species irrigation affinity (expressed as the spearman rank correlation coefficient) and species moisture indicator value. For orthopterans no significant relationship was found, however this may be a result of the low number of $N = 7$ species (Fig. 5a). The combined number of species of conservation concern of all three taxa did not differ between irrigated (3.6±0.6) and non-irrigated (2.4±0.3) meadows ($z = 1.853$, $P = 0.064$).

Discussion

Unexpectedly, traditional flood irrigation had no significant effect on diversity and species of conservation concern of orthopterans, carabids, and spiders in lowland meadows. However, flood irrigation and the associated environmental parameter humidity influenced the species composition of all taxa and shifted species assemblages towards more moisture-dependent species.

Table 1. Effects of management and environmental variables on species composition of orthopterans, carabids, and spiders in irrigated and non-irrigated meadows in the Queichtal, Germany.

	Orthopterans		Carabids		Spiders	
	R^2	P	R^2	P	R^2	P
Factors						
Irrigation (yes or no)	0.078	**0.019**	0.050	0.171	0.080	**0.041**
Fertilization (yes or no)	0.028	0.451	0.012	0.976	0.023	0.710
Habitat parameters						
Humidity	0.085	**0.011**	0.088	**0.024**	0.060	0.102
Nitrogen	0.023	0.568	0.029	0.582	0.033	0.451
Landscape parameters						
Distance to water	0.032	0.370	0.048	0.175	0.012	0.924
Distance to forest	0.031	0.405	0.027	0.613	0.061	0.121

Significance was tested by permutational multivariate ANOVA (command 'adonis' in R package vegan). Significant results (P<0.05) are shown in bold.

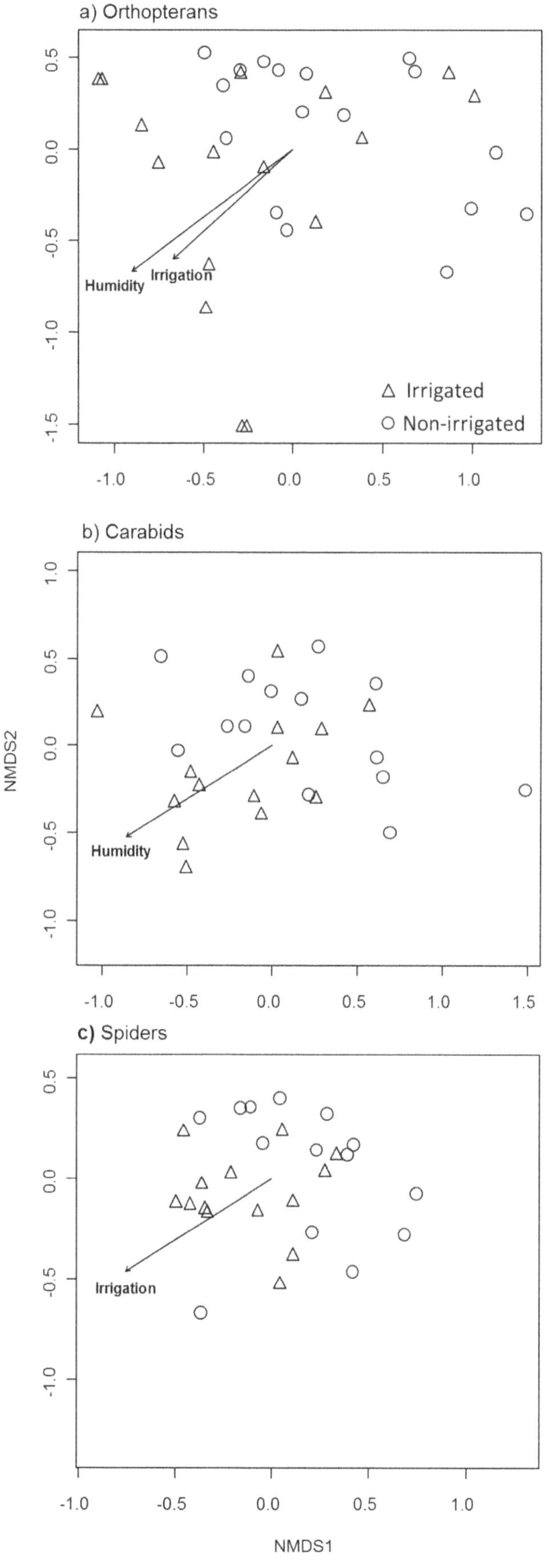

Figure 4. NMDS ordinations of a) orthopterans, b) carabid and c) spider species composition of irrigated and non-irrigated meadows. Only significant environmental variables are shown (permutational multivariate ANOVA, for statistics see Table 1).

Local diversity

We assumed that traditional flood irrigation leads to small-scale differences in moisture and sediment conditions and consequently higher microhabitat and vegetation heterogeneity resulting in increased local diversity [16,48]. However, in contrast to our hypothesis, traditionally flood irrigated meadows did not contain a higher local diversity (in terms of species richness and Simpson diversity) of orthopterans, carabids, and spiders compared to non-irrigated meadows. Possibly, the traditional irrigation system in our study area - on average four flooding events with a maximum of twelve irrigation days between April and August - is not sufficient to induce (measurable) heterogeneity effects. Riedener et al. [16] assumed that effects of the irrigation technique on diversity are only effective in combination with other management factors such as mowing and grazing regimes. Additionally, landscape variables such as patch isolation can influence species [32]. We accounted for possible confounding effects of management (mowing frequency, fertilization) and landscape parameters (distance to permanent water, distance to forest), none of which differed significantly between irrigated and non-irrigated meadows. However, as in any observational study, we cannot rule out that additional unmeasured management or environmental parameters have influenced our results.

In the literature, several studies in riparian habitats were able to detect positive effects of (natural) flooding on diversity [48–51]. However, in contrast to our study system with no (non-irrigated) and low intensive (irrigated) flooding, flood intensities in these studies are mostly studied along gradients containing higher intensities. Gerisch et al. [50] found positive effects for carabids at the river Elbe (Germany) and explained this by higher resource diversity in frequently flooded habitats. At the river Meuse (Belgium/Netherlands), Lambeets [50] could show that flooding initially had a positive effect on carabid diversity, which peaked at intermediate flooding degrees. This is in line with findings of Pollack et al. [48] in riparian meadows where plant species richness was highest at intermediately flooded river banks because of increased microhabitat heterogeneity.

Similar to diversity, densities of orthopterans and carabids did not differ between meadow types. However, spider densities were higher in irrigated meadows. One explanation might be enhanced food availability, because short time flooding can enhance soil organisms [52], which present important food source especially for linyphiid spiders [53].

Community differentiation (beta diversity)

Irrigation did not influence community differentiation and, in contrast to our hypothesis, beta diversity of orthopterans, carabids, and spiders was not higher in irrigated compared to non-irrigated meadows. Although flood irrigation between irrigated meadows differed in time and intensity, these differences were obviously too weak to result in more diverse species assemblages. Moreover, the traditional flooding method in the region - where the dammed river water slowly streams into the meadows and back into the river through a system of open ditches - leads to a relatively homogenous water flow. This prevents stagnant moisture [12] and moisture conditions on irrigated sites might be more uniform than expected. Human-altered repetitive flood events are known to result in uniform species compositions due to a homogenization of habitat structure [54,55].

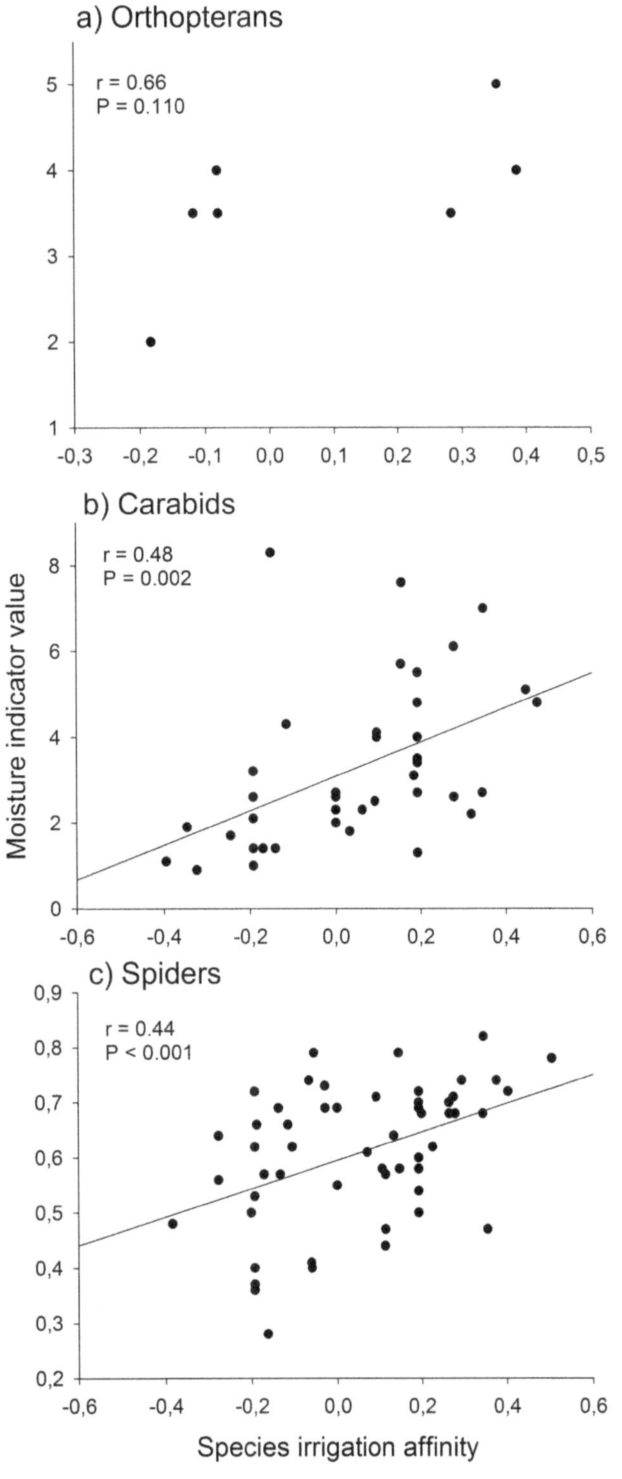

Figure 5. Relationship between species irrigation affinity (spearman rank correlation coefficient of species densities to irrigation) and moisture indicator value of a) orthopterans, b) carabids, and c) spiders.

Species composition

As hypothesized, meadow irrigation and the associated altered humidity conditions influenced species composition of orthopterans, carabids, and spiders. Irrigation may therefore increase beta diversity at the landscape scale and contribute to diverse grassland communities. Assemblages of irrigated meadows contained more moisture-dependent species compared to non-irrigated ones. This was reflected for carabids and spiders by the positive relationships between species irrigation affinity and their moisture indicator value. For orthopterans this effect was not significant (most likely because of the low number of species), but the two species with the highest moisture indicator value – *Mecostethus parapleurus* and *Stetophyma grossum* – were significantly more abundant in irrigated meadows (Table S3). For all three taxa, humidity is known to be one of the most influencing environmental parameter structuring species compositions [27,43,56]. Impacts of (natural) flood disturbance on species and trait composition of orthopterans were previously shown by Dziock et al. [57]. Bonn et al. [48] found that flood regime strongly influenced carabid species assemblages, which was also found by Lambeets et al. [50] for carabid and spider communities. Although fertilization can strongly influence arthropods [58,59] we could not detect an effect of fertilization. In general, fertilization application rates were low in the study area (0 to ~60 kg N•ha-1•yr-1), a range in which also plants showed no significant decrease in species richness (unpublished data). Similar to fertilization, the landscape parameters, distance to forest and to permanent water, did not differ between irrigated and non-irrigated meadows and had no effect on species compositions, respectively.

In contrast to our hypothesis, species compositions of flood irrigated meadows did not contain more species of conservation concern than non-irrigated meadows. This is in contrast to Bonn et al. [49] and Lambeets et al. [51], where anthropogenic alterations in flooding regimes not only have a strong influence on arthropod communities but also on the distribution of rare (and often endangered) riparian species. Again, this difference may be due to the low irrigation intensity in our study system.

Conclusion

Flood irrigation had no significant effect on local and beta diversity of orthopterans, carabids, and spiders in lowland meadows. In contrast, flood irrigation clearly changed species assemblages towards moisture-dependent species and probably increased beta diversity at the landscape scale. However, these species were mostly common species and assemblages of irrigated meadows did not contain more species of conservation concern compared to non-irrigated ones. More variable and intensive (higher duration and/or frequency) flooding regimes are likely to provide much stronger conservation benefits. Moreover, beneficial effects of flood irrigation might be more pronounce along the irrigation infrastructures (open ditches, drains, weirs) than in the open meadow, which will be tested in further studies.

Supporting Information

Table S1 Site characteristics of irrigated and non-irrigated meadows in the 'Queichtal', Germany. For explanations see text.

Table S2 Moisture indicator values of orthopterans based on information in Maas et al. (2002). Information was coded numerically as follows: Strongly xerophilic = 1, xerophilic = 2, mesophilic = 3, hygrophilic = 4, strongly hygrophilic = 5.

Table S3 Abundances (mean and SE) of orthopterans, carabid, and spider species in irrigated and non-

irrigated meadows in the Queichtal, Rhineland-Palatine (Germany). Differences were tested with Poisson or, in case of overdispersion, quasi-Poisson GLM's for count data (only species with ≥10 individuals). n.t. = not tested. Significant results ($P < 0.05$) are shown in bold.

Acknowledgments

We thank Lisa Gundall, Melanie Hilbert, Larissa Hermann, Franziska Möller, Max Pfirrmann, Stefan Pietrusky Tessa Schardt, Lisa Weigel and Kim Wilsdorf for help with the field work and species determination.

Author Contributions

Conceived and designed the experiments: JS MHE. Performed the experiments: JS MA IR MHE. Analyzed the data: JS. Contributed reagents/materials/analysis tools: JS MA MHE. Contributed to the writing of the manuscript: JS MHE.

References

1. Muller S (2002) Appropriate agricultural management practices required to ensure conservation and biodiversity of environmentally sensitive grassland sites designated under Natura 2000. Agric Ecosys. Environ 89: 261–266.
2. Poschlod P, WallisDeVries MF (2002) The historical and socioeconomic perspective of calcareous grasslands - lessons from the distant and recent past. Biol Conserv 104: 361–376.
3. Isselstein J, Jeangros B, Pavlu V (2005) Agronomic aspects of biodiversity targeted management of temperate grasslands in Europe – A review. Agron Res 3: 139–151.
4. Duelli P, Obrist MK (2003) Regional biodiversity in an agricultural landscape: the contribution of seminatural habitat islands. Basic Appl Ecol 4: 129–138.
5. Strijker D (2005) Marginal lands in Europe – causes of decline. Basic Appl Ecol 6: 99–106.
6. Rounsevell MDA, Ewert F, Reginster I, Leemans R, Carter TR (2005) Future scenarios of European agricultural land use - II. Projecting changes in cropland and grassland. Agric Ecosyst Environ 107: 117–135.
7. Baur B, Cremene C, Groza G, Rakosy L, Schileyko AA, et al. (2006) Effects of abandonment of subalpine hay meadows on plant and invertebrate diversity in Transylvania, Romania. Biol Conserv 132: 261–273.
8. Ellenberg H, Leuschner C (2012) Die Vegetation Mitteleuropas mit den Alpen. Stuttgart: UTB. 1334 p.
9. Rosenthal G, Hölzel N (2009) Renaturierung von Feuchtgrünland, Auengrünland und mesophilem Grünland. In: Zerbe S, Wiegleb G, editors. Renaturierung von Ökosystemen in Mitteleuropa. Heidelberg: Spektrum Akademischer Verlag. pp. 283–316.
10. Leibundgut C (2004) Historical meadow irrigation in Europe - a basis for agricultural development. In: Rodda JC, Ubertini L, editors. The Basis of Civilization - Water Science? IHAS Publication 286: 77–87.
11. Schellberg S (2005) Meadow irrigation in the federal state Baden-Württemberg. Portrayal of a nearly forgotten land use system. Integrated Land and Water Resources Management in History, Schriften der DWhG, Sonderband 2: 123–131.
12. Hassler DM, Glaser K-H (1995) Wässerwiesen - Geschichte, Technik und Ökologie der bewässerten Wiesen, Bäche und Gräben in Kraichgau, Hardt und Bruhrain. Ubstadt-Weiher: Verlag Regionalkultur. 432 p.
13. Leibundgut C (2011) Die Wässermatten des Oberaargaus. Ein regionales Kulturerbe als Modell für Europa? Jahrb des Oberaargaus 54: 121–144.
14. Kleijn D, Baquero RA, Clough Y, Díaz M, De Esteban J, et al. (2006) Mixed biodiversity benefits of agri-environment schemes in five European countries. Ecol Lett 9: 243–254.
15. Kleijn D, Sutherland WZ (2003) How effective are European agri-environment schemes in conserving and promoting biodiversity? J Appl Ecol 40: 947–969.
16. Riedener E, Rusterholz H-P, Baur B (2013) Effects of different irrigation systems on the biodiversity of species-rich hay meadows. Agric Ecosyst Environ 164: 62–69.
17. MacArthur RH, MacArthur JW (1961) On bird species diversity. Ecology 42: 594–598.
18. MacArthur RH, Wilson EO (1967) The theory of island biogeography. Princeton: Princeton University Press. 224 p.
19. Sousa WP (1984) The role of disturbance in natural communities. Ann Rev Ecol Syst 15: 353–391.
20. Benton TG, Vickery ZA, Wilson JD (2003) Farmland biodiversity: is habitat heterogeneity the key? Trends Ecol Evol 18: 182–188.
21. Curry JP (1994) Grassland Invertebrates – Ecology, influence on soil fertility and effects on plant growth. London: Chapman and Hall. 437 p.
22. Thiele HU (1977) Carabid beetles in their environments. A study on habitat selection by adaptations in physiology and behaviour. Heidelberg, Berlin, New York: Springer. 369 p.
23. Wise D (1993) Spiders in ecological webs. Cambridge: Cambridge University Press. 328 p.
24. Báldi A, Kisbenedek T (1997) Orthopteransn assemblages as indicators of grassland naturalness in Hungary. Agric Ecosyst Environ 66: 121–129.
25. Schirmel J, Mantilla-Contreras J, Blindow I, Fartmann T (2011) Impacts of succession and grass encroachment on heathland Orthopterans. J Insect Conserv 15: 633–642.
26. Rainio J, Niemelä J (2003) Ground beetles (Coleoptera: Carabidae) as bioindicators. Biodivers Conserv 12: 487–506.
27. Schirmel J, Buchholz S (2011) Response of carabid beetles (Coleoptera: Carabidae) and spiders (Araneae) to coastal heathland succession. Biodivers Conserv 20: 1469–1482.
28. Scott AG, Oxford GS, Selden PA (2006) Epigeic spiders as ecological indicators of conservation value for peat bogs. Biological Conserv 127: 420–428.
29. Horvath R, Magura T, Szinetar C, Tothmeresz B (2009) Spiders are not less diverse in small and isolated grasslands, but less diverse in overgrazed grasslands: A field study (East Hungary, Nyirseg). Agric Ecosyst Environ 130: 16–22.
30. Briem E, Geiger M (2008) Das Fließgewässer Queich. In: Geiger M, editor. Haardt, Weinstraße und Queichtal: Ein Geo-Führer. Neustadt/Weinstraße: Pollichia. pp. 114–116.
31. Ssymank A, Hauke U, Rückriem C, Schröder E (1998) Das europäische Schutzgebietssystem NATURA 2000– BfN Handbuch zur Umsetzung der Fauna-Flora-Habitat-Richtlinie und der Vogelschutz-Richtlinie. Schriftenr Landschaftspfl Natursch 53: 1–560.
32. Bailey D, Schmidt-Entling MH, Eberhart P, Herrmann JD, Hofer G, et al. (2010) Effects of habitat amount and isolation on biodiversity in fragmented traditional orchards. J Appl Ecol 47: 1003–1013.
33. Google (2011) Google Earth 6.1.0.5001. Google Inc. Available: http://www.google.com/earth/index.html. Accessed 2013 Oct 19.
34. Gardiner T, Hill J, Chesmore D (2005) Review of the methods frequently used to estimate the abundance of Orthopterans in grassland ecosystems. J Insect Conserv 9: 151–173.
35. Bellmann H (2006). Der Kosmos Heuschreckenführer. Die Arten Mitteleuropas sicher bestimmen. Stuttgart: Kosmos. 350 p.
36. Müller-Motzfeld G (2006) Die Käfer Mitteleuropas. Band 2 Adephaga 1: Carabidae (Laufkäfer). Heidelberg/Berlin: Spektrum Akademischer Verlag. 536 p.
37. Roberts MJ (1996) Collins field guide. Spiders of Britain and Northern Europe. Berkshire: D & N Publishing. 383 p.
38. Pfeifer MA, Niehuis M, Renker C (2011) Die Fang- und Heuschrecken in Rheinland-Pfalz. Flora und Fauna in Rheinland-Pfalz 41: 1–678.
39. Schüle P, Persohn M (2000) Rote Liste der in Rheinland-Pfalz gefährdeten Laufkäfer (Coleoptera: Carabidae). Ministerium für Umwelt und Forsten. Mainz: Rhein Main Druck. 29 p.
40. Nährig D, Harms KH (2003) Rote Listen und Checklisten der Spinnen in Baden-Württemberg. Mannheim: Landesanstalt für Umweltschutz Baden-Württemberg, JVA Mannheim-Druckerei. 203 p.
41. Maas S, Detzel P, Staudt A (2002) Gefährdungsanalyse der Heuschrecken Deutschlands. Verbreitungsatlas, Gefährdungseinstufung und Schutzkonzepte. Münster: Landwirtschaftsverlag. 401 p.
42. Irmler U, Gürlich S (2004) Die ökologische Einordnung der Laufkäfer (Coleoptera: Carabidae) in Schleswig-Holstein. Faunistisch-Ökologische Mitt Suppl 32: 1–115.
43. Entling W, Schmidt MH, Bacher S, Brandl R, Nentwig W (2007) Niche properties of Central European spiders: shading, moisture and the evolution of the habitat niche. Global Ecol Biogeogr 16: 440–448.
44. Zuur AF, Ieno IN, Walker NJ, Saveliev AA, Smith GM (2009) Mixed effects models and extensions in ecology with R. Berlin, Heidelberg: Springer. 574 p.
45. Anderson MJ, Ellingsen KE, McArdle BH (2006) Multivariate dispersion as a measure of beta diversity. Ecol Lett 9: 683–693.
46. Oksanen J, Blanchet FG, Kindt R, Legendre P, O'Hara RB, et al. (2011) vegan: Community Ecology Package. R package version 1.17-11. http://CRAN.R-project.org/package=vegan.
47. R Development Core Team (2010) R: A language and environment for statistical computing. Vienna: R Foundation for Statistical Computing. www.R-project.org.

48. Pollock MM, Naiman RJ, Hanley TA (1998) Plant species richness in riparian wetlands - a test of biodiversity theory. Ecology 79: 94–105.
49. Bonn A, Hagen K, Wohlgemuth-von Reiche D (2002) The significance of flood regimes for carabid beetle and spider communities in riparian habitats – a comparison of three major rivers in Germany. River Res Appl 18: 43–64.
50. Gerisch M, Agostinelli V, Henle K, Dziock F (2012) More species, but all do the same: contrasting effects of flood disturbance on ground beetle functional and species diversity. Oikos 121: 508–515.
51. Lambeets K, Vandengehuchte ML, Maelfait JP, Bonte D (2008) Understanding the impact of flooding on trait-displacements and shifts in assemblage structure of predatory arthropods on river banks. J Animal Ecol 77: 1162–1174.
52. Plum N (2005) Terrestrial invertebrates in flooded grassland: a literature review. Wetlands 25: 721–737.
53. Nyffeler M (1999) Prey selection of spiders in the field. J Arachnol 27: 317–324.
54. Bonn A, Kleinwächter M (1999) Microhabitat distribution of spider and ground beetle assemblages (Araneae, Carabidae) on frequently inundated river banks of the River Elbe. Z Ökologie Naturschutz 8: 109–123.
55. Vanbergen AJ, Woodcock BA, Watt AD, Niemelä J (2005) Effect of land-use heterogeneity on carabid communities at the landscape scale. Ecography 28: 3–16.
56. Ingrisch S, Köhler G (1998) Die Heuschrecken Mitteleuropas. Magdeburg: Westarp-Wissenschaften. 460 p.
57. Dziock F, Gerisch M, Sieger M, Hering I, Scholz M, et al. (2011) Reproducing or dispersing? Using trait based habitat templet models to analyse Orthopterans response to flooding and land use. Agric Ecosyst Environ 145: 85–94.
58. Siemann E (1998) Experimental tests of effects of plant productivity and diversity on grassland arthropod diversity. Ecology 79: 2057–2070.
59. van Wingerden WKRE, van Kreveld AR, Bongers W (1992) Analysis of species composition and abundance of grasshoppers (Orth., Acrididae) in natural and fertilized grasslands. J Appl Entomol 113: 138–152.

22

Genetic Diversity and Structure of *Sinopodophyllum hexandrum* (Royle) Ying in the Qinling Mountains, China

Wei Liu, Dongxue Yin, Jianjun Liu*, Na Li

College of Forestry, Northwest A & F University, Yangling, China

Abstract

Sinopodophyllum hexandrum is an important medicinal plant whose genetic diversity must be conserved because it is endangered. The Qinling Mts. are a *S. hexandrum* distribution area that has unique environmental features that highly affect the evolution of the species. To provide the reference data for evolutionary and conservation studies, the genetic diversity and population structure of *S. hexandrum* in its overall natural distribution areas in the Qinling Mts. were investigated through inter-simple sequence repeats analysis of 32 natural populations. The 11 selected primers generated a total of 135 polymorphic bands. *S. hexandrum* genetic diversity was low within populations (average $H_e = 0.0621$), but higher at the species level ($H_e = 0.1434$). Clear structure and high genetic differentiation among populations were detected by using the unweighted pair group method for arithmetic averages, principle coordinate analysis and Bayesian clustering. The clustering approaches supported a division of the 32 populations into three major groups, for which analysis of molecular variance confirmed significant variation (63.27%) among populations. The genetic differentiation may have been attributed to the limited gene flow ($N_m = 0.3587$) in the species. Isolation by distance among populations was determined by comparing genetic distance versus geographic distance by using the Mantel test. Result was insignificant ($r = 0.212$, $P = 0.287$) at 0.05, showing that their spatial pattern and geographic locations are not correlated. Given the low within-population genetic diversity, high differentiation among populations and the increasing anthropogenic pressure on the species, in situ conservation measures were recommended to preserve *S. hexandrum* in Qinling Mts., and other populations must be sampled to retain as much genetic diversity of the species to achieve ex situ preservation as a supplement to in situ conservation.

Editor: Raffaella Balestrini, Institute for Sustainable Plant Protection, C.N.R., Italy

Funding: This work was supported by the program from the Forestry Research Foundation for the Public Service Industry of China (200904004). The funders had no role in study design, data collection and analysis, decision to publish, or preparation of the manuscript.

Competing Interests: The authors have declared that no competing interests exist.

* Email: ljj@nwsuaf.edu.cn

Introduction

Sinopodophyllum hexandrum (Royle) Ying, family Berberidaceae, the only species of this genus in China, commonly known as Himalayan mayapple, is an endangered and medicinal perennial herb native to the Himalayan regions at elevations ranging from 2 700 m to 4 500 m [1–4]. Plants provide us with many important medicaments, including anticancer and antiinfective agents [5], and traditional Chinese medicine has contributed to identifying these substances [6]. *S. hexandrum* is a traditional Chinese medicine that has been used in folk medicine [7]. The roots and rhizomes of *S. hexandrum* contain large amounts of lignans. The most important lignan for human health is arguably the most active cytotoxic aryltetralin lignan, podophyllotoxin, with three times the podophyllotoxin levels compared to the American species *Podophyllum peltatum* [8,9,10], as a precursor for the semi-synthesis of the anticancer pharmaceuticals, such as etoposide (VP-16), teniposide (VM-26), GP-7, NK-611, etopophos, GL-331 and TOP-53 [9–15]. The destructive harvest of these plants added *S. hexandrum* to the endangered species list of the Convention on International Trade in Endangered Species of Wild Fauna and Flora [16]. *S. hexandrum* was classified as an endangered species (grade 3) in 1987 by the Chinese Plant Red Book [4].

Currently, with the enhanced awareness of its medicinal value and superior efficacy in clinical applications, the wild *S. hexandrum* populations in China has been noted to be very small and to be rapidly declining. The availability of podophyllotoxin from plants has become increasingly limited due to intense collection, habitat fragmentation, low natural regeneration rate, and the lack of organized cultivation. Wild *S. hexandrum* populations could become extinct without timely and effective protective measures, costing humans an ideal drug against cancer. Knowledge of the genetic diversity at intraspecific levels is an important prerequisite for species conservation and a rational exploitation program planning. However, previous studies have mainly focused on the identification and separation of the chemical components [17–25], biological properties [26–32], and micropropagation of *S. hexandrum* [33]. Only studies on genetic diversity in *S. hexandrum* from the Northwestern Himalayan region are available [34,35,36]. In particular, two areas where *S. hexandrum* is grown, namely, western Sichuan Province and Himalaya–Hengduan Mt. region in China, were investigated to identify genetic diversity of this species through inter-simple sequence repeats (ISSR) and amplified fragment length polymorphism (AFLP) markers [37,38,39], respectively. These reports with same results showed that *S. hexandrum* populations had relatively high genetic diversity ($H_e = 0.2944 \sim 0.3377$). However, there is a consensus that alpine plants are faced with pollinator restriction [40]. The unclear extent of the species dispersal mechanisms makes it interesting to study the relationships between populations. Furthermore, many reports about genetic diversity of other medicinal plants, such as *Calamagrostis porteri ssp. insperata* [41], *Aegiceras corniculatum* [42], *Sonneratia alba* [43], *Coscinium fenestratum* [44], *and Lilium pumilum* [45], have been published based on the ISSR approach.

Molecular markers are very useful tools for genetic diversity studies. ISSR markers are molecular markers especially suited to genetic polymorphisms analyses of species without available sequence information [46,47]. Studies on the population relationships, genetic diversity and conservation of *S. hexandrum* in Qinling Mts. are requisite because climate change and local overexploitation may cause unknown endangering mechanisms. The present study aims to establish management strategies for the conservation genetics of *S. hexandrum* by (1) examining the levels of genetic variability within and among *S. hexandrum* populations sampled from Qinling Mts. where historical records showed *S. hexandrum* having grown naturally, (2) assessing the possible factors that affect the genetic variation observed, and (3) comparing these within and among *S. hexandrum* populations with data published for itself or other plant taxa with similar characteristics.

Materials and Methods

Ethics statement

The endangered species were collected, and research activities were scientifically conducted under the permits issued by the local forestry department. A detailed description of the experimental material collection and procedures is provided. The experimental procedures were approved by the Ethics Committee for Plant Experiments of Northwest A & F University and the State Forestry Administration, P. R. China. The names of the authorities that issued the permit for each location were listed in Table S1 in Text S1.

Study area

This study was performed in the Qinling Mts. (32°41′ to 34°59′ N, 103°54′ to 110°34′ E), which are located in central of China (Figure 1). The Qinling Mts., a 1 500 Km mountain chain, run east–west and act as an important watershed divider between two great Chinese rivers, the Yangtze River and the Yellow River, which constitute a transitional zone between the northern subtropical zone and warm-temperate zone. The Qinling Mts. were considered to be a biodiversity hotspot in China [48]. As one of the distribution areas of *S. hexandrum*, Qinling Mts. have unique environmental features which have high impact on the evolution of the species.

Plant materials

S. hexandrum distribution pattern and extent of in Qinling Mts. were investigated from 2010 to 2011. The *S. hexandrum* populations are distributed in small and scattered patches. According to the field survey information, a total of 32 wild *S. hexandrum* populations were sampled for DNA analysis between July 19, 2012, and September 17, 2012, to ensure collection period consistency (Table 1). *S. hexandrum* has a wide geographic distribution throughout the Qinling Mts. (Figure 1). The altitude of the sample sites ranged from 1 013 m to 2 883 m (Table1). Geographical distances between populations ranged from 5.5 km to 276.8 km. 20 plants were sampled from each population. The horizontal and vertical distances between sampled plants within each population were over 20 and 5 m, respectively, to increase the likelihood of sampling inter-individual variation within each population [37,38]. About 2 g to 10 g of fresh young leaves per plant was immediately frozen in liquid nitrogen and then kept at − 80 °C until DNA isolation. The key information on *S. hexandrum* populations in all sampling sites is summarized in Table 1.

DNA extraction

Total genomic DNA was extracted from frozen leaves by using a plant genomic DNA rapid extraction kit (Spin-column) (BioTek Corporation, Beijing, China; http://bioteke.biogo.net/). The extracted DNA was quantified by comparing with known DNA of standard quantity (Lambda DNA) through electrophoresis in ethidium bromide-stained 1.0% agarose gels (Gene Genius Bio Imaging System; Synegene), and the extracted DNA was diluted in TE buffer to a final concentration of 50 ng/μL and stored at− 20°C before PCR amplification.

Primer screening and ISSR-PCR amplification

A total of 100 ISSR primers (synthesized by Shanghai Sheng Gong Biotechnology CO. LTD, China) were screened based on the primer set published by the Biotechnology Laboratory, University of British Columbia, Canada (UBC set No. 9) and the studies on Himalayan mayapple [34–39]. An optimum reaction system was obtained by screening DNA, Mg^{2+}, dNTP, primer (UBC900 was used for preliminary test), and Taq DNA polymerase concentrations and annealing temperature, and reaction conditions. The optimization showed that 20 μL of reaction system is ideal. Each 20 μL amplification reaction consisted of 1×PCR buffer (10 mM Tris-HCl at pH 8.3, 50 mM L^{-1} KCl, 0.001% gelatin, and 1.5 mmol L^{-1} $MgCl_2$), 1.6 mmol L^{-1} dNTP mix, 0.6 μmol L^{-1} primer (UBC900 was used for preliminary test), 15 ng of template DNA, and 1.0 U Taq DNA polymerase (TaKaRa Biotechnology, Dalian, China), using a cycling profile of initial 5 min at 94°C, followed by 45 cycles of 30 s at 94°C, 45 s annealing at 50°C, and 90 s extension at 72°C, ending with a final extension of 7 min at 72°C.

The optimized PCR experiment conditions were applied for primer screening in a PTC100TM Programmable Thermal Controller (MJ Research, Waltham, MA, USA). Six populations (TCG, LJG, BG, MDS, YMG and CTG) with observable variations (morphology, habitat, etc) were selected to initially screen 100 primers by using 10 samples for each population. Primers that generated scorable bands and high levels of polymorphisms were selected by genotyping all populations. The amplification products were electrophoresed on 1.0% agarose gels buffered with 1.0×TBE for 2.5 h at 100 V and were detected through ethidium bromide staining, and the gels were imaged in the Gene Genius Bioimaging System. Band size was estimated from a 0.1 kbp DNA ladder (TaKaRa Biotechnology, Dalian, China). Each primer was amplified in triplicate to confirm reliability and reproducibility. A reaction without DNA was used as negative control.

Data analyses

Amplification results were scored according to the positions of the DNA bands from electrophoresis, being labeled "1" for presence of the bands and "0" for absence of the bands in the data

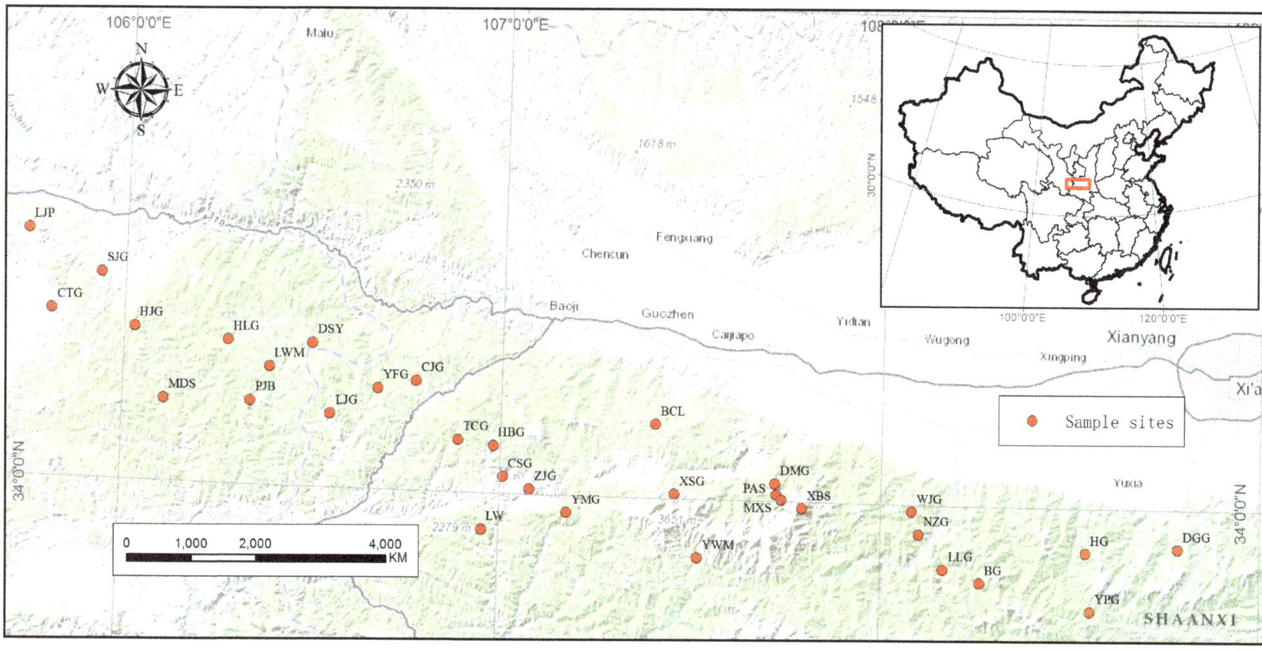

Figure 1. Locations of the 32 *S. hexandrum* populations in Qinling Mts. sampled for this study.

matrix. Only stable bands with repeatable differences were considered valid for polymorphism loci.

The resulting ISSR phenotype data matrix (binary matrix from 0 to 1) was analyzed using the POPGENE software (version 1.31) [49] to compute genetic diversity parameters, such as allele frequencies, percentage of polymorphic bands (PPB), number of alleles per locus (A_o), effective number of alleles per locus (A_e), total gene diversity (H_t), the level of gene flow (N_m), Nei's gene diversity (H_e), gene distance (GD), Shannon's information index (H_o), within-population diversity (H_s), and mean coefficient of gene differentiation (G_{st}). An unweighted pair groups mean arithmetic (UPGMA) mean dendrogram was constructed using PowerMarker 3.23 to examine the genetic relationship at the species level [50]. A bootstrap (resampling) test was performed 1 000 times to determine distances between the populations using PHYLIP version 3.69 (PHYLogeny Inference Package) programs [51]. Bayesian analysis of population structure was performed as implemented in STRUCTURE (version 2.2) to infer the most likely number of population genetic clusters (K) in the ISSR dataset [52]. K ranged from 1 to 10, with 10 replicate runs for each K, and a burn-in period of 2×10^5 and 5×10^4 iterations. The "no admixture model" and independent allele frequencies were chosen for this analysis. The most likely number of clusters was estimated according to the model values (ΔK) based on the second order rate of change, with respect to K, of the likelihood function [53]. To detect within-group structure, subsequent runs were performed for each obtained clusters using the same settings as previous. Population similarity was also explored and visualized through principle coordinate analysis (PCoA) using NTSYSpc 2.10e [54]. Analysis of molecular variance (AMOVA), which partitions total phenotypic variance within and among populations, was performed using WIN AMOVA (version 1.55), which was provided by the Genetics and Biometry Laboratory, University of Geneva, Geneva, Switzerland [55]. The AMOVA input files with the Euclidean distance matrix were created using AMOVA-PREP 1.01 [56]. Significance level was tested by comparing the frequency distributions from the original data and the data generated by a set of 1 000 computer simulations. A Mantel test was performed using Tools for Population Genetic Analysis (TFPGA) [56] for computing 5 000 permutations to test the isolation by distance (IBD) among populations by comparing genetic distance between all pairwise combinations of populations versus geographic distance. Geographical distance, L_{ab}, was computed using the following formula: $L_{ab}=\mathrm{Arccos}\ [\cos(LAT_a)\text{-}COS(LONG_a)\cos(LAT_b)\cos(LONG_b) + \cos(LAT_b)\sin(LONG_b)\cos(LAT_b)\sin(LONG_b) + \sin(LAT_a)\sin(LAT_b)]\times R$ [57]. $LONG_a$, LAT_a and $LONG_b$, LAT_b are the longitudes and latitudes (in radians) of sampling sites a and b, respectively; R indicates the radius of the earth, which is 6 378 km; and L_{ab} represents the geographic distance between sampling sites a and b.

Results

Genomic DNA amplification results

For polymorphism testing of the 640 *S. hexandrum* individuals from 32 populations, forty-eight ISSR primers amplified visible bands and were chosen from the initial set of 100 primers to screen for reproducible markers. PCR amplification results (Table 2) show that 11 primers produced 241 clear and replicated bands (250 bp to 2 000 bp), of which 135 were polymorphic (56.02%) with 100% reproducibility. Individual primers detected between 19 (UBC825) and 29 (UBC900) loci can amplify clear bands, with an average of 21.91. The percentage of polymorphism ranged from 42.86% (UBC834) to 68.97% (UBC900), indicating that the selected primers are highly polymorphic across *S. hexandrum* populations. The H_t was 0.1434, whereas H_s was found to be 0.0599. The G_{st} value of 0.5823 indicated that 41.77% of the genetic diversity resided within the populations. The N_m among the sampled populations was calculated as 0.3587 using G_{st} through the formula $(0.5(1-G_{st})/G_{st})$.

Population genetic diversity

Detailed statistical analyses were performed on the ISSR amplification results (Table 3). The numbers of polymorphic

Table 1. Sample information of the 32 sampling sites.

No.	Population	Prefecture	Coordinates	N	Altitude/m	SO/SD/°	Vegetation type
1	Caotangou	Maiji	E105°48′N34°21′	20	1 899	SE70	Shrub-grass
2	Tancaogou	Fengxian	E106°53′N34°6′	20	1 728	SW45	Shrub-grass
3	Chunshugou	Fengxian	E107°1′N34°2′	20	1 512	SW65	Shrub-grass
4	Hougou	Fengxian	E108°33′N33°55′	20	1 439	N75	Shrub-grass
5	Xiaoshagou	Taibai	E107°27′N34°1′	20	2 332	NW60	Forest-edge, Shrub-grass
6	Doumugong	Meixian	E107°43′N34°2′	20	2 748	NW75	Forest-edge, Shrub-grass
7	Pinganssi	Meixian	E107°43′N34°1′	20	2 815	NW75	Forest-edge, Shrub-grass
8	Mingxingsi	Meixian	E107°44′N34°0′	20	2 637	E75	Forest-edge, Shrub-grass
9	Xiabansi	Meixian	E107°47′N33°59′	20	2 883	SE50	Forest-edge, Shrub-grass
10	Laojungou	Liangdang	E106°33′N34°9′	20	1 902	SE60	Cliff, edge of forest
11	Youfanggou	Fengxian	E106°40′N34°12′	20	1 545	SW70	Shrub-grass
12	Baicaoling	Taibai	E107°24′N34°9′	20	1 373	S60	Forest-edge, Shrub-grass
13	Nianzigou	Zhouzhi	E108°6′N33°56′	20	1 821	SW75	Meadow, edge of forest
14	Wenjiagou	Zhouzhi	E108°5′N33°59′	20	1 765	S80	Meadow, edge of forest
15	Liulingou	Zhouzhi	E108°10′N33°52′	20	1 013	NW75	Meadow, edge of forest
16	Beigou	Zhouzhi	E108°16′N33°51′	20	1 579	NW85	Meadow, edge of forest
17	Dagangou	Huxian	E108°47′N33°55′	20	1 335	NW65	Shrub-grass
18	Yaowangmiao	Taibai	E107°31′N33°53′	20	1 779	SE40	Forest-edge, Shrub-grass
19	Yingpangoukou	Huxian	E108°34′N33°47′	20	1 487	SE60	Shrub-grass
20	Maiduoshigou	Liangdang	E106°7′N34°10′	20	1 520	SE70	Cliff, edge of forest
21	Panjiaba	Liangdang	E106°20′N34°10′	20	1 976	SW45	Cliff, edge of forest
22	Zhangjiagou	Fengxian	E107°5′N34°1′	20	1 469	SW75	Shrub-grass
23	Longwangmiao	Fengxian	E106°57′N33°55′	20	1 475	SW75	Shrub-grass
24	Huangbaigou	Fengxian	E106°59′N34°6′	20	1 745	SW80	Shrub-grass
25	Dashuiyugou	Maiji	E106°30′N34°18′	20	1 878	W75	Shrub-grass
26	Chenjiagou	Fengxian	E106°46′N34°13′	20	1 515	SW45	Shrub-grass
27	Longwangmiao	Maiji	E106°23′N34°15′	20	1 736	NW60	Shrub-grass
28	Yinmagou	Taibai	E107°11′N33°58′	20	2 252	S60	Forest-edge, Shrub-grass
29	Hualingou	Maiji	E106°17′N34°18′	20	1 793	SW55	Shrub-grass
30	Huojigou	Maiji	E106°2′N34°19′	20	1 718	SE65	Shrub-grass
31	Shijiagoucun	Maiji	E105°56′N34°25′	20	1 366	SE60	Shrub-grass
32	Liujiaping	Maiji	E105°44′N34°30′	20	1 690	SW30	Shrub-grass

Note: N means sample size; SO/SD means slope orientation/degree.

Table 2. Description of the11 selected primers used for ISSR amplification.

No.	Primer	Sequences 5′→3′	N_t	N_p	P_r %	H_t	H_s	G_{st}
1	UBC825	$(AC)_8T$	19	12	63.16	0.1715	0.0835	0.5132
2	UBC834	$(AG)_8YT$	21	9	42.86	0.1035	0.0440	0.5748
3	UBC844	$(CT)_8AGC$	22	11	50.00	0.1160	0.0517	0.5543
4	UBC845	$(CT)_8AGG$	23	12	52.17	0.1192	0.0494	0.5855
5	UBC853	$(CT)_8AGT$	20	9	45.00	0.1121	0.0352	0.6858
6	UBC855	$(AC)_8YT$	21	11	52.38	0.1347	0.0645	0.5212
7	UBC857	$(AC)_8CTG$	20	13	65.00	0.1763	0.0763	0.5672
8	UBC867	$(GGC)_6$	19	10	52.36	0.1251	0.0564	0.5492
9	UBC873	$(GACA)_4$	22	11	50.00	0.1165	0.0441	0.6214
10	UBC895	AGAGTTGGTAGCTCTTGATC	25	17	68.00	0.1875	0.0765	0.5920
11	UBC900	ACTTCCCCACAGGTTAACACA	29	20	68.97	0.2150	0.0773	0.6404
Sum			241	135				
Mean			21.91	12.27	56.02	0.1434	0.0599	0.5823

Note: N_t-No. of total amplified bands; N_p-No. of polymorphic bands; P_r-Polymorphism rate; H_t- Total gene diversity among population; H_s-within population diversity; G_{st}-Mean coefficient of gene differentiation.

bands are different among populations. The highest PPB (39.17%) was observed in the DMG population. However, only 20.73% of the bands were polymorphic in the SJG population. The PPB was 56.02% at the species level, whereas those of the single populations ranged from 20.73% (SJG) to 39.17% (DMG), with an average of 27.33%. A_e ranged from 1.0182 (HJG) to 1.2643 (BG), with 1.1630 at the population level and 1.3732 at the species level. H_e within populations was lower (0.0621) than that of the species level (0.1434). Within each population, the H_e of most populations ranged from 0.0226 to 0.0971. Only in the population HG, a high H_e (0.1229, Table 3) was observed. H_o ranged from 0.0244 to 0.1038, with an average of 0.0637 at the population level and 0.2362 at the species level. Allele frequencies calculated using Popgene software were shown in Table S2 in Text S2.

Genetic structure and differentiation of the populations

UPGMA cluster analysis. UPGMA clustering analysis defined three major groups among 32 populations (Figure 2). Group 1 included 11 populations, which were further divided into three subgroups. Populations TCG, CSG and HG were in subgroup 1a; populations XSG, DMG, PAS, MXS and XBS were in subgroup 1b; and populations LJG, YFG, and BCL were in subgroup 1c. Populations NZG, WJG, LLG and BG (Zhouzhi County), population DGG (Huxian County), population YWM (Taibai County) and population YPG (Huxian County) were grouped in cluster 2. Group 3 contained 14 populations, which were sampled from two adjacent cities, Baoji and Tianshui, and were subdivided into two clusters. Populations MDS, PJB, ZJG, LW, and HBG (Baoji) and DSY (Tianshui) were in subgroup 3a. Populations LWM, HLG, HJG, SJG, CTG, and LJP (Tianshui) and CJG, YMG (Baoji) were in subgroup 3b. Seven populations including DSY, LWM, HLG, HJG, SJG, CTG, and LJP in Group 3 were from Gansu Province. The other populations clustered in Groups 1, 2, and 3 were from Shaanxi Province. The *S. hexandrum* population was not clustered on the UPGMA tree according to geographic distance, which may indicate no obvious correlation between geographic distribution and genetic distance.

Principle coordinate analysis. PCoA was used for ordination and exploration of the similarity between populations. All samples were clearly separated into three major groups on the first (PCo1) and second (PCo2) principal coordinates (Figure 3). PCo1 explained 13.6% of the total variance, and PCo2 explained 8.7% of the total variance. Groups 1 and 2 are clustered within each other's vicinity even though clearly separated, indicating higher similarity between these two groups. The clustering of the populations was in agreement with the UPGMA dendrogram.

Mantel test for IBD. A Mantel test for IBD was performed to assess the correlation between the genetic distance matrix and the corresponding geographic distance matrix of the wild *S. hexandrum* populations. The correlation coefficient *r* of genetic distance and geographic distance was 0.212 ($P = 0.287$), and the correlation analysis diagram (Figure 4) was comprised of many disordered and scattered points, indicating that the IBD of the wild *S. hexandrum* populations in Qinling Mts. was not significant at the level of 0.05. The Mantel test did not indicate correlation between genetic distance and geographical provenance.

Bayesian clustering. The genetic structure of the *S. hexandrum* samples were further analyzed using Bayesian clustering algorithm in the STRUCTURE software. The ΔK method indicated that the most likely K value was 3 (Figure 5). Sharp divisions were observed for the three clusters (Figure 6). The assignments of the populations to Groups 1, 2, and 3 were stable and consistent with UPGMA and PCoA clustering.

Table 3. Descriptive statistics summary of the *S. hexandrum* populations.

Population	Code	PPB (%)	A_o	A_e	H_e	H_o
Tancaogou	TCG	23.63	1.3546	1.1684	0.0559	0.0737
Chunshugou	CSG	22.62	1.2687	1.1556	0.0337	0.0689
Hougou	HG	25.66	1.5642	1.2357	0.1229	0.1038
Xiaoshagou	XSG	25.32	1.5721	1.1957	0.0777	0.0733
Doumugong	DMG	39.17	1.3905	1.1404	0.0296	0.0255
Pinganssi	PAS	33.95	1.4397	1.1833	0.0567	0.0442
Mingxingsi	MXS	26.94	1.4720	1.1667	0.0381	0.0667
Xiabansi	XBS	23.51	1.4986	1.231	0.0542	0.0614
Laojungou	LJG	28.52	1.2199	1.1937	0.0426	0.0566
Youfanggou	YFG	21.95	1.3688	1.1846	0.0533	0.0755
Baicaoling	BCL	28.84	1.5579	1.1795	0.0603	0.0715
Nianzigou	NZG	25.18	1.473	1.2166	0.0868	0.0508
Wenjiagou	WJG	31.99	1.4238	1.1737	0.0797	0.0321
Liulingou	LLG	26.85	1.5053	1.2000	0.0882	0.0733
Beigou	BG	29.01	1.5319	1.2643	0.0670	0.0680
Dagangou	DGG	27.07	1.5912	1.2128	0.0448	0.0781
Yaowangmiao	YWM	32.03	1.5309	1.2024	0.0728	0.0972
Yingpangoukou	YPG	24.93	1.6054	1.2290	0.0778	0.0799
Maiduoshigou	MDS	24.63	1.1576	1.0445	0.0226	0.0656
Panjiaba	PJB	32.95	1.1872	1.1293	0.0252	0.0277
Zhangjiagou	ZJG	26.17	1.2364	1.1722	0.0523	0.0464
Longwangmiao	LW	24.46	1.2953	1.2199	0.0498	0.0636
Huangbaigou	HBG	24.07	1.3276	1.1913	0.0684	0.0994
Dashuiyugou	DSY	31.74	1.2435	1.0573	0.0804	0.0704
Chenjiagou	CJG	28.75	1.4232	1.2048	0.0643	0.0544
Longwangmiao	LWM	33.62	1.2577	1.0735	0.0422	0.0722
Yinmagou	YMG	23.52	1.4565	1.2381	0.0971	0.061
Hualingou	HLG	23.34	1.1842	1.1088	0.0387	0.0603
Huojigou	HJG	30.96	1.0761	1.0182	0.0541	0.0244
Shijiagoucun	SJG	20.73	1.1253	1.0611	0.0412	0.0431
Caotangou	CTG	26.42	1.1088	1.0826	0.0315	0.0533
Liujiaping	LJP	25.93	1.2165	1.0802	0.0573	0.0961
Average		27.33	1.3645	1.1630	0.0621	0.0637
Total		56.02	1.7801	1.3732	0.1434	0.2362

Note: A_o, observed number of alleles per locus; A_e, effective number of alleles per locus; He, Nei's gene diversity; H_o, Shannon's information index; PPB, percentage of polymorphic bands.

Analysis of molecular variance. The pairwise GD values (Table 4) were small and ranged from 0.0467 (XSG and DMG) to 0.3997 (BG and LW), indicating low differentiation within populations. Low population differentiation indicated that gene flow within each clustered group may be high or that isolation time was recent.

AMOVA (Table 5) was also performed for population differentiation to further evaluate genetic structure. Highly significant ($P<0.000$ 2) genetic variance was expectedly observed among the populations and explained 63.27% of the total variance, supporting the results from the hierarchical and Bayesian clustering. Only 36.73% of the total genetic variance occurred within populations, indicating higher genetic differentiation between the populations than within each population and the emergence of genetic differentiation among populations. The G_{st} value (0.5823) (Table 2) showed there were more variation among populations than that within populations, confirming the AMOVA results.

Discussion

Genetic diversity of *S. hexandrum* in Qinling Mts

The ISSR markers developed in this study effectively revealed low genetic diversity within the *S. hexandrum* populations sampled in Qinling Mts. Populations are isolated given the population differentiation and clear clustering. The comparison of the average genetic diversity to that of *S. hexandrum* from Northwestern Himalayan region [34,36,38] and other Berberidaceae species [58,59] based on the ISSR approach showed that *S. hexandrum* populations of Qinling Mts. have low genetic diversity (average $H_e = 0.0621$). The *S. hexandrum* in the Northwestern Himalayan

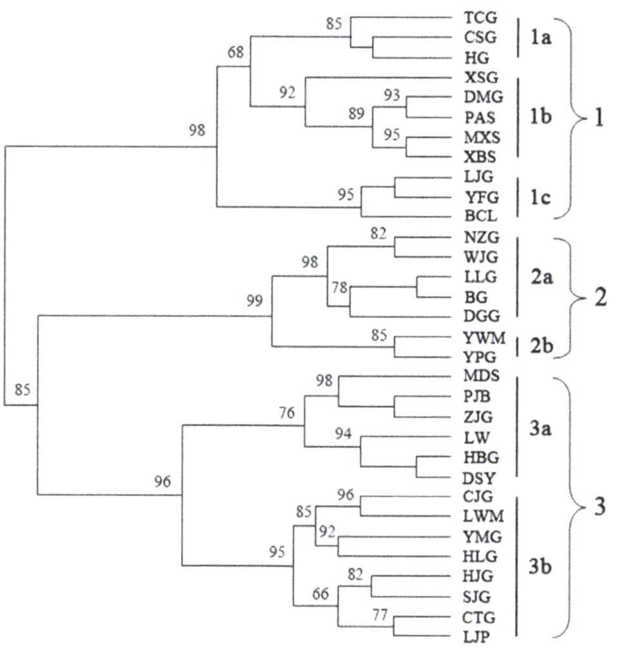

Figure 2. UPGMA clustering of *S. hexandrum* populations in Qinling Mts.

region [34,36,38] showed high genetic variation ($H_e = 0.2944$, 0.092, respectively). *S. hexandrum* is native to the Himalayan region, growing in valleys with secondary vegetation, or under shrubs or around trees [2]. Their habitat is significantly different from that of Qinling Mts. *S. hexandrum* reproduces through vegetative reproduction and seeds. Insects and birds are limited in the high altitude regions, implying that *S. hexandrum* pollination is easier than in the Qingling Mts. The sizes of wild populations of *S. hexandrum* are very small and declines each year in Qinling Mts. because of habitat fragmentation and deterioration caused by human disturbance (overcollection due to economic interests). The rapid decrease in individuals in wild populations may also cause loss of the genetic diversity of this endangered species. Another Berberidaceae species, *Dysosma versipellis*, is an endangered species endemic to China and has been listed as a key protected wild plant in China due to habitat fragmentation or destruction. Qiu et al. found this species had high level of genetic diversity in China ($H_e = 0.378$) [58]. *Dysosma pleiantha*, a threatened medicinal plant species distributed in southeastern China, sexually and asexually reproduces. High H_e (0.364) was observed in this species [59]. The relatively high level of genetic variation observed within two species suggested that the balance between vegetative reproduction and sexual reproduction was more in favor of sexual reproduction in the populations *D. versipellis* and *D. pleiantha* than in the *S. hexandrum* populations. The H_e found in this study was 0.1434 at the species level, lower than those of some strictly self-pollinating soybean species ($H_e = 0.1714$) [60] and the self-pollinating *Oryza granulata* ($H_e = 0.210$) [61], which also indicated that potential selfing system in these populations reduced genetic diversity of *S. hexandrum* populations. Historical events are also responsible for the variation in genetic diversity [62].

Genetic diversity is affected by multiple factors, such as geographical distribution, mating system, life form, pollen and seed dispersal [63,64]. Low genetic variation within populations could be attributed to seed dispersal and the predominant clonal reproduction in *S. hexandrum* in this high mountainous area. However, the genetic diversity of *S. hexandrum* is not much lower than other endangered species analyzed using ISSR markers, such as *Leontice microrhyncha* (Berberidaceae) ($H_e = 0.021$), which is a polycarpic perennial herb found in deciduous or coniferous forests

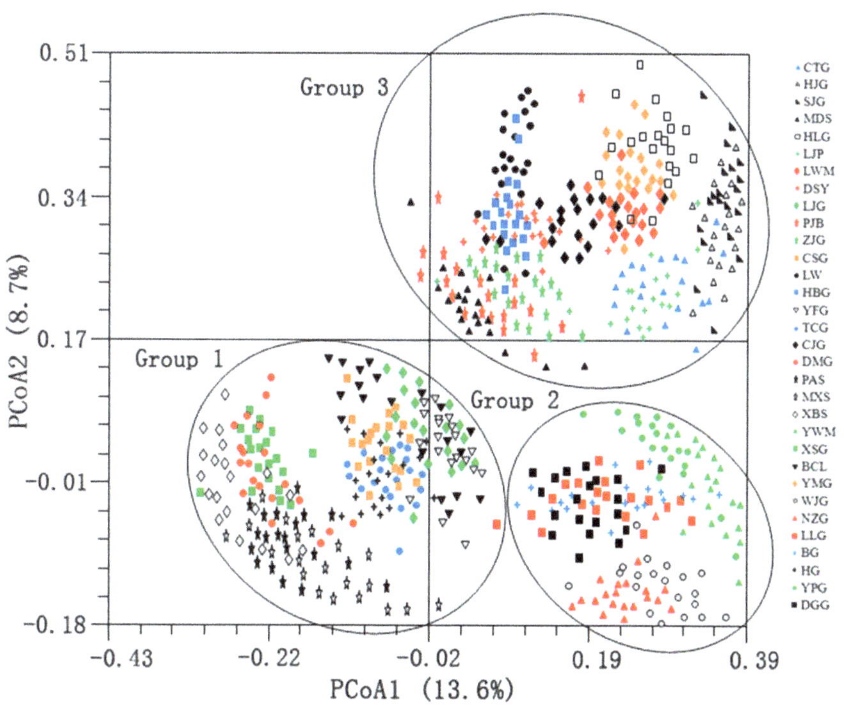

Figure 3. Distribution of individuals of the 32 *S. hexandrum* populations from Qinling Mts., according to the first (PCo1) and second (PCo2) principal coordinates. PCo1 and PCo2 account for 13.6 and 8.7% of the total variation, respectively.

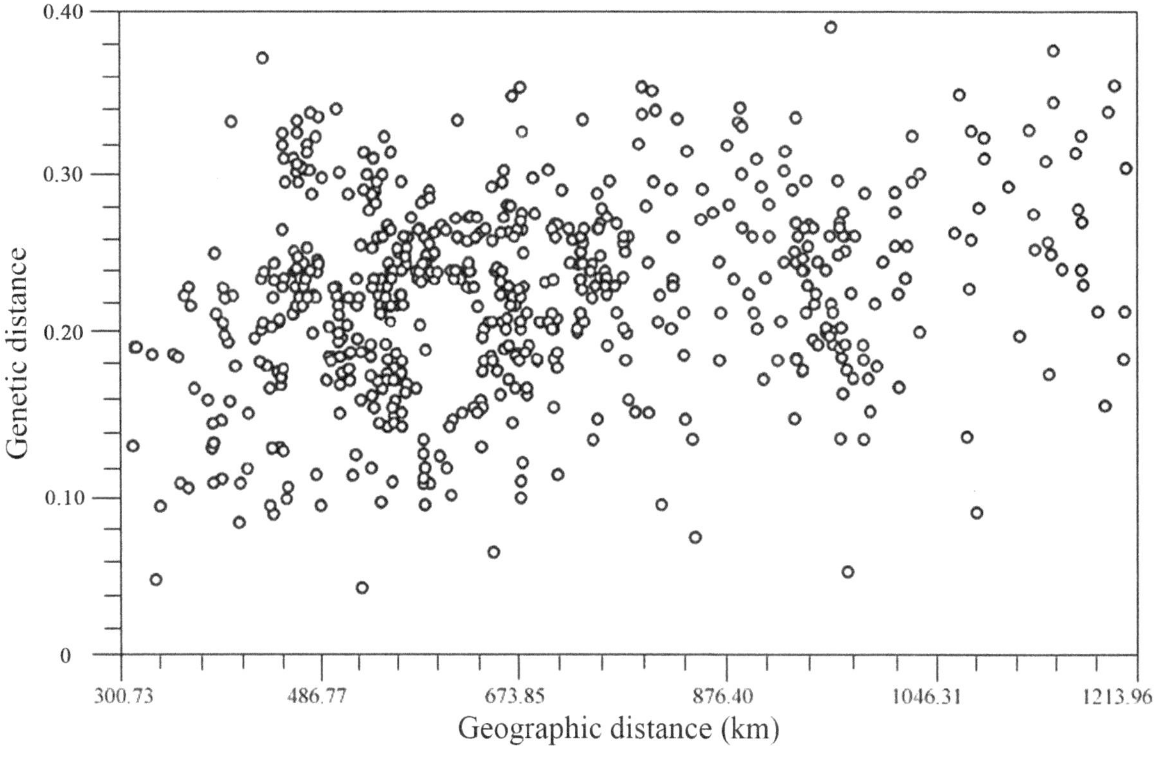

Figure 4. Mantel regression of the pairwise relationship between genetic and geographical distances for *S. hexandrum* populations.

in Korea and Northeast China [65]. For species *L. microrhyncha,* each pollinated flower produces an 8 mm berry and seed dispersal is restricted due to its heavy berry [66]. The genetic diversity of *S. hexandrum* was also higher than those of two other species, the endangered *Pinus squamata* ($H_e = 0.020$) [67] and the first-degree endangered species *Manglietia decidua* ($H_e = 0.0637$) [68]. *P. squamata* and *M. decidua* are extremely rare and endangered tree species in China. The extremely low genetic diversity of this two species could have resulted from the severe bottleneck effect during their evolutionary process. The gene drift and inbreeding may further decrease their genetic diversity in the shrinking populations. The weak competitive ability against broad - leaved trees and human activities may also accelerated the decrease of genetic variation.

S. hexandrum is reasonably long lived because its rhizomes easily reproduce, which could slow down the loss of genetic diversity. Pollen dispersal is generally restricted to a small region due to the large pollen size, which limits gene flow to increase or

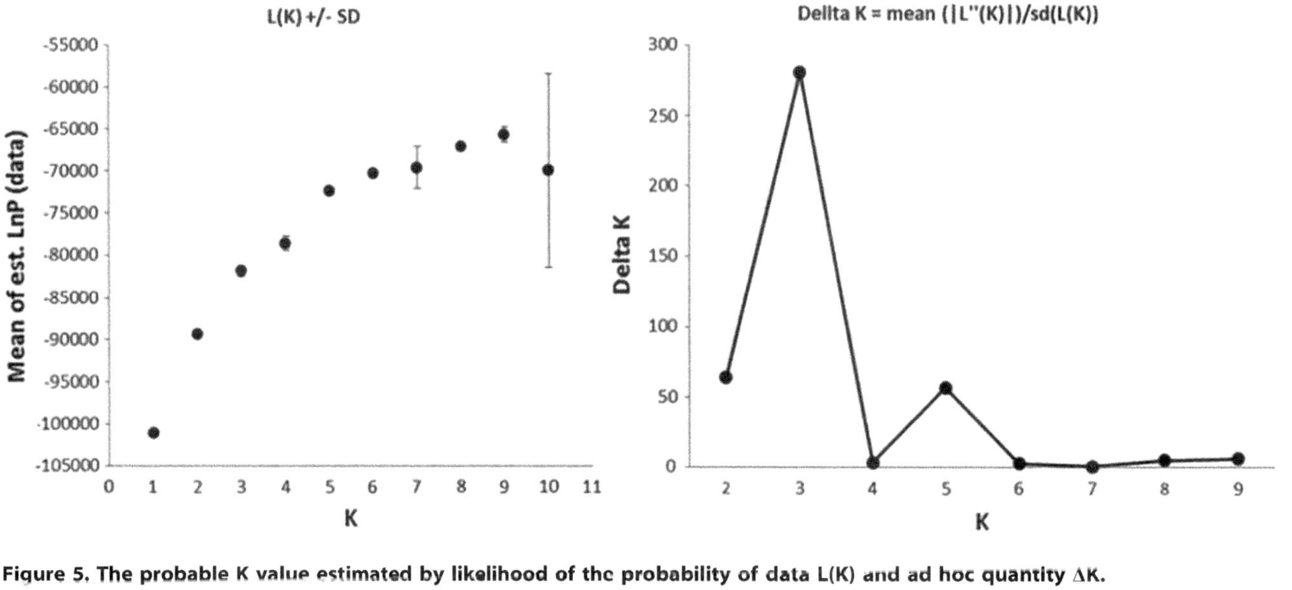

Figure 5. The probable K value estimated by likelihood of the probability of data L(K) and ad hoc quantity ΔK.

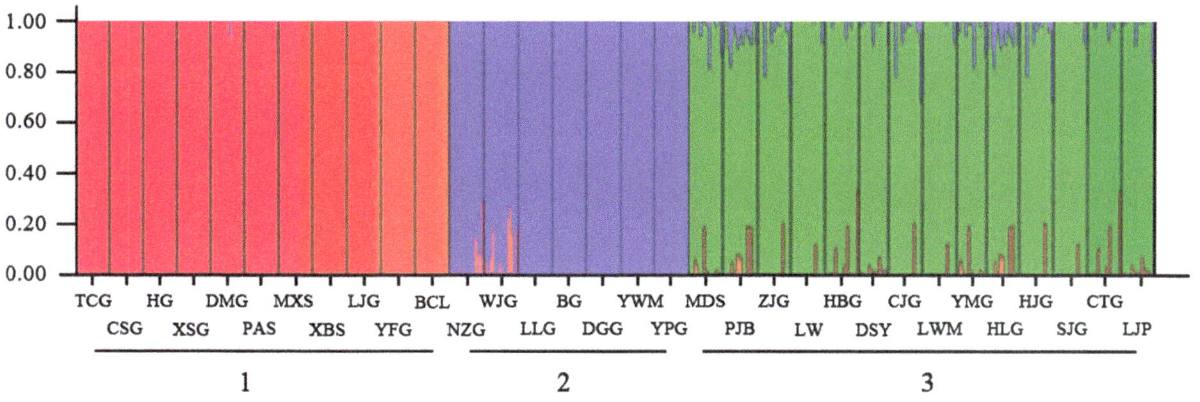

Figure 6. Bayesian clustering for infering population structure of *S. hexandrum* populations from Qinling Mts.

maintain genetic diversity. *S. hexandrum* is native to the Himalayan region, including China, India, Nepal, and Myanmar. No other genetic diversity studies of *S. hexandrum* exist in other locations. Further research should include more populations in other regions of China and countries.

High genetic differentiation and distinct genetic structure

High level of genetic differentiation and clear population structure was detected in this study. K = 3 in the Bayesian clustering as a meaningful value because group 1, 2 and 3 could be divided by detecting the within-group substructure. This result is also supported by distance-based clustering and PCoA. Groups 1 and 2 are genetically close but significantly different. Estimation of the number of clusters K should be treated with care because it is computationally difficult to obtain accurate estimates and the method merely provides an ad hoc approximation [52].

All sampled individuals were strongly assigned to their original populations, and all data strongly support the conclusion that the 32 *S. hexandrum* populations distributed in the Qinling Mts. are clustered into three major groups. These methods consistently showed that high genetic differentiation existed among *S. hexandrum* populations, which is consistent with genetic variation studies in certain selfing species [69]. This would mean that *S. hexandrum* should be a selfing species or a selfing predominant species, which is consistent with previous studies on *S. hexandrum* by Ma et al. [70]. Aside from the breeding system, the high genetic differentiation across populations may also be caused by genetic drift [71]. Wright [72] noted that genetic drift would lead a small population to emerge with a distinct genetic differentiation when the N_m value is lower than 1.0. The N_m of *S. hexandrum* (0.3587) determined using the POPGENE software was lower than 1.0 in the present study, which suggested that some genetic drift may have emerged among the populations of this species. The distribution of *S. hexandrum* populations obviously tend to fragment based on the field investigation, which is consistent with the possibility of genetic drift.

Migration of plant populations can occur through dispersal of pollen and seed [73]. But a number of factors such as fragmented geographical distribution, lack of pollinators or seed dispersers can be a barrier to gene flow between populations [74,75]. Limited gene flow among *S. hexandrum* populations may be related to inbreeding of the species and limited seed propagation distance. Some studies have found that seed dispersal is the primary factor influencing variation of gene flow and population structure [76]. Heavy mature berries of *S. hexandrum* usually drop to the ground because of rain or wind, settling some seeds in the soil, whereas others are dispersed by cattle, birds, or humans. Therefore, the short distance of seed dispersal of *S. hexandrum* probably resulted in limited gene flow among populations. Mountain ranges and rivers are possible barriers to either dispersal of pollen or rhizomes of *S. hexandrum*, reproductively isolating the populations. The restriction of gene flow associated with geographical distance is consistent with the results of previous studies on this species [34,35,36].

Implications for conservation

S. hexandrum is a rare and threatened species [4] in China. Assessment of genetic diversity is important for designing conservation strategies for threatened and endangered species [77,78]. The results of this study showed that there was low genetic diversity among *S. hexandrum* populations and genetic differentiation among populations was higher than within populations. Genetic diversity loss has deleterious effects on species fitness and threatens the population survival and could be the key reason that explains the endangerment of *S. hexandrum* in Qinling Mts. [79,80]. The estimation of the genetic diversity and population genetic structure could provide bases for *S. hexandrum* conservation and its reasonable utilization. The results will help determine what to conserve and where and how to conserve this species.

The field survey showed that the habitats of some populations have been destroyed by human disturbance for great medical value. Damage to natural habitats would led to a decrease in population sizes and probably a subsequent increase in inbreeding, decreasing its genetic diversity. In situ conservation effectively and sustainably prevents this problem. The establishment of *S. hexandrum* reserves should be the primary method because the Qinling Mts. are situated in state forest conservation areas, where cutting and hunting are restricted. The management for the conservation of genetic variability in this species should aim to preserve not only large populations but also as many of the small populations outside nature reserves as possible. Reduced levels of genetic variation, especially in the smaller populations, will affect the species' ability to adapt to changes in its habitat [81]. Positive correlations between population size, expected heterozygosity, and plant fitness were found in *Gentiana pneumonanthe* [82] and *Arnica Montana* [81]. Thus, policy plans should also be developed to stimulate seedling recruitment in the small populations (e.g., PAS and MXS). It may be dangerous to mix highly divergent populations because it could cause loss of adaptive diversity [45]. Therefore, it is necessary to improve gene flow among populations

Table 4. Pairwise genetic distance estimates from 241 ISSR markers in *S. hexandrum*.

Population	TCG	CSG	HG	XSG	DMG	PAS	MXS	XBS	LJG	YFG	BCL	NZG	WJG	LLG	BG	DGG	YWM	YPG	MDS	PJB	ZJG	LW	HBG	DSY	CJG	LWM	YMG	HLG	HJG	SJG	CTG	LJP
TCG	***																															
CSG	0.1362	***																														
HG	0.0712	0.1379	***																													
XSG	0.1756	0.1444	0.1064	***																												
DMG	0.1833	0.1433	0.1343	0.0467	***																											
PAS	0.1654	0.1373	0.1312	0.0513	0.1514	***																										
MXS	0.1532	0.1920	0.1255	0.0568	0.2222	0.0634	***																									
XBS	0.1743	0.1636	0.1168	0.0764	0.3867	0.0754	0.0822	***																								
LJG	0.1503	0.1520	0.1205	0.0608	0.1290	0.0879	0.1075	0.0732	***																							
YFG	0.1853	0.1585	0.1484	0.0654	0.1640	0.0819	0.1015	0.0797	0.1755	***																						
BCL	0.1897	0.1574	0.1453	0.0709	0.1684	0.1575	0.1262	0.0786	0.1799	0.2085	***																					
NZG	0.1974	0.1514	0.1396	0.0905	0.1461	0.1786	0.1278	0.3726	0.1876	0.1907	0.1586	***																				
WJG	0.2095	0.2061	0.1309	0.0680	0.3582	0.1546	0.1093	0.1273	0.1697	0.1790	0.1555	0.3824	***																			
LLG	0.1673	0.1777	0.2213	0.1457	0.3891	0.1896	0.1158	0.0989	0.1830	0.1994	0.1498	0.2020	0.1616	***																		
BG	0.2448	0.1592	0.2445	0.1568	0.3483	0.1940	0.1147	0.3804	0.1885	0.1907	0.1411	0.3772	0.2163	0.2304	***																	
DGG	0.2727	0.1657	0.2449	0.1368	0.1980	0.2017	0.1087	0.0869	0.2081	0.1829	0.1422	0.1818	0.1879	0.2020	0.2035	***																
YWM	0.2696	0.1646	0.2185	0.1356	0.2179	0.2138	0.1334	0.0858	0.1815	0.1772	0.1384	0.1873	0.1763	0.1835	0.1948	0.2076	***															
YPG	0.2639	0.1586	0.3831	0.1431	0.1844	0.1716	0.1350	0.0798	0.1861	0.1685	0.2015	0.2069	0.1828	0.1900	0.3893	0.2272	0.2040	***														
MDS	0.1552	0.2133	0.3086	0.0830	0.2136	0.2014	0.1142	0.1345	0.1624	0.1612	0.2136	0.1784	0.1817	0.1889	0.2172	0.2047	0.2251	0.2114	***													
PJB	0.2520	0.1849	0.3223	0.1026	0.2202	0.1462	0.1207	0.1061	0.1453	0.1891	0.1696	0.3726	0.2137	0.1829	0.2141	0.2093	0.1991	0.2318	0.2197	***												
ZJG	0.1799	0.1641	0.3050	0.0741	0.1530	0.1127	0.0896	0.3760	0.1674	0.1860	0.1694	0.0781	0.1853	0.2376	0.2084	0.1729	0.2341	0.2231	0.3793	0.1879	***											

Table 4. Cont.

Population	TCG	CSG	HG	XSG	DMG	PAS	MXS	XBS	LJG	YFG	BCL	NZG	WJG	LLG	BG	DGG	YWM	YPG	MDS	PJB	ZJG	LW	HBG	DSY	CJG	LWM	YMG	HLG	HJG	SJG	CTG	LJP
LW	0.2768	0.1706	0.3384	0.0787	0.1413	0.0843	0.1136	0.0971	0.1927	0.1803	0.1524	0.0977	0.1405	0.2092	0.3997	0.2079	0.2385	0.2474	0.2108	0.1822	0.2160	***										
HBG	0.1778	0.1695	0.1430	0.0842	0.1617	0.0753	0.1683	0.3744	0.1667	0.2150	0.1747	0.0729	0.2006	0.1884	0.1905	0.2123	0.2162	0.1946	0.2154	0.1735	0.2206	0.2416	***									
DSY	0.1574	0.1635	0.3868	0.1394	0.1530	0.0818	0.2399	0.0540	0.2017	0.1510	0.1711	0.0775	0.1307	0.1949	0.2184	0.2200	0.2283	0.2011	0.3709	0.1772	0.2261	0.2208	0.2214	***								
CJG	0.1925	0.2182	0.1617	0.1110	0.3773	0.0807	0.1154	0.3891	0.2061	0.1556	0.1624	0.1916	0.1769	0.1938	0.2153	0.2021	0.2161	0.2000	0.2405	0.2051	0.2457	0.2273	0.2461	0.0884	***							
LWM	0.1963	0.1898	0.1586	0.0865	0.1169	0.0747	0.1219	0.0929	0.1838	0.1611	0.1569	0.2112	0.1676	0.1878	0.2096	0.1899	0.2372	0.1940	0.2139	0.2020	0.2087	0.2262	0.2177	0.0624	0.0477	***						
YMG	0.2042	0.1653	0.1529	0.1514	0.3719	0.1294	0.0908	0.1008	0.1959	0.1807	0.1848	0.1836	0.1757	0.2425	0.2009	0.2110	0.2145	0.2487	0.3815	0.1963	0.2152	0.2202	0.1631	0.0974	0.0756	0.1038	***					
HLG	0.1864	0.1718	0.1442	0.1410	0.1563	0.1010	0.1148	0.0930	0.1837	0.1651	0.1817	0.1882	0.1880	0.2141	0.1936	0.1870	0.1941	0.2203	0.2240	0.1876	0.2141	0.2749	0.1910	0.1018	0.0725	0.0772	0.1090	***				
HJG	0.1747	0.1707	0.1369	0.1481	0.1640	0.0825	0.1395	0.0919	0.2048	0.1697	0.1760	0.3937	0.1935	0.1896	0.2215	0.2220	0.2292	0.2148	0.2436	0.1844	0.2081	0.2465	0.0774	0.0795	0.0668	0.0818	0.0874	0.1415	***			
SJG	0.0991	0.1647	0.3648	0.1470	0.1761	0.0890	0.1111	0.0859	0.1821	0.1752	0.1673	0.2133	0.2131	0.1961	0.2184	0.2264	0.2330	0.2344	0.2159	0.2123	0.2628	0.2220	0.0717	0.0916	0.0581	0.0873	0.0939	0.1131	0.0940	***		
CTG	0.1864	0.1894	0.1617	0.1722	0.1339	0.1366	0.1021	0.1106	0.1617	0.1948	0.1581	0.1622	0.1975	0.1950	0.2127	0.2341	0.2409	0.2096	0.3624	0.2092	0.2344	0.2285	0.0839	0.0794	0.0526	0.1069	0.0928	0.0886	0.0880	0.0853	***	
LJP	0.2107	0.1610	0.1560	0.1686	0.3655	0.1082	0.1086	0.3822	0.1968	0.1723	0.1860	0.1687	0.2021	0.1890	0.3834	0.2462	0.2231	0.2142	0.2213	0.2153	0.2700	0.2274	0.0894	0.1005	0.0805	0.0793	0.0868	0.0951	0.0847	0.0918	0.0907	***

Table 5. Analysis of molecular variance (AMOVA) for 640 individuals in 32 populations of *S. hexandrum* using 11 selected ISSR primers.

Source of variation	df	Sum of squares	Mean squares	Variance component	Total variance (%)	*P*-value
Among population	31	4,028.326	129.946	3.9128	63.27	<0.000 2
Within population	608	950.912	1.564	1.564	36.73	<0.000 2

within each group through some artificial means, such as transplanting individuals (by seed, rhizomes from one population to another). Furthermore, to avoid human overcollection, greater awareness for *S. hexandrum* protections must be emphasized, and related forest departments should be encouraged to undertake conservation through an integrated conservation strategy based on demographic, ecological, and genetic aspects.

As a supplement to in situ conservation, ex situ conservation would also be feasible as underlined by other studies on endangered species [83,84,85]. Populations may be partially preserved through seed banks or in vitro germplasm collections. *S. hexandrum* has favorable sexual reproduction. Each plant produces approximately 60 seeds, with a maximum of approximately 180 seeds [70]. Seed collection is easier for *S. hexandrum* than other endangered species. Thus, a strategy involving extensive collection to ensure full sampling of genetic diversity, subsequent cultivation in a garden at least 1000 m above sea level [7], and reintroduction into their original wild habitats seems feasible, although *S. hexandrum* mainly grows wild on high altitude mountain ranges. For *S. hexandrum* populations in Qinling Mts., there are some preserved forest farms which could be used for relocation. However, ex situ conservation has many drawbacks because it is impossible to recreate the habitat as a whole. The new environment may have important ecological differences compared with the original habitat, and the approach is technically challenging and is often expensive. Therefore, ex situ conservation is recommended only to supplement in situ conservation or as a last resort. In vitro techniques are also proven to be an effective alternative means of propagation that facilitates the recovery of the rare and endangered *S. hexandrum* [33]. At present, an effective protocol of in vitro propagation, involving multiple shoot formation from zygotic embryos and subsequent rooting, could be available for *S. hexandrum*. In vitro propagation may well be used as a means to rescue zygotic embryos for this species. In vitro techniques induce variability, but plants raised from tissue cultures may be screened for useful somaclonal variants and exploited to obtain plants or cultures with high podophyllotoxin contents, which possibly reduces the pressure on natural *S. hexandrum* populations.

Most of the genetic diversity of the important medicinal and endangered species *S. hexandrum* in Qinling Mts. must be guaranteed with these combined and sustained efforts.

Acknowledgments

The authors are also grateful to Genlu Bai, Guowei Xia and all the colleagues in the same laboratory for the assistance in the work.

Author Contributions

Conceived and designed the experiments: WL JJL. Performed the experiments: WL DXY NL. Analyzed the data: WL. Contributed reagents/materials/analysis tools: JJL. Contributed to the writing of the manuscript: WL.

References

1. Editorial committee, Chinese Academy of Sciences (2010) Vol. 29, Flora of China. Beijing: Science Press. 249–251 p. (In Chinese.).
2. Ying TS (1979) Study on Dysosma Woodson and Sinopodophyllum Ying (new genus) of the Berberidaceae. Acta Phytotax Sin 17: 17–23. (In Chinese with English abstract.).
3. Chatterjee R (1952) Indian podophyllum. Econ Bot 6: 342–354.
4. Fu LG (1992) Plant red book of China: Rare threatened plant. Beijing: Science Press. (In Chinese.).
5. Cragg GM, Newman DJ, Snader KM (1997) Natural products in drug discovery and development. J Nat Prod 60: 52–60.
6. Dong JE, Ma XH, Wei Q, Peng SB, Zhang SC (2011) Effects of growing location on the contents of secondary metabolites in the leaves of four selected superior clones of *Eucommia ulmoides*. Ind Crop Prod 34: 1607–1614.
7. Li GM (1975) Introduction a medicine plant: *Sinopodophyllum emodi Wall. var. Chinense Sprague*. Journal of Botony 2: 28. (In Chinese with English abstract.).
8. Fay DA, Ziegler HW (1985) Botanical source differentiation of Podophyllum resin by high performance liqid chromatography. J Liq Chromatogr 8: 1501–1506.
9. Giri A, Narasu ML (2000) Production of podophyllotoxin from *Podophyllum hexandrum*: a potential natural product for clinically useful anticancer drugs. Cytotechnology 34: 17–26.
10. Stahelin HF, Von Warhurg A (1991) The chemical and biological route from podophyllotoxin glucoside to etoposide: ninth cain memorial avard lecture. Cancer Res 51: 5–11.
11. Kamil WM, Dewick PM (1986) Biosynthetic relationship of aryltetralin lactone lignans to dibenzylbutyrolactone lignans. Phytochemistry 25: 2093–2102.
12. Holthuis JJM (1988) Etoposide and teniposide. Pharm Weekbl 10: 101–116.
13. Imbert TF (1998) Discovery of podophyllotoxin. Biochimie 80: 207–222.
14. Moraes RM, Lata H, Bedir E, Maqbool M, Cushman K (2002) The American mayapple and its potential for podophyllotoxin production. In: Janick J, Whipkey A, editors. Trends in new crops and new uses. Alexandria, VA: ASHS Press. pp. 527–532.
15. Yousefzadi M, Sharifi M, Behmanesh M, Moyano E, Bonfill M, et al. (2010) Podophyllotoxin: current approaches to its biotechnological production and future challenges. Eng Life Sci 4: 281–292.
16. Lata H, Moraes RM, Bertoni B, Pereira Ana MS (2010) In vitro germplasm conservation of *Podophyllum peltatum* L. under slow growth conditions. In Vitro Cell Dev Biol–Plant 46: 22–27.
17. Lin MC, Lin JH, Chen SK, Cheng YW, Cheng HW (2008) Simultaneous determination of podophyllotoxin, quercetin and kaempferol in podophyllin by liquid chromatography tandem mass spectrometry. J Food Drug Anal 16: 29–40.
18. Zhou Y, Jiang SY, Ding LS, Cheng SW, Xu HX, et al. (2008) Chemical fingerprinting of medicinal plants "Gui-jiu" by LC-ESI multiple-stage MS. Chromatographia 68: 781–789.
19. Purohit MC, Bahuguna R, Maithani UC, Rawat MSM (1999) Variation in podophylloresin and podophyllotoxin content in different populations of *Podophyllum hexeandrum*. Curr Sci India 77: 1078–1079.
20. Zhao CQ, Cao W, Nagatsu A, Ogihara Y (2001) Three new glycosides from *Sinopodophyllum emodi* (Wall.) Ying. Chem Pharm Bull 49: 1474–1476.
21. Kong Y, Xiao JJ, Meng SC, Dong XM, Ge YW, et al. (2010) A new cytotoxic flavonoid from the fruit of *Sinopodophyllum hexandrum*. Fitoterapia 81: 367–370.

22. Zhao CQ, Zhu YY, Chen SY, Ogihara Y (2011) Lignan glucoside from *Sinopodophyllum emodi* and its cytotoxic activity. Chinese Chem Lett 22: 181–184.
23. Sun YJ, Li ZL, Chen H, Liu XQ, Zhou W, et al. (2011) Three new cytotoxic aryltetralin lignans from *Sinopodophyllum emodi*. Bioorg Med Chem Lett 21: 3794–3797.
24. Qin Y, Gui MY, Yu LN, Ru H, Jin YR, et al. (2009) RP-HPLC determination of lignans in *Sinopodophyllum emodi* Wall. Chin J Pharm Anal 29: 1490–1493. (In Chinese with English abstract.).
25. Huang K, Jiang W, Zhao JF, Wang CH, Liu X, et al. (2012) Determination of podophyllotoxin and total lignans in *Sinopodophyllum emodi*. China journal of Chinese material medicia 37: 1360–1365. (In Chinese with English abstract.).
26. Inamori Y, Kubo M, Tsujibo H, Ogawa M, Baba K, et al. (1986) The biological activities of podophyllotoxin compounds. Chem pharm Bull 34: 3928–3932.
27. Goel HC, Prasad J, Sharma A, Singh B (1998) Antitumour and radioprotective action of *Podophyllum hexandrum*. Indian J Exp Biol 36: 583–587.
28. Chattopadhyay S, Bisaria VS, Panda AK, Srivastava AK (2004) Cytotoxicity of invitro produced podophyllotoxin from *Podophyllum hexandrum* on human cancer cell line. Nat Prod Res 18: 51–57.
29. Reddy PB, Paul DV, Agrawal SK, Saxena AK, Kumar HMS, et al. (2008) Design, synthesis, and biological testing of 4b-[(4-Substituted)-1,2,3-triazol-1-yl] podophyllotoxin analogues as antitumor agents. Arch Pharm Chem Life Sci 341: 126–131.
30. Zhu CG, Yang J, Xiong Y (2004) Research progress in natural antineoplastic podophyllotoxin and its derivatives. Drug Eval 1: 306.
31. Wang DW, Guo FX, Ma XY (1997) The antitumor activity of *Sinopodophyllum emodi*. Chinese medicine matria 20: 571–574. (In Chinese with English abstract.).
32. Li GY (2005) Effects of ethanol extracts from *Podophylum emodi var.* on proliferation and apoptosis of breast careinoma cell line MCF-7. Chinese Journal of New Drugs 26: 2185–2188. (In Chinese with English abstract.).
33. Nadeem M, Palni LMS, Purohit AN, Pandey H, Nandi SK (2000) Propagation and conservation of *Podophyllum hexandrum* Royle: an important medicinal herb. Bio Conserv 92: 121–129.
34. Alam MA, Naik PK, Gulati P, Gulati AK, Mishra GP (2008) Characterization of genetic structure of *Podophyllum hexandrum* populations, an endangered medicinal herb of Northwestern Himalaya, using ISSR-PCR markers and its relatedness with podophyllotoxin content. Afr J Biotechnol 7: 1028–1040.
35. Alam MA, Gulati P, Aswini KG, Gyan PM, Pradeep KN (2009) Assessment of genetic diversity among *Podophyllum hexandrum* genotypes of Northwestern Himalayan region for podophyllotoxin production. Indian J Biotechnol 8: 391–399.
36. Naik PK, Alam MA, Singh H, Goyal V, Parida S, et al. (2010) Assessment of genetic diversity through RAPD, ISSR and AFLP markers in *Podophyllum hexandrum*: a medicinal herb from the Northwestern Himalayan region. Physiol Mol Biol Plants 16: 135–148.
37. Xiao M, Li Q, Guo L, Luo T, Duan WX, et al. (2006) AFLP analysis of genetic diversity of the endangered species *Sinopodophyllum hexandrum* in the tibetan region of Sichuan province, China. Biochem Genet 44: 47–59.
38. Xiao M, Li Q, Wang L, Guo L, Li J, et al. (2006) ISSR analysis of the genetic diversity of the endangered species *Sinopodophyllum hexandrum* (Royle) Ying from Western Sichuan province, China. J Integr Plant Bio 48: 1140–1146.
39. Li Y, Zhai SN, Qiu YX, Guo YP, Ge XJ, et al. (2011) Glacial survival east and west of the 'Mekong–Salween Divide' in the Himalaya–Hengduan mountains region as revealed by AFLPs and cpDNA sequence variation in *Sinopodophyllum hexandrum* (Berberidaceae). Mole Phylogenet Evol 59: 412–424.
40. Totland Ø (1999) Effects of temperature on performance and phenotypic selection on plant traits in alpine *Ranunculus acris*. Oecologia 120: 242–251.
41. Esselman EJ, Li J, Crawford DJ, Winduss JL, Wolfe AD (1999) Clonal diversity in the rare *Calamagrostis porteri ssp. insperata* (Poaceae): Comparative results for allozymes and random amplified polymorphic DNA (RAPD) and inter-simple sequence repeat (ISSR) marker. Mol Ecol 8: 443–451.
42. Ge XJ, Sun M (1999) Reproductive biology and genetic diversity of a cryptoviviparous mangrove *Aegiceras corniculatum* (Myrsinaceae) using allozyme and inter-simple sequence repeat (ISSR) analysis. Mol Ecol 8: 2061–2069.
43. Li HS, Chen GZ (2004) Genetic diversity of *Sonneratia alba* in China detected by inter-simple sequence repeat (ISSR) analysis. Acta Bot Sin 46: 515–521.
44. Thriveni HN, Sumangala RC, Shivaprakash KN, Ravikanth G, Vasudeva R, et al. (2014) Genetic structure and diversity of *Coscinium fenestratum*: a critically endangered liana of Western Ghats, India. Plant Syst Evol 300: 403–413.
45. Tang N, Mo G, van Tuyl JM, Arens P, Liu JJ, et al. (2014) Genetic diversity and structure of *Lilium pumilum* DC. in southeast of Qinghai–Tibet plateau. Plant Syst Evol 300: 1453–1464.
46. Bornet B, Branchard M (2001) Nonanchored inter simple sequence repeat (ISSR) markers: reproducible and specific tools for genome fingerprinting. Plant Mol Biol Rep 19: 209–215.
47. Nagaoka T, Ogihara Y (1997) Applicability of inter-simple sequence repeat polymorphisms in wheat for use as DNA markers in comparison to RFLP and RAPD markers. Theor Appl Genet 94: 597–602.
48. Li JJ, Zhu ZC, Min ZL (1989) Comprehensive survey of the Taibai Mountain preserve (1st ed). Xi'an: Shaanxi Normal University Press. 1–9 p. (In Chinese.).
49. Yeh FC, Yang RC, Boyle T, Ye ZH, Mao JX (1997) POPGENE: The user friendly shareware or population genetic analysis. Molecular Biology and Biotechnology Center, University of Alberta, Edmonton.
50. Liu KJ, Muse SV (2005) PowerMarker: an integrated analysis environment for genetic marker analysis. Bioinformatics 21: 2128–2129.
51. Felsenstein J (1985) Confidence limits on phylogenies: an approach using the bootstrap. Evolution 39: 783–791.
52. Pritchard JK, Stephens M, Donnelly P (2000) Inference of population structure using multilocus genotype data. Genetics 155: 945–959.
53. Evanno G, Regnaut S, Goudet J (2005) Detecting the number of clusters of individuals using the software Structure: a simulation study. Mol Ecol 14: 2611–2620.
54. Jensen RJ (1989) Ntsys-Pc-numerical taxonomy and multivariate analysis system-version 1.40. Q Rev Biol 64: 250–252.
55. Exeoeffr L, Smouse PE, Quattro JM (1992) Analysis of molecular variance inferred from metric distances among DNA haplotypes: applications to human mitochondrial DNA restriction data. Genetics 131: 479–491.
56. Miller MP (1997) Tools for Population Genetic Analysis (TFPGA), Vesion 1.3. USA: Department of Biological Sciences, Northern Arizona University, Ariozna.
57. Larson SR, Palazzo AJ, Jensen KB (2003) Identification of western wheatgrass cultivars and accessions by DNA fingerprinting and geographic provenance. Crop Sci 43: 394–401.
58. Qiu YX, Li JH, Liu HL, Chen YY, Cheng XF (2006) Population structure and genetic diversity of *Dysosma versipellis* (Berberidaceae), a rare endemic from China. Biochem Syst Ecol 34: 745–752.
59. Zong M, Liu HL, Qiu YX, Yang SZ, Zhao MS, et al. (2008) Genetic diversity and geographic differentiation in the threatened species *Dysosma pleiantha* in China as revealed by ISSR analysis. Biochem genet 46: 180–196.
60. JinY, Zhang WJ, Fu D X, Lu BR (2003) Sampling strategy within a wild soybean population based on its genetic variation detected by ISSR markers. Acta Bot Sin 45: 995–1002. (In Chinese with English abstract.).
61. Wu CJ, Chen ZQ, Huang XQ, Yin SH, Cao KM, et al. (2004) Genetic diversity among and within populations of *Oryza granulata* from Yunnan of China revealed by RAPD and ISSR markers: implications for conservation of the endangered species. Plant Sci 167: 35–42.
62. Karron JD (1991) Patterns of genetic variation and breeding systems in rare plant species. In: Falk DA, Holsinger, KE, editors. Genetics and conservation of rare plants. New York: Oxford University Press. pp. 87–98.
63. Hamrick JL, Godt MJW, Sherman-Broyles SL (1992) Factors influencing levels of genetic diversity in woody plant species. New Forest 6: 95–124.
64. Ohsawa T, Saito Y, Sawada H, Ide Y (2008) Impact of altitude and topography on the genetic diversity of *Quercus serrata* populations in the Chichibu Mountains, central Japan. Flora 203: 187–196.
65. Lee TB (1980) Illustrated flora of Korea. Seoul: Hyang-mun Pub. Co. (In Korean.).
66. Chang CS, Kim H, Park TY, Maunder M (2004) Low levels of genetic variation among southern peripheral populations of the threatened herb, *Leontice microrhyncha* (Berberidaceae) in Korea. Biol Conserv 119: 387–396.
67. Zhang ZY, Li DZ (2003) Conservation genetics of extremely endangered pine, *Pinus squamata*. Acta Botanica Yunnanica 25: 544–55. (In Chinese with English abstract.).
68. Liao WF, Xia NH, Deng YF, Zheng QY (2004) Study on genetic diversity of *Manglietia decidua*. Acta Botanica Yunnanica 26: 58–64. (In Chinese with English abstract.).
69. Nybom H, Bartish IV (2000) Effects of life history traits and sampling strategies on genetic diversity estimates obtained with RAPD markers in plants. Plant Ecol Evol Syst 3: 93–114.
70. Ma SB, Xu ZY, Hu ZH (1997) A contribution to the reproductive biology of *Sinopodophyllum hexandrum* (Role) Ying (Berberidaceae). Acta Bot Boreal – Occident Sin 17: 49–55. (In Chinese with English abstract.).
71. Hamrick JL, Godt MJW (1989) Allozyme diversity in plant species. In: Brown AHD, Clegg MT, Khaler AL, Weir BS, editors. Plant population genetics, breeding and genetic resources. Sunderland: Sinauer Associates, Inc., Publisher. pp. 43–63.
72. Wright S (1965) The interpretation of population structure by F-statistics with special regard to systems of mating. Evolution 19: 395–420.
73. Ennos RA (1994) Estimating the relative rates of pollen and seed migration among plant-populations. Heredity 72: 250–259.
74. Slatkin M (1985) Gene flow in natural-populations. Annu Rev Ecol Syst 16: 393–430.
75. Zhou TH, Qian ZQ, Li S, Guo ZG, Huang ZH, et al. (2010) Genetic diversity of the endangered Chinese endemic herb *Saruma henryi Oliv.* (Aristolochiaceae) and its implications for conservation. Popul Ecol 52: 223–231.
76. Kalisz S, Hanzawa FM, Tonsor SJ, Thiede DA, Voigt S (1999) Ant-mediated seed dispersal alters pattern of relatedness in a population of *Trillium grandifloru*. Ecology 80: 2620–2634.
77. Hamrick JL (1983) The distribution of genetic variation within and among natural plant populations. In: Schonewald-Cox CM, Chambers SM, McBryde B, Thomas WL, editors. Genetics and conservation. Menlo Park: Benjamin Cummings Publishing Company. pp. 335–348.
78. Francisco-Ortega J, Santos-Guerra A, Kim SC, Crawford DJ (2000) Plant genetic diversity in the Canary Islands: a conservation perspective. Am J Bot 87: 909.
79. Malone CL, Knapp CR, Taylor JF, Davis SK (2003) Genetic consequences of pleistocene fragmentation: isolation, drift, and loss of diversity in rock iguanas (Cyclura). Conserv Genet 4: 1–15.

80. Reed DH (2003) Correlation between fitness and genetic diversity. Conserv Biol 17: 230–237.
81. Luijten SH, Dierick A, Oostermeijer JGB, Raijmann LEL, Den Nijs JCM (2000) Population size, genetic variation and reproductive success in the rapidly declining, self-incompatible perennial (*Arnica montana*) in the Netherlands. Conserv Biol 14: 1776–1787.
82. Oostermeijer JGB, Van Eijck MW, Van Leeuwen NC, Den Nijs JCM (1995) Analysis of the relationship between allozyme heterozygosity and fitness in the rare *Gentiana pneumonanthe* L. J Evol Biol 8: 739–757.
83. Bunn E (2005) Development of in vitro methods for ex situ conservation of *Eucalyptus impensa*, an endangered mallee from southwestern Western Australia. Plant Cell Tiss Org 83: 97–102.
84. Cochrane JA, Crawford AD, Monks LT (2007) The significance of ex situ seed conservation to reintroduction of threatened plants. Aust J Bot 55: 356–361.
85. Li QM, Xu ZF, He TH (2002) Ex situ genetic conservation of endangered *Vatica guangxiensis* (Dipterocarpaceae) in China. Biol Conserv 106: 151–156.

PERMISSIONS

All chapters in this book were first published in PLOS ONE, by The Public Library of Science; hereby published with permission under the Creative Commons Attribution License or equivalent. Every chapter published in this book has been scrutinized by our experts. Their significance has been extensively debated. The topics covered herein carry significant findings which will fuel the growth of the discipline. They may even be implemented as practical applications or may be referred to as a beginning point for another development.

The contributors of this book come from diverse backgrounds, making this book a truly international effort. This book will bring forth new frontiers with its revolutionizing research information and detailed analysis of the nascent developments around the world.

We would like to thank all the contributing authors for lending their expertise to make the book truly unique. They have played a crucial role in the development of this book. Without their invaluable contributions this book wouldn't have been possible. They have made vital efforts to compile up to date information on the varied aspects of this subject to make this book a valuable addition to the collection of many professionals and students.

This book was conceptualized with the vision of imparting up-to-date information and advanced data in this field. To ensure the same, a matchless editorial board was set up. Every individual on the board went through rigorous rounds of assessment to prove their worth. After which they invested a large part of their time researching and compiling the most relevant data for our readers.

The editorial board has been involved in producing this book since its inception. They have spent rigorous hours researching and exploring the diverse topics which have resulted in the successful publishing of this book. They have passed on their knowledge of decades through this book. To expedite this challenging task, the publisher supported the team at every step. A small team of assistant editors was also appointed to further simplify the editing procedure and attain best results for the readers.

Apart from the editorial board, the designing team has also invested a significant amount of their time in understanding the subject and creating the most relevant covers. They scrutinized every image to scout for the most suitable representation of the subject and create an appropriate cover for the book.

The publishing team has been an ardent support to the editorial, designing and production team. Their endless efforts to recruit the best for this project, has resulted in the accomplishment of this book. They are a veteran in the field of academics and their pool of knowledge is as vast as their experience in printing. Their expertise and guidance has proved useful at every step. Their uncompromising quality standards have made this book an exceptional effort. Their encouragement from time to time has been an inspiration for everyone.

The publisher and the editorial board hope that this book will prove to be a valuable piece of knowledge for researchers, students, practitioners and scholars across the globe.

LIST OF CONTRIBUTORS

Oswaldo Cruz Neto and Ariadna V. Lopes
Departamento de Botânica, Universidade Federal de Pernambuco, Recife, Pernambuco, Brazil

Antonio V. Aguiar
Department of Wildlife Ecology and Conservation, University of Florida, Gainesville, Florida, United States of America

Alex D. Twyford, Linda E. Neaves and R. Toby Pennington
Royal Botanic Garden Edinburgh, Edinburgh, United Kingdom

Nerea Abrego
Department of Biological and Environmental Science, University of Jyväskylä, Jyväskylä, Finland
Department of Plant Biology and Ecology, University of the Basque Country (UPV/EHU), Bilbao, Spain

Gonzalo García-Baquero and Isabel Salcedo
Department of Plant Biology and Ecology, University of the Basque Country (UPV/EHU), Bilbao, Spain

Panu Halme
Department of Biological and Environmental Science, University of Jyväskylä, Jyväskylä, Finland
Natural History Museum, University of Jyväskylä, Jyväskylä, Finland

Otso Ovaskainen
Department of Biosciences, University of Helsinki, Helsinki, Finland

Rui Ding, Xu-Hui Chen, Li-Jun Zhang, Xiao-Dan Yu, Bo Qu, Ru Duan and Yu-Feng Xu
College of Bioscience and Biotechnology, Shenyang Agricultural University, Shenyang, Liaoning, People's Republic of China

Li Zhang, Caihong Wei, Lingyang Xu, Fuping Zhao and Lixin Du
National Center for Molecular Genetics and Breeding of Animal, Institute of Animal Sciences, Chinese Academy of Agricultural Sciences, Beijing, People's Republic of China

Jian Lu, Gang Liu, Zhigang Wang, Xu Han and Chousheng Liu
National Center of Preservation & Utilization of Genetic Resources of Animal, National Animal Husbandry Service, Beijing, People's Republic of China

Amanda V. Rocha and Renato Caparroz
Laboratório de Genètica e Biodiversidade, Instituto de Ciências Biológicas, Universidade de BrasÍ lia, BrasÍ lia, Distrito Federal, Brazil

Luis O. Rivera
Centro de Investigación y Transferencia de Jujuy, Universidad Nacional Jujuy - CONICET, Jujuy, Argentina

Jaime Martinez
Projeto Charão - Associação Amigos do Meio Ambiente, Carazinho, Rio Grande do Sul, Brazil

Nêmora P. Prestes
Projeto Charão - Associação Amigos do Meio Ambiente, Carazinho, Rio Grande do Sul, Brazil
Instituto de Ciências, Universidade de Passo Fundo, Passo Fundo, Rio Grande do Sul, Brazil

Wen-Hong Chen
Key Laboratory for Plant Diversity and Biogeography of East Asia, Kunming Institute of Botany, Chinese Academy of Sciences, Kunming, Yunnan, China
University of the Chinese Academy of Sciences, Beijing, China

Yu-Min Shui and Hong Wang
Key Laboratory for Plant Diversity and Biogeography of East Asia, Kunming Institute of Botany, Chinese Academy of Sciences, Kunming, Yunnan, China

Jun-Bo Yang and Zhi-Rong Zhang
Plant Germplasm and Genomics Center, Germplasm Bank of Wild Species, Kunming Institute of Botany, Chinese Academy of Sciences, Kunming, Yunnan, China

Kanae Nishii and Michael Möller
Science Division, Royal Botanic Garden Edinburgh, Edinburgh, Scotland, United Kingdom

Fang Wen
Guangxi Institute of Botany, Guangxi Zhuang Autonomous Region and Chinese Academy of Sciences, Guilin, Guangxi, China

Yi Yu, Qiang Fan, Rujiang Shen, Jianhua Jin and Wenbo Liao
Guangdong Key Laboratory of Plant Resources and Key Laboratory of Biodiversity Dynamics and Conservation of Guangdong Higher Education Institutes, School of Life Sciences, Sun Yat-Sen University, Guangzhou, China

Wei Guo
Department of Horticulture and Landscape Architecture, Zhongkai University of Agriculture and Engineering, Guangzhou, China

Dafang Cui
College of Forestry, South China Agriculture University, Guangzhou, China

Kathrin A. Witzenberger
Zoo Hoyerswerda, Hoyerswerda, Germany
Trier University, Department of Biogeography, Trier, Germany

Axel Hochkirch
Trier University, Department of Biogeography, Trier, Germany

Wei Ji, Gui-Rong Zhang and Kai-Jian Wei
Key Laboratory of Freshwater Animal Breeding, Ministry of Agriculture, College of Fisheries, Huazhong Agricultural University, Wuhan, P. R. China
Freshwater Aquaculture Collaborative Innovation Centre of Hubei Province, Wuhan, P. R. China

Wei Ran and Wei-Min Wang
Key Laboratory of Freshwater Animal Breeding, Ministry of Agriculture, College of Fisheries, Huazhong Agricultural University, Wuhan, P. R. China

Jonathan P. A. Gardner
Key Laboratory of Freshwater Animal Breeding, Ministry of Agriculture, College of Fisheries, Huazhong Agricultural University, Wuhan, P. R. China
Freshwater Aquaculture Collaborative Innovation Centre of Hubei Province, Wuhan, P. R. China
School of Biological Sciences, Victoria University of Wellington, Wellington, New Zealand

Gui-Wei Zou
Key Laboratory of Freshwater Biodiversity Conservation, Ministry of Agriculture, Yangtze River Fisheries Research Institute, Chinese Academy of Fishery Sciences, Wuhan, P. R. China

Holly B. Ernest
Wildlife Health Center, School of Veterinary Medicine, University of California Davis, Davis, California, United States of America
Wildlife and Ecology Unit, Veterinary Genetics Laboratory, School of Veterinary Medicine, University of California Davis, Davis, California, United States of America
Department of Veterinary Sciences, University of Wyoming, Laramie, Wyoming, United States of America

T. Winston Vickers and Walter M. Boyce
Wildlife Health Center, School of Veterinary Medicine, University of California Davis, Davis, California, United States of America

Scott A. Morrison
The Nature Conservancy, San Francisco, California, United States of America

Michael R. Buchalski
Wildlife Health Center, School of Veterinary Medicine, University of California Davis, Davis, California, United States of America
Wildlife and Ecology Unit, Veterinary Genetics Laboratory, School of Veterinary Medicine, University of California Davis, Davis, California, United States of America

Martina Grdiša, Klaudija Carović-Stanko, Ivan Kolak and Zlatko Satovic
Department of Seed Science and Technology, Faculty of Agriculture, University of Zagreb, Zagreb, Croatia

Zlatko Liber and Ivan Radosavljević
Department of Botany, Division of Biology, Faculty of Science, University of Zagreb, Zagreb, Croatia

Debjani Roy Choudhury, Nivedita Singh, Amit Kumar Singh, Sundeep Kumar and Rakesh Singh
Division of Genomic Resources, National Bureau of Plant Genetic Resources, New Delhi, 110 012, India,

Kalyani Srinivasan and R. K. Tyagi,
Division of Germplasm Conservation, National Bureau of Plant Genetic Resources, New Delhi, 110 012, India

Altaf Ahmad
Department of Botany, Faculty of Science, Jamia Hamdard (Hamdard University), New Delhi, 110062, India

N. K. Singh
National Research Centre on Plant Biotechnology, Indian Agricultural Research Institute, New Delhi, 110 012, India

Wei Zhao and Wei Guo
National and Local United Engineering Laboratory for Chinese Herbal Medicine Breeding and Cultivation, School of Life Science, Jilin University, Changchun, Jilin, China
School of Life Science, Jilin University, Changchun, Jilin, China
Institute of Botany, Changbai Mountain Academy of Sciences, Erdao, Jilin, China

Xiaozheng Shi
Key Laboratory for Molecular Enzymology and Engineering, the Ministry of Education, Jilin University, Changchun, Jilin, China
School of Life Science, Jilin University, Changchun, Jilin, China

Jiangnan Li, Chengbai Liu and Xia Chen
National and Local United Engineering Laboratory for Chinese Herbal Medicine Breeding and Cultivation, School of Life Science, Jilin University, Changchun, Jilin, China
School of Life Science, Jilin University, Changchun, Jilin, China

Elisa Boscari and Leonardo Congiu
Department of Biology, University of Padova, Padova, Italy

Jose Martin Pujolar
Department of Bioscience, Aarhus University, Aarhus C, Denmark

Isabelle Dupanloup
Institute of Ecology and Evolution, University of Berne, Berne, Switzerland
Swiss Institute of Bioinformatics, Lausanne, Switzerland

Riccardo Corradin
Department of Statistics, University of Padova, Padova, Italy

Michael Hellmair
Department of Fisheries Biology, Humboldt State University, Arcata, California, United States of America
FISHBIO, Chico, California, United States of America

Andrew P. Kinziger
Department of Fisheries Biology, Humboldt State University, Arcata, California, United States of America

Qing Zhang and Jianming Niu
School of Life Sciences, Inner Mongolia University, Hohhot, China
Sino-US Center for Conservation, Energy and Sustainability Science, Inner Mongolia University, Hohhot, China

Xiangyang Hou and Yong Ding
Grassland Research Institute of Chinese Academic of Agricultural Science, Hohhot, China

Frank Yonghong Li
AgResearch Grasslands Research Centre, Palmerston North, New Zealand

Yanlin Zhou, Liqing Zhao, Wenjing Ma and Sarula Kang
School of Life Sciences, Inner Mongolia University, Hohhot, China

Xin Li
Sino-US Center for Conservation, Energy and Sustainability Science, Inner Mongolia University, Hohhot, China

Wei-wei Zhou, Hong-man Chen, Jie-qiong Jin, Jun-xiao Yang, Ke Jiang, Yun-yu Wang and Jing Che
State Key Laboratory of Genetic Resources and Evolution, and Yunnan Laboratory of Molecular Biology of Domestic Animals, Kunming Institute of Zoology, Chinese Academy of Sciences, Kunming, China

Bao-lin Zhang
State Key Laboratory of Genetic Resources and Evolution, and Yunnan Laboratory of Molecular Biology of Domestic Animals, Kunming Institute of Zoology, Chinese Academy of Sciences, Kunming, China
Laboratory for Conservation and Utilization of Bio-resources, Yunnan University, Kunming, China

Robert W. Murphy
State Key Laboratory of Genetic Resources and Evolution, and Yunnan Laboratory of Molecular Biology of Domestic Animals, Kunming Institute of Zoology, Chinese Academy of Sciences, Kunming, China
Centre for Biodiversity and Conservation Biology, Royal Ontario Museum, Toronto, Ontario, Canada

Ya-ping Zhang
Laboratory for Conservation and Utilization of Bio-resources, Yunnan University, Kunming, China

Stephanie L. Greene
National Center for Genetic Resource Preservation, United States Department of Agriculture-Agricultural Research Service, Fort Collins, Colorado, United States of America

Theodore J. Kisha and Long-Xi Yu
Western Regional Plant Introduction Station, United States Department of Agriculture-Agricultural Research Service, Pullman, Washington, United States of America

Mauricio Parra-Quijano
Departamento de Biologıa Vegetal, Universidad Politecnica de Madrid, Madrid, Spain

Mohar Singh, Ishwari Singh Bisht, Sandeep Kumar, Manoranjan Dutta, Kailash Chander Bansal and Moreshwar Karale
National Bureau of Plant Genetic Resources, Pusa, New Delhi, India

Ashutosh Sarker
International Centre for Agricultural Research in Dry Areas, South Asia and China Regional Programme, NASC Complex, DPS Marg, Pusa, New Delhi, India

Ahmad Amri
International Centre for Agricultural Research in Dry Areas, Tunisia, Tunis

Shiv Kumar
International Centre for Agricultural Research in Dry Areas, Rabat, Morocco

Swapan Kumar Datta
Division of Crop Science, Indian Council of Agricultural Research, Krishi Bhawan, New Delhi, India

Rafe M. Brown, Karen V. Olson and Anthony J. Barley
Biodiversity Institute and Department of Ecology and Evolutionary Biology, University of Kansas, Lawrence, Kansas, United States of America

Jennifer A. Weghorst
Biodiversity Institute, University of Kansas, Lawrence, Kansas, United States of America

Mariano R. M. Duya
Institute of Biology, University of the Philippines Diliman, Quezon City, Philippines

Melizar V. Duya
Diliman Science Research Foundation Inc., University of the Philippines Diliman, Quezon City, Philippines

Myron Shekelle
Research Institute of EcoScience, Ewha Womans University, Seoul, Republic of Korea

Irene Neri-Arboleda
Thomas Glen Erin Animal Hospital, Mississauga, Ontario, Canada

Jacob A. Esselstyn
Museum of Natural Science & Department of Biological Sciences, Louisiana State University, Baton Rouge, Louisiana, United States of America

Nathaniel J. Dominy
Department of Anthropology, Dartmouth College, Hanover, New Hampshire, United States of America,

Perry S. Ong and Adrian Luczon
Institute of Biology, University of the Philippines, Diliman, Quezon City, Philippines

Gillian L. Moritz
Department of Biological Sciences, Dartmouth College, Hanover, New Hampshire, United States of America

Mae Lowe L. Diesmos
University of Santo Tomas, Espana, Manila, Philippines

Arvin C. Diesmos
Herpetology Section, Zoology Division, Philippine National Museum, Ermita, Manila, Philippines

Cameron D. Siler
Sam Noble Oklahoma Museum of Natural History and Department of Biology, University of Oklahoma, Norman, Oklahoma, United States of America

Jens Schirmel, Martin Alt, Isabell Rudolph and Martin H. Entling
Institute of Environmental Science, University of Koblenz-Landau, Landau, Germany

Wei Liu, Dongxue Yin, Jianjun Liu and Na Li
College of Forestry, Northwest A & F University, Yangling, China

Index